Progress in
Inorganic Chemistry

Volume 26

Advisory Board

PROGRESS IN
INORGANIC CHEMISTRY

Edited by

STEPHEN J. LIPPARD

DEPARTMENT OF CHEMISTRY
COLUMBIA UNIVERSITY
NEW YORK, NEW YORK

VOLUME 26

AN INTERSCIENCE® PUBLICATION
JOHN WILEY & SONS, New York · Chichester · Brisbane · Toronto

An Interscience® Publication

Library of Congress Catalog Card Number: 59–13035
ISBN 0–471–04944–1

Printed in the United States of America

10 9 8 7 6 5 4 3 2 1

Preface

Two of the chapters in this volume, that by Lagow and Margrave on direct fluorination and that by Gerloch on magnetochemistry, depart somewhat in scope and content from reviews published in recent years. In particular, these chapters are in-depth summaries of work conducted primarily by the authors, rather than comprehensive and/or critical surveys of a field in general. (The series of articles by Kepert on stereochemistry, published in previous volumes and to be continued in future ones, also falls into this category.) *Progress in Inorganic Chemistry* will continue to provide a forum for such discussions of work from individual laboratories when, in the view of the Editor and the Advisory Board, they are likely to impact on the thinking or experimental procedures of a sufficiently broad cross section of our readership. As always, constructive comments about this and other policies are welcomed.

STEPHEN J. LIPPARD

New York, New York
July 1979

Contents

Progress in
Inorganic Chemistry

Volume 26

A Local View in Magnetochemistry

M. GERLOCH

University Chemical Laboratories
Cambridge, England

CONTENTS

I. INTRODUCTION

Surely an essential feature of any technique used in chemistry is that it should relate to the wider area and scope of chemistry and not be limited to one specific area. About 10 years ago, Nyholm made some wide-ranging criticizms (42) of magnetochemistry based, I believe, on a correct observation that the techniques used were very often failing to meet this basic requirement. The last few years, however, have seen an enormously valuable reworking of the techniques of magnetochemistry and I hope in this chapter to demonstrate that contemporary results might again capture the interest of magnetochemists and workers in other areas of chemistry as well.

I illustrate the new approaches by reference to work in two broad classes of magnetic systems, one concerned with single-center problems — "magnetically dilute" "distortion" studies — and the other concerned with the question of magnetic exchange. The literature on paramagnetic and antiferromagnetic systems has traditionally been separated essentially because of the rather different theoretical techniques required by these types of systems. However, as I am interested in describing how magnetochemical studies might yield information related to chemical bonding, I have attempted to draw the different themes together by discussing recent work on the antiferromagnetism of some binuclear transition metal complexes. I begin by sketching the nature of the problems that must be solved in the exchange area and argue that a satisfactory treatment of the two-center problems can only be made following a detailed analysis of the appropriate one-center systems. Accordingly the discussion is interrupted at this point with a review of the current state of the art in paramagnetism. Finally I return to the exchange problem in a little more detail.

II. THE HEISENBERG-DIRAC-VAN VLECK MODEL FOR EXCHANGE

Within the chemical literature, at least, a large number of the studies of binuclear antiferromagnetism have relied on interpretation from the Heisenberg-

Dirac-Van Vleck (HDVV) model of exchange (48). Heisenberg and Dirac independently provided an explanation of the phenomenon of ferromagnetism in terms of the electrostatic Coulomb operator e^2/r_{12}, their arguments being most simply illustrated by an interaction between two hydrogen atoms. The neglect of overlap between the orbitals of the interacting atoms leads to a spin-triplet ground state and excited spin-singlet state for the pair, separated by an energy determined solely by the quantum mechanical exchange integral

$$K = \langle \, \phi_a(1)\phi_b(2) \left| \frac{e^2}{r_{12}} \right| \phi_a(2)\phi_b(1) \, \rangle \tag{1}$$

Dirac and Heisenberg showed how the same state of affairs could be reproduced by an *effective spin-Hamiltonian*

$$\mathscr{H}' = - \mathscr{J}_{ab} s_a \cdot s_b \tag{2}$$

acting upon the spin functions of each center, the parameter $\mathscr{J}_{ab}$ being simply related to the true exchange integral K. The spin-Hamiltonian formalism caused considerable excitement at the time, because it provided the rationale for the dipolar coupling of spins previously ascribed to enormous internal magnetic fields in ferromagnets. Van Vleck's contribution was to generalize the treatment to include interaction between many-electron systems with spin angular momenta S_A, S_B, namely,

$$\mathscr{H}_{HDVV} = - \mathscr{J}_{AB} S_A \cdot S_B \tag{3}$$

with the proviso that the interacting systems were devoid of any *orbital* angular momentum. The "exchange parameter" $\mathscr{J}_{AB}$ is now a function of various two-electron exchange integrals, K, in the system. Recent contributions to the HDVV model, notably those by Löwdin (40), have shown that formalism 3 remains valid even when the interacting systems overlap. However, in these cases the $\mathscr{J}$ parameter also involves one-electron integrals representing overlap or covalency effects that normally reverse the sign of $\mathscr{J}$ and so account for the more commonly observed phenomenon of antiferromagnetism. Although a misnomer, the name "exchange integral" is still used for $\mathscr{J}$ in these circumstances.

Two main aspects of the HDVV approach are important for the present discussion. One aspect is that the model is valid only in the absence of orbital angular momentum contributions to the system, so that its use has been broadly restricted to (say) binuclear complexes of iron(III), octahedral nickel(II), or distorted octahedral copper(II). The other aspect concerns the fact that the parameter is a "global" parameter of the system sequestering many contributions from various ligands and orbitals and, in this respect, may be compared with how Dq in ligand-field theory represents the net effect of electrostatic and covalent (σ and π) effects in the $t_{2g} - e_g$ orbital splitting. Attempts to localize our conception of the exchange phenomenon have been made by various physicists, notably Kanamori (38), Goodenough (35), and Anderson (1, 2), and from them have emerged our notion of "exchange pathways."

III. EXCHANGE PATHWAYS

Perhaps the coarsest subdivision we can make here is the separation of "direct exchange" and "superexchange" relating to the absence or presence of any participation by bridging groups in the metal–metal interaction processes. Many papers on the exchange process have been concerned with demonstrating the dominance of one or other of these pathways in particular molecules. The arguments invariably revolve around the sort of pathway analysis discussed by Anderson (2), the essential features of which are summarized in Fig. 1. Fig. 1*a* shows *d* orbitals of two metals interacting through a single *p* orbital of some bridging group. The stages by which we deduce that the best mutual arrangement of the spins on the two metals is antiparallel are shown, in a purely pedagogic fashion, in *1–3*. In *1* we envisage a filled bridging orbital overlapping with the left-hand metal orbital containing an α spin. As a result of overlap, some β spin from the bridge pairs (*2*) with this α spin, leaving an equal amount of bridge α spin available to pair with the spin on the right-hand metal, which therefore tends to align as a β spin (*3*). An alternative viewpoint is provided by conceiving of a contribution from the molecular orbital shown in Fig. 1*b*. In contrast, parallel orientation of the spins, associated with ferromagnetism, may arise when two (or more) *orthogonal* bridging orbitals are involved, as shown in Fig. 1*c*. The central feature here (step 4) is the effective application of Hund's rule in the vicinity of the bridging atom. Overall, the figure shows that ferromagnetism is associated with points of orthogonality in the exchange pathway, while antiferromagnetism results from overlap, electron delocalization, insipient bond formation, or equivalent descriptions. Figure 1*c* indicates a further point, namely, that should the metal orbitals significantly overlap *directly*, then, by analogy with Fig. 1*a* and the summary above, the coupling would tend to be *anti*ferromagnetic. In such cases, there arises the question of which pathway, direct metal–metal or superexchange by way of the orthogonal bridge orbitals, will dominate: certainly the result will be some combination of both effects. Some guide towards resolving this question is afforded by arguments given by Anderson, the main thrust of which is that a small degree of overlap is most effective (*2*); that is, whether we are concerned with direct or superexchange mechanisms, a pathway involving no orthogonalities and leading to antiferromagnetism will tend to predominate over any ferromagnetic routes.

The relevance of this brief pathway discussion is to note that this kind of argument is often the most that is brought to bear on the interpretation of systems studied within the HDVV formalism. Thus we observe many empirical correlations between the purely phenomenological HDVV $\mathscr{J}$ parameters, obtained

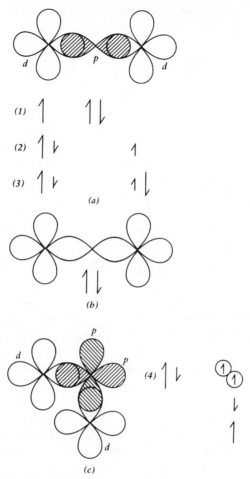

Fig. 1.   Exchange pathways: (a) antiferromagnetic superexchange by way of a single bridge orbital, (b) a molecular orbital viewpoint for a, (c) ferromagnetic superexchange by way of two orthogonal bridge orbitals.

by fitting the magnetic susceptibilities of powdered samples, and the structural nature of the system. How do $\mathscr{J}_{HDVV}$ values relate to the presence or absence of bridging groups, to the number of bridging groups, to the linearity or otherwise of the bridges, to their chemical nature, and to the metal electron configuration? The intellectual content and chemical reward of this empiricism are akin to those enjoyed in the early days of "one-center" paramagnetism interpreted by the "spin-only" formula. It is valid so far as it goes, but surely we can do better.

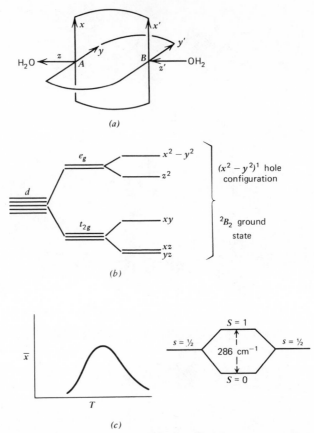

Fig. 2.   Magnetic exchange in copper acetate: (*a*) coordinate frames, (*b*) one-center orbital splitting diagram, (*c*) variation of susceptibility with temperature fitted within the HDVV model.

IV. COPPER ACETATE

Chemists, too, have tried to localize our perception of the exchange process. Building on similar notions of pathways, Figgis and Martin (25) proposed the existence of a weak δ bond in the binuclear copper acetate molecule. Actually, much of the physics of the antiferromagnetic exchange in this system had been explained previously by Bleaney and Bowers (6) in their famous ESR study — and at a time before the binuclear nature of the molecule had been proved by crystallographers — but within the chemical literature, probably the contribution best remembered is that by Figgis and Martin, and this because of the *chemical* feature of the δ bond, supposedly demonstrated by magnetic studies. It is doubtful if many chemists still accept this interpretation today, as is shown later, but that does not detract from the central point: that magneto-chemistry is about chemistry.

Let us review the origins of the δ bond in copper acetate. Figure 2 shows a sketch of the binuclear system, together with the conventional coordinate frames and symmetry designations of the d orbitals. The five-coordinate struc-ture around each copper atom is close to C_{4v} symmetry and we expect to stabilize d_{z^2} and d_{xz}, d_{yz} with respect to the precursor octahedral e_g and t_{2g} functions, and the single hole in the d^9 system resides in the $x^2 - y^2$, or δ orbital on each center. The two orbitally nondegenerate $^2B_2$ ground states couple antiferromagnetically and the behavior of the powder susceptibility with tempe r-ature, shown in Fig. 2c, is well fitted by the HDVV model, with the interaction parameter $\mathcal{J}_{HDVV} = -286 \text{ cm}^{-1}$. In other words, unpaired electrons in the $d_{x^2 - y^2}$ orbitals of each copper atom couple together in an antiferromagnetic sense. In view of the closeness of the copper ions (2.64 Å), Figgis and Martin proposed a direct interaction between the δ-type orbitals to form a *weak* δ bond.

V. ORBITAL ANGULAR MOMENTUM AND EXCHANGE

About 6 years ago, I and my colleagues in Cambridge began to apply ourselves to the general problem of magnetic exchange. For about the previous 10 years we had confined our attention to the paramagnetism of single-center systems, partly out of a reluctance to become limited by the "global" phenomenology of the HDVV model and partly from a timorous respect for more detailed exchange microtheory. Our experience within the area of mag-netically dilute systems had taught us, as I discuss later, that any worthwhile understanding of paramagnetism requires detailed quantum-mechanical models applied to the more comprehensive magnetic anisotropy data obtainable from single-crystal studies. In relation to magnetic exchange systems, we anticipate that magnetic anisotropy might arise from the exchange processes themselves, but such anisotropy may not simply be additive with that arising from low one-center symmetry; rather, it may be multiplicative with it in some complex way. Instead of studying systems that, in principle, may be tractable within a modified HDVV model, or by using a more general spin Hamiltonian, which in our view only disguises the role played by various exchange pathways, we considered that a frontal attack on the orbital problem was timely. The point here is that in paramagnetism, anisotropy arises from formal orbital degeneracy. If we are to measure the magnetic anisotropy in exchange systems, we recognise that it may arise in whole or in part from similar formal orbital degeneracy. In turn, it is under just these circumstances that the validity of the HDVV model vanishes. Accordingly, and for reasons of chemical stability, variation, and existence, we embarked on a study of five- and six-coordinate binuclear cobalt(II) systems. Among these are the cobalt benzoates.

The molecular structures of these benzoates (4, 13, 16, 18) are analogous to those of the copper carboxylates, including copper acetate, and our discussion makes use of the coordinate frames shown in Fig. 2. A full description of our approach (8, 7) to the magnetism of these systems is outside the scope of this review, but the following sketch illustrates some of the more important features. We consider a Hamiltonian acting upon the binuclear entity,

$$\mathcal{H} = \mathcal{H}^A + \mathcal{H}^B + \mathcal{H}^{AB} \tag{4}$$

in which

$$\mathcal{H}^A = \sum_{i<j} \frac{e^2}{r_{ij}^2} + \zeta \sum_i l_i \cdot s_i + V_{lf} \tag{5}$$

represents the perturbations of "interelectron repulsion," spin—orbit coupling and the ligand field on the functions of one cobalt atom, center A. $\mathcal{H}^{AB}$, the exact form of which need not be specified here, represents an interaction Hamiltonian acting on *product* functions made up from the *two* centers; this may be called, loosely, the "exchange Hamiltonian." We are concerned now with evaluating matrix elements of the form

$$\langle \chi_a^A \chi_b^B | \mathcal{H}^{AB} | \chi_c^A \chi_d^B \rangle \tag{6}$$

in which the superscripts refer to the cobalt centers and the subscripts to different many-electron functions made up from basis d orbitals. We may anticipate that the various exchange energy splittings, calculated below, will be smaller than the primary ligand field effects but perhaps comparable with those from any low-symmetry ligand field and with spin—orbit coupling. Two inter-related questions arise at the outset. How large a basis do we need for our work? The d^7 configuration is 120-fold degenerate and would imply diagonalization of a 14,400 x 14,400 complex matrix — clearly an impractical proposition in terms of computer store and time. If we restrict ourselves to a basis spanning the complete spin-quartet terms of d^7, the 40 functions of $^4F + {}^4P$ imply 1600 product functions. Even limiting discussion to the formal $^4T_{1g}$ (O_h parentage) ground term involves 144 products. This size basis is just tractable, as we discuss later, but a description simply in terms of this $^4T_{1g}$ set would not be appropriate, for the manner in which the ligand field perturbation is treated can be expected to have important repercussions on the utility of any (necessarily restricted) basis chosen for the two-center problem. Thus the second question concerns the nature of V_{lf} in Eq. 5. As exchange energies might be comparable with various components of the ligand field energies, how accurately need the one-center problem be solved? Indeed, any realistic assessment of the exchange phenomenon in these systems can only proceed with an awareness of the general background of the magnetism of the *paramagnetic* one-center systems. So let us leave this discussion here and return after we have reviewed the state of the art in paramagnetism.

Paramagnetism

Paramagnetism has often been reviewed along historical lines illustrating the advancing and regressing nature of successive approximations that have been entertained over the years. The approach taken here has been chosen to highlight the failures and weaknesses of various older methods in magnetochemistry that have led to the construction of successful new theoretical and practical techniques relating to the subject. An early achievement of quantum theory was the interpretation of the magnetic moments of the lanthanides by Hund's formula

$$\mu = g\sqrt{J(J+1)} \tag{7}$$

in which the Landé splitting factor is given by

$$g = 1 + \frac{J(J+1) + L(L+1) - S(S+1)}{2J(J+1)} \tag{8}$$

The broadly successful application of these formulas to the paramagnetism of lanthanide complexes was due to the wide multiplet widths in the f block metals (large spin–orbit coupling coefficients λ) and to the small effect of the ligand field on the deep-lying f orbitals. No comparably useful formula for the magnetic moments of d block complexes exists, except perhaps for the spin-only formula:

$$\mu_{so} = \sqrt{4S(S+1)} = \sqrt{n(n+2)} \tag{9}$$

but this, of course, cannot reproduce the temperature variation of transition metal complex moments frequently observed. It was probably not until about 1960 that a *generally* useful model emerged.

VI. MAGNETISM OF IONS WITH
FORMAL ORBITAL TRIPLET GROUND TERMS

Building upon the work of Kotani, Bleaney and his group, and Stevens, Figgis described an approach to the interpretation of the paramagnetism of ions with formally orbitally degenerate ground terms (20, 21, 23, 24). Although a number of systems had been studied and interpreted in great detail previously, Figgis' model was the first to be applied widely throughout the transition metal block, with respect to variation in metal, coordination number, and ligand. The average powder susceptibilities of about 60 complexes and their variation with temperature were measured and "fitted" within the range 80–300°K by Figgis et al. (20, 21, 23, 24). The approach may be illustrated by its application to ions with formal $^2T_2$ ground terms (20), for example, tetrahedral copper(II) complex ions.

The model describes the simultaneous perturbation of the sixfold degenerate $^2T_2$ basis by the effects of spin–orbit coupling and an axially symmetric ligand field distortion,

$$\mathcal{H}' = \lambda L \cdot S + V_{\mathrm{axial}} \tag{10}$$

The exact nature of the axial field is irrelevant within this model – for example, it could be any of the distortions illustrated in Fig. 3 – and its effect is represented parametrically by a splitting between the orbital doublet and orbit singlet arising from the triplet $^2T_2$ term. The splitting, Δ in Fig. 4 (defined conventionally as positive if the singlet lies lowest), is thus defined, not in terms of the operator

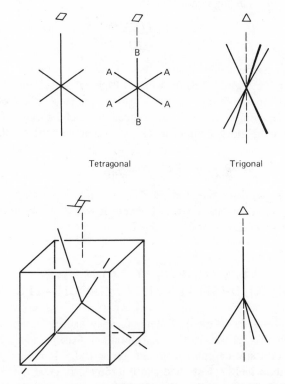

Fig. 3.   Examples of axially distorted octahedra and tetrahedra.

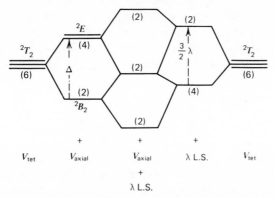

Fig. 4.   Perturbation of a $^2T_2$ cubic field term by spin–orbit coupling and an axial distortion.

V_{axial}, but by the energy difference

$$\Delta = \epsilon(^2E) - \epsilon(^2B_2) \tag{11}$$

The model effectively grew from the recognition that low-symmetry splittings can be comparable with those from spin–orbit coupling previously studied by Kotani, and in the middle of Fig. 4 are shown the three Kramers' doublets that arise from these two effects together. The energies and composition of these doublets are functions of two parameters, Δ and λ. The model involves a further variable, namely, k, Stevens' orbital reduction factor (32, 46) appearing in the magnetic moment operator,

$$\mu_\alpha = \beta_0(kL_\alpha + 2S_\alpha); \qquad \alpha = \parallel, \perp \tag{12}$$

This parameter, representing in effect the first-order failure of ligand field theory, was hopefully identified at the time with the degree of covalency in the metal–ligand bonds. Altogether then, the fitting of experimental susceptibilities by variation of the three parameters Δ, λ, k could be expected to yield information about the magnitude and sign of the axial-field splitting and, in so far as $\lambda_{complex}$ differed from $\lambda_{free\text{-}ion}$ and k differed from unity, they could comment on the covalent nature or otherwise of the bonding in the complex. In this way chemical aims for magnetism were kept in view.

Unfortunately, despite its successful reproduction of the detailed temperature variation of the susceptibilities of so many transition metal complexes, a generally acceptable interpretation of the desired parameter values did not emerge. A number of difficulties were recognized by Figgis and Lewis from the beginning. Orbital reduction factors sometimes, but not always, completely failed to correlate with accepted notions of π bonding and covalency, although since then arguments have been presented (32) to show why direct correlation may not obtain. Similarly empirical relationships between λ values and covalency or ligand field strength were not apparent. However, the most serious problems with the model began to emerge a little later, and interpretation (26) of the magnetism of nominally tetrahedral copper(II) ions illustrates this rather well.

VII. LARGER BASES

Crystals of Cs_2CuCl_4 contain nominally tetrahedral $CuCl_4^{2-}$ ions, shown by x-ray crystallographic analysis to have the tetragonally compressed approximate D_{2d} symmetry (Fig. 5a). It is likely, therefore, that the orbitals containing the z label are stabilized with respect to the tetrahedral energies, as shown in Fig. 5b, and given rise to a $^2B_2$ ground term.

Application of the $^2T_2$ model, described above, to the powder suscepti-bility of Cs_2CuCl_4 leads to a value for Δ, the tetragonal field splitting, of about 5000 cm^{-1}, in agreement with an orbital singlet ground term. Furthermore, the single-crystal polarized spectrum of this complex, reported by Ferguson (19), confirms the size and magnitude of Δ as shown in Fig. 5c.

At this stage, therefore, the crystal structure, crystal spectrum, and

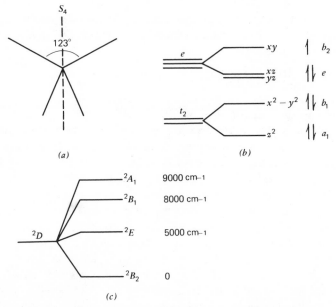

Fig. 5.   (a) The compressed tetragonal distortion in $CuCl_4^{2-}$. (b) Orbital splitting diagram expected from a. (c) Summary of term splittings from polarized spectrum (19).

powder susceptibilities are compatible within the $^2T_2$ model. However, that model involves axially distorted systems and, as indicated in Ref. 12, requires the computation of magnetic moments parallel and perpendicular to the axis of distortion. While fitting the powder susceptibilities above, the model simultaneously predicts that the parallel moment should be less than that perpendicular to the approximate S_4 axis. Experimentally, measurements (2) of the crystal magnetic anisotropy unambiguously yield the opposite result:

$$^2T_2 \text{ theory: } \mu_\| < \mu_\perp \qquad \text{expt: } \mu_\| > \mu_\perp$$

The explanation of this contradiction, which seemed unlikely at the time it was proposed, is that the energetically distant levels $^2B_1$, and $^2A_1$ must be included in the calculation: they contribute through spin—orbit coupling and in the second-order Zeeman effect. Their effect (22), though small, is important and can be seen from Fig. 6.

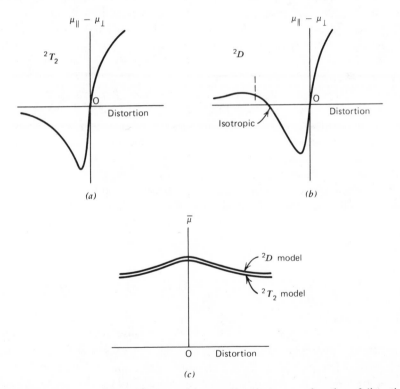

Fig. 6.   Comparison of $^2T_2$ and ^2D bases: (a) magnetic anisotropy as function of distortion with T_2 basis, (b) the same with ^2D basis, (c) mean moments as functions of distortion for both models. All curves are typical for about 300°K for Cs_2CuCl_4, based on those given in Ref. 26 (elongation to the right, compression to the left).

A typical plot of anisotropy, $\mu_\| - \mu_\perp$, versus distortion, calculated within the restricted $^2T_2$ basis, is shown in Fig. 6a. The curve lies in two diametrically opposite quadrants, which indicates a direct relation between the sign of the anisotropy and the sign of the distortion: for the compressed geometry of the $CuCl_4^{2-}$ tetrahedron, the model uniquely predicts $\mu_\| < \mu_\perp$. Without distortion the system is isotropic, of course, as required by symmetry. The inclusion of the high-lying $^2A_1$ and $^2B_1$ terms, corresponding to a diagonalization of the complete free-ion term ^2D, shifts the trace by a small, but important, amount, as shown in Fig. 6b. No longer does the curve lie only in two quadrants: no longer do we observe a unique relation between geometry and the sign of the magnetic anisotropy. Indeed at one point on the abscissa the anisotropy vanishes. This "accidental" isotropy does not indicate structural isotropy, of course; in the present example, it occurs when the ground-state splitting Δ is about 3500 cm^{-1}. The double reversal in the sign of $\mu_\| - \mu_\perp$ permits the "fitting" of the experimental anisotropy in $CuCl_4^{2-}$ ions near the dashed line in Fig. 6b when account is taken of all properties, crystal structure, crystal spectrum, and crystal magnetism. It is also interesting to see the behavior of *powder* or *average* magnetic moments with respect to the bases chosen. From Fig. 6c we note, in this particular example, that the difference between the models is almost negligible in terms of average moments alone.

Two important generalizations can be made; they do not derive from this study alone but are confirmed time and again from studies involving other metals, geometries, and ligands. The first generalization is that calculations must include a sufficiently complete basis set of functions: today we use the complete set of terms with maximum spin multiplicity (in d block systems) as a matter of course, but occasionally even this may be insufficient. Secondly, we conclude that it is impossible to rely on average magnetic moments to provide unambiguous fitting parameters, so that a complete set of principal susceptibilities from a single crystal study constitutes a basic minimum data set for any useful study of this kind.

VIII. FULLER SPECIFICATIONS FOR AXIAL FIELDS

The way in which various higher lying excited states may contribute to the "ground-state" magnetism depends on the geometry of the system. Let us consider the average powder moments at room temperature for ferrous Tutton salt, $FeSO_4 \cdot (NH_4)_2 SO_4 \cdot 12H_2O$ and ferrous fluorosilicate $FeSiF_6 \cdot 6H_2O$. The values are 5.5 and 5.2, respectively. An earlier fitting of these magnetic moments and their temperature variations within the cubic field ground term $^5T_{2g}$ had yielded values of Stevens' orbital reduction factor k of about 1.0 and about 0.7,

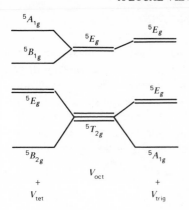

Fig. 7.   Splitting of ^5D term in trigonally and tetragonally distorted octahedral symmetry.

respectively. The contemporary understanding of Stevens' factor implied that the M–L bonds in the fluorosilicate were much more covalent and π bonding than those in the Tutton salt. However, the coordination geometry in both complexes involves the formally octahedral hexaquo ions $Fe(H_2O)_6^{2+}$, the essential difference between the systems being concerned with the counterions and the different crystal lattices. In view of the lessons learned from the tetrahedral copper systems, an obvious first step towards understanding this problem was to perform calculations within the complete spin-quintet manifold, namely, ^5D, but then of course, it is still difficult to see why the moments of the two apparently virtually identical ions should be so different, regardless of the size of the basis chosen for the calculations. A clue is provided by the different point group symmetries adopted in the two lattices. In the fluorosilicate lattice the iron complex possesses $\bar{3}$ symmetry, while in the Tutton salt there is a very approximate tetragonal distortion. The term splitting diagrams for the two geometries are shown in Fig. 7.

The eventual successful interpretation (29) of the paramagnetism of these two salts originates from this difference in molecular symmetry. In the tetragonal case, the higher components of the cubic field ^5E term contribute only through spin–orbit coupling, while in the fluorosilicate they also mix into the lower $^5E(^5T_{2g})$ term by way of the trigonal component of the ligand field. The final analysis yields approximately equal orbital reduction factors (ca. 0.7) for both systems. The lesson to be learned from this study is that the nature of an axial distortion is relevant for the magnetism, a description in terms of a splitting like that in Eq. 11 being quite inadequate. The splitting Δ of Eq. 11 represents a difference between diagonal terms in the low-symmetry matrix; in the trigonal system there is also an off-diagonal element connecting the two ^5E terms.

IX. RHOMBIC SYMMETRY

Actually the contribution from upper states may not always be very important. A dominance by a few low-lying states can yield surprising results as the magnetism of *trans*-dimesityl bis(diethylphenylphosphine)cobalt(II) shows. This nominally square planar low-spin cobalt(II) molecule shown in Fig. 8*a*, in which the α-methyl groups of the *trans*-mesityl ligands effectively block the fifth and sixth octahedral coordination sites around the cobalt, was studied by Bentley et al. (5). Although the symmetry of the first coordination shell is

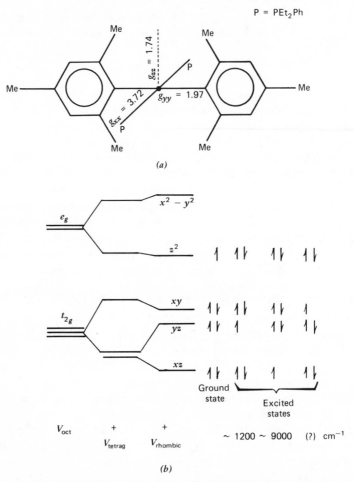

(a)

(b)

Fig. 8.   *trans*-Dimesityl bis(diethylphenylphosphine)cobalt(II): (*a*) coordination geometry and principal *g* values, (*b*) orbital splitting diagram and configurations used (5) to interpret *a*.

rhombic, we might expect to be able to regard the molecular geometry as approximately tetragonal with physical properties defining the out-of-plane direction as essentially unique. The ESR g values, however, tell a different story. The g tensor *is* approximately axial, in that g_{zz} (1.74) is similar to g_{yy} (1.97), but the "unique" g_{xx} (3.72) lies parallel to the P····P vector and not perpendicular to the coordination plane. The eventual interpretation of this phenomenon is based on the splitting diagram shown in Fig. 8*b*.

The primary distortion *is* represented by a tetragonal field with respect to the coordination plane, but recognition of the inequivalence of the phosphines and mesityls superimposes a rhombic component on the field. The ground and a few low-lying configurations are indicated in the figure, the relative energies of these having been deduced from the g values. The optical spectrum shows a single absorption in the visible region, at about 9000 cm^{-1}, in agreement with this scheme. The central point here is that the magnetic properties (g values and susceptibilities) are dominated by the ground and lowest excited configurations, $(xz)^2(yz)^2(yx)^2(z^2)^1$ and $(xz)^2(yz)^1(xy)^2(z^2)^2$, alone. These configurations carry much more information relating to the rhombic component of the low-symmetry ligand field than to the overall tetragonal distortion and so I suggest two important generalizations. Firstly, we should not expect, in general, that different techniques, being more sensitive to different subsets of the energy manifold, should equally reflect the *overall* symmetry and electronic structure of the system. Secondly, small departures from axial symmetry, which may be regarded as insignificant from a structural or general chemical point of view, may utterly dominate the magnetic properties for which they must therefore not be ignored nor unduly approximated.

X. LOW-SYMMETRY FIELDS

The problem of very low symmetry is central to a proper understanding of the paramagnetism of most transition metal complexes. We illustrate this point with a discussion of the relation between crystal and molecular susceptibilities (39). In magnetically dilute systems we may associate with each paramagnetic center a molecular magnetic tensor K, describable, for example, by three orthogonal principal molecular susceptibilities and their orientation in space. Susceptibility measurements, however, involve the determination of the *crystal* susceptibility tensor χ, which is given by an appropriate tensorial sum of all *molecular* susceptibility tensors in the unit cell. An example of this is indicated for a monoclinic system in Fig. 9*a*. The monoclinic crystal class is defined by there being a unique symmetry axis, conventionally the *b* axis, parallel to which is a twofold rotation or screw axis and/or perpendicular to which is a mirror or glide plane. For clarity only two symmetry-related molecules are shown and

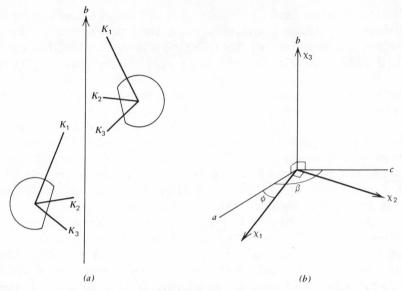

Fig. 9.   Relationship of two symmetry-related molecular susceptibility tensors to mono-clinic crystal axes. (b) Definitions of principal crystal susceptibilities in monoclinic crystals.

these are related by a screw axis. The crystal magnetic property must have at least the point group symmetry of the crystal plus, in the case of centrosymmetric susceptibility tensors, a center of inversion. The susceptibility symmetry must therefore be compatible with the Laue group $2/m$. The relative orientation of the principal crystal susceptibilities with respect to the monoclinic crystal axes are shown in Fig. 9b: by convention χ_3, the symmetry-determined principal susceptibility, lies parallel to b. The remaining χ_1 and χ_2 lie in the ac plane in directions that are not determined by symmetry. Experiment yields values for χ_1, χ_2, and χ_3 (and possibly ϕ) and, provided orientations of the molecular susceptibilities K with respect to the crystal axes a, b, c are known, the magnitudes of the principal molecular susceptibilities may be calculated by reversing and decomposing the tensor transformation that relates K and χ, namely,

$$\chi = \sum_{T}^{\substack{\text{equivalent} \\ \text{molecules}}} TKT^{\dagger} \tag{13}$$

where T represents the transformation matrix expressing the orientation of the molecular tensor in the crystal. Formerly such was common practice. The systems studied involved axially distorted octahedral or tetrahedral entities where the orientation of the principal molecular susceptibilities presumably could be confidently inferred by comparison with the known crystal structure (the days of structure determination from magnetism were already long past). But consider the case of the cobalt Tutton salt, $CoSO_4 \cdot (NH_4)_2 SO_4 \cdot 12H_2O$ containing nominally octahedral $Co(H_2O)_6^{2+}$ ions. A presumption that the principal molecular susceptibilities lie parallel, or nearly parallel, to the Co—O bond vectors yielded (33) a negative value for one of the molecular susceptibilities — an extremely unlikely situation. Furthermore, rotation of the two in-plane susceptibilities (a slight tetragonal elongation obtains in the system) by 45° to position them between the appropriate Co—O bonds yields very different magnitudes for the principal molecular susceptibilities. Clearly the presumed orientation of the molecular tensor was untenable. It might be argued here that $Co(H_2O)_6^{2+}$ was bound to furnish a difficult case. Approximate structural isotropy (i.e., near O_h symmetry) should yield an isotropic (spherical) tensor property, so that small distortions would dominate the tensor orientation; indeed, second-order crystal lattice effects may dominate here. However, more or less the same behavior is found (31) for the much more obviously tetragonally distorted molecule *trans*-Co(pyridine)$_4$(NCS)$_2$ for which some calculated molecular susceptibilities have very low and unacceptable values. The problem here, of course, is what criterion exists by which we may judge the calculated molecular susceptibilities to be "reasonable"? There is none, and so our practice today, except where molecular tensor orientations are *rigorously* defined by symmetry, is to refrain from calculating molecular magnetic susceptibilities from the observed crystal properties. Instead, comparison between observed and quantum mechanically calculated magnetic properties takes place in the *crystal* frame. That principle aside, this discussion serves to emphasize the importance of low-symmetry ligand fields and demonstrates the necessity of finding a suitable quantum mechanical model in which to accommodate them.

XI. SYMMETRY-BASED LIGAND FIELD PARAMETERS

The splittings of the d function manifold are conventionally dealt with by the formalism and techniques we call ligand field theory. The representation of the ligand effects by a one-electron potential is the central feature of the theory, conferring upon it great simplicity and hence utility. Recent studies (28) have shown that the approach is based on a much sounder footing than a brief acquaintance with quantum chemistry would indicate.

It is common to express a general ligand field potential as a superposition of spherical harmonics (34) and the expressions for the octahedral potential,

V_{oct}, in tetragonal and trigonal quantization are well-known:

tetragonal:
$$V_{oct}^{(4)} = Y_4^0 + \sqrt{\frac{5}{14}}\,(Y_4^4 + Y_4^{-4}) \tag{14}$$

trigonal:
$$V_{oct}^{(3)} = Y_4^0 + \sqrt{\frac{10}{7}}\,(Y_4^3 - Y_4^{-3}) \tag{15}$$

The expansion coefficients $\sqrt{5/14}$ and $\sqrt{10/7}$ are determined entirely by the O_h symmetry and the normalized forms of spherical harmonics. Calculations of energy involve only one parameter, Dq, and this may be summarized by the relationship

$$\langle d\,|\,V_{oct}\,|\,d\rangle \Rightarrow Dq \tag{16}$$

Axially distorted octahedra, having lower symmetry, leave three expansion coefficients undetermined, namely,

$$V_{D_{4h}} = aY_2^0 + bY_4^0 + c(Y_4^4 + Y_4^{-4}) \tag{17a}$$

$$\equiv aY_2^0 + b'Y_4^0 + c'V_{oct}^{(4)} \tag{17b}$$

for tetragonal distortions, and

$$V_{D_{3d}} = \alpha Y_2^0 + \beta Y_4^0 + \gamma(Y_4^3 - Y_4^{-3}) \tag{18a}$$

$$\equiv \alpha Y_2^0 + \beta'Y_4^0 + \gamma'V_{oct}^{(3)} \tag{18b}$$

for trigonal symmetry. These ligand field potentials may be parameterized simply by the coefficients a, b, c and so forth, but if so, little chemical or detailed structural insight will be gained. Some workers prefer the "descent-in-symmetry" approach implied by Eqs. 17b and 18b to maintain a constant definition for Dq, and when evaluating matrix elements, they refer to the tetragonal parameters Dt, Ds or the trigonal ones $D\sigma, D\tau$, that is,

	tetragonal	trigonal	
$\langle d\,|\,Y_2^0\,|\,d\rangle \Rightarrow$	Ds	$D\sigma$	(19)
$\langle d\,|\,Y_4^0\,|\,d\rangle \Rightarrow$	Dt	$D\tau$	

Many spectroscopic and magnetic studies have been concerned with empirical correlations between these parameters and features of structural and chemical interest in the molecules. It should be noticed, however, that these symmetry-based parameters are "global" (like $\mathscr{I}_{HD\,v\,v}$ which is discussed earlier), referring to the field of all ligands *as a whole*. [The same is true of recent more comprehensive symmetry-defined parameters proposed by Donini et al. (17).] Being based on the minimum assumptions of ligand field theory, and hence, for some, preferred as more "basic," these parameters lack possibilities for immediate chemical relevance and appeal.

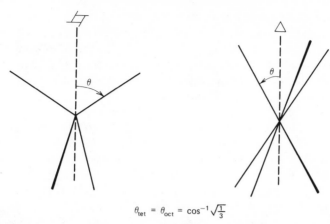

$$\theta_{tet} = \theta_{oct} = \cos^{-1}\sqrt{\tfrac{1}{3}}$$

Fig. 10.   The angular parameter θ in trigonally distorted octahedra and tetragonally distorted tetrahedra.

XII. THE POINT-CHARGE MODEL

As in our discussion of exchange pathways earlier, several attempts have been made to localize our perception of the ligand field, though not all have been fully successful. Our own studies, as well as those of many others, were concerned for some time with the point-charge model (34). In this, as in the very closely related point-dipole approach, ligands are represented as point charges (or dipoles) located at their donor atoms. The technique was much used for angularly distorted systems and can be reviewed very briefly by reference to trigonally distorted octahedra and tetragonally distorted tetrahedra, as in Fig. 10. In both geometries we may consider the cases where all ligands are identical, the departure from cubic symmetry being represented by the angle θ. The point-charge model provides a degree of separation between the angular parameter θ and radial parameters Dq and Cp, defined as

$$Dq = \tfrac{1}{6}ze^2 \int R(d)\, \frac{r^4}{a^5}\, R(d) r^2 \; dr \tag{20}$$

$$Cp = \tfrac{2}{7}ze^2 \int R(d)\, \frac{r^2}{a^3}\, R(d) r^2 \; dr \tag{21}$$

The second- and fourth-order radial integrals (Cp and Dq, respectively) are designed not to involve the angular coordinate of the ligands and so hopefully may refer directly to features of the M—L bonds or interactions. The angle θ must be treated as a *parameter*, even though it should take a value near that determined for the molecular structure, because of the gross approximation involved in representing ligands as point charges. It was further hoped that sensible relationships between Dq and Cp values between different geometries (e.g., those shown in Fig. 10) should emerge as the "chemically unimportant" angular or distortion features are factorized off from the radial parameters in this model. The precise nature of these relationships is not very clear however, when we realize that the definition of Dq in Eq. 20, for example, only coincides with that in Eq. 14 for exact O_h or T_d symmetry. In passing, it is worth mentioning that much effort has been invested in seeking "transferability" of parameters between ligands with common central metals or vice versa, the desire to do so deriving from the approximate "factorizability" of the spectrochemical series. The really important aspect of this factorizability is that it is only *approximate*; otherwise there would be little point in attempting to derive values for the ligand field parameters in more than a few systems.

Let us return then to the problem of low symmetry in transition metal complexes. The most direct and unassuming approach would be to write a symmetry-based expansion of the ligand field potential in terms of spherical harmonics. For a completely unsymmetrical molecule (C_1) this would be written as

$$V_{C_1} = aY_2^0 + bY_2^{\pm 1} + cY_2^{\pm 2} + dY_4^0 + eY_4^{\pm 1} + fY_4^{\pm 2} + gY_4^{\pm 3} + hY_4^{\pm 4} \qquad (22)$$

for d electrons, involving eight expansion coefficients as parameters. Even if this number of parameters is acceptable in zero-symmetry situations, there is little chemical utility in this approach. Alternatively we might employ the point-charge model, assigning Dq and Cp values to each ligand (or ligand type), and parameterize all bond angles. This method seems even worse. At this level of discussion, therefore, we can only identify a need. Either we use the chemically uninformative symmetry-defined parameters of Eq. 22 or we seek an approach that is equally applicable to molecules of any geometry and that relates as directly as possible to the local M—L interaction or bonding features in the system. I describe such an approach shortly.

XIII. DIMINISHING RETURNS IN MAGNETOCHEMISTRY

I mention in the opening paragraph of this chapter that the utility of magnetochemistry, that is, its contribution to the wider area of chemistry, was questioned a few years ago by Nyholm. It is especially significant that he, who was probably the chemist most responsible for the inclusion of magneto-chemistry as a routine tool in inorganic chemistry research laboratories through-out the world, should feel obliged, in his Plenary Lecture (42) at the 1967 International Conference on Coordination Chemistry meeting in Japan, to doubt the technique so close to his heart. Yet he was surely right, as must be apparent from the background sketched above. Let us summarize the problems in magnetochemistry that existed in 1967:

1. There were too many ligand field parameters without obvious chemical relevance.

2. Little relationship was evident between the models and parameters used for molecules of different geometry.

3. There was an inability to cope with very low or nonexistent symmetry.

In addition we may identify two problems not mentioned so far:

4. No effective method for the determination of the complete susceptibility tensors of triclinic crystals was available. Many papers had been published on this subject, yet, surprisingly, no reliable technique had been devised.

5. At that time, it was still generally difficult to measure principal crystal susceptibilities at temperatures lower than that of liquid nitrogen. As susceptibilities vary most rapidly at temperatures lower than this, studies performed within the 80–300°K range often were not sufficiently exacting for the quantum models proposed.

Overall, therefore, too much theoretical and computational effort, often used with insufficient data, was producing increasing technical complexity and minimal chemical insight. It is this conclusion that prompted Nyholm's criticizms and, for that matter, the opening sentence of this chapter. Over the past few years, however, the situation has been enormously improved in each respect as I now try to demonstrate.

XIV. CURRENTLY AVAILABLE TECHNIQUES

A. Apparatus

On the practical side, apparatus design and availability are much better now. At the "high-technology" end of the market we have seen the introduction of magnetometers (referred to by the makers as "susceptometers"!) built around the SQUID (semiconducting quantum interference devices) that are capable of the detection and measurement of extremely feeble magnets. As such these techniques are ideally (though not exclusively) suited to magnetic investigations in biological systems. Along more traditional lines, comprehensive commercial Faraday magnetometers capable of measurements on powders and single crystals down to 1.6°K, are now available. These apparatus usually employ a superconducting solenoid whose axis lies parallel to the sample suspension, and thus they have the disadvantage that crystal susceptibilities can only be measured in one direction at a time. Nevertheless, the technical expertise built into these systems frees the user of most of the difficulties previously experienced by nonexperts with liquid helium cryogenics. Our own low-temperature Faraday

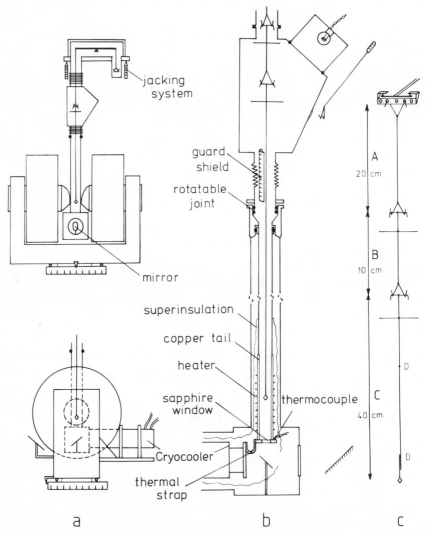

Fig. 11.   A single-crystal Faraday balance (12) for measurements in the temperature range 20–300°K: (a) general view, (b) the cryostat, (c) the crystal suspension.

balance (12), though less sophisticated than many commercial systems, can be constructed from commercially available units and its simplicity may deserve a brief review.

This device is shown schematically in Fig. 11. The overall height of the apparatus is about 1 m. Crystals, typically weighing 1–5 mg, are glued to a quartz suspension that is constructed so as to prevent sample rotation with respect to the microbalance arm and at the same time be unlinked in various places for ease of servicing and for the replacement of samples. The crystal is contained in an atmosphere of helium gas within a simple cryostat. Cooling to a base temperature of about 20°K is effected by a commercial closed-cycle heat pump, and any temperature between 20 and 300°K is maintained by balancing the cooling with a simple noninductively-wound Nichrome heater. The cooler is operated simply by electrically activating compressor pumps within the system. There are no cryogens, and no cryogenic problems can arise provided a suitable vacuum is maintained in the outer chamber. The complete cryostat and cooler are clamped to a 10 cm electromagnet free to rotate in the horizontal plane. Hence susceptibilities may be measured in all directions in the plane perpendicular to the suspension – a great advantage over the vertical superconducting solenoid systems. If temperatures below 20°K are required, an equivalent system can be constructed in which the heat pump is replaced by a cold-finger continuous-flow cryostat – a system requiring only a little cryogenic expertise. Overall, therefore, these balances permit the measurement of single-crystal susceptibilities down to 20°K (or to 3°K, with the continuous-flow device) in a routine manner. The apparatus is relatively robust and inexpensive.

B. Triclinic Crystals

Many equations have been published (though rarely applied) over the years giving the relationships between the principal susceptibilities and their orientations in triclinic crystals, and various measurements that may be made. In this sense, procedures for the determination of triclinic crystal susceptibilities have been known for a long time. However, except in cases of special molecular pseudosymmetry (much exploited several years ago for planar aromatic diamagnets), all these methods suffer from an inadequate treatment of the question of experimental accuracy. This is not a pedantic point, for the practical fact is that susceptibility measurements are often subject to errors that are sufficiently large to render the theoretically exact equations for triclinic systems insoluble. We recently developed practical and theoretical techniques for the measurement of the susceptibilities of triclinic crystals and successfully applied them to the cobalt benzoate systems referred to earlier (7). The obvious disadvantage of dealing with triclinic systems is the considerable amount of practical measurement and crystal manipulation required. The great advantage, however, is that, once measured, the crystal susceptibility tensor usually yields the molecular susceptibility tensor directly (by identity), thus providing six independent pieces of data (rather than three) with which to compare theory.

C. General Susceptibility Tensor Components

Turning from the more practical side of recent advances, let us consider the "routine" aspect of the calculation of molecular susceptibilities from the eigenvalues and eigenvectors produced by some ligand field model. The usual procedure, using Van Vleck's equation

$$K = \frac{N_A \sum_i (W_i^{12}/kT - 2W_i^{II}) e^{-W_i^0/kT}}{\sum_i e^{-W_i^0/kT}} \tag{23}$$

in conventional notation, is useful only if we know beforehand, that is, by symmetry, the directions in which the principal molecular susceptibilities lie. In other words, the formula is used only for the diagonal components of the susceptibility. Since we wish to consider molecules of low or nonexistent symmetry, then, however that unsymmetric ligand field is to be represented quantum mechanically, we do need a formula for all elements $K_{\alpha\beta}$ of the molecular susceptibility tensor. Such a formula was recently derived (30) and is a central feature of the present-day susceptibility calculations.

D. A General Program

Examples have been given to demonstrate the need for variable, and perhaps extensive, basis sets of functions in the calculation of magnetic moments. Similarly we have seen the necessity for an adequate treatment of (wholly or partly) unsymmetrical molecules. With these requirements in mind, I and my colleagues in Cambridge have constructed a system of programs (9, 30), using standard FORTRAN language, for the calculation and fitting of magnetic susceptibilities, electronic spectra, and ESR g values using bases for any d^n or f^n configuration in terms of any combination of free-ion terms and/or states for molecules of any (or no) symmetry. In this way all paramagnetic molecules, regardless of geometry, coordination number, metal, or ligand, can be treated in a common manner. All interelectron repulsion and spin—orbit coupling effects are included; but the issue of central importance for our present discussion is the parameterization of the ligand field. In the end, this is the main concern of magnetochemistry. We have chosen to represent the ligand field by the local "bonding-orientated" method, known as the *angular overlap model* (AOM).

XV. THE ANGULAR OVERLAP MODEL

The original angular overlap model (AOM), devised and developed by Schäffer and Jørgensen (37, 43–45), was based on the notion that the antibonding energy of a metal orbital was proportional to the square of the overlap integral between that metal orbital and an appropriate ligand function. Reappraisals of the AOM by Schäffer (43) have demonstrated the relationships among the AOM, perturbation theory and ligand field potentials expressed by Eqs. 13–18. A recent examination (28) of the nature of ligand field theory in general showed that such antibonding energies may not be exactly proportional to S^2_{ML} but confirms that the formalism of the method remains intact as a sensible and justifyable procedure. The most important aspect of the AOM is the way the ligand field may be parameterized in terms of local metal–ligand interaction, taking due regard of the local symmetry of these interactions. The essence of the approach, summarized below, is shown in Fig. 12.

Within the local frame of each M–L moiety the AOM method usually

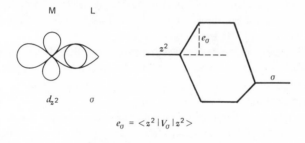

$$e_\sigma = \langle z^2 | V_\sigma | z^2 \rangle$$

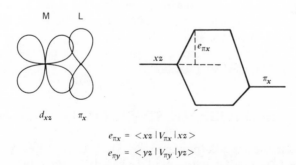

$$e_{\pi x} = \langle xz | V_{\pi x} | xz \rangle$$
$$e_{\pi y} = \langle yz | V_{\pi y} | yz \rangle$$

Fig. 12.   Representations of AOM parameters. The scheme is not necessarily based on an overlap criterion, as in mo theory, unless V_σ, V_π are associated uniquely with ligand σ and π functions (see Ref. 28).

presumes that the interactions are diagonal. By this is meant that the d_z^2 metal function interacts only with a ligand σ function, the d_{xz} only interacts with a ligand π_x function, and the d_{zy} only interacts with a π_y function. It is usual to neglect δ interaction (d_{xy} or $d_{x^2-y^2}$ with ligand δ functions), but these can be included within the formalism. (Actually, depending on origin definitions, "neglect" of δ interactions involves no approximation whatever.) The locally diagonal nature of these interactions is removed if the local M—L symmetry is not sufficiently close to C_2 symmetry or higher. This problem can be dealt with but is not discussed here.

It is important to note that the e parameters are defined as energy shifts, so that the detailed nature of V_σ, $V_{\pi x}$ and so forth need not be specified. Thus no separation between electrostatic or true covalent interactions is attempted, a philosophy that lies behind my present use of the word interaction rather than bond. The model is a ligand field model, not some *ab initio* SCF or many-body theory. Nevertheless, it appears reasonable to associate the sign of the e parameters fairly closely with the notions of ligand donor and acceptor properties; positive for donors and negative for acceptors. The final main assumption of the AOM approach is that energy shifts from different modes of bonding (σ π . . .) and from each ligand are presumed to be additive. Again, a recent study of ligand field theory supports the viability of this idea (28).

Overall, therefore, instead of representing the ligand field *as a whole* by parameters like Dq, Ds, and Dt, the AOM expresses it as a sum over all ligands and bonding modes, so conferring upon the local e parameters an immediate chemical "transparency." Molecules of any geometry are at once equally tractable, for the interaction at any one ligand site and orientation are related to a suitable (arbitrary) global reference frame simply by rotation matrices constructed from the known structural coordinates of the complex. The technicalities of the procedure are well-documented and need not concern us here. It is more appropriate and direct to look at the results of some studies that have used this model. In doing so, we note that many studies, often by the originators Schäffer and Jørgensen, and by their colleagues, have been concerned with the interpretation of electronic spectra. This chapter is concerned with magnetochemistry, so I hope the spectroscopic work is complemented by the mainly magnetochemical studies I now discuss.

XVI. SELECTED RECENT STUDIES IN PARAMAGNETISM

The molecule *trans*-Co(pyridine)$_4$(NCS)$_2$ is discussed earlier to illustrate the inadequacy of an axial approach to the ligand field distortion. The molecule is centrosymmetric, one immediate consequence of which is that the packing of the four pyridine groups around the central metal does not follow the expected

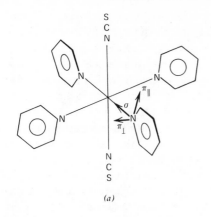

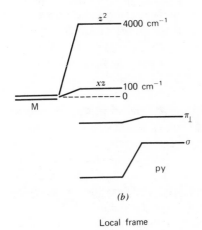

Fig. 13.   *trans*-Co(pyridine)$_4$(NCS)$_2$:
(*a*) orientation of pyridine π bonding parameters, (*b*) relative sizes of e_σ and $e_{\pi\perp}$ parameters for local Co-pyridine frame.

"propeller" fashion. As all donor atom—Co—donor atom bond angles are close to 90°, this feature constitutes the major structural departure from tetragonal symmetry. In passing, we may note that a point-charge representation would necessarily be incapable of reflecting this distortion, although a point-dipole scheme might. Within the angular overlap model, the e parameters refer to directions shown in Fig. 13, the orientation of $\pi_\parallel$ and $\pi_\perp$ being related directly to the known structure coordinates. Calculations were performed within the complete spin-quartet basis $^4P + ^4F$, and the parameter set comprised B, ζ, k for interelectron repulsion, spin—orbit coupling, and Stevens' orbital reduction factor, respectively; $e_\sigma(\text{py})$, $e_{\pi\perp}(\text{py})$, $e_{\pi\parallel}(\text{py})$, $e_\sigma(\text{NCS})$, $e_{\pi\,\text{av}}(\text{NCS})$.

This large list of parameters immediately raises the central *practical* difficulty with low-symmetry systems. Can we hope to determine unambiguous values for each parameter by fitting to the susceptibilities, perhaps together with any electronic spectral information? In general there can be no confident answer to this question, but it seems, from several studies so far undertaken, that, while each system presents its own difficulties, the experimental data are frequently determined by a relatively small subset of these parameters. One immediate simplification, perhaps, is that the Racah parameter B, which is so important for high-lying excited states, only affects magnetic moments a little, in that contributions from these states are roughly inversely proportional to their energies relative to the ground state. Furthermore, in some cases, correlations between the "fitting abilities" of (say) two parameters may be observed, without there being any significant weakening of the model's response to other parameters. The fact remains that low-symmetry systems beget a large number of parameters and involve the investigator in difficult walks through polyparameter space! Incidentally, the number of AOM parameters in any one system is not simply related to the number of coefficients in an expansion of the ligand field potential as spherical harmonics; however, the number of e parameters need not be more, and is often less, than the number of such expansion coefficients.

Returning to the study (31) of *trans*-Co(py)$_4$(NCS)$_2$, the model was fit to the principal crystal susceptibilities (to within 2%) and several main conclusions were reached.

1. The principal *molecular* susceptibilities, calculated as the penultimate step in reproducing the experimental *crystal* susceptibilities, were found to lie far from any M–L bond directions or from any suitable bisectors of these vectors; nor did the magnitudes of the principal molecular susceptibilities describe anything like an axially symmetric susceptibility tensor. It is clear why so much difficulty was experienced earlier in the interpretation of the crystal susceptibilities.

2. The e_σ values for the pyridine and thiocyanate ligands, though not sensitively determined, are between 3000 and 4000 cm^{-1}. In strict octahedral symmetry,

$$10Dq = 3e_\sigma - 4e_\pi \qquad (24)$$

and so with increasing experience with the AOM model, these e_σ values appear wholly unexceptional.

3. As expected, we found no M–L interaction ($e_{\pi\parallel} = 0$) in the plane of each pyridine ligand.

4. However, perpendicular to the pyridine planes, a small positive π interaction was unambiguously defined: 50 cm^{-1} < $e_{\pi\perp}$ < 130 cm^{-1}. Any value for $e_{\pi\perp}$ outside this range, produced hopelessly inadequate descriptions of the magnetic properties, and it is certainly this small amount of π interaction that is responsible for the low symmetry of the ligand field and the "unusual" orientation of the susceptibility tensor. This is an example of a sharply fitting parameter.

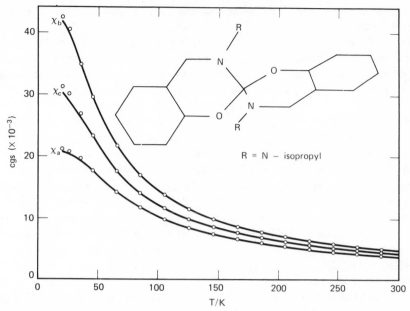

Fig. 14.   Bis(N-isopropylsalicylidene)nickel(II): The chemical structure shows coordination geometry, and the graph shows observed (o) and calculated (—) principal susceptibilities.

We might compare the e_σ(py) and $e_{\pi\perp}$(py) values pictorially (Fig. 13b) as a forceful demonstration, in the local M–L frames, of how small this π donation is. The important point for the moment is that this detailed study unambiguously demonstrates that the e parameters *immediately and obviously* reveal sensible chemical conclusions; in the present system that pyridine acts as a weak π donor.*

A similar study has been performed on the isomorphous iron(II) analogue of this system (31). The experimental susceptibilities were different, of course. The basis functions for the calculations were different also: in *trans*-Fe(py)$_4$ (NCS)$_2$ the d^6 configuration gives a ^5D term as the only level of maximum spin multiplicity. Otherwise the procedures followed those adopted for the cobalt analogue. Except that in this case the principal molecular susceptibilities were calculated to lie in entirely different but uninformative directions, all the results from the ferrous system virtually coincided with those from the cobalt one. This

*An identification of $e_\pi = 0$ with absolutely no π bonding appears to be a matter more of definition than of theoretical proof. In practical terms, however, the notion appears defensible by reference to an effective "calibration" in that e_π(NH$_3$) ~ 0; furthermore, the results described in this chapter lend strong support to an overlap criterion for AOM e parameters. See also Ref. 28.

is surely strong support for, and a vindication of, the general approach we have examined.

Another study (12) concerns the nominally tetrahedral complex, bis(N-isopropylsalicylidene)nickel(II) shown in Fig. 14. The principal crystal suscepti-bilities over the temperature range 20–300°K have been fitted, as shown in Fig. 14, within the complete spin-triplet basis $^3F + ^3P$ (further calculations demon-strated the unimportance of including spin-singlet terms in the model) using the parameter set $B \zeta k, e_\sigma(O), e_\sigma(N), e_{\pi\|}(N), e_{\pi\perp}(N), e_{\pi\|}(O), e_{\pi\perp}(O)$.

The main results to emerge from this study are that $e_\sigma(O) \sim e_\sigma(N) \sim$ 4000 cm^{-1}, an unexceptional value; that $e_{\pi\|}(N)$, defined parallel to the plane of the sal groups, is zero – a chemically sensible result; and that $e_{\pi\perp}(N) \sim$ 900 cm^{-1}. This last figure shows that the nominally sp^2 hybridized imine N atom enters into substantial π interaction with the nickel atom, serving also to put the 100 cm^{-1} π interaction of pyridine with Co/Fe above into perspective.

Closely related to this "tetrahedral" nickel molecule is the five-coordinate complex shown in Fig. 15. Bis(N-β-diethylamine-ethyl-5-chlorosalisyl-aldiminato)nickel(II) differs from the previous system in three main features: N-isopropyl groups are replaced by tertiary amines, the phenyl rings are substi-tuted with chlorine atoms para to the oxygen donor atoms, and the coordination is approximately square pyramidal rather than distorted tetrahedral. Crystal

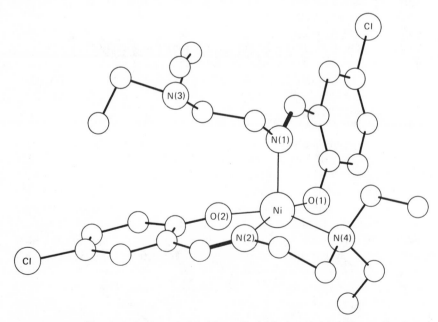

Fig. 15.   Bis(N-β-diethylamine-ethyl-5-chlorosalicylaldiminato)nickel(II) coordination geometry.

susceptibilities in the 20–300°K range have again been accurately reproduced within the $^3F + {}^3P$ basis (with a complete d^8 basis check). The parameter set was the same as for the "tetrahedral" system plus $e_\sigma(N_T)$ for the tertiary amine (only one of which coordinates to the nickel atom). The main conclusions from this study (11) are given below.

1. $e_\sigma(N) \sim e_\sigma(O) \sim 4000$ cm^{-1}, as for the "tetrahedral" analogue.

2. $e_\sigma(N_T) \sim 2500$ cm^{-1}. The much lower value accords well with the long Ni–N$_T$ bond lengths of 2.2 Å compared with Ni–N(imine) ~ 1.96 Å.

3. $e_{\pi\parallel}(N) = 0$; $e_{\pi\perp}(N) \sim 900$ cm^{-1}, again.

4. The π interaction from the oxygen atoms in this five-coordinate system is much larger than that in the four-coordinate one. $e_{\pi\perp}(O)$ this system $\sim$ 1500 cm^{-1} and $e_{\pi\perp}(O)$ tetrahedral system ~ 800 cm^{-1}.

We take this last result as an indication of the para substituent on the benzene ring. Whether one regards the effect as being due to a mesomeric donation of π charge from the chlorine to the ring and then to the oxygen or as the result of the chlorine substitution raising the energy of the ligand π orbitals to better match that of the metal is still open to question. One thing is clear, however: this sort of magnetochemical conclusion, referring to electronic effects originating far out on a ligand being detected by magnetism on the central metal, is not the sort arrived at using earlier models in which ligand field effects are represented by "global-type" parameters. There is no doubt that the present finding requires support from similar studies on related molecules, but the point here is that magneto*chemistry* is beginning to justify its name.

Finally, let us consider results from similar studies (15) of the series of nominally tetrahedral molecules, $M(PPh_3)_2X_2$, where M = Co^{II}, Ni^{II} and X = Cl, Br. The Ni(PPh$_3$)$_2$Cl$_2$ is famous as the first-reported tetrahedrally coordinated nickel(II) molecule. A recently completed study of the single-crystal magnetic susceptibilities of all four compounds together with a detailed analysis of the previously reported polarized single-crystal optical spectra of M(PPh$_3$)$_2$-Cl$_2$ (M = Ni, Co) yielded one main result. The $e_\pi(P)$ parameters were large and negative, while e_π(halogen) values were larger than expected and positive. This result seems to be understandable immediately in terms of π acceptance by the phosphines tending to promote π donation by the halogens. Recently Mason and Meek argued (41) from bond-additivity relationships and very accurate x-ray structure analyses that there is no evidence for M–L π bonding with phosphines. The magnetic study certainly argues for substantial π bonding in the present complexes *so far as the metal* d *orbitals are concerned*. The italicized phrase provides one means, of course, to make the two viewpoints compatible. Anyway, for our part, my colleagues and I are currently engaged in similar studies on a wide variety of phosphine complexes to explore the point further. These are being coupled with spin-polarized neutron diffraction studies in the hope of obtaining more direct evidence of spin delocalization.

Overall, therefore, these few examples demonstrate that the sort of questions now being asked in studies of paramagnetism lie within the wider area and interest of chemistry. We have hopefully reversed the situation of only a few years ago when the problems were essentially technical and inward-looking, having an everdecreasing relevance for chemists interested in molecules and bonding. Much of this was in our minds when we embarked on our study of antiferromagnetism in cobalt benzoates. Now that we have completed our study of paramagnetism, the time has come to return to the problem of magnetic exchange.

Magnetic Exchange

XVII. ONE-CENTER PROBLEMS IN COBALT BENZOATE

We left our discussion of antiferromagnetism in the binuclear cobalt benzoates at the point of inquiring how carefully the one-center problem at each cobalt atom must be treated with respect to both basis and geometry. However much we may have to approximate any treatment of the exchange phenomenon, it is surely proper to deal more thoroughly with other parts of the problem whenever possible. It is virtually certain that inadequate treatment of the one-center system will not be compensated for by any necessary limitations in the exchange model; yet this simple fact is often ignored. We begin, therefore, with an analysis of each square pyramidal five-coordinate cobalt center within the angular overlap model. The local symmetry is not exactly C_4, partly owing to small displacements of the nitrogen and oxygen donor atoms and partly owing to any π interaction with the axial quinoline ligands presumably not being axially symmetric with respect to the Co—N vector. While this effect may be small (and so it turns out), our experience with $M(py)_4(NCS)_2$ warns us that such effects must not be ignored at the outset. All these details are routinely incorporated within the AOM procedure from the known ligand coordinates and orientations. Calculations were performed within the complete spin-quartet basis spanned by $^4P + {}^4F$ free-ion terms. The parameter set comprises B, ζ, $e_\sigma(Q)$, $e_\sigma(Bz)$, $e_{\pi\perp}(Q)$, $e_{\pi\perp}(Bz)$, and, in the more detailed studies not discussed here, $e_{\pi\parallel}(Bz)$, where Q = axial quinoline, Bz = bridging benzoate. In an attempt to remove some of these parameters from the overall binuclear problem subsequently studied, the spectrum of the binuclear unit was fitted (7) on the assumption that the relatively weak exchange phenomenon would not significantly affect the spin-allowed quartet transitions. Only the "major" spectral parameters, B, $e_\sigma(Q)$, $e_\sigma(Bz)$ could be estimated in this way, there being

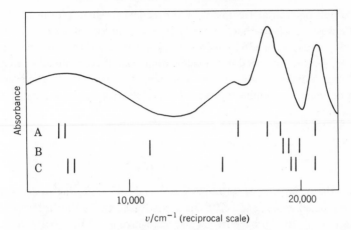

Fig. 16. Reflectance spectrum of $Co_2(benzoate)_4(quinoline)_2$ fitted with
(A) $B = 850$ cm^{-1}, $e_\sigma(Q) = 2500$ cm^{-1}, $e_\sigma(Bz) = 4000$ cm^{-1};
(B) $B = 850$ cm^{-1}, $e_\sigma(Q) = 4000$ cm^{-1}, $e_\sigma(Bz) = 2500$ cm^{-1};
(C) $B = 850$ cm^{-1}, $e_\sigma(Q) = 3200$ cm^{-1}, $e_\sigma(Bz) = 3700$ cm^{-1}.

insufficient dependence of transition energies on spin—orbit coupling or on the various π parameters. All features of the unpolarized spectrum in Fig. 16 are adequately reproduced by the model for $B = 850$ cm^{-1}, $e_\sigma(Q) = 2500$ cm^{-1}, $e_\sigma(Bz) = 4000$ cm^{-1}, which are reasonable values in view of earlier experience with one-center systems. Also included in the figure are two other sets of transition energies calculated with other parameter values to indicate the nature of variations considered (7).

XVIII. THE TWO-CENTER, PRODUCT BASIS

The eigenfunctions deriving from the one-center AOM calculations are expressed as linear combinations of the 40 basis functions from $^4F + {}^4P$, namely,

$$\psi_I = \sum_i^{40} c_{iI}\, \chi_i \tag{25}$$

in which it is convenient to regard the basis functions χ as three-hole Slater determinants built from the real d orbitals. Now, as is discussed in the introduction, it is computationally impractical to construct and diagonalize (perhaps thousands of times as are usually required in a many-parameter model) a product basis matrix for the binuclear system using all 40 basis functions (and hence yielding 1600 products). The central strategy of our two-center calculations is to build product functions of a few of the lowest-energy combinations (25) left after the full one-center treatment. The practical maximum for us was inclusion of the lowest 12 levels that, in a rough way, correspond to the formal $^4T_{1g}$ functions of an octahedral precursor. However, even though only these 12 functions, ψ, were included in the calculation, each is constructed from all 40 of the original basis determinants and contains a great deal of the relevant information defining the detailed ligand field, interelectron repulsion, and spin–orbit perturbations.

The two-center calculation is based on the construction and diagonalization of the 144-fold product basis under a binuclear Hamiltonian

$$\mathscr{H} = \mathscr{H}^A + \mathscr{H}^B + \mathscr{H}^{AB} \qquad (26)$$

in which the one-center Hamiltonians $\mathscr{H}^A$ and $\mathscr{H}^B$, referring to centers A and B, are discussed earlier and $\mathscr{H}^{AB}$ accounts for all two-center interactions. The justification of this procedure, while recognizing overlap between the two centers, the important problem of antisymmetrizing, and the detailed form of the effective Hamiltonian $\mathscr{H}^{AB}$, are matters of some complexity and are not reviewed here; the reader is referred to the original paper (8). Our present concern is with localized exchange pathways and for this it is sufficient to sketch the breakdown of the "exchange" matrix elements.

$$\langle \psi_I^A \psi_J^B | \mathscr{H}^{AB} | \psi_K^A \psi_L^B \rangle = \sum_i^{40} \sum_j^{40} \sum_k^{40} \sum_l^{40} c_{iI}^* c_{jJ}^* c_{kK} c_{lL} \langle \chi_i^A \chi_j^B | \mathscr{H}^{AB} | \chi_k^A \chi_l^B \rangle \qquad (27)$$

Note that the origin of some (but by no means all) of the computational complexity lies in products like 40^4 (~2.5 M) which occur over 10,000 times in any one matrix!

XIX. INTERACTION PARAMETERS

Omitting the bulk of the technical argument, matrix elements (Eq. 27) may be broken down into linear combinations of surviving orbital matrix elements of the type:

$$\langle \phi_a^A \phi_b^B | \mathscr{H}^{AB} | \phi_b^A \phi_a^B \rangle \qquad (28)$$

in which the ϕ are one-hole orbitals on the respective centers. In turn we may define exchange or, better, interaction parameters such as the following:

$$J_{\zeta\zeta} \equiv \langle xy^A(1)xy^B(2) | \mathscr{H}^{AB} | xy^A(2)xy^B(1) \rangle$$

$$J_{\epsilon\epsilon} \equiv \langle x^2 - y^{2\,A}(1)x^2 - y^{2\,B}(2) | \mathscr{H}^{AB} | x^2 - y^{2\,A}(2)x^2 - y^{2\,B}(1) \rangle$$

$$J_{\theta\theta} \equiv \langle z^{2\,A}(1)z^{2\,B}(2) | \mathscr{H}^{AB} | z^{2\,A}(2)z^{2\,B}(1) \rangle \qquad (29)$$

$$J_{\zeta\epsilon} \equiv \langle xy^A(1)x^2 - y^{2\,B}(2) | \mathscr{H}^{AB} | x^2 - y^{2\,A}(2)xy^B(1) \rangle$$

$$J_{\zeta\theta} \equiv \langle xy^A(1)z^{2\,B}(2) | \mathscr{H}^{AB} | z^{2\,A}(2)xy^B(1) \rangle$$

The J parameters refer to local exchange pathways, although they implicitly involve contributions from all four bridging benzoates simultaneously. $J_{\zeta\zeta}$ relates to a pathway in which the exchange processes couple d_{xy} functions on each metal center. This may involve direct metal—metal interaction but is more likely to involve the superexchange mechanism through the delocalized benzoate π orbitals that overlap directly with each d_{xy} orbital. Using overlap as a criterion, we might expect this pathway to be more effective than those represented by $J_{\epsilon\epsilon}$ or $J_{\theta\theta}$ involving $d_{x^2-y^2}$ and d_{z^2}, respectively. Again direct interaction may contribute, but any superexchange process would involve routes through the benzoate σ framework, which is not expected to furnish good delocalization. The "cross terms" $J_{\zeta\epsilon}$ and $J_{\zeta\theta}$ have almost identically no effect on the exchange processes in coupled d^7 systems in C_4 symmetry, a group-theoretical result that does not carry over to the d^9 systems, which are discussed later. Other J parameters, which may be written like Eq. 27, involve orthogonal pathways and their relevance in the present antiferromagnetic systems is probably minimal.

XX. RESULTS

With this background, let us summarize the problems and results of our studies on three binuclear cobalt benzoates. The systems involve benzoate, o-nitrobenzoate, or o-methylbenzoate (toluate) bridges with the axial ligands being either 4-methyl quinoline or 6-methyl quinoline, the appropriate combinations being determined by the ability to prepare suitable single crystals. A practical difficulty obtains with all three benzoate systems studied in that they form triclinic crystals (4, 13, 16). The problems associated with triclinic crystals are outlined above and their recent solution is discussed. In recompense for all the considerable experimental effort involved in the determination of these crystal susceptibility tensors we have, with only one molecule in the asymmetric unit as here (actually one-half of the centrosymmetric dimer), an experiment yielding the molecular magnetic tensor directly. In each case the tensors are rhombic with the smallest principal susceptibility lying nearly parallel to the Co—Co vector. Based on the maxima in the susceptibility versus temperature curves, the antiferromagnetic effect increases along the bridging series in the order nitrobenzoate, benzoate, toluate. Running with this ordering is a small trend towards greater axiality of the susceptibility tensor and a closer alignment of the minimum susceptibility with the Co—Co vector (7, 14).

An additional problem with triclinic systems is concerned with the process of fitting theory to experiment, for here it is necessary to reproduce all six independent components of the magnetic tensor simultaneously rather than, perhaps, just the three principal moments in more symmetric crystals. Coupled

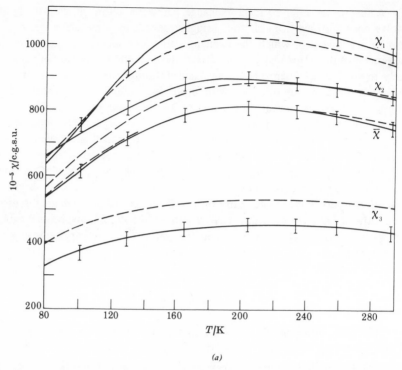

(a)

Fig. 17. $Co_2(benzoate)_4(4\text{-Me-quinoline})_2$. *(a)* Comparison between observed (——) and calculated (– – –) principal and average susceptibilities over the temperature range $80-300°K$.

with this is a problem in the presentation of the results, which, though not fundamental, is worth mentioning. After a great number of parameter variations, involving a computational investment in excess of 50 times that required for an analogous one-center investigation, acceptable agreement between theory and experiment has been achieved. The quality of fit shown in Fig. 17*a* for $Co_2(benzoate)_4(4\text{-Me quinoline})_2$ [not as good as that obtained in the, as yet, incomplete study of the nitrobenzoate system (14)] is not as exact as that now commonly obtained in one-center studies. It is, however, easily better than any others based on different parameter values. However, a more revealing representation of the theoretical agreement is shown in Fig. 17*b*, in which projections of observed and calculated susceptibility ellipsoids are compared at a representative temperature (205°K); I suggest that the agreement is quite good.

The fit shown in these figures was obtained with the B, $e_\sigma(Q)$, $e_\sigma(Bz)$ values cited above, together with $e_{\pi\perp}(Q) \sim 100$ cm^{-1}, $e_{\pi\perp}(Bz) \sim 375$ cm^{-1}, and $J_{\zeta\zeta} = 1200$ cm^{-1}. The important feature of the calculations, however, is that *no fit is possible at all for other J parameters* either in isolation or in conjunction

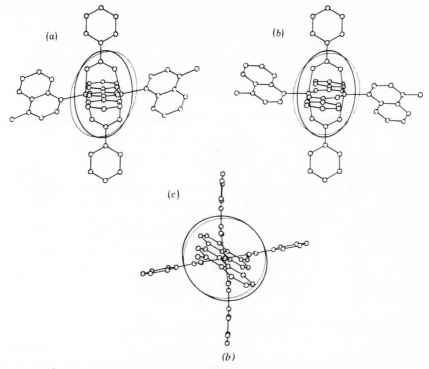

(b) At 205°K projections of the observed (thick lines) and calculated susceptibility ellipsoids down three mutually orthogonal directions (7).

with others. Values of $J_{\epsilon\epsilon}$ or $J_{\theta\theta}$, for example, in excess of 40 cm^{-1} taken in conjunction with $J_{\zeta\zeta}$ near 1200 cm^{-1} radically alter the magnitude of magnetic anisotropies, the absolute value of χ, the temperature dependencies of all χ, or, indeed, all of these features. The study (7, 8) therefore affords an *unambiguous selection of the* $\text{xy}^A\text{--xy}^B$ *exchange pathway* in this molecule, and this presumably involves the delocalized benzoate π orbitals. Preliminary results on $Co_2(o\text{-nitrobenzoate})_4(6\text{-Me quinoline})_2$ and $Co_2(\text{toluate})_4(4\text{-Me quinoline})_2$ yield similar conclusions with $J_{\zeta\zeta}$ values near 900 and 1600 cm^{-1}, respectively. The trend in $J_{\zeta\zeta}$ values appears to reflect the electron-withdrawing nature of the nitro substituent and electron-donating properties of the methyl group. Finally, I might remark that the magnitude of these J parameters, so large in comparison with typical $\mathscr{J}_{HDVV}$ values in other systems, need not cause concern. Firstly, they refer to the simultaneous participation in the exchange process by all four bridging groups, and secondly, they may be compared, in some very loose sense, with the one-center AOM parameters. Insofar as $J_{\zeta\zeta}$ refers to superexchange through the benzoate π bonds, a degree of "insipient" bond formation by way of this network is implied: against that background an exchange pathway parameter of 1200 cm^{-1} for four π linkages need not seem excessive. Furthermore, the J parameters are defined with respect to orbitals on the metal atoms rather than states and so the overall exchange effect is less than the present $J_{\zeta\zeta}$ value might imply in that the participation of d_{xy} in the ground states is considerably less than total.

XXI. COPPER ACETATE

Our pathway through this review thus comes full circle, for, as the xy–xy exchange route has been demonstrated in the cobalt benzoates, it is natural to inquire if a similar mechanism can also account for the antiferromagnetism in the structurally analogous copper acetate discussed at the outset. There is no question but that the magnetism of copper acetate is characterized by an $(x^2 - y^2)^A (x^2 - y^2)^B$ product ground state ($\delta\delta$) split into a spin-singlet lying 286 cm^{-1} below a spin-triplet. The point of interest here is whether this splitting of δ functions can originate solely in a coupling between $\delta'(xy)$ functions in the exchange process.

We let the lowest energy orbital on each copper atom be written as the linear combination

$$\psi_\delta = a(x^2 - y^2) + b(xy) \tag{30}$$

Let us further presume that $J_{\zeta\zeta}$ and $J_{\zeta\epsilon}$ (Eq. 29) represent the only important "connection," for present purposes, between the two centers. The "cross-term" $J_{\zeta\epsilon}$, whose effect does *not* vanish in d^9 systems, might be expected to take a value rather less than $J_{\zeta\zeta}$; the exact relationship between them need not concern us here. As an indicator we focus attention of the smaller exchange parameter $J_{\zeta\epsilon}$. We now ask what value of $J_{\zeta\epsilon}$ is required to reproduce the HDVV splitting $\mathscr{J}_{\text{HDVV}} = -286 \text{ cm}^{-1}$ in the ground state as a result of various contributions to the mixing coefficient b in Eq. 30.

We first consider spin–orbit mixing between $(x^2 - y^2)$ and (xy) functions of the copper atom, a process that has been suggested (36) to provide a suitable means of making a $\delta'(xy)$ pathway viable. Taking the maximum likely λ value of -830 cm^{-1} from the free ion, the trivial degree of mixing that results leads to an estimate of $J_{\zeta\epsilon}$ around $20,000 \text{ cm}^{-1}$. A closer approach derives from considering the effects of spin–orbit coupling and the low-symmetry ligand field together. Mixing by means of this latter effect arises if the local one-center symmetry is less than C_{2v}. Using the ligand coordinates and orientations given by the early x-ray structural analysis (47), together with the usual AOM approach, we find the mixing implies a required $J_{\zeta\epsilon}$ value of about 5000 cm^{-1}: perhaps this might just be acceptable. However, a similar calculation based on the more recent and accurate coordinates from neutron diffraction analysis (10), which shows the molecule to be more symmetrical than first supposed, puts $J_{\zeta\epsilon}$ back up to about 9000 cm^{-1}. Overall, therefore, the "usual" one-center perturbations cannot produce a ground state with sufficient (xy) character for the $xy-xy$ exchange pathway to provide a viable explanation of the observed antiferromagnetism.

Nevertheless, I do not consider the δ bond, originally proposed, to provide the most likely origin of the exchange in copper acetate. The $xy-xy$ pathway through the acetate π orbitals can still account for the experimental results, but essentially by a "second-order" phenomenon. Thus, as discussed by Ballhausen and Hansen (3), the "symmetric" product functions for the dimer, for example, $(xy^A)(xy^B)$ and $(x^2 - y^2)^A(x^2 - y^2)^B$ all transform as the totally symmetric representation of the binulcear point group and may therefore mix. It has been proposed (27) that such mixing can occur by means of the total Hamiltonians of the system and arising essentially from the closeness of the two centers. The interaction proposed need not necessarily imply direct metal–metal exchange as opposed to superexchange coupling, for it takes place between the two monomeric units as a whole. We have shown elsewhere that the extent of such mixing need only be quite small to make the $xy-xy$ superexchange pathway relevant once again. The same approach simultaneously allows the electronic spectrum, the ESR g values and, perhaps, the zero-field splitting factors, as well as the single-crystal magnetism, all to be satisfactorily reproduced. Undoubtedly the "second-order" nature of our explanation (27) of the magnetism of copper acetate, which extends to all similar carboxylate dimers, lacks the directness of the conclusions reached for the cobalt analogue. Yet it shows a relationship among all these binuclear carboxylates, as well as reminds us of the subtle nature of exchange processes. In the d^7 systems, of course, the participation of the important xy orbitals in the ground state is first order and so an analogous second-order correction would be relatively unimportant and, in any case, sequestered in the primary mechanism.

XXII. CONCLUSIONS

I have reviewed a number of studies of the magnetism of one- and two-center systems to show an important difference between magnetic and spectroscopic properties. Provided we are interested only in band positions in an electronic spectrum, a great deal can be learned merely from qualitative interpretations essentially because the splitting patterns often directly reflect the group theoretical representations. A similarly qualitative treatment of spectral intensities (beyond a simple spin-selection rule) yields virtually nothing. Magnetic susceptibility studies are also rather like that: "10%" theories often yield 90% rubbish. We must set up models for magnetism that are reasonably complete, certainly including spin—orbit coupling and a true picture of the actual geometry in the system. Some parameters may not be of prime interest in any given study [for some, the one-center parameters B, $e_\sigma(Q)$, $e_\sigma(Bz)$ in the benzoate study would be of little interest], but to obtain reliable values for those parameters that *are* of interest, a comprehensive study must be made; and in this context, single-crystal susceptibility measurements are surely mandatory.

Historically one can see why small distortions have been neglected for so long. When crystal field theory was first proposed, physicists were concerned that small details of a crystal lattice might cause energy splittings that could markedly affect magnetic properties, and if they did this would seem to remove the essential simplicity and utility of the crystal-field method so far as magnetism was concerned. However, Van Vleck (48) proved a theorem showing that, provided such splittings were not larger than kT, their effects on *average* susceptibilities are very small. Accordingly the physicists neglected low-symmetry components for many years and the chemists followed suit. It is now realized that no useful chemical information may generally emerge from mean moments alone, but for measurements of crystal anisotropies, Van Vleck's theorem ceases to be useful and the original concerns come to the fore once more. The point is that now there are methods to deal with these problems in a way that is beneficial to our understanding of chemical bonding.

The difference between magnetism and magnetochemistry is that the goal for the latter is to see some *chemically meaningful* results. The global treatment of the ligand field, though occasionally more rigorous from a symmetry point of view, does not provide chemical insight. *Local* AOM parameters and pathway treatments in exchange problems seem to offer the best chance of providing that chemical interest. The models described here are parametric ones. This is inevitable in view of the wide range of systems studied and the present state of the art with *ab initio* calculations of any sort. In the end, a parametric theory is only useful to us, however, if the chemical significance of these parameters is immediately obvious.

References

1. P. W. Anderson, *Phys. Rev., 181,* 25 (1969).
2. P. W. Anderson, *Solid State Phys.*, *14*, 99 (1963).
3. C. J. Ballhausen and A. E. Hansen, *Trans. Faraday Soc.*, *61*, 631 (1965).
4. S. Bellard and G. M. Sheldrick, in preparation.
5. R. B. Bentley, F. E. Mabbs, W. R. Smail, M. Gerloch, and J. Lewis, *J. Chem. Soc. A.*, *1970*, 3003.
6. B. Bleaney and K. D. Bowers, *Proc. Roy. Soc. (Lond.)*, *A214*, 451 (1952).
7. P. D. W. Boyd, J. E. Davies, and M. Gerloch, *Proc. Roy. Soc. (Lond.)*, *A360*, 191 (1978).
8. P. D. W. Boyd, M. Gerloch, J. H. Harding, and R. G. Woolley, Proc. Roy. Soc. *(Lond.)*, *A360*, 161 (1978).

9. "CAMMAG," FORTRAN program by D. A. Cruse, J. E. Davies, J. H. Harding, M. Gerloch, D. J. Mackey, and R. F. McMeeking.
10. R. Chidambaram and G. M. Brown, *Acta. Crystallogr.*, *B29*, 2393 (1973).
11. D. A. Cruse and M. Gerloch, *J. Chem. Soc. Dalton Trans.*, *1977*, 1613.
12. D. A. Cruse and M. Gerloch, *J. Chem. Soc. Dalton Trans.*, *1977*, 152.
13. J. E. Davies, S. Bellard, and G. M. Sheldrick, in preparation.
14. J. E. Davies and M. Gerloch, in preparation.
15. J. E. Davies, M. Gerloch, and D. J. Phillips, in press. (*J. Chem. Soc. Dalton.*)
16. J. E. Davies, V. Rivera, and G. M. Sheldrick, *Acta, Crystallogr.*, *B33*, 156 (1977).
17. J. C. Donini, B. R. Hollebone, and A. P. B. Lever, *Prog. Inorg. Chem.*, *22*, 225 (1977).
18. J. Drew, M. B. Hursthouse, P. Thornton, and A. J. Welch, *Chem. Commun.*, *1973*, 52.
19. J. Ferguson, *J. Chem. Phys.*, *40*, 3406 (1964).
20. B. N. Figgis, *Trans. Faraday Soc.*, *57*, 204 (1961).
21. B. N. Figgis, M. Gerloch, J. Lewis, F. E. Mabbs, and G. A. Webb, *J. Chem. Soc. A.*, *1968*, 2086.
22. B. N. Figgis, M. Gerloch, J. Lewis, and R. C. Slade, *J. Chem. Soc. A.*, *1968*, 2028.
23. B. N. Figgis, J. Lewis, F. E. Mabbs, and G. A. Webb, *J. Chem. Soc. A.*, *1967*, 442.
24. B. N. Figgis, J. Lewis, F. E. Mabbs, and G. A. Webb, *J. Chem. Soc. A.*, *1966*, 1411.
25. B. N. Figgis and R. L. Martin, *J. Chem. Soc.*, *1956*, 3837.
26. M. Gerloch, *J. Chem. Soc. A.*, *1968*, 2023.
27. M. Gerloch and J. H. Harding, *Proc. Roy. Soc. (Lond.)*, *A360*, 211 (1978).
28. M. Gerloch, J. H. Harding, and R. G. Woolley, in preparation.
29. M. Gerloch, J. Lewis, G. G. Phillips, and P. N. Quested, *J. Chem. Soc. A.*, *1970*, 1941.
30. M. Gerloch and R. F. McMeeking, *J. Chem. Soc. Dalton Trans.*, *1975*, 2443.
31. M. Gerloch, R. F. McMeeking, and A. M. White, *J. Chem. Soc. Dalton Trans.*, *1975*, 2452.
32. M. Gerloch and J. R. Miller, *Prog. Inorg. Chem.*, *10*, 1 (1968).
33. M. Gerloch, and P. N. Quested, *J. Chem. Soc. A.*, *1971*, 2308.
34. M. Gerloch and R. C. Slade, *Ligand Field Parameters*, Cambridge University Press, London, 1973.
35. J. B. Goodenough, *Phys. Chem. Solids*, *6*, 281 (1958).
36. A. K. Gregson, R. L. Martin, and S. Mitra, *Proc. Roy. Soc. (Lond.)*, *A320*, 473 (1970).
37. C. K. Jørgensen, *Modern Aspects of Ligand Field Theory*, North-Holland, Amsterdam, 1971.
38. J. Kanamori, *Phys. Chem. Solids*, *10*, 87 (1959).
39. K. S. Krishnan and K. Lonsdale, *Proc. Roy. Soc. (Lond.)*, *A156*, 597 (1936).
40. P. O. Löwdin, *Rev. Mod. Phys.*, *34*, 80 (1962).
41. R. Mason and D. W. Meek, *Angew. Chem. Int. Ed. Engl.*, *17*, 183 (1978).
42. R. S. Nyholm, *IUPAC 10th Int. Conf. Coord. Chem., Jap. 1967.*
43. C. E. Schäffer, *Struct. Bonding*, *5*, 68 (1968).
44. C. E. Schäffer, and C. K. Jørgensen, *J. Inorg. Nucl. Chem.*, *8*, 143 (1958).
45. C. E. Schäffer, and C. K. Jørgensen, *Mat. Fys. Medd.*, *34*, 13 (1965).
46. K. W. H. Stevens, *Proc. Roy. Soc. (Lond.)*, *A219*, 542 (1954).
47. J. N. van Niekerk and F. R. L. Schoening, *Acta Crystallogr.*, *6*, 227 (1953).
48. J. H. Van Vleck, *Theory of Electric and Magnetic Susceptibilities*, Oxford University Press, London, 1932.

UV Photoelectron Spectroscopy in Transition Metal Chemistry

ALAN H. COWLEY

Department of Chemistry, The University of Texas at Austin, Austin, Texas

I. INTRODUCTION

A. Preliminaries and Scope of Review

When a photon of sufficient energy impinges on a neutral molecule, M, it is possible to eject an electron, leaving behind a radical cation, $M^{\dagger}$, in one of several possible electronic states. It is well-known that energy is conserved during the photoionization event; hence if $h\nu$ represents the energy of the incident photons in Eq. 1 it is obvious that measurement of the kinetic energies of the

$$h\nu = IE_x + E_k \qquad (1)$$

photoemitted electrons, E_k, provides experimental access to the various ionization energies, IE_x. Historically the field of photoelectron spectroscopy (PES) grew in a bifurcated manner, x-ray photoelectron spectroscopy (XPS or ESCA) being used primarily for the ionization of core electrons and the lower energy, higher resolution technique of ultraviolet photoelectron spectroscopy (UPS) being employed for the study of valence electrons. Even though recent conceptual and technical developments have tended to unify the XPS and UPS areas, the present work is concerned only with the latter in the interest of maintaining the chapter within manageable proportions.

Following the pioneering work of Turner et al. (273) most ultraviolet photoelectron spectrometers are designed to utilize He(I) photons with an energy of 21.22 eV as the source of monochromatic radiation. Subsequently the observational range of the UPS experiment was extended by the development of He(II) photon sources of energy 40.81 eV. The UPS method is employed generally to study photoionization in the vapor phase; however, considerable progress has been made recently in applying the UPS technique to solids (68, 250) and solutions (8, 222, 223). A discussion of the interesting field of chemisorbed species would in itself constitute an entire review article. No attempt is made here to consider the various technical details pertaining to the UPS experiment; for this aspect the reader is referred to the various treatises on the subject (111, 242, 273).

The basic objective of the present chapter is to discuss the vapor-phase UPS results and allied theoretical work for transition metal compounds in as comprehensive a manner as possible. Even though some 1978 articles are covered here, the literature search should be regarded as being complete only to the end of 1977. Some aspects of this particular subject have been presented previously (111, 128, 133, 165, 166, 231, 242, 272, 273); however, heretofore there has not been a comprehensive, up-to-date review article on the utility of UPS in transition metal chemistry.

B. Molecular Orbitals and Ionic States

Consider a model system M that features three singly degenerate molecular orbitals (MOs), 1, 2, and 3 and one virtual orbital, 4 (Fig. 1). Electron ejection from MOs 1, 2, and 3 affords three distinct doublet states of the cation M^+ as shown in Figs. $1b-1d$. Thus the UPS of M should exhibit three peaks, the ionization energies (IEs) for which correspond to the differences in energy between that of the molecular ground state and those of the three ionic states. There are, in fact, additional possible consequences of the impingement of photons on M. In the "shake-up" phenomenon (Figs. $1e$ and f) the photo-ionization is accompanied by the promotion of one of the remaining electrons into the virtual orbital, thus producing excited configurations of the cation M^+ containing doublet and quartet states. In the "shake-off" process the system is sufficiently disturbed that the excited electron is eliminated, thereby producing the configuration depicted in Fig. $1g$.

In view of the foregoing discussion it is clear that UPS data should be interpreted with reference to the appropriate electronic states rather than

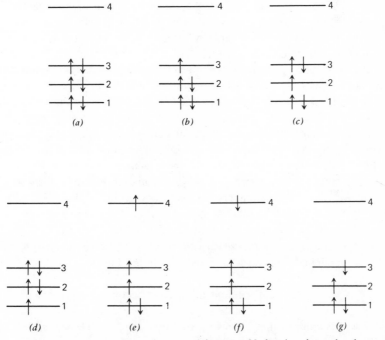

Fig. 1. Orbital occupancy scheme for a model system M showing the molecular ground state, three doublet ionic states, and "shake-up" and "shake-off" phenomena.

molecular orbitals. The preferred interpretational procedure is to perform, for example, Hartree—Fock self-consistent field (SCF) calculations on the molecular ground state and the relevant ionic states and to compare these computed energy differences with the experimental ionization energies (IEs). It is seen below that so-called ΔSCF calculations have been performed on a few transition metal compounds. However, this procedure involves considerable expenditure of computer time for many-electron systems. Two general approaches have been adopted to minimize the computational expense. The most common short cut is to invoke Koopman's theorem (196), which equates the negatives of the SCF-computed eigenvalues of the molecular ground state with the ionization energies of the various molecular orbitals (Eq. 2). While this theorem provides an

$$IE_x = -\epsilon_j^{SCF} \tag{2}$$

inexpensive and, in some respects, conceptually satisfying connection between photoelectron spectroscopy and molecular orbital theory, it should be recalled that the "frozen orbital" approach neglects *inter alia* orbital rescaling and electron correlation effects. It has become clear that Koopmans' theorem is a bad approximation for the photoionization of MOs with predominantly metal character. ΔSCF calculations have revealed that for metal-localized MOs considerable charge migration toward the metal occurs on proceeding from the molecular ground state to the corresponding ionic state.

As a compromise between the time-consuming ΔSCF calculations and the severe approximations inherent in Koopmans' theorem, an increasing number of authors are turning their attention to SCF—X_α scattered wave calculations (188, 189, 265). This type of calculation can be implemented for many-electron systems at considerably less cost than for the Hartree—Fock methods. Equally importantly, the X_α method permits the direct calculation of ionization energies by the transition-state method. In this approach the difference of two state energies is computed by means of a single calculation in the "transition state" that involves occupation numbers halfway between the initial and final states. Generally the results of X_α transition-state IEs are in reasonable agreement with those computed by the more laborious ΔSCF method. One disadvantage of the X_α method, however, is the sensitivity of the results to the nature of the potentials employed. This aspect is elaborated in Section II.

C. Photoionization Cross Sections

There is considerable interest in the intensities of UPS bands both theoretically and because of their utility in making spectral assignments. Nearly all theoretical approaches to the calculation of photoionization cross sections, σ, rely on three basic assumptions: (1) the interaction of the radiation field with the molecule is weak, (2) the electric dipole approximation is valid, implying that the momentum of the ionizing photon is effectively neglected, and (3) the electrons are treated nonrelativistically. On the basis of the foregoing assumptions

$$\sigma = \frac{2\pi e^2 h^2}{m^2 c \nu} \left| \mu \cdot \left\langle A \left| \sum_i \nabla \right| B \right\rangle \right|^2$$

where ν is the frequency of the radiation field, μ is the polarization vector of the radiation, A is the wave function appropriate to the final state of the system, and B is the wave function that describes the initial state of the system. For polyatomic molecules the initial state is described by a single determinantal wave function calculated in the Hartree—Fock scheme, and the final state usually involves the plane wave approximation for the ejected electron (98, 115,254, 269). The theoretical problem is rendered more tractable by invoking the frozen-orbital approximation.

At this point photoionization cross sections have been computed mostly for diatomic molecules, π-electron systems, and other relatively small molecules [see Rabalais (242) for a summary of this work up to 1976]. Very few photoionization cross section calculations have been performed (108) on transition metal systems and the agreement with experimental intensities is rather poor. For the most part, therefore, one must rely on empirical trends when dealing with the photoionization of metal-containing molecules. A number of such trends have now emerged and are useful for spectral assignment.

1. For the photoionization of predominantly metal nd orbitals there is generally a significant increase in cross section as the principal quantum number n increases (172). This phenomenon is referred to as the "heavy atom effect."

2. Generally the cross sections for the ionization of MOs with appreciable metal valence d character increased markedly compared to ligand MOs on proceeding from He(I) to He(II) radiation (108, 172, 174). This effect is most pronounced when the metal atom is attached to a ligand with a principal quantum number $\geqslant 3$, such as S, Cl, Br, and I.

3. The He(I) photoionization cross sections for nf orbitals are rather small and thus nf contributions are difficult to detect in He(I) UPS (232). However, the photoionization cross sections are somewhat larger for He(II) radiation (107).

4. Regarding the photoionization of predominantly ligand MOs, many authors proceed on the assumption that there is a crude relationship between spectral intensity and orbital degeneracy. This assumption should be viewed with caution, however, since the generalization is accurate only for the ionization of electrons with similar localization properties.

D. The Photoionization of Open-Shell Systems

Complexities can arise when dealing with the UPS of open-shell molecules because photoionization of either the open-shell itself or of subsequent closed shells can generate a large number of ionic states. As an example of these complexities, consider two ionizations of the paramagnetic carbonyl $V(CO)_6$ (121). Ionization of the open-shell, namely, $(t_{1u})^6 (t_{2g})^5 \rightarrow (t_{1u})^6 (t_{2g})^4$, leads to four ionic states: $^1A_{1g}$, ^1E_g, $^1T_{2g}$, and $^3T_{1g}$, while the ionization process $(t_{1u})^6 (t_{2g})^5 \rightarrow (t_{1u})^5 (t_{2g})^5$ results in a total of eight ionic states: $^{1,3}T_{2u}$, $^{1,3}T_{1u}$, $^{1,3}E_u$, and $^{1,3}A_{2u}$. The rules for predicting the ionic states formed and the relative intensities of the UPS peaks have been derived (85, 86) and are summarized briefly below.

1. If a closed shell is ionized, all states arising from the coupling of the positive hole with the open-shell state will be realized, the relative cross sections for the production of these states being in proportion to the total (i.e., spin-orbital) degeneracies. For example, the relative photoionization cross sections for the production of $^3T_{2u}$, $^1T_{2u}$, ^3E_u, ^1E_u, $^3A_{2u}$, and $^1A_{2u}$ states would be in the ratio 9:3:6:2:3:1.

2. If the orbitals belonging to different subshells are assumed to have the same one-electron cross sections, the integrated ionization cross section of a particular subshell is simply proportional to the occupancy of that orbital in the subshell in the molecule.

3. If the open-shell is ionized, the relative probabilities of producing different ionic states will reflect the fractional parentage coefficients (44, 155, 156), which may, but will not in general, be proportional to spin-orbital degeneracies.

4. If a molecule contains two or more open-shells, it is necessary to consider the coupling that already exists between the different open-shells in the molecule. The probability of ionization is usually expressed in terms of Racah coefficients and fractional parentage coefficients.

II. METAL CARBONYLS

The electronic structures of the metal carbonyls have been investigated vigorously over the past few years because the interaction between CO and metals (76, 137) is essential for an understanding of both organometallic chemistry and CO chemisorption phenomena. There is continuing debate concerning the relative importance of σ-donation from the carbon "lone pair" of carbon monoxide and "back donation" from metal d electrons into vacant π^* ligand orbitals. The most frequently cited data in favor of π back donation are the trends of CO stretching frequencies (109, 192, 193) of tetracoordinate and hexacoordinate metal carbonyls.

A. Nickel Carbonyl

A good starting point for a discussion of the bonding in metal carbonyls is to consider a very qualitative "back of the envelope" scheme for $Ni(CO)_4$ in which only σ bonding between $Ni(3d)$ and 5σ (carbon "lone pair") carbon monoxide MOs are considered (Fig. 2). The essential features of the scheme are the splitting of the $Ni(3d)$ AOs into t_2 and e sets on account of the ligand field of T_d symmetry, and σ-bonding interactions between symmetry-adapted combinations of 5σ MOs of CO molecules and $Ni(3d)/Ni(4s)$ AOs, which produce a_1 and t_2 MOs in $Ni(CO)_4$. When other valence shell interactions are included, the MO scheme becomes significantly more complex. For instance, the 4σ MOs of the CO ligands produce an additional set of t_2 and a_1 symmetry-adapted MOs and the 1π and 2π MOs of CO each result in $t_1 + t_2 + e$ sets of orbitals. Discussions of π back bonding would revolve primarily around the interaction of the t_2 and e symmetry-adapted MOs from the 2π MO of CO with the metal $3d$ orbitals. Obviously in such a complex system it is necessary to turn to some form of MO calculation to delineate the sequence of MOs and to estimate the relative magnitudes of the various interactions.

Several approximate MO calculations have appeared (175, 185, 225, 253,

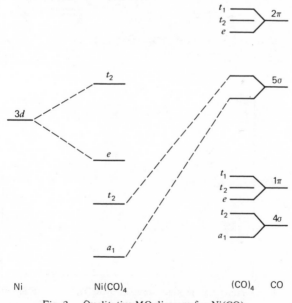

Fig. 2.   Qualitative MO diagram for $Ni(CO)_4$.

257) for Ni(CO)$_4$. However, these calculations are at considerable variance, not only with respect to the orderings and magnitudes of orbital energies, but also with respect to charge distributions. Since such questions as the relative importances of σ and π bonding were not settled by approximate calculations, attention has been turned recently to more elaborate calculations. As mentioned in the Introduction, one very important theoretical development has been the calculation of the electronic structures of transition metal compounds by means of the Hartree—Fock—Slater or X$_\alpha$ method. There are, in fact, two types of X$_\alpha$ calculations — the multiple scattering (MS—X$_\alpha$) method of Slater and Johnson

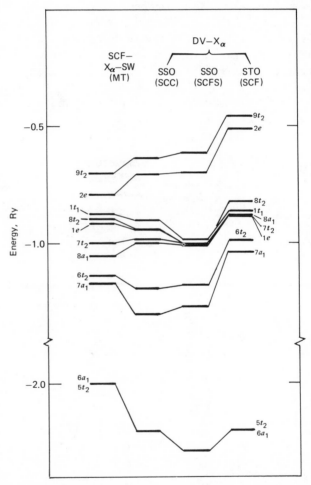

Fig. 3.   MOs for the molecular ground state of Ni(CO)$_4$ from different types of X$_\alpha$ calculation. (From Ref. 194.)

(189, 265) and the discrete variational (DV–X_α) method of Ellis and Painter (114). Furthermore, the MS–X_α method can be subdivided into cases where the so-called "muffin tin" approximation is used and those where a system of overlapping spheres is employed. For the metal carbonyls it has been found that the conclusions based on X_α calculations vary considerably according to the particular method. Thus, in the MS–X_α calculation on $Ni(CO)_4$, it was concluded (190) that the molecular stability derives from a strong σ-bonding interaction between the 5σ MOs of the CO ligands and Ni $3d$ and $4s$ orbitals, and that π back bonding is not important. On the other hand, DV–X_α (12) and approximate Hartree–Fock–Slater calculations (194) indicated the importance of π back bonding. The differences in conclusions from these X_α calculations derive primarily from the accuracy of the potentials employed. Figure 3 indicates rather dramatically the sensitivity of the MOs of $Ni(CO)_4$ in the molecular ground state to the particular type of X_α calculation.

The Hartree–Fock calculations (97, 179, 180) on $Ni(CO)_4$ agree with each other quite well in terms of the predicted sequence of MOs in the molecular ground state (Table I). Both calculations support the postulate of some π back bonding. Calculations of the first two ionic states of $Ni(CO)_4^+$ have also been done (177) (Table I), revealing that substantial relaxation energies (>5 eV) are associated with the predominantly metal $9t_2$ and $2e$ MOs.

The UPS of $Ni(CO)_4$ (152, 177, 211) consists of two low-energy peaks (Fig. 4 and Table I) that can be assigned with confidence to the $^2T_2$ and ^2E ionic states on the basis of all the Hartree–Fock and X_α calculations. The spectral region beyond 14 eV, however, consists of several closely spaced ionizations. In view of the differences in the various calculations both with respect to

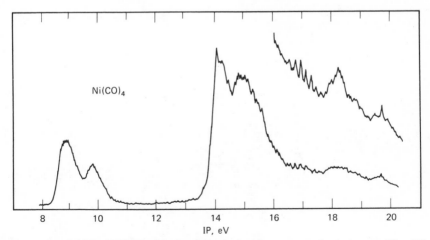

Fig. 4.   He(I) UPS of $Ni(CO)_4$. The bands marked with arrows are due to traces of free CO. (From Ref. 177.)

TABLE I

Hartree–Fock Eigenvalues, Computed Ionization Energies, and Experimental UPS Data for Ni(CO)$_4$

MO	Nature	Hartree–Fock		Computed IEs, eV				Experimental IE[c]
		Energy, eV[a]	Energy, eV[b]	ΔSCF[c]	X_α (DV)[d]	X_α (approx.)[e]		
$9t_2$	Ni (3d)	−10.75	−11.69	7.0	9.3	11.7		8.9
$2e$	Ni (3d)	−12.82	−13.53	7.8	10.3	13.0		9.8
$8t_2$	N–Cσ + COπ	−17.52	−18.51		14.1	17.0		14.1
$1t_1$	N–Cσ + COπ	−17.77	−18.71		14.6	16.7		14.9
$7t_2$	N–Cσ + COπ	−18.01	−19.00		15.0	17.0		.
$1e$	N–Cσ + COπ	−18.37	−19.00		15.1	17.0		
$8a_1$	N–Cσ + COπ	−18.86	−19.64		15.0	17.6		15.7
$6t_2$	N–Cσ + COπ	−21.58	−22.61		16.5	19.3		18.2
$7a_1$	N–Cσ + COπ	−22.01	−23.26		17.2	20.4		19.7

[a] Reference 97.
[b] Reference 180.
[c] Reference 177.
[d] Reference 12.
[e] Reference 194.

orbital composition and in regard to the sequencing of MOs and the corresponding ionic states, no further assignments are possible at the present time.

B. Iron Pentacarbonyl

In D_{3h} geometry the Fe($3d$) AOs transform as $a_1{}'(d_{z^2})$, $e''(d_{xz}, d_{yz})$, and $e'(d_{x^2-y^2}, d_{xy})$. Simple crystal field considerations indicate the energy sequence is $a_1{}' < e' < e''$ and the valence electron count dictates that the $a_1{}'$ orbital is vacant.

Discrete variational X_α MO calculations (10) (Table II) confirm this ordering and thus permit assignment of the first two UPS peaks of Fe(CO)$_5$ (Fig. 5) (10, 211). As with the other metal carbonyls the MO calculations reveal that the predominantly metal nd MOs are followed by a series of closed spaced MOs. A similar situation has been observed for the hydride Fe(CO)$_4$H$_2$ (159).

TABLE II
Calculated[a,b] and Experimental[a] Ionization Energies for Fe(CO)$_5$

MO		Calculated IE, eV	Experimental IE, eV
10e'	Fe($3d$)	7.7	8.6
3e''		9.0	9.9
10$a_1{}'$–13$a_1{}'$	CO	17.7–13.7	>13.5
6e'–9e'	CO	15.9–13.4	>13.5
6$a_2{}''$–8$a_2{}''$	CO	16.3–12.8	>13.5
1e''–2e''	CO	15.0–14.2	>13.5
1$a_2{}'$	CO	13.9	>13.5

[a] Reference 10.
[b] Transition state method; discrete variational X_α calculations.

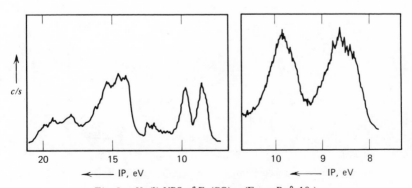

Fig. 5. He(I) UPS of Fe(CO)$_5$. (From Ref. 10.)

C. Group VIB Hexacarbonyls

The high symmetry of the metal hexacarbonyls renders them ideal for molecular orbital—photoelectron spectroscopic study (1) because of the equivalent orbital interactions at each site and (2) because of the orthogonality of metal (nd)—σ and metal (nd)—π interactions in O_h geometry. A very qualitative scheme depicting the σ bonding in Group VIA hexacarbonyls is presented in Fig. 6. The principal interaction is between the metal orbitals and the t_{1u}, e_g, and a_{1g} symmetry-adapted carbon monoxide 5σ (carbon "lone pair") MOs. The predominantly metal nd orbital of t_{2g} symmetry, which is the highest occupied molecular orbital (HOMO), is capable of interacting with the t_{2g} combination of 2π virtual orbitals to effect π back bonding. The various interactions of the 1π and 4σ MOs of carbon monoxide are not shown since they would complicate the MO scheme unduly.

There have been many quantum mechanical calculations on the Group VIB hexacarbonyls at both the semiempirical (25, 52, 256) and *ab initio* (179,

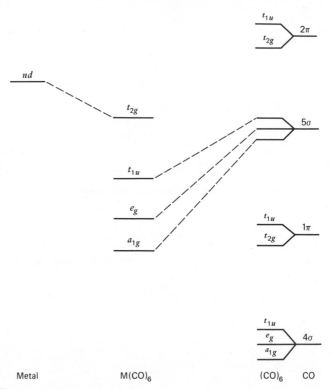

Fig. 6.   Qualitative scheme for σ bonding in Group VIA hexacarbonyls.

180, 185) levels. More recently an SCF–X_α MO calculation (12, 187) was performed on $Cr(CO)_6$. As in the case of $Ni(CO)_4$ there are some significant differences between the DV–X_α and MS–X_α calculations. Not only are the sequences of computed IEs different (see the $5e_g$ and $8a_{1g}$ MOs in Table III), but also the bonding interpretation differs in the sense that the DV calculation indicated significant π back bonding, while the MS calculation indicated that this type of bonding was insignificant.

Turning to the He(I) and He(II) UPS of the $M(CO)_6$ species (M = Cr, Mo, W) (172, 211, 273), three spectral regions, A, B, and C are apparent (Fig. 7), a feature that is evident in all the MO calculations. Peak A is uniformly attributed to electron ejection from the $2t_{2g}$ MO of principally metal nd character, thus generating the $^2T_{2g}$ ionic state. The ionization energies for band A exhibit the trend $Cr(CO)_6 < Mo(CO)_6 < W(CO)_6$ as expected on the basis of other photoionization studies.

In each of the carbonyls, region B comprises a narrow band, B_1, ~13.3 eV, which, in the consensus view, is attributable to ionization of the $8t_{1u}$ MO, which is primarily CO 5σ in nature. The rest of the B band consists of an extended system of overlapping bands. Detailed assignments are difficult in this region, partly because the sequences of eigenvalues are dependent on the method of calculation (Table III) and partly because the details of the band structure are slightly different in the various published spectra. Furthermore, as evidenced by the large changes in spectral intensity upon He(II) irradiation, it is difficult to make arguments based on relative band intensities. One can say, however, that the complex region, B_2 to B_7, is due to ionization of the $1t_{1g}$, $1t_{2u}$, $7t_{1u}$, and $1t_{2g}$ MOs composed mainly of carbon monoxide 1π character and of $5e_g$ and $8a_{1g}$ MOs derived principally from the ligand 5σ orbital. Finally, region C is due to electron ejection from the $4e_g$, $6t_{1u}$, and $7a_{1g}$ MOs, which are derived largely from the ligand 4σ orbital. Two additional comments are in order concerning peak A in the metal hexacarbonyls. First, note that there is an appreciable increase of intensity of this peak, particularly in the He(I) UPS, in the order $Cr(CO)_6 < Mo(CO)_6 < W(CO)_6$. This is an example of the so-called "heavy atom effect," the origins of which remain obscure. Second, note that in the case of $W(CO)_6$ band A actually consists of two bands separated by 0.26 eV. This arises because of splitting of the $^2T_{2g}$ cation state into U' and E'' components under the action of spin–orbit coupling. In the O^* double group the difference in energy between the U' and E'' multiplets is $3/2\zeta_{t_{2g}}$ where $\zeta_{t_{2g}}$ is the spin–orbit coupling constant for the t_{2g} orbital in question (156). The experimental splitting of 0.26 eV implies >30% delocalization of the tungsten $5d$ electrons into the ligands in the $^2T_{2g}$ radical cation. Additional poorly resolved fine structure on the A bands of $W(CO)_6$ is due to that vibrational (possibly C–O stretching) progression associated with the spin–orbit split components.

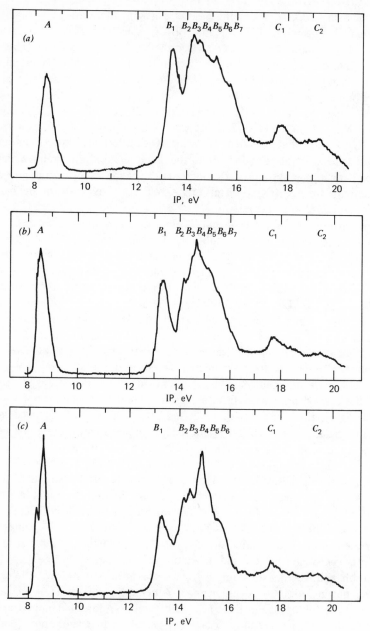

Fig. 7. He(I) UPS of the Group VIB hexacarbonyls: (a) $Cr(CO)_6$; (b) $Mo(CO)_6$; (c) $W(CO)_6$. The lettering system (A, B, etc.) is discussed in the text. (From Ref. 172.)

TABLE III

Hartree–Fock Eigenvalues and Computed Ionization Energies for $Cr(CO)_6$, and Experimental UPS Data for $M(CO)_6$, $M = Cr, Mo, W$

MO	Hartree–Fock eigenvalues, eV[a]	Computed IEs, eV		Experimental IEs, eV[d]			Spectral region[d]
		$X_\alpha(DV)$[b]	$X_\alpha(MS)$[c]	$Cr(CO)_6$	$Mo(CO)_6$	$W(CO)_6$	
$2t_2g$	−10.7	8.9	8.6	8.40	8.50	8.56[e]	A
$8t_1u$	−17.5	12.9	12.1	13.38	13.32	13.27	B_1
$1t_1g$	−18.5	14.6	12.9	14.21	14.18	14.20	B_2-B_7
$1t_2u$	−18.7	14.8	13.1	14.40	14.4	14.42	B_2-B_7
$7t_1u$	−19.3	15.0	13.5	–	14.66	14.88	B_2-B_7
$1t_2g$	−19.3	15.2	13.4	15.12	15.2	15.2	B_2-B_7
$5eg$	−17.9	13.9	13.6	15.60	15.6	15.54	B_2-B_7
$8a_1g$	−20.2	14.7	15.0	16.2	16.2	–	B_2-B_7
$4eg$	−22.4	16.1	17.8	17.82	17.70	17.84	C
$6t_1u$	−22.7	16.2	17.9	19.7	19.7	20.2	C
$7a_1g$	−24.4	17.7	17.9	–	–	–	C

[a] Reference 180.
[b] Reference 12.
[c] Reference 187.
[d] Reference 172.
[e] Band split due to spin–orbit coupling. This is the U' component. See text.

D. Open-Shell Metal Carbonyls

As pointed out earlier in this chapter, the ionization of open-shell systems can produce a large number of ionic states. An excellent example of this phenomenon is provided by the paramagnetic hexacarbonyl $V(CO)_6$, which has

TABLE IV
Ionic States Resulting From the First Two Ionizations of $V(CO)_6$

Ground state	Ion configuration	Ionic states produced	Ligand field energies, eV	Experimental IE, eV[a]
$^2T_{2g}(t_{1u}^6 t_{2g}^5)$	$t_{1u}^6 t_{2g}^4$	$^3T_{1g}$	0	7.52
		$^1T_{2g}$	$6B + 2C$	7.88
		^1E_g	$6B + 2C$	7.88
		$^1A_{1g}$	$15B + 5C$	
	$t_{1u}^5 t_{2g}^5$	$^3T_{2u}$		
		$^1T_{2u}$		
		$^3T_{1u}$		
		$^1T_{1u}$		
		^3E_u		
		^1E_u		
		$^3A_{2u}$		
		$^1A_{2u}$		

[a] Data and assignments from Ref. 121.

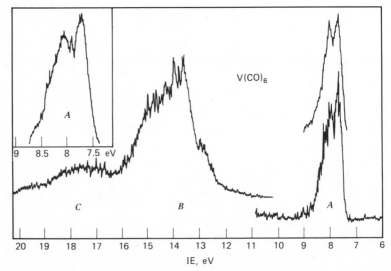

Fig. 8.   He(I) UPS of $V(CO)_6$. (From Ref. 121.)

the ground-state electronic configuration $t_{1u}^6 t_{2g}^5 (^2T_{2g})$. Table IV illustrates that, theoretically, the first two ionizations of $V(CO)_6$ should give rise to a total of 12 ionic states! As anticipated, the UPS of $V(CO)_6$ (121) is considerably more complex than that of the diamagnetic hexacarbonyl $Cr(CO)_6$ (Fig. 8). This is particularly apparent in the low IE region, which corresponds to ionization of essentially pure metal $3d$ orbitals of t_{2g} symmetry. Thus, whereas $Cr(CO)_6$ exhibits a single, sharp band at 8.40 eV, $V(CO)_6$ exhibits at least two peaks at 7.52 and 7.88 eV. From Table IV it follows that the 7.52 eV peak corresponds to the $^3T_{1g}$ state, and the 7.88 eV peak is attributed to the $^1T_{2g}$ and ^1E_g states, which are expected to lie close in energy. The weak ionization corresponding to the $^1A_{1g}$ state is not discernible and may represent part of the 7.88 eV band.

III. SUBSTITUTED METAL CARBONYLS

Three general classes of complexes are considered in this section: the Group VIIB carbonyl derivatives with one-electron donor ligands, the Group VIB carbonyl derivatives featuring two-electron donor ligands, and $LM(CO)_4$ compounds.

A. $LM(CO)_5$ Compounds of Group VIIB

Replacement of one of the CO groups of a metal hexacarbonyl by another ligand reduces the symmetry from O_h to C_{4v}.

O_h	C_4
$t_{2g}(d_{xy}, d_{xz}, d_{yz})$	$e(d_{xz}, d_{yz}) + b_2(d_{xy})$
$e_g(d_{x^2-y^2}, d_{z^2})$	$a_1(d_{z^2}) + b_1(d_{x^2-y^2})$

If only the metal nd electrons are considered, the above abbreviated correlation table indicates that the t_{2g} orbitals split into e and b_2 components upon lowering the symmetry from O_h to C_{4v}. Removal of a z axial CO ligand from a metal hexacarbonyl would be expected to lead to a molecular orbital sequence $b_2 < e$ on the grounds that the metal d_{xz} and d_{yz} orbitals interact less strongly with the π^* MOs of the CO groups than does the d_{xy} orbital. However, considerable electronic redistribution may accompany the replacement of the CO group by the incoming ligand L such that, a priori, it is not possible to predict the relative ordering of the e and b_2 MOs. Additionally, metal–ligand or ligand MOs may appear in the same region as the metal MOs. It is precisely these points, together with the fact that the early UPS data were of lower resolution, that have caused considerable discussion regarding the interpretation of the UPS of $LM(CO)_5$ compounds. It is discussed below how ab initio MO/Koopmans' theorem calculations have tended to add to the confusion. On the other hand, the acquisition of He(II) data and the study of spin—orbit coupling effects in the UPS of the heavier species have proved to be very useful for spectral interpretation. Of course, drawing analogies between the UPS of lighter and heavier elements implies its own set of assumptions. Nevertheless, it is shown below that, taking the data as a whole, a coherent, self-consistent series of spectral assignments can be made.

Let us illustrate some of these points by considering the simplest species that feature hydride ligands. In the early work (116) only two low-energy ionizations were detected in the UPS of $HMn(CO)_5$ and these were attributed to electron ejection from the manganese e and b_2 MOs. However, subsequent studies (87, 161) have shown that, in fact, the first band involves two closely spaced ionizations (Fig. 9), which these authors ascribed to the manganese e and b_2 MOs. The third band (originally considered to be the second) was attributed to the Mn—H σ bond of a_1 symmetry. *Ab initio* MO calculations (157, 158) suggested a reversed ordering for the MOs, that is, $b_2 < e < a_1$ (Table V); however, it is now known that, in conjunction with Koopmans' theorem, MO

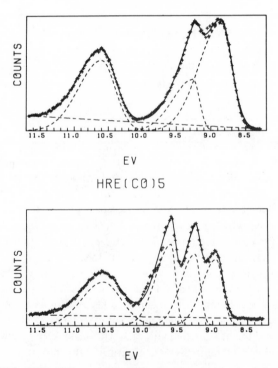

Fig. 9. He(I) UPS of $HMn(CO)_5$ and $HRe(CO)_5$. Reprinted with permission from M. B. Hall, *J. Am. Chem. Soc.*, **97**, 2057 (1975). Copyright by the American Chemical Society.

TABLE V

Calculated Ionization Energies (eV)[a,b] for the First Three Ionic States of HMn(CO)$_5$
and CH$_3$Mn(CO)$_5$

Ionic state	Koopmans' theorem				ΔSCF	
	Ab initio	CNDO(A)[c]	CNDO(B)[d]	Fenske et al.	Ab initio	CNDO(A)[c]
			HMn(CO)$_5$			
^2E	12.2	8.9	9.3	9.8	8.2	4.2
$^2B_2$	12.3	8.1	9.1	10.36	8.6	4.3
$^2A_1$	10.4	9.7	7.0		8.4	8.1
			CH$_3$Mn(CO)$_5$			
^2E	11.9	9.2	10.3	8.7	8.0	4.4
$^2B_2$	12.2	8.4	10.1	9.3	8.6	4.5
$^2A_1$	9.4	10.1	7.6	(9.7)	7.5	9.0

[a] All data except those for Fenske et al. from Ref. 159.
[b] Data for Fenske et al. from Refs. 134 and 162.
[c] 3d, 4s, 4p Mn basis sets.
[d] 3d only Mn basis set.

calculations of this type tend to overemphasize the stability of the metal MOs (81, 124, 248). Approximate MO calculations (134, 159), on the other hand, have produced various eigenvalue sequences (Table V). When calculations of the individual ionic states are made and the differences between the energies of these and that of the molecular ground state are computed (ΔSCF), it is clear that large Koopmans' theorem deviations arise when metal-localized MOs are ionized. This is true for both *ab initio* and CNDO wave functions. The correctness of the sequence of ionic states based on the CNDO ΔSCF calculations has been confirmed by studying the effect of spin–orbit coupling on the UPS of the rhenium analogue HRe(CO)$_5$ (87, 161, 174). To consider the effects of spin–orbit coupling it is necessary to employ the double group C_{4v}^*.

The correctness of the sequence of ionic states based on the CNDO ΔSCF calculations has been confirmed by studying the effect of spin–orbit coupling on the UPS of the rhenium analog HRe(CO)$_5$ (87, 161, 174). To consider the effects of spin–orbit coupling it is necessary to employ the double group C_{4v}^*.

C_{4v}	C_{4v}^*
$^2A_1$	E'
$^2A_2$	E'
$^2B_1$	E''
$^2B_2$	E''
^2E	$E' + E''$

It is immediately apparent that only the ^2E states of $HRe(CO)_5^+$ are split by spin–orbit coupling. However, it is also obvious that subsequent interactions can take place between MOs of the same (E' or E'') symmetry. Considering only the predominantly metals orbitals, Hall (161) has presented three schemes that differ in the relative energies of the b_2 and e MOs.

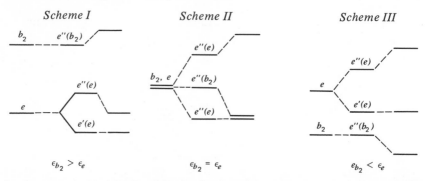

In the experimental UPS of $HRe(CO)_5$ (87, 161, 174) the splitting of the first and second bands is 0.31 eV and that between the second and third bands is 0.34 eV (Fig. 9). These relative spacings are consistent only with Scheme III, thus confirming that the e MO is less stable than the b_2 MO. The spin–orbit coupling parameter, $\zeta_{5d}(Re)$, of 0.25 eV deduced from the spectra compares with the range 0.26–0.34 eV estimated for the Re^{2+} cation (136, 156). The reduction of $\zeta_{5d}(Re)$ in $HRe(CO)_4$ is consistent with the delocalization of metal $5d$ electrons into the 2π MOs of the CO ligands.

The history of the UPS of the methylated species $CH_3M(CO)_5$, $M = Mn$, Re is somewhat similar. In the first UPS work on $CH_3Mn(CO)_5$, only two low-energy peaks were discernible (116). Furthermore, since the relative intensities of these peaks were inverted compared to those of $HMn(CO)_5$, it was suggested that the sequence of MOs was $e < b_2$. The inverted order was attributed to a π-type (hyperconjugative) interaction between the e MOs on the CH_3 moiety and the metal $3d$ AOs of e symmetry (116). However, subsequent reexamination (87, 202) of the UPS of $CH_3Mn(CO)_5$ under higher resolution revealed that the second band is due to two closely spaced ionizations. The results of the various MO calculations (134, 159, 162), both with and without orbital relaxation, are summarized in Table V. The sequence of ionic-state energies $^2A_1 < {}^2B_2 < {}^2E$ has been confirmed by studies of the UPS of $CH_3Re(CO)_5$ (87, 161, 174) that revealed that this species, like $HRe(CO)_5$, was only interpretable according to Scheme III (161). Interestingly, although the earlier UPS assignments were incorrect, there is evidence for the operation of hyperconjugative effects in the methylated compounds. This is provided by the fact that the gap between the b_2 and e ionizations is larger for $CH_3M(CO)_5$ than for $HM(CO)_5$, which is a consequence of destabilization of the metal MOs of e symmetry.

In the case of the SiH_3 and GeH_3 analogues it is evident (Table VI) that the average IEs pertaining to the ^2E and $^2B_2$ ionic states are in the order $Si > Ge > C$. This contrasts with the order $C > Si \approx Ge$, which would be anticipated on the basis of most scales of electronegativity. Previously this anomalous trend was attributed (88, 90, 91, 141) to conjugative interactions

TABLE VI

Lower Ionization Energies (eV) and Assignments for $LMn(CO)_5$ and $LRe(CO)_5$ Complexes

Ligand, L	Metal		a_1 (metal–ligand bond)	e (ligand)	Ref.
	e	b_2			

a. $LMn(CO)_5$[c]

Ligand, L	e	b_2	a_1 (metal–ligand bond)	e (ligand)	Ref.
H	8.85	9.25	10.60	–	161
CH_3	8.65	9.12	9.49	–	161, 202[d]
SiH_3	8.99	9.38	–	–	87
GeH_3	8.90	9.26	–	–	87
Me_3Si	9.0	9.3	–	–	87
Cl	8.94	9.56	11.18	10.56	161, 174, 203[e]
Br	8.86	9.56	10.11	10.81	161, 174, 203[e]
I	9.69	9.69	10.44	8.44[a]	161, 174, 203[e]
				8.74[a]	
CF_3	9.17	9.51	10.53	–	174, 203[f]
$SiCl_3$		(9.36)[b]			174
		9.58[b]			174
$SnMe_3$		8.63[b]			174
		(9.01)[b]			174
		9.66[b]			174

Ligand, L	Metal			Metal–ligand bond	Ligand		Ref.
	$e''(e)$	$e'(e)$	$e''(b_2)$	$e'(a_1)$	$e''(e)$	$e'(e)$	

b. $LRe(CO)_5$[c]

Ligand, L	$e''(e)$	$e'(e)$	$e''(b_2)$	$e'(a_1)$	$e''(e)$	$e'(e)$	Ref.
H	8.94	9.25	9.59	10.59	–	–	161
CH_3	8.72	8.98	9.53	9.53	–	–	161
SiH_3	8.9	9.1	9.5, 9.6	–	–	–	87
GeH_3	8.9	9.13	9.4, 9.6	–	–	–	87
Cl	8.80	9.04	9.86	11.21	10.76	10.76	161
Br	8.80	9.04	9.94	10.91	10.37	10.64	161
I	9.75	10.08	9.75	10.52	8.32	8.77	161
CF_3CO[a]	9.40	9.69	9.97	8.80			161

TABLE VI (*continued*)

Ligand, L	a_1 (metal–metal σ bond)	Metal–metal π bonds		e_2 (nonbonding)	Ref.
		e_3	e_1		
		c. $M_2(CO)_{10}$; M = Mn, Re			
$Mn_2(CO)_{10}$	8.02	8.35	9.03	9.03	174
$Re_2(CO)_{10}$	8.06	8.56	9.28	9.60	161
		8.86	9.60	9.60	161

[a] Splitting due to spin–orbit coupling of I.
[b] Spectra not sufficiently resolved for detailed assignment.
[c] The MOs are labeled according to the double group C_{4v}.* The precursor MOs in C_{4v} symmetry are indicated in parentheses.
[d] Reference 202 for data; Ref. 161 for interpretation.
[e] Reference 203 for data; Refs. 161 and 174 for interpretation.
[f] Reference 203 for data; Ref. 174 for interpretation.

between filled orbitals of π symmetry and vacant receptor d orbitals on Si or Ge. For the transition metal complexes, at least, this view would appear to be oversimplified, since the b_2 MO of the metal is only capable of δ bonding to Si or Ge. It has been suggested, therefore, that the order of IEs is due to the σ-acceptor capabilities of the ligands, which, inferentially, are in the order $SiH_3 > GeH_3 > CH_3$. In the case of $Me_3SiMn(CO)_5$ no evidence was obtained for σ–π mixing (236) between the MOs of the Me_3Si moiety and the metal MOs of e symmetry. The bonding in $SiF_3Mn(CO)_5$ has also been discussed on the basis of UPS measurements (87a).

The dimeric species $M_2(CO)_{10}$ can be considered to be special cases of the $LM(CO)_5$ class of compounds. In the solid phase $Mn_2(CO)_{10}$ is known to adopt the staggered conformation (94). Two $M(CO)_5$ fragments can thus be combined in D_{4d} symmetry to afford the MOs of the dimer as indicated below:

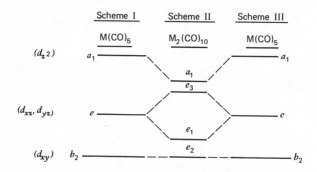

The UPS of $Mn_2(CO)_{10}$ and $Re_2(CO)_{10}$ can be assigned satisfactorily (116, 161, 174) according to the above scheme. In the case of $Re_2(CO)_{10}$ (Fig. 10) each of the ^2E states split into two components as a result of the operation of spin–orbit coupling (161, 174). The fact that the gap between the e MOs is larger for $Re_2(CO)_{10}$ than for $Mn_2(CO)_{10}$ may be taken to indicate stronger π bonding in the Re–Re bond than in the Mn–Mn bond. This factor may contribute to the proclivity of Re and other third row transition elements to form cluster compounds. It should be noted, perhaps, that the adoption of the eclipsed conformation (D_{4h}) in the vapor phase cannot be excluded either on theoretical (163) or spectral interpretation grounds. In fact, the only major difference in the D_{4h} and D_{4d} models would be that the b_2 MO, which is nonbonding in D_{4d} symmetry, would split slightly (estimated 0.15 eV) in D_{4h} symmetry into bonding and antibonding MOs.

The halides $XM(CO)_5$, have also attracted considerable attention (53, 116, 161, 174, 203) on account of their spectral complexity (Fig. 11). In part this arises because the halogens possess a_1 and e orbitals, and the latter can interact with the metal d_{xz} and d_{yz} orbitals. Furthermore, the MO schemes for the various halides are dependent on the relative electronegativities of the halogens and metals as shown by Hall (161) and Higginson et al. (174).

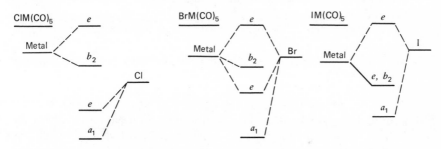

Even more complexities develop when the metal and/or halogen exhibits significant spin–orbit coupling. Although other assignments have been proposed (46, 53, 134, 203), the above schemes due to Hall (161) and Higginson et al. (174) seem best able to explain the various spectroscopic observations, namely,

1. Of the manganese halides only the first band of $IMn(CO)_5$ is split into two spin–orbit components.

2. Only the first band of $ClRe(CO)_5$ is split as a result of spin–orbit coupling. No splitting is evident on the third band.

3. For $BrRe(CO)_5$ spin–orbit coupling effects are evident on both the first and third bands. The near equality of these splittings (0.27 eV) is between the spin–orbit coupling parameters for Re (0.25 eV) and Br (0.31 eV) and suggests that the e MOs are approximately equal mixtures of $Re(5d)$ and $Br(4p)$ contributions.

RE2(CO)10

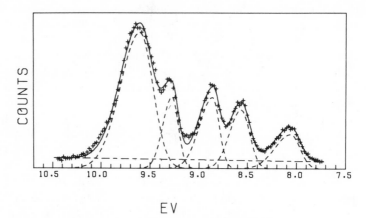

E V

Fig. 10. He(I) UPS of $Re_2(CO)_{10}$. Reprinted with permission from M. B. Hall, *J. Am. Chem. Soc.*, **97**, 2057 (1975). Copyright by the American Chemical Society.

4. The b_2 MO, which is noninteracting, is almost the same energy in all the Mn and Re systems.

5. The IEs of the a_1 MO decrease with decreasing electronegativity of the halogen.

6. The assignments are consistent with He(I)/He(II) relative intensity considerations (174).

Inspection of the IE data in Table VI reveals that the b_2 MO, which is highly localized on the metal, responds to changes of electronegativity of the ligand, L, in the anticipated manner. In the conventional view of bonding in metal carbonyl derivatives, increasing the electronegativity of L should promote more σ donation and less π acceptance (into the antibonding C–O $2\pi^*$ MOs) on the part of the CO ligands. Both factors should lead to increases in the C–O bond strength. It is rather interesting, therefore, to note that there is a good correlation between the squares of the C–O stretching frequencies and the IEs of the b_2 MO (Fig. 12) (174).

B. LM(CO)₅ Compounds of Group VIB

A number of Group VIB pentacarbonyl compounds of the general type LM(CO)₅, L = amine (173), pyridine (277), phosphine (173, 278), sulfide (278), or isonitrile (173), have been investigated in an effort to probe the relative σ-donor or π-acceptor abilities of the various ligands.

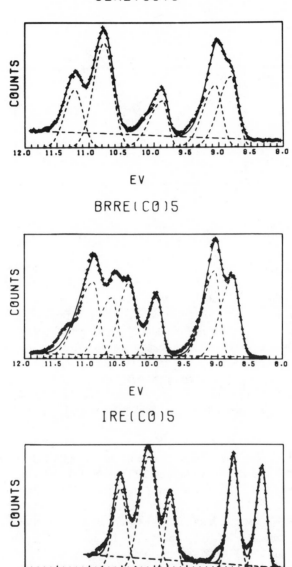

Fig. 11. He(I) UPS of ClRe(CO)$_5$, BrRe(CO)$_5$, and IRe(CO)$_5$. Reprinted with permission from M. B. Hall, *J. Am. Chem. Soc.*, **97**, 2057 (1975). Copyright by the American Chemical Society.

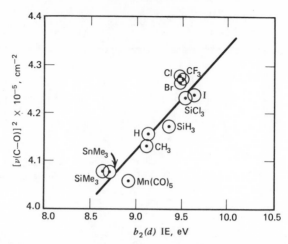

Fig. 12.　Correlation between C–O stretching frequency and $^2B_2$ ionic state energy ("d orbital ionization energy") for $LMn(CO)_5$ compounds. (From Ref. 174.)

As with the Group VIIB pentacarbonyls discussed above, the metal d orbitals transform as b_2 and e. However, it is not clear what the relative orderings of these MOs should be in every case. Intensity considerations, together with spin–orbit coupling effects in the UPS of the tungsten compounds (173), suggest that, in general, the energies of the cation states are in the order $^2E < {}^2B_2$. By contrast, *ab initio* MO calculations on $Cr(CO)_5NH_3$ and $Cr(CO)_5PH_3$ indicate that the ground-state eigenvalues are in the reverse order (173). This disparity has been attributed to the large relaxation energies associated with the ionization of MOs of predominantly metal character. The ^2E and $^2B_2$ ionic-state assignments in Table VII should therefore be viewed as provisional.

In principle, the energies of the e and b_2 metal MOs should be sensitive to the nature of the ligand, L, since, to a first approximation, the b_2 MO interacts only with the π MO of CO, while the e MO is capable of interacting with both L and CO π orbitals. Thus a π-acceptor ligand should stabilize the e MO and diminish the $e-b_2$ separation (providing, of course, that the e MO is the HOMO). The $e-b_2$ separations (Table VII) are in the order amine $\sim$ pyridine $>$ sulfide $>$ phosphine, which correlates with the π-acceptor capabilities of these classes of ligands (146). However, this result should be viewed with caution, not only because of the uncertainty of the UPS assignments, but also because alkyl substitution at nitrogen in the amine–tungsten compounds does not affect the $e-b_2$ separation.

The shifts in the statistically weighted average e/b_2 IE values compared to the IE value for the t_{2g} MOs of $Cr(CO)_6$ (8.40 eV) are in the order pyridine $>$ amine $>$ sulfide $>$ phosphine. This order is similar to that observed for the shifts

TABLE VII
Lower Ionization Energies (eV) and Assignments[a] for $LCr(CO)_5$ and $LW(CO)_5$ Complexes and the Corresponding Free Ligands

Ligand, L	Metal				Ligand		Ref.
	e	b_2	a_1 (metal–ligand)	b_1 (metal–ligand)	a_1	b_1	
			$LCr(CO)_5$				
NH_3	7.56	7.85	–	–	–	–	173
$(CH_3)_3N$	7.45	7.76	10.57	–	8.53	–	173
Pyr[b]	7.30	7.59	–	–	–	–	277
4-CH_3Pyr	7.22	7.48	–	–	–	–	277
4-$(CH_3)_3$CPyr	7.17	7.47	–	–	–	–	277
4-ClPyr	7.42	7.66	–	–	–	–	277
4-BrPyr	7.37	7.64	–	–	–	–	277
4-CH_3OPyr	7.18	7.45	–	–	–	–	277
4-CH_3COPyr	7.5	7.8	–	–	–	–	277
CH_3NC	7.61 (sh)[c]	7.77	–	–	–	–	277
PH_3	7.90 (sh)[c]	8.03	11.43	–	10.59	–	173
$(CH_3)_3P$	7.58 (sh)[c]	7.72	10.00	–	8.60	–	173
$(C_2H_5)_3P$	7.44 (sh)[c]	7.58	9.67	–	8.31	–	278
$(C_2H_3)_3P$	7.52 (sh)[c]	7.66	9.60	–	8.51	–	278
$(CH_3)_2S$	7.59	7.79	10.00	12.10	8.69	11.20	278
$(C_2H_5)_2S$	7.45	7.67	9.71	11.48	8.44	10.70	278
$(C_2H_3)_2S$	7.6_1	7.8_0	9.48	–	8.42	–	278
$(CH_3)(CH_2Cl)S$	7.74	7.90	10.27	–	9.17	–	278

Ligand, L	Metal		a_1 (metal–ligand)	a_1 (ligand)	Ref.
	e	b_2			
		$LW(CO)_5$			
NH_3	7.54[d]	8.06	–	–	173
	7.75[d]				173
$(CH_3)_2NH$	7.41[d]	7.95	11.14	8.93	173
	7.62[d]				173
$(CH_3)_3N$	7.41[d]	7.96	10.75	8.45	173
	7.62[d]				173

[a] Assignments for ionizations of e and b_2 metal MOs tentative. See text.

[b] Pyr = pyridine.

[c] sh = shoulder.

[d] Splitting due to spin–orbit coupling.

in the $Cr(2p)$ core IEs of $Cr(CO)_6$, $Cr(CO)_5NH_3$, and $Cr(CO)_5PMe_3$ (19) and is presumably a consequence of the ability of the ligand to place a charge at the metal.

The "lone pair" MO of the free amine and phosphine possesses a_1 symmetry in C_{3v} local symmetry. As anticipated, this MO undergoes considerable stabilization as coordination to the metal occurs. The "lone pair stabilization energies" of amines have been found to be greater than those of phosphines with respect to main group acceptors such as BH_3, O, and S (32, 113, 210). The sulfide ligands feature two MOs with significant lone pair character.

$$b_1 \qquad\qquad\qquad\qquad a_1$$

Both of these MOs become stabilized upon coordination; however, the b_1 MO is stabilized ~50% more than the a_1 MO.

Several carbene complexes of the general formula $Cr(CO)_5C(X)Y$ have been examined by UPS (31). These interesting derivatives can be regarded as further examples of the $LM(CO)_5$ class of compounds. The carbene group can function as a ligand by σ donation from a σ* MO, which is essentially a filled sp^2 carbon orbital, and by accepting charge from the metal into a vacant π* level, which is essentially a vacant carbon $2p$ AO. X-ray structure data (67, 183, 216, 217) indicate that the plane of the carbene moiety lies between that of the cis carbonyl groups, thus conferring a C_s skeletal symmetry on these complexes. In C_s symmetry the degeneracy of the $Cr(3d)$ AOs is lifted, since the t_{2g} representation splits into $2a' + a''$, while the five Cr—CO σ bonds span the irreducible representations $3a' + 2a''$, and the Cr—carbene σ bond transforms as a'. A very qualitative scheme for the interaction of CO and carbene σ MOs with the $Cr(3d)$ AOs is presented in Fig. 13. Although the π* level of the carbene is in a better position energywise for the interaction with the $Cr(3d)$ orbitals, structural studies (67, 183, 216, 217) indicate that carbene ligands are less effective π acceptors than CO. This apparent anomaly is caused by two factors: (1) the less favorable geometry for π back bonding in the case of the carbenes and (2) the fact that the π* receptor MO of the carbene is only singly degenerate, while the corresponding 2π MO of CO is doubly degenerate.

The UPS of each of the nine carbene compounds studied (Table VIII and Fig. 14) show a band at low IE that can be ascribed to electron ejection from the predominantly $Cr(3d)$ orbitals of symmetries $a'(2)$ and a''. The band often exhibits shoulders and can be curve-fitted with two or three Gaussian peaks. The

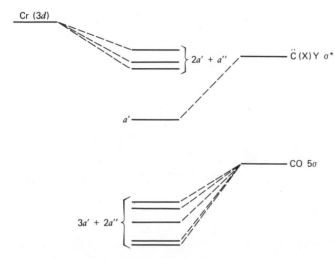

Fig. 13. Qualitative scheme depicting the interaction of Cr($3d$) orbitals with Co and carbene MOs in $Cr(CO)_5C(X)Y$ complexes. The individual sequencing of a' and a'' MOs is not shown since MO calculations indicate that this varies with the type of carbene. (From Ref. 31.)

TABLE VIII

Ionization Energy Data (eV) for Predominantly Metal MOs and Metal–Carbene σ-bonding MOs of $Cr(CO)_5C(X)Y$ Complexes[a]

X	Y	Cr($3d$) MOs			Cr–carbene σ-bonding MO
OMe	Me	7.47	7.89		9.89
SMe	Me	7.35	7.59	7.79	9.91
NMe$_2$	Me	7.12	7.35	7.61	9.72
NH$_2$	Me	7.45	7.80		10.31
OMe	C$_4$H$_3$O	7.37	7.68		9.92
NH$_2$	C$_4$H$_3$O	7.22	7.52		10.30
OMe	C$_6$H$_5$	7.39	7.78		9.26
NH$_2$	C$_6$H$_5$	7.25	7.52	7.73	9.80
NMe$_2$	C$_6$H$_5$	7.02	7.26	7.54	9.49

[a] Data from Ref. 31.

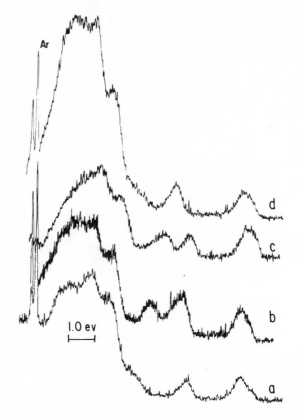

Fig. 14.   He(I) UPS of (a) $(CO)_5CrC(OCH_3)CH_3$, (b) $(CO)_5CrC(SCH_3)CH_3$, (c) $(CO)_5CrC-(N(CH_3)_2)CH_3$, (d) $(CO)_5CrC(NH_2)CH_3$. Reprinted with permission from *J. Am. Chem. Soc.*, **99**, 4321 (1977). Copyright by the American Chemical Society.

second UPS band is assignable to the $^2A'$ ionic state, which arises from ionization of the metal–carbene σ bond. Beyond this point the spectra become more complex as a result of the various ligand and C–X π-bond ionizations. The most significant aspect of the data is the fact that for all the carbene complexes the average of the IEs pertaining to the predominantly Cr(3d) MOs is less than the IE corresponding to the $^2T_{2g}$ ionic state of $Cr(CO)_6$ (8.40 eV). This observation is in accord with the poorer π-acceptor capability of the carbene ligand.

In a UPS study of thiocarbonyl complexes such as $M(CO)_5CS$, M = Cr, W (202a) it was found that the ionizations associated with the σ and π levels of coordinated CS are clearly separated from the other ionizations. It was concluded that CS is a better π acceptor than CO.

C. $LM(CO)_4$ Compounds

Relatively few compounds of this type have been investigated by UPS
(87). Structural studies (for example, Ref. 245) have established that the local
geometry around the $M(CO)_4$ moiety is C_{3v}. The orbitals of predominantly
metal d character, therefore, will transform as $e + e$, and the metal–ligand bond
will transform as a_1. The UPS of some $LCo(CO)_4$ complexes (Table IX) confirm
these expectations.

TABLE IX
Lower Ionization Energies (eV) for $LCo(CO)_4$ Complexes[a]

Ligand, L	Metal		a_1 (metal–ligand)
	e	e	
H	8.90	9.90	11.5
SiH_3	8.85	9.90	–
GeH_3	8.80	9.80	–

[a] Data from Ref. 87.

IV. π-ALLYL, ARENE, AND OLEFIN COMPLEXES

A. π-Allyl Complexes

In a sense the π-allyl compounds of the transition metals can be regarded
as the simplest of the "sandwich" molecules. Bis(π-allyl)nickel, the best known
of such complexes, has been shown by x-ray crystallography (104, 105) to have
a staggered arrangement of π-allyl moieties and hence a C_{2h} molecular conformation. The electronic structure of the ground state of bis(π-allyl)nickel has
been investigated by both semiempirical (47) and *ab initio* (274, 275) methods,
and a semiempirical computation has been performed on bis(π-allyl)palladium
(47).

The interpretation of the UPS of bis(π-allyl)nickel remains controversial.
In the initial publication (209) on this matter it was suggested that the first three
spectral bands, which exhibit an approximately 2:2:1 intensity ratio (Fig. 15),
are due to the ionization of essentially pure Ni($3d$) MOs; hence from the
standpoint of Koopmans' theorem the sequence of the higher lying MOs is
$d_{z^2} < d_{xz} = d_{yz} < d_{xy} = d_{x^2-y^2}$. Similar arguments were used to interpret the
UPS of bis(π-allyl)palladium (Fig. 15) (209). Subsequently the bis(π-allyl)nickel
interpretation was questioned by Veillard and co-workers, who calculated the

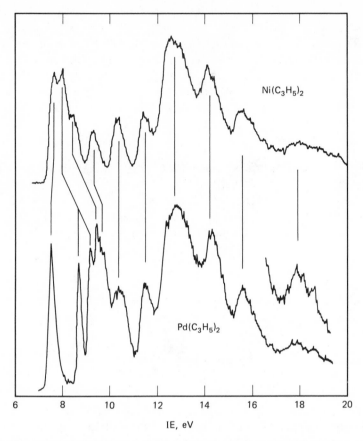

Fig. 15.   He(I) UPS of bis (π-allyl)nickel and bis(π-allyl)palladium. (From Ref. 209.)

energies of the molecular ground state and some of the ionic states of $(C_3H_5)_2Ni^+$ using the minimal (248) and double-zeta (247) quality basis sets. While the two basis sets produced somewhat different sequences of molecular ground-state MOs, both calculations indicated that very large (as much as 10 eV) relaxation energies are associated with the ionization of the $3b_g$, $9a_g$, $10a_g$, and $11a_g$ MOs, which involve large percentages of Ni($3d$) character (Table X). By contrast, the nearly pure ligand MOs (e.g., $7a_u$, $11b_u$, $6a_u$, and $10b_u$) exhibit little, if any, relaxation effects. Accordingly Veillard et al. concluded that Koopmans' theorem breaks down badly for the metal-localized MOs but is a reasonable approximation for the ligand MOs. Specifically these calculations imply that the very stable $9a_g$ MO in the molecular ground state of $(C_3H_5)_2Ni$ [the thirteenth MO in one calculation (248) and the twelfth in the other (247)]

TABLE X

Orbital Energies (eV), Population Analyses, and Ionization Energies (eV) for Bis(π-allyl)nickel and Some of Its Ions[a]

MO	Energy	% Ni(3d)	% ligand π	% ligand σ	Corresponding ionic state	% Ni (3d)	% ligand π	% ligand σ	Computed IE	Experimental IE
$7a_u$	9.52	—	94	6	^2A_u	—	94	6	8.92 (7.30)[b]	9.48
$6b_g$	11.07	34	56	10	—	—	—	—	—	
$13a_g$	13.85	27	53	20	—	—	—	—	—	
$11b_u$	14.18	—	93	7	^2B_u	—	94	6	13.6	
$12a_g$	15.75	8	26	66	—	—	—	—	—	
$5b_g$	15.97	42	1	57	—	—	—	—	—	
$6a_u$	15.97	—	—	100	^2A_u	—	—	100	15.7	
$10b_u$	16.05	—	3	97	^2B_u	—	4	96	15.8	
$11a_g$	16.35	83	4	13	^2A_g	98	—	2	8.21	8.17
$4b_g$	16.52	25	—	75	—	—	—	—	—	
$10a_g$	17.50	69	2	29	^2A_g	99	1	—	8.52	8.59
$5a_u$	17.88	—	—	100	—	—	—	—	—	
$9a_g$	18.26	62	12	26	^2A_g	96	2	2	7.92 (5.71)[b]	7.85
$3b_g$	18.83	38	—	62	^2B_g	99	—	1	8.03	8.17
$9b_u$	19.29	—	—	100	—	—	—	—	—	
$8a_g$	20.90	38	—	62	—	—	—	—	—	

[a] Experimental data from Ref. 209; interpretation and computations from Refs. 247 and 248.
[b] Values in parentheses refer to extended basis set computations. See Ref. 247.

77

TABLE XI

Experimental Ionization Energy Data (eV) for Bis(π-allyl)nickel and Related Compounds[a]

UPS band	$(\diagup\diagdown)_2$Ni	$(\diagup\diagdown)_2$Ni	$(\diagdown\diagup)_2$Ni	$(\diagdown\diagup)_2$Ni
1	7.7_6	7.5_3	7.5_3	7.2_2
2	8.1_9	7.9_1	8.0_0	7.6_8
3	8.5_8	8.3_2	8.4_0	8.1_0
4	9.4_0	9.2_2	9.1_3	8.7_8
5	10.3_8	9.8_6	10.1_0	9.7_3
6	11.5_5	10.9_3	11.1_5	10.7_0
7	12.7	12.2	12.3	12.2
8	14.2	12.7	12.8	13.4
9	15.6	15.0	15.0	14.0

[a] Data from Ref. 23.

is responsible for the first UPS ionization. A complete listing of the proposed assignments of Veillard et al. is presented in Table X.

Even though only some of the ionic states were calculated and some of the relaxation energies are much larger than any others ever reported, the bis(π-allyl)nickel case is very frequently cited as the example *par excellence* where Koopmans' theorem is violated. In the most recent investigation of the UPS of bis(π-allyl)nickel and cognates, Batich (23) suggested that the ^2A_u ionic state, which arises from the ionization of the $7a_u$ ligand π orbital, is responsible in part for the second UPS peak. This conclusion was reached on the basis of intensity, alkyl substitution, and He(II) irradiation arguments. It must be concluded that the bis(π-allyl)nickel question is still open and more investigation is warranted. In view of this uncertainty the UPS data for the methylated derivatives of bis(π-allyl)nickel are given in Table XI without interpretation.

B. Arene Complexes

A rather similar situation regarding the breakdown of Koopmans' theorem exists for the metallocenes. The electronic structures of these interesting compounds have attracted the attention of both theoreticians and experimentalists for several years. Ferrocene is the metallocene that has been studied the most extensively. From the UPS standpoint this stems from the diamagnetism of the neutral molecule and the stability of the ferricenium cation.

To probe further the bonding descriptions of the transition metal metallocenes it is instructive to assemble a "back of the envelope" MO scheme. The available structural evidence (249, 280) indicates that the energies of the staggered (D_{5d}) and eclipsed conformations are rather close. Thus, in the solid state, ferrocene adopts the D_{5d} conformation (106), while the eclipsed arrangement is favored in the vapor phase (34) as it is in the solid state of ruthenocene (167). However, the irreducible representations pertaining to the D_{5d} and D_{5h} point groups are quite similar. The present discussion is made with reference to D_{5d} symmetry in conformity with the majority of prior theoretical discussions. The π-type MOs of the cyclopentadienide rings are well-known (70) and, by virtue of inter-ring interactions, can be combined into symmetric pairs of MOs as shown on the left-hand side of Fig. 16. The D_{5d} point group effects a differentiation of the metal nd AOs into the symmetries $e_{1g}(d_{xz}, d_{yz})$, $e_{2g}(d_{xy}, d_{x^2-y^2})$, and $a_{1g}(d_{z^2})$. The metal nd AOs are then allowed to interact with the ligand π combinations according to the prescriptions of qualitative pertubation molecular orbital (PMO) theory. Thus the metal e_{1g} and ligand e_{1g} MOs are destabilized and stabilized, respectively; the reverse is true for the metal and ligand e_{2g} MOs, and the metal a_{1g} MO is stabilized by virtue of interaction with the metal $(n + 1)s$ AO. For ferrocene and its heavier congeners the total of 18 ligand π plus metal electrons is accommodated by filling the MO scheme up to and including the e_{2g} metal MO. Initially the UPS of ferrocene (123, 130, 243, 273) was interpreted according to Fig. 16 with the tacit assumption of the validity of Koopmans' theorem. Thus the first, somewhat broad band (Fig. 17) was attributed to ionization of the metal-localized e_{2g} MO and the sharper second UPS band was ascribed to electron ejection from the essentially nonbonding a_{1g} metal MO. The next two spectral bands were assigned to ionization of the ligand π MOs of symmetries e_{1u} and e_{1g}. Interestingly, in one study (243) [but not reproducible in another (123)] structure was discernible on the first three ionizations with an interval of 275 cm^{-1}. This corresponds to the ring-metal stretching frequency ν_4, which occurs at 303 cm^{-1} in the molecular ground state of ferrocene (205, 206). In summary, and on the basis of Koopmans' theorem, the UPS data on ferrocene imply the MO hierarchy

$$e_{1g}(\pi\text{-Cp}) < e_{1u}(\pi\text{-Cp}) < a_{1g}(3d) < e_{2g}(3d)$$

in the molecular ground state. A similar sequence of ferrocene MOs, namely,

$$a_{1g}(\pi\text{-Cp}) < e_{1u}(\pi\text{-Cp}) < a_{1g}(3d) \sim e_{2g}(3d)$$

has been deduced on the basis of ESR studies of substituted ferricenium cations (214, 239, 241). Various orbital sequences have emerged in numerous semiempirical MO calculations (5, 7, 36, 93, 139, 176, 195, 252, 263, 267). In stark contrast to the foregoing considerations, however, *ab initio* Hartree–Fock calculations with either minimal (81) or extended (14) basis sets indicate that in the molecular ground state of ferrocene the sequence of MOs is

$$a_{1g}(3d) < e_{2g}(\sigma\text{-Cp}) \sim a_{2u}(\pi\text{-Cp}) \sim e_{2u}(\sigma\text{-Cp}) < e_{2g}(3d) < e_{1g}(\pi\text{-Cp}) \sim e_{1u}(\pi\text{-Cp})$$

Apart from the appearance of σ-type ring MOs, which have not been considered hitherto, the striking difference between the *ab initio* calculations and the

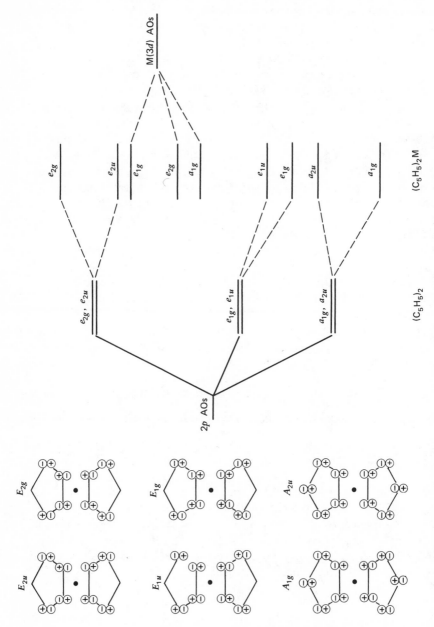

Fig. 16. A qualitative MO scheme for the transition metal metallocenes. Only ring π and metal d orbitals are included. The staggered D_5d conformation is assumed.

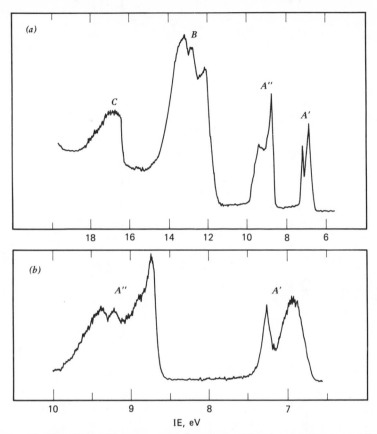

Fig. 17.   The He(I) UPS of ferrocene: (a) full spectrum, (b) high-resolution scan of the first two band systems. (From Ref. 123.)

Koopmans' theorem interpretation of the MOs in the molecular ground state of ferrocene is that the metal-localized e_{2g} and e_{1g} orbitals are stabilized appreciably. As in the case of bis(π-allyl)nickel, which is discussed earlier, the discrepancies arise because of relaxation effects. From Table XII it is clear that only modest electronic rearrangement accompanies the ionization of the delocalized ligand π-type MOs and the localized metal MOs undergo marked electronic rearrangement with the percent metal character increasing sharply on proceeding from the neutral molecule to the appropriate ionic state. Both Hartree–Fock calculations predict the energies of the $^2E_{1g}$ and $^2E_{1u}$ ligand-like ionic states to be very close; accordingly, it is impossible to be completely definitive about these UPS assignments.

As is discussed earlier for $Ni(CO)_4$, the results of SCF–X_α calculations depend on the nature of the potential employed. The two X_α calculations on

TABLE XII

Ab initio Eigenvalues for the Ground State of Ferrocene, Computed Ionization Energies (eV) and Experimental Ionization Energies (eV)

Molecular ground state (ab initio)[a]		Cation state	Computed IEs				Experimental IEs[e]
MO	Energy (eV)		Δ(SCF)[a]	Δ(SCF)[b]	X_α(DV)[c]	X_α(MT)[d]	
$e_{2g}(3d)$	14.42	$^2E_{2g}$	8.3	5.69	6.7	8.5	6.858
$a_{1g}(3d)$	16.57	$^2A_{1g}$	10.1	7.46	6.7	7.9	7.234
$e_{1u}(\pi\text{-Cp})$	11.67	$^2E_{1u}$	11.1	8.85	8.1	9.3	8.715
$e_{1g}(\pi\text{-Cp})$	11.89	$^2E_{1g}$	11.2	8.79	8.6	9.7	9.38
Bands resulting from ionization of σ and π-type ring MOs	—		—	—	—	—	12.2
							13.6
							16.4

[a] Data from Ref. 81. Near minimal basis set.
[b] Data from Ref. 14.
[c] Data from Ref. 11.
[d] Data from Ref. 246.
[e] Data from Ref. 243. However, the assignments for the $^2E_{1u}$ and $^2E_{1g}$ ionic states are not certain. See text.

ferrocene (11, 246) provide a further illustration of this phenomenon. Thus, in the muffin tin approximation, the X_α calculations of Rösch and Johnson (246) predict, rather surprisingly, the $^2A_{1g}$ state of the ferricenium cation to be the ground state, and the discrete variational method (11) indicates that the $^2A_{1g}$ and $^2E_{2g}$ Fe 3d-like states are isoenergetic.

Substitution in the ferrocene rings necessarily reduces the effective molecular symmetry to at least C_{2v}, thereby reducing each doubly degenerate MO to two singly degenerate MOs. The anticipated splittings, however, are too small to resolve in the UPS of 1,1′-dimethyl- or 1,1′-dichloroferrocene (123) (Table XIII). In the UPS of the latter two additional peaks appeared in the ~11 eV region. These are due to the ionization of chlorine 3p lone pair MOs. Since the chlorine ligands are probably too far apart to interact directly, it has been suggested that, as in, for example, chlorobenzene, this is due to the differentiation of in-plane and out-of-plane interactions between the Cl(3p) orbitals and the π system of the aromatic ring (273).

The UPS of the heavier metallocenes $(C_5H_5)_2$Ru and $(C_6H_4Me)_2$Ru are similar to those of ferrocene in many respects and have been assigned in a comparable manner (123) (Table XIII). The spectrum of 1,1′-dimethylosmocene

TABLE XIII
Experimental IE Data (eV) and Assignments for Closed-Shell Metallocenes[a]

Compound	Ionic state produced and MO ionized			
	$^2E_{2g}[e_{2g}(nd)]$	$^2A_{1g}[a_{1g}(nd)]$	$^2E_{1u}[e_{1u}(\pi\text{-Cp})]$	$^2E_{1g}[e_{1g}(\pi\text{-Cp})]$
$(C_5H_5)_2$Fe	6.858	7.234	8.715	9.38
$(C_5H_4Me)_2$Fe	6.72	7.06	8.53	9.17
$(C_5H_4Cl)_2$Fe	7.03	7.37	8.71	9.49
$(C_5H_5)_2$Ru	7.45	7.63	8.51	9.93
$(C_5H_4Me)_2$Ru	7.25	7.25	8.24	9.76
$(C_5H_4Me)_2$Os	6.93[b,c] 7.55[b,c]	7.21[c]	8.26	9.90
$(C_5H_5)_2$Mg	–	–	9.03[d] (9.26)[d]	8.11[d] 8.23[d] 8.44[d]
$(C_5H_4Me)_2$Mg	–	–	8.62[e] (8.86)[e]	7.78[e] (7.90)[e] (8.10)[e]

[a] Data for $(C_5H_5)_2$Fe from Ref. 243. Assignments discussed in text. Other data and assignments from Ref. 123.
[b] Splitting due to spin−orbit coupling.
[c] Assignments uncertain; see text.
[d] Splittings due to Jahn-Teller effect.
[e] Splittings due to low symmetry.

is somewhat more complex. The observation of three, rather than two, peaks in the low IE region is consistent with the spin–orbit splitting of the $^2E_{2g}$ ionic state into the 5/2 and 3/2 components. However, it is not possible to assign the $^2A_{1g}$, $^2E_{2g}$ (5/2), and $^2E_{2g}$ (3/2) ionic states unequivocally on purely empirical grounds (123).

The main group metallocenes, $(C_5H_5)_2Mg$ and $(C_5H_4Me)_2Mg$, have also been investigated by UPS (123). In these molecules the d AOs are of relatively high energy; hence the qualitative LCAO synthesis is concerned only with interaction between the valence $s(a_{1g})$ and $p(a_{2u} + e_{1u})$ AOs and the $(C_5H_5)_2$ ring π MOs. Under these conditions the e_{1g} and e_{1u} π-type MOs no longer cross over (Fig. 16), the d-type AOs are removed from consideration, and the HOMO becomes the (nonbonding) ring π MO. According to this simple view the e_{1u} ring MO should possess some bonding character by virtue of its interaction with the higher-lying metal p AOs of the same symmetry. However, in the UPS of both $(C_5H_5)_2Mg$ and $(C_5H_4Me)_2Mg$, the widths of the peaks corresponding to the production of the $^2E_{1g}$ and $^2E_{1u}$ ionic states are similar. In crude terms, therefore, one could view the bonding in magnesium metallocenes as largely "ionic."

The appropriately weighted IE data pertaining to the production of metal-localized $^2E_{2g}$ and $^2A_{2g}$ ionic states increase in the order $(C_5H_4Me)_2Os$ $\sim (C_5H_4Me)_2Ru > (C_5H_4Me)_2Fe$ (123). This trend is anticipated on the basis of ionization energy data for the free atoms and also by analogy with other triads of organometallic compounds such as $Cr(CO)_6$, $Mo(CO)_6$, and $W(CO)_6$. The observation that these average IEs are essentially identical for the Os and Ru compounds is consistent with the general chemical similarity of these elements. The decrease in the $^2E_{2g}$–$^2A_{1g}$ separation on proceeding from Fe to Ru is attributable to an increase in metal–ligand covalency in the sense that an increased $e_{2g}(M)$–$e_{2g}(\pi^*)$ interaction (Fig. 16) will stabilize the $e_{2g}(M)$ MO, and an increased $a_{1g}(M)$–$a_{1g}(\pi)$ interaction will destabilize the $a_{1g}(M)$ MO. The other significant general trend in the transition metal metallocenes is the increase of the ionization cross sections for the metal-centered MOs relative to those of ligand MOs with increasing atomic number. This is a further example of the "heavy atom effect."

The UPS of monocyclopentadienyl iron complexes of the types $[(C_5H_5)Fe(CO)_2]_2$, $[(C_5H_5)Fe(CO)]_4$, and $C_5H_5Fe(CO)X$, X = Cl, Br, I, or Me can be treated in terms of a local C_{5v} symmetry for the C_5H_5–Fe interaction (268). However, as pointed out subsequently (150), Symon and Waddington used an incorrect assignment for ferrocene. Furthermore, there is some spectral evidence that the sample of $[(C_5H_5)Fe(CO)]_4$ underwent decomposition to ferrocene and CO.

Bent bis(cyclopentadienyl)metal complexes have been of interest for several years. In the initial theoretical model Ballhausen and Dahl (16) proposed that in $(C_5H_5)_2MoH_2$ the two d electrons reside in a lone pair MO situated between the two metal–hydrogen bonds. Subsequent structural studies on this class of compound, summarized in Refs. 148 and 151, have revealed that such a location for the lone pair is highly unlikely, and an alternative bonding model has been proposed. A convenient way to visualize this particular bonding scheme is to start with a conventional D_{5d} metallocene structure and bend the metal-ring axes, thereby producing a C_{2v} skeletal geometry. (Note that in this correlation diagram the x axis, unconventionally, becomes the twofold axis.) Under these conditions the doubly degenerate, predominantly metal and metal-ring MOs split into singly degenerate MOs as indicated in Fig. 18. When, for example, CO binds to the $(C_5H_5)_2M$ moiety the metal–carbon σ bond is formed by interaction of the $5\sigma(a_1)$ MO of CO and the a_1 metal orbital. In C_{2v} symmetry the 2π virtual orbitals of CO span the irreducible representations

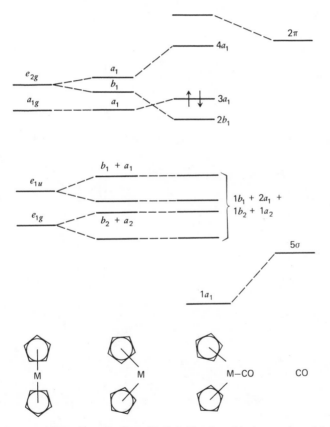

Fig. 18.   Proposed MO scheme for bent $(C_5H_5)_2M$ unit and its interaction with CO. (From Ref. 151.)

TABLE XIV

Experimental Ionization Energies (eV) and Proposed Assignments for Carbonyl, Ethylene, and π-allyl Complexes of Bent $(C_5H_5)_2M$ Units[a]

$(C_5H_5)_2MoCO$	$(C_5H_5)_2Mo(C_2H_4)$	$(C_5H_5)_2W(C_2H_4)$	$(C_5H_5)_2Nb(C_3H_5)$	Assignment[b,]	
				C_{2v}	C_s
5.9	6.0	6.0	5.7	$3a_1$	$4a'$
6.8	6.9	7.1	8.0	$2b_1$	$3a''$
8.8	8.8	9.0	8.6	Olefin—metal bonding[c]	
9.3	9.2	9.3	9.2	$2a_1 + 1b_2 + 1b_1 + 1a_2$	$2a' + 3a' + 1a' + 2a'$
9.6		9.5		$2a_1 + 1b_2 + 1b_1 + 1a_2$	$2a' + 3a' + 1a' + 2a'$
	11.3	11.3		$2a_1 + 1b_2 + 1b_1 + 1a_2$	$2a' + 3a' + 1a' + 2a'$

[a] Data and interpretations from Ref. 151.
[b] See Fig. 18 for orbital numbering scheme.
[c] No distinction possible between olefin-metal ionizations.

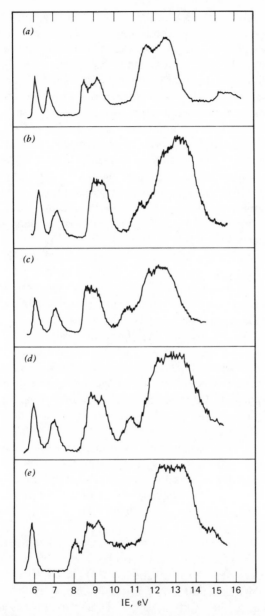

Fig. 19. He(I) UPS of (a) $[(C_5H_5)_2MoCo]$, (b) $[(C_5H_5)_2Mo(C_2H_4)]$, (c) $[(C_5H_5)_2W$-$(C_2H_4)]$, (d) $[(C_5H_5W(C_3H_5)]$, (e) $[(C_5H_5)_2Nb(C_3H_5)]$. (From Ref. 151.)

$b_1 + b_2$. Since there is no metal orbital of b_2 symmetry, π back bonding takes place by interaction of the b_1 orbitals on the metal and carbon monoxide. The same general scheme can be used to describe the bonding of C_2H_4 to the $(C_5H_5)_2M$ group. Thus for the CO and C_2H_4 complexes (Table XIV and Fig. 19) the first UPS peak is due to ionization of the $3a_1$ MO. The narrowness of this peak is consistent with the nonbonding nature of the $3a_1$ MO. Note that the second UPS peak of these complexes is distinctly broader, since the $2b_1$ MO is involved in a π back bonding interaction with the CO or C_2H_4 ligand. In the C_2H_4 complexes an additional peak is apparent at $\sim$11.0 eV. This is most reasonably ascribed to the ionization of an orbital that is an admixture of the $1a_1$ olefin π MO and a metal a_1 MO. Since the first IE of uncoordinated C_2H_4 is at 10.51 eV (273) it is clear that the olefin π electrons are stabilized by at least 0.5 eV upon complexation to the metal. Since the symmetry of the metal orbital involved in back bonding to the olefin is b_1, the C_2H_4 ligand is orientated in the xy plane (as defined above).

The treatment of the UPS of the π-allyl complex $(C_5H_5)_2Nb(C_3H_5)$ is similar to that of the ethylene complexes. One significant difference between the ethylene and π-allyl complexes, however, is that the symmetry is reduced to C_2 in the latter. The correlations between the appropriate orbital symmetries in the C_{2v} and C_s point groups are indicated in Table XIV.

If the hydride ligands are considered to reside in the xy plane of the $(C_5H_5)_2M$ moiety, the appropriate, symmetry-adapted combinations of H($1s$) AOs are a_1, $a_1 + b_1$, and $2a_1 + b_1$ in $(C_5H_5)_2MH$, $(C_5H_5)_2MH_2$, and $(C_5H_5)_2MH_3$, respectively. If bonding to the appropriate number of H atoms replaces the bonding of CO to the bent $(C_5H_5)_2M$ unit in Fig. 18, it follows that the numbers of metal lone pairs in the mono-, di-, and trihydrides are two $(a_1 + b_1)$, one (a_1), and none, respectively. The UPS of some representative hydrides (Table XV) support the foregoing bonding description. Thus $(C_5H_5)_2ReH$ exhibits two low-energy ionizations, the dihydrides, $(C_5H_5)_2MH_2$, M = Mo, W, show one such ionization, and $(C_5H_5)_2TaH_3$ does not possess a UPS band below 8.0 eV (Fig. 20). As expected, the UPS of the dimethyl compounds, $(C_5H_5)_2MMe_2$, M = Mo, W are remarkably similar to those of the corresponding dihydrides.

The UPS of the bent bis(cyclopentadienyl) metal chlorides $(C_5H_5)_2MCl_2$ and $(C_5H_4Me)_2MCl_2$, M = Ti, V have also been measured and interpreted with the aid of Fenske–Hall MO calculations (235). In the Ti complexes the $3a_1$ metal-localized MO is vacant, while in the V complexes this orbital is singly occupied, leading to the additional spectral complexities that are expected for the ionization of an open-shell system comparable studies have been made of analogous Zr and Hf compounds (64a).

As pointed out in the section dealing with paramagnetic metal carbonyls, the ionization of open-shell molecules can give rise to additional spectral com-

TABLE XV
Experimental Ionization Energies (eV) and Proposed Assignments for Hydride and Methyl Complexes of Bent $(C_5H_5)_2$ M Units[a]

$(C_5H_5)_2$ ReH	$(C_5H_5)_2$ MoH$_2$	$(C_5H_5)_2$ WH$_2$	$(C_5H_5)_2$ TaH$_2$	$(C_5H_5)_2$ MoMe$_2$	$(C_5H_5)_2$ WMe$_2$	Assignment[b,c]
6.4	6.4	6.4	8.1	6.1	6.0	$3a_1$
7.0	8.9	8.9	8.7	8.3	8.3	$2b_1$
8.8	9.5	9.6	9.6	8.9	8.8	$2a_1 + 1b_2 + 1b_1 + 1a_2$
9.2	—	—	—	—	9.0	$2a_1 + 1b_1 + 1b_1 + 1a_2$
9.9	—	—	—	—	—	$2a_1 + 1b_1 + 1b_1 + 1a_2$

[a] Data and interpretation from Ref. 151.
[b] See Fig. 2 for orbital numbering scheme.
[c] No distinction possible between higher-lying ionizations.

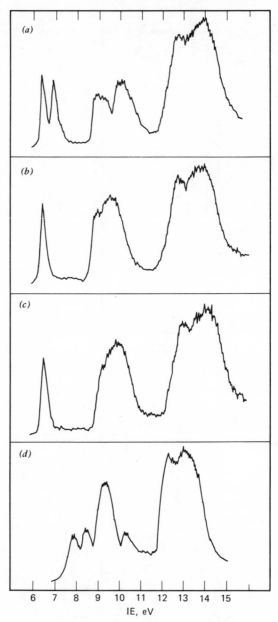

Fig. 20. He(I) UPS of (a) $[(C_5H_5)_2ReH]$, (b) $[(C_5H_5)_2MoH_2]$, (c) $[(C_5H_5)_2WH_2]$, (d) $[(C_5H_5)_2TaH_2]$. (From Ref. 151.)

plexities, since, in general, each electronic configuration of the ion will afford more than one ionic state. The calculation of all the ionizations under these circumstances would be a formidable task, particularly for systems involving two or more open shells. For the most part, therefore, the UPS assignments of the open-shell metallocenes are based on ligand field considerations and the correlation between predicted ionic states and observed spectral bands. Each of these approaches can be severely limited, however. For instance, ligand field data are only widely available for molecular ground states and the effects of orbital rescaling in proceeding to a given ionic state are difficult to assess in a quantitative manner. Furthermore, the theory of relative cross sections discussed earlier does not take account of configuration interaction; moreover, different MO subshells have been found to possess different one-electron cross sections. These difficulties are further compounded by uncertainties concerning the ground-state electronic configurations of chromocene and manganocene in the vapor phase (Table XVI). For example, in the solid state of manganocene, the observed paramagnetism corresponding to five unpaired electrons is clearly consistent with the $^6A_{1g}$ ground state (154). However, in the vapor phase low-spin states such as $^2E_{2g}$ or $^2A_{1g}$ (or even intermediate quartet spin states) cannot be

TABLE XVI
Ground-State Electronic Configurations of Open-Shell Metallocenes

Metallocene	d^n	Ground state
$(C_5H_5)_2V$	d^3	$^4A_{2g}(a^1_{1g}e^2_{2g})$
$(C_5H_5)_2Cr$	d^4	$^3A_{2g}(a^2_{1g}e^2_{2g})$ or $^3E_{2g}(a^1_{1g}e^3_{2g})$
$(C_5H_5)_2Mn$	d^5	$^6A_{1g}(a^1_{1g}e^2_{2g}e^2_{1g})$ or $^2E_{2g}(a^2_{1g}e^3_{2g})$ or $^2A_{1g}(a^1_{1g}e^4_{2g})$
$(C_5H_5)_2Co$	d^7	$^2E_{1g}(a^2_{1g}e^4_{2g}e^1_{1g})$
$(C_5H_5)_2Ni$	d^8	$^3A_{2g}(a^2_{1g}e^4_{2g}e^2_{1g})$

excluded with certainty. Indeed, it is well established that the pseudo-isoelectronic species $(C_5H_5)(C_6H_6)Cr$ and $(C_5H_5)(C_7H_7)V$ possess low-spin ground states in the condensed phases (138, 171, 244). With the foregoing considerations as a background it is not surprising that uncertainty surrounds the UPS assignments of the open-shell metallocenes.

To proceed it is convenient to assemble a tabulation of the ionic states that are produced by electron ejection from the metal-centered MOs (Table XVII). Along with these ionic states, which can be arrived at by simple group theoretical considerations, are listed the relative ligand field energies (122). Correct to first order, these were established with respect to the following orbital sequence.

TABLE XVII

Ionic States Resulting From Lower Energy Ionizations of Open-Shell Metallocenes, Ligand Field Energies and Experimental Ionization Energies (eV)

Ground state	Ion configuration	Ionic states produced	Ligand field energies[a]	Experimental ionization energies[a]
$(C_5H_5)_2V$				
$^4A_{1g}(a_{1g}^1e_{2g}^2)$	$a_{1g}^1e_{2g}^1$	$^3E_{2g}$	$A - 8B + \Delta_2$	6.78 (6.60)[b]
	e_{2g}^2	$^3A_{2g}$	$A + 4B$	6.78 (6.60)[b]
$(C_5H_5)_2Cr$				
$^3A_{2g}(a_{1g}^2e_{2g}^2)$	$a_{1g}^2e_{2g}^1$	$^2E_{2g}$	$3A + 8B + 2\Delta_2$	
	$a_{1g}^1e_{2g}^2$	$^4A_{2g}$	$3A - 12B + \Delta_2$	
	$a_{1g}^1e_{2g}^2$	$^2A_{2g}$	$3A + 3C + \Delta_2$	
$^3E_{2g}(a_{1g}^1e_{2g}^3)$	$a_{1g}^1e_{2g}^2$	$^4A_{2g}$		$-^c$
	$a_{1g}^1e_{2g}^2$	$^2A_{2g}$		$-^c$
	$a_{1g}^1e_{2g}^2$	$^2A_{1g}$	$3A - 8B + 5C + \Delta_2$	$-^c$
	$a_{1g}^1e_{2g}^2$	$^2E_{1g}$	$3A - 8B + 3C + \Delta_2$	$-^c$
	e_{2g}^3	$^2E_{2g}$	$3A + 12B + 4C$	$-^c$
$(C_5H_5)_2Mn$				
$^6A_{1g}(a_{1g}^1e_{2g}^2e_{1g}^2)$	$a_{1g}^1e_{2g}^2e_{1g}^1$	$^5E_{1g}$	$6A - 21B + 2\Delta_2 + \Delta_1$	$-^c$
	$a_{1g}^1e_{2g}^1e_{1g}^2$	$^5E_{2g}$	$6A - 21B + 3\Delta_2 + 2\Delta_1$	$-^c$
	$e_{2g}^2e_{1g}^2$	$^5A_{1g}$	$6A - 21B + 2\Delta_2 + 2\Delta_1$	$-^c$

$^2E_{2g}(a_{1g}^2e_{2g}^3)$	$a_{1g}^2e_{2g}^2$	$^1E_{1g}$	$6A - 16B + 7C + 2\Delta_2$	$-^c$
	$a_{1g}^2e_{2g}^2$	$^1A_{1g}$	$6A - 16B + 9C + 2\Delta_2$	$-^c$
	$a_{1g}^2e_{2g}^2$	$^3A_{2g}$	$6A - 16B + 5C + 2\Delta_2$	$-^c$
	$a_{1g}^1e_{2g}^3$	$^3E_{2g}$	$6A - 8B + 5C + \Delta_2$	$-^c$
	$a_{1g}^1e_{2g}^3$	$^1E_{2g}$	$6A + 7C + \Delta_2$	$-^c$
$^2A_{1g}(a_{1g}^1e_{2g}^4)$	$a_{1g}^1e_{2g}^3$	$^3E_{2g}$		$-^c$
	$a_{1g}^1e_{2g}^3$	$^1E_{2g}$		$-^c$
	e_{2g}^4	$^1A_{1g}$	$6A + 24B + 8C$	

$(C_5H_5)_2Co$

$^2E_{1g}(a_{1g}^2e_{2g}^4e_{1g}^1)$	$a_{1g}^2e_{2g}^4$	$^1A_{1g}$	$15A - 20B + 15C + 2\Delta_2$	5.56 (5.37)
	$a_{1g}^2e_{2g}^3e_{1g}^1$	$^3E_{1g}$	$15A - 26B + 12C + 3\Delta_2 + \Delta_1$	7.18 (6.97)
	$a_{1g}^2e_{2g}^3e_{1g}^1$	$^1E_{1g}$	$15A - 14B + 14C + 3\Delta_2 + \Delta_1$	
	$a_{1g}^2e_{2g}^3e_{1g}^1$	$^3E_{2g}$	$15A - 32B + 12C + 3\Delta_2 + \Delta_1$	7.18 (6.97)
	$a_{1g}^2e_{2g}^3e_{1g}^1$	$^1E_{2g}$	$15A - 32B + 14C + 3\Delta_2 + \Delta_1$	

$(C_5H_5)_2Ni$

$^3A_{2g}(a_{1g}^2e_{2g}^4e_{1g}^2)$	$a_{1g}^2e_{2g}^4e_{1g}^1$	$^2E_{1g}$	$21A - 31B + 18C + 3\Delta_2 + \Delta_1$	6.51 (6.36)
	$a_{1g}^2e_{2g}^3e_{1g}^2$	$^4E_{2g}$	$21A - 43B + 14C + 4\Delta_2 + 2\Delta_1$	Overlapping bands in region ~8.0–10.5
	$a_{1g}^2e_{2g}^3e_{1g}^2$	$^2E_{2g}$	$21A - 34B + 17C + 4\Delta_2 + 2\Delta_1$	Overlapping bands in region ~8.0–10.5
	$a_{1g}^1e_{2g}^4e_{1g}^2$	$^4A_{2g}$	$21A - 31B + 14C + 3\Delta_2 + 2\Delta_1$	
	$a_{1g}^1e_{2g}^4e_{1g}^2$	$^2A_{2g}$	$21A - 28B + 17C + 3\Delta_2 + 2\Delta_1$	

^{a} Data from Ref. 48.
^{b} Values in parentheses refer to the 1,1'-dimethyl metallocenes.
^{c} Assignments controversial; see text.

$$e_{1g}(d_{xz}, d_{yz}) \quad \overline{}$$

$$\Big\updownarrow \Delta_1$$

$$a_{1g}(d_{z^2}) \quad \overline{}$$

$$\Big\updownarrow \Delta_2$$

$$e_{2g}(d_{xy}, d_{x^2-y^2}) \quad \overline{}$$

In the cases of chromocene and manganocene, alternate sets of ionic states are presented because of the ambiguities regarding the molecular ground state.

Only one band appears in the low-energy region of the UPS of vanadocene and its 1,1-dimethyl analogue (122). Since two ionic states are predicted in this region, due to the ionization of a_{1g} or e_{2g} metal MOs, it was suggested that the $^3A_{2g}$ and $^3E_{2g}$ ionic states are too close in energy to permit spectral resolution. Such an idea would be consistent with the small value estimated for the $^2A_{1g}$–$^2E_{2g}$ separation on ligand field grounds ($12B - \Delta_2 \approx 0.19$ eV. See Table XVI). The band moves to lower IE by about 0.2 eV in 1,1'-dimethyl-vanadocene. This is typical of the shifts exhibited by the predominantly metal MO ionizations of the other metallocenes.

For chromocene none of the various UPS assignments is completely satisfactory (85, 122, 243). It is agreed, however, that the molecular ground state is $^3E_{2g}$ rather than $^3A_{2g}$. This conclusion evolves from the fact that the $^4A_{2g}$ ground state affords only three low-energy ionic states, while the $^3E_{2g}$ ground state leads to five. Experimentally the low-energy region (<8 eV) of the UPS of $(C_5H_5)_2$Cr comprises at least four ionizations (122, 243). The assignments for these four ionizations remain equivocal, since none of the four assignments advanced thus far (Table XVIII) is completely satisfactory from the standpoint of either relative peak intensities or simple ligand field considerations. A more elaborate theoretical analysis encompassing configuration interaction is clearly needed.

TABLE XVIII

Proposed Assignments for the First Four UPS Bands of Chromocene

Assignment	Band I	Band II	Band III	Band IV	Ref.
	$(5.71)^a$	$(7.04)^a$	$(7.30)^a$	$(7.58)^{a,b}$	
1	$^4A_{2g} + {}^2E_g$	$^2E_{1g}$	$^2A_{1g}$	$^2A_{2g}$	85
2	$^2E_{2g}$	$^4A_{2g}$	$^2E_{1g}$	$^2A_{1g} + {}^2A_{2g}$	243
3	$^4A_{2g}$	$^2E_{1g}$	$^2A_{1g}$	$^2A_{2g}$	122
4	$^4A_{2g}$	$^2E_{1g} + {}^2E_{2g}$	$^2A_{1g}$	$^2A_{2g}$	122

^{a} Data from Ref. 122.
^{b} Shoulder.

The UPS of manganocene and the related 1,1'-dimethyl compound pose particularly difficult assignment problems (122). As pointed out above, it is not clear whether the molecular ground state is $^6A_{1g}$, one of the low spin states $^2E_{2g}$ or $^2A_{1g}$, or even one of the intermediate quartet spin states. The UPS assignments have, in fact, been made with respect to both the $^6A_{1g}$ and the $^2A_{1g}$ ground state (122, 243). Thus Evans et al. (122) attribute the 6.91, 10.10, and 10.51 eV bands (Fig. 21) to the production of the $^5E_{1g}$, $^5A_{1g}$, and ^5E_g ionic states, respectively, arising from the high-spin ground state; on the other hand, Rabalais et al. (243), favoring the $^2A_{1g}$ ground state, attribute the first UPS band of $(C_5H_5)_2Mn$ to the formation of the closely spaced $^1A_{1g}$ (6.70 eV), $^3E_{2g}$ (6.85 eV), and $^1E_{2g}$ (7.10 eV) ionic states. The latter assignment has been questioned on the basis of ligand field considerations (276). Furthermore, INDO-type calculations (59) indicate that the ground state of manganocene and the computed ionization energies are in reasonable agreement with the interpretation of Evans et al. (122). The UPS of 1,1'-dimethylmanganocene is especially interesting in the sense that at least three additional peaks are discernible at low IE compared to that of $(C_5H_5)_2Mn$ (Fig. 21). It has been suggested (122) that

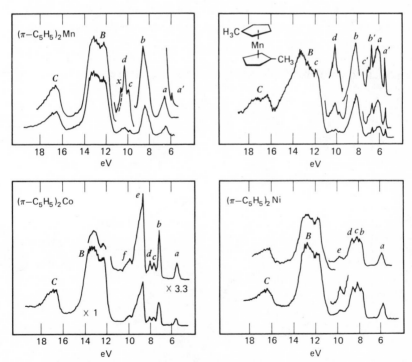

Fig. 21. He(I) UPS of manganocene, 1,1'-dimethylmanganocene, cobaltocene, and nickelocene. (From Ref. 122.)

the additional peaks arise because the methylated compound exists as a mixture of spin states in the vapor phase, possibly on account of increased splitting of the metal-localized orbitals. Since the $^6A_{1g}$ and $^2E_{2g}$ states of manganocene are computed by the INDO method (59) to have equilibrium energies differing by only ~800 cm^{-1}, a mixture of spin states is not unreasonable.

The UPS of $(C_5H_5)_2Co$ and $(C_5H_4Me)_2Co$ are particularly rich in bands in the low IE region (122). Fortunately the ground-state electronic configuration of cobaltocene is not in doubt (Table XVI) and the first peak can be attributed to the ionization $a_{1g}^2 e_{2g}^4 e_{1g}^1 \rightarrow a_{1g}^2 e_{2g}^4$ (Fig. 21). The observation that this peak is broader than that of the metal-localized MOs of ferrocene is consistent with the view that the e_{1g} metal MO is antibonding. The next peak is assigned to the production of the $^3E_{1g}$ and $^3E_{2g}$ ionic states, which arise by ionization of the e_{2g} metal MOs. However, beyond this point the assignments are not definitive and, in fact, four possible assignments have been suggested for the remaining $^{1,3}E_{1g}$ and $^1E_{2g}$ ionic states (122).

Finally, nickelocene exhibits only one low-energy band, which is assignable to ionization of the e_{1g} metal-centered MO, leading to the $^2E_{1g}$ ionic state (122, 243). The asymmetry of this band is presumably due to the Jahn-Teller instability of the $^2E_{1g}$ state. The remaining ionic states emanating from the ionization of the metal MOs of e_{2g} and a_{1g} symmetry presumably appear in a heavily overlapped region at higher IE.

The trends in the IEs pertaining to the predominantly d MOs of the open-shell metallocenes are rather complex. These irregularities arise because of the

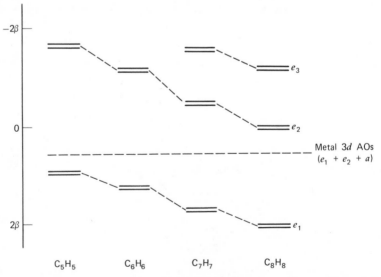

Fig. 22.   Hückel MO energies for e MOs of carbocyclic systems of general formula C_nH_n.

subtle effects of electron repulsion in the various open-shell configurations. If, however, an average ionization energy for the metal a_{1g} and e_{2g} MOs is defined by weighting each state from an ionized configuration according to its relative theoretical cross section, it is found that, with the exception of $(C_5H_5)_2Mn$, the average IE displays the anticipated increase in proceeding across the transition series from vanadocene to nickelocene.

The effects of changing the ring sizes of the arene ligands have been discussed in two publications (117, 119). At the outset let us consider the symmetries of the various species involved. In the case of dibenzenechromium there has, in fact, been considerable uncertainty regarding the skeletal symmetry. However, the current view (224) is that it is D_{6h}. The true symmetries of the methylated analogues and the mixed-ring sandwich compounds are obviously lower than D_{6h}. However, if these species are regarded as freely rotating "discs" the effective symmetries are $D_{\infty h}$ or $C_{\infty v}$. As shown in the following correlation table, the splitting of, for example, the nd AOs is unaffected by a change of symmetry from D_{6h} to $D_{\infty h}$ or $C_{\infty v}$.

nd AOs	D_{6h}	$D_{\infty h}$	$C_{\infty v}$
d_{z^2}	A_{1g}	$\Sigma_g^+ \ (\equiv A_{1g})$	$\Sigma^+ \ (\equiv A_1)$
(d_{xz}, d_{yz})	E_{1g}	$\Pi_g \ (\equiv E_{1g})$	$\Pi \ (\equiv E_1)$
$(d_{xy}, d_{x^2-y^2})$	E_{2g}	$\Delta_g \ (\equiv E_{2g})$	$\Delta \ (\equiv E_2)$

It is perhaps instructive to consider the relative energies of the doubly degenerate Hückel MOs of the C_nH_n carbocyclic systems as a function of n (Fig. 22). The relative energies of the free metal $3d$ AOs have also been indicated; for the early transition metals these orbitals are expected to lie between the e_1 and e_2 ring MOs. As a consequence of the stabilization of all the e-type ring MOs with increasing ring size, the e_1-e_1 metal–ring interaction should decrease progressively with increasing ring size, while the opposite should be true for the e_2-e_2 metal–ring interaction. One important consequence of this trend is a reversal of the order of the $a_{1(g)}$ and $e_{2(g)}$ metal–localized MOs on proceeding from the bis(cyclopentadienide) complexes to sandwich molecules involving larger rings.

The various approximate MO calculations on the ground state of dibenzenechromium (5, 139, 176, 261, 262, 264) are not in agreement regarding the sequence of MOs; however, these computations do agree regarding the relative ordering of the a_{1g} and e_{2g} metal-localized MOs. Ab initio MO calculations (160) on $(C_6H_6)_2Cr$ indicate that in the molecular ground state the $4e_{2g}$ MO is less stable than the $8a_{1g}$ MO (Table XIX). However, when ΔSCF computations were made on the first two ionic states it was demonstrated that much larger relaxation energies are associated with the ionization of the a_{1g} MO on account of the high percentage of metal character. While the computed relaxation

TABLE XIX

Ab initio Eigenvalues, Computed Ionization Energies, and Experimental UPS Data
for $(C_6H_6)_2Cr^a$

MO	Nature	Energy, eV	ΔSCF computed IE, eV	Experimental IE, eV
$4e_{2g}$	53% Cr(3d)	−7.5	4.9	6.46
$8a_{1g}$	92% Cr(3d)	−11.3	5.1	5.45
$4e_{1g}$ ⎱	Benzene π	−12.0		9.56(sh)
$6e_{1u}$ ⎰	$(1e_{2g})$	−12.9		9.80
$3e_{2u}$ ⎱	Benzene σ	−15.0		11.39
$3e_{2g}$ ⎰	$(3e_{2g})$	−15.2		11.39
$6a_{2u}$ ⎱	Benzene π	−15.5		11.19(sh)
$7a_{1g}$ ⎰	$(1a_{1u})$	−17.1		12.7(sh)
$5e_{1u}$		−18.1		14.23
$3e_{1g}$		−18.2		14.23
$1b_{1g}$		−18.5		14.23
$1b_{2u}$		−18.5		14.23
$2b_{2g}$		−18.6		14.2(sh)
$2b_{1u}$		−18.8		14.2(sh)
$5a_{2u}$		−21.0		16.83
$6a_{1g}$		−21.5		16.83

^{a} Data and assignments from Ref. 160.

energies did not quite reverse the ordering of the $^2E_{2g}$ and $^2A_{1g}$ ionic states, it was suggested that this reversal would occur if a more flexible Cr(3d) basis set were employed. For the assignment of ionizations higher than 8 eV it was assumed that Koopmans' theorem is valid, since it is only the first two MOs of $(C_6H_6)_2$Cr that involve appreciable metal character. ESR and electronic spectral studies (110, 132, 238, 240) indicate that the ground state for the dibenzenechromium cation is $^2A_1$. The UPS assignments for dibenzene-chromium, dibenzenemolybdenum, and various methylated derivatives are presented in Table XX.

The mixed-ring compounds $(\pi\text{-}C_6H_6)(\pi\text{-}C_5H_5)$Mn and $(\pi\text{-}C_7H_7)$ $(\pi\text{-}C_5H_5)$Cr are formally isoelectronic with dibenzenechromium and, as expected, the UPS of all three compounds (117, 119) are quite similar in the spectral region below 8 eV. Above 8 eV, however, the spectra are more complex than those of the symmetrical sandwich compounds as a result of the presence of ionizations from two types of ring. The mixed-ring titanium compound $(\pi\text{-}C_7H_7)(\pi\text{-}C_5H_5)$Ti is also diamagnetic but possesses two valence electrons less than chromocene. The UPS of this species (119) is in accord with the view that the $a_{1(g)}$ metal-centered MO is unoccupied in the molecular ground state. Similar mixed-ring compounds of Zr, Nb, and Mo have also been investigated by UPS (156a).

One striking feature of the UPS data for chromocene and its cognates (Table XX) is the very low value for the first IE. In fact, the first IE of

TABLE XX

Ionization Energies (eV) and Assignments for Closed-Shell Arene Complexes[a]

Compound	Ionic state			
	$^2A_{1\,(g)}$(metal)	$^2E_{2\,(g)}$(metal)	$^2E_{2\,(u)}$(ligand)	$^2E_{1\,(g)}$(ligand)
$(\pi\text{-}C_6H_6)_2Cr$	5.4	6.4	9.6	9.6
$(\pi\text{-}C_6H_5Me)_2Cr$	5.24	6.19	9.16	9.53
$(\pi\text{-}C_6H_3Me_3)_2Cr$	5.01	5.88	8.90	8.90
$(\pi\text{-}C_6H_6)_2Mo$	5.52	6.59	9.47	10.15
$(\pi\text{-}C_6H_5Me)_2Mo$	5.32	6.33	9.05	9.75
$(\pi\text{-}C_6H_3Me_3)_2Mo$	5.13	6.03	8.63	9.31
$(\pi\text{-}C_7H_7)\,(\pi\text{-}C_5H_5)Cr$	5.59	7.19	$-c$	$-c$
$(\pi\text{-}C_6H_6)\,(\pi\text{-}C_5H_5)Mn$	6.36	6.72	$-c$	$-c$
$(\pi\text{-}C_7H_7)\,(\pi\text{-}C_5H_5)Ti$	$-b$	6.83	$-c$	$-c$

[a] Data and assignments from Refs. 117 and 119.
[b] a_1 metal MO unoccupied. See text.
[c] Assignments of ligand ionizations not attempted for mixed-ring systems.

$(C_6H_3Me_3)_2Cr$ (5.01 eV) is the lowest value reported for a transition metal compound. Undoubtedly this factor is responsible for the ease of oxidation of chromocene (280).

The trends in the first two IEs within the isoelectronic series $(\pi\text{-}C_6H_6)_2Cr$, $(\pi\text{-}C_6H_6)(\pi\text{-}C_5H_5)Mn$, and $(\pi\text{-}C_5H_5)_2Fe$ are displayed in Fig. 23. This clearly shows the crossover of the $^2A_{1(g)}$ and $^2E_{2(g)}$ ionic states, which arise from the ionization of MOs of predominantly metal $3d$ character. Note that the crossover

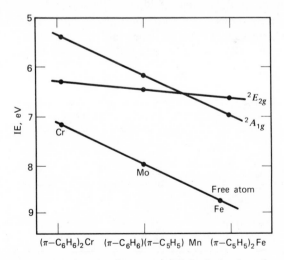

Fig. 23.   Ionization energy data for free atoms and for ionization of metal-localized a_{1g} and e_{2g} molecular orbitals. (Data from Refs. 117 and 119.)

is a consequence of the relative steepness of the $^2A_{1(g)}$ plot. The fact that this plot parallels that of the first IEs for free Cr, Mn, and Fe atoms rather closely is indicative of the high degree of metal character in the $a_{1(g)}$ MO.

Several open-shell arene—metal complexes have also been investigated by UPS (117, 119). they all feature one less valence electron than dibenzenechromium. As with the open-shell metallocenes, difficulties arise *inter alia* because of uncertainties regarding the molecular ground state. There are two reasonable possibilities: $^2E_2(e_2^3a_1^2)$ or $^2A_1(e_2^4a_1)$. Where ESR data are available (171, 244) the $^2A_1$ ground state is favored; hence this state is taken to be the preferred one here. Ionization of the a_1 electron in the $^2A_1$ ground state affords the $^1A_1$ ionic state, while ionization of an e_2 electron leads to $^1E_2$ and $^3E_2$ ionic states. The ionization energy data and tentative assignments for these molecules are given in Table XXI.

The electronic and molecular structures of the bis(cyclo-octatetraene)-actinides have attracted significant recent attention (15). However, despite these efforts some uncertainty remains regarding the nature and extent of f-orbital participation in the metal—ring bonding. In an effort to provide additional insight into this question Clark and Green (61, 62) have measured and interpreted the UPS of $(C_8H_8)_2$Th and $(C_8H_8)_2$U. A qualitative bonding scheme (Fig. 24) can be assembled by first considering the Hückel sequence of ring orbitals, allowing for ring—ring interaction, and finally by considering the symmetry-allowed interactions between the $(C_8H_8)_2$ MOs and metal $5f$ and $6d$ AOs. Addition of the appropriate numbers of electrons to the scheme then reveals that the ground-state electronic configurations of thorocene and uranocene are $a_{1g}^2a_{2u}^2e_{1g}^4e_{2u}^4e_{2g}^4e_{1u}^4$ and $a_{1g}^2a_{2u}^2e_{1g}^4e_{2u}^4e_{2g}^4e_{1u}^4f^2$, respectively. Consideration of the latter electronic configuration would imply the $^3A_{2u}$ ground state for uranocene. Unfortunately, however, the UPS of $(C_8H_8)_2$U provides little information on this point. If, in fact, the molecular ground state were $^3A_{2u}$, the ionic state produced, namely, $^2E_{3u}$, would be split into $J_z = \frac{5}{2}$ and $\frac{7}{2}$ components on

TABLE XXI
Ionization Energy Data (eV) and Tentative Assignments for
Open-Shell Arene—Metal Compounds[a]

Compound	Ionic state		
	$^1A_1$	$^3E_2$	$^1E_2$
$(\pi\text{-}C_8H_8)(\pi\text{-}C_5H_5)$Ti	5.67	7.62	7.62
$(\pi\text{-}C_6H_3Me_3)_2$V	5.61	5.33	6.08
$(\pi\text{-}C_7H_7)(\pi\text{-}C_5H_5)$V	6.42	6.77	7.28
$(\pi\text{-}C_6H_6)(\pi\text{-}C_5H_5)$Cr	6.20	6.20	7.15

[a] Data and assignments from Refs. 117 and 119.

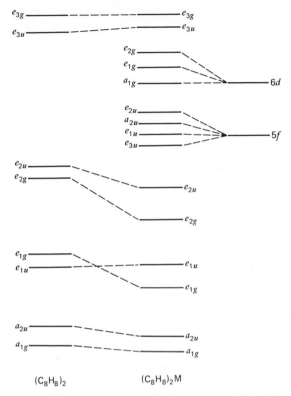

Fig. 24. Qualitative MO scheme for bis(cyclooctatetraene) actinides. (From Ref. 61.)

account of spin—orbit coupling, the estimated peak separation being about 0.6 eV. However, uranocene exhibits only one sharp peak at low IE (6.20 eV) (Fig. 25). It could, of course, be argued that the other component is obscured beneath the second UPS peak; however, intensity arguments render this possibility unlikely. The UPS of thorocene and the remaining peaks of uranocene are readily assigned (Table XXII and Fig. 25) by reference to the qualitative MO scheme in Fig. 24. Interestingly the first spectral peak of uranocene underwent a significant increase in intensity in changing from He(I) to He(II) radiation, thus indicating that the relative cross section for $5f$ orbitals increases relative to that of carbon. The fact that the band attributed to ionization of the e_{2u} MOs of thorocene and uranocene also underwent an increase of intensity in changing from He(I) to He(II) radiation implies that these orbitals have a significant contribution from the metal $5f$ AOs. That the differences in IE pertaining to the ionizations of the e_{2u} and e_{2g} MOs are larger than those of the e_{1u} and e_{1g} MOs can be taken to imply that metal $5d$ orbital participation is also important.

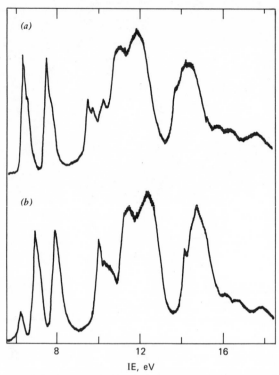

Fig. 25. He(I) UPS of (a) $(C_8H_8)_2$ Th and (b) $(C_8H_8)_2$ U. (From Ref. 61.)

TABLE XXII
Experimental Ionization Energies (eV) and Assignments for
Bis(cyclooctatetrene) actinides[a]

$(C_8H_8)_2$ Th	$(C_8H_8)_2$ U	Assignment
	6.20	f^2
6.79	6.90	e_{2u}
7.91	7.85	e_{2g}
9.90	9.95	e_{1u}, e_{1g}
10.14	10.28	e_{1u}, e_{1g}
10.65	10.56	e_{1u}, e_{1g}

[a] Data and assignments from Ref. 61.

102

C. Olefin Complexes

An electron diffracton study (95) of $(C_2H_4)Fe(CO)_4$ has established that the C_2H_4 ligand substitutes for one of the equatorial CO groups of $Fe(CO)_5$. The molecular symmetry of the ethylene complex is therefore C_{2v}. Discrete variational X_α MO calculations have been performed (10) on this molecule, producing the calculated IEs shown in Table XXIII. Note that, as with $Fe(CO)_5$ (10, 211), the lowest energy UPS peaks are associated with electron ejection from primarily metal $3d$ MOs. Lowering the symmetry from D_{3h} to C_{2v} causes splitting of the e' MO into a b_2 and an a_1 MO, while the e'' MO becomes an a_2 and a b_1 MO; hence there are four low-energy bands in the UPS of (C_2H_4)-$Fe(CO)_4$, while $Fe(CO)_5$ only exhibits two such ionizations (Fig. 5). Interestingly the average IEs for the predominantly metal orbitals of $Fe(CO)_5$ and $(C_2H_4)Fe(CO)_4$ are almost identical, suggesting similar charges on iron in both compounds. Surprisingly the IEs of free and coordinated ethylene are extremely similar even though the C=C bond distance of uncomplexed C_2H_4 (1.335 Å) is very different from that of coordinated C_2H_4 (1.462 Å). [Compare the IEs of the $1b_u$ (π_{CC}) and $1b_{1g}$ (σ_{CH}) MOs of free C_2H_4 with the corresponding $21a_1$ and $3a_2$ MOs of uncomplexed C_2H_4 in Table XXIII.] Overlap population analyses (10) indicate that this situation arises because of the interplay of π donation from the $1b_{1u}$ MO and back donation into the $1b_{3g}$ MO of ethylene as it undergoes coordination. However, it is possible that the complex undergoes dissociation in the vapor phase.

A study of the mode of bonding of butadiene, cyclobutadiene, and trimethylenemethane to the iron tricarbonyl moiety by UPS and *ab initio* MO

TABLE XXIII

Calculated and Experimental Ionization Energies (eV) for $(C_2H_4)Fe(CO)_4$ and C_2H_4

| MO | | $(C_2H_4)Fe(CO)_4{}^a$ | | C_2H_4 | |
		Calculated IE	Experimental IE	MO	Experimental IE
$12b_2$ (d_{xy})	Fe($3d$)	7.6	8.4–8.6		
$22a_1$ $(d_{x^2-y^2})$	Fe($3d$)	8.0	9.2		
$4a_2$ (d_{xy})	Fe($3d$)	8.8	9.6		
$11b_1$ (d_{xz})	Fe($3d$)	9.1	9.8		
$21a_1$	C_2H_4	10.1	10.5	$1b_{1u}$	10.5
$3a_2$	C_2H_4	11.9	12.3	$1b_{1g}$	12.45
$14a_1 - 20a_1$	$CO + C_2H_4$	17.4–13.3	>13.5		
$1a_2 - 2a_2$	$CO + C_2H_4$	15.0–13.8	>13.5		
$6b_1 - 10b_1$	$CO + C_2H_4$	16.4–12.5	>13.5		
$7b_2 - 11b_2$	$CO + C_2H_4$	17.6–13.5	>13.5		

^{a} Data from Ref. 10.

calculations has been presented in a series of papers by Hillier et al. (65, 66, 164). Low-resolution UPS studies of these compounds were reported earlier by Dewar and Worley (99, 100, 281). More recently the UPS of various cyclic diene complexes of $Fe(CO)_3$ and $Ru(CO)_3$ have been presented by Green et al. (153).

The structures of (butadiene)$Fe(CO)_3$ and (trimethylenemethane)$Fe(CO)_3$ are known to be C_s and C_{3v} in the solid and vapor states, respectively (3, 218). The *ab initio* MO calculations (65, 66, 164) indicate that there is an appreciable negative charge on each of the three organic ligands due to metal → ligand π-donation. For example, in the case of (butadiene)$Fe(CO)_3$ π donation into the butadiene LUMO($2b_2$) is greater than π back bonding from the HOMO ($1a_2$) into vacant Fe orbitals. For (trimethylenemethane)$Fe(CO)_3$ the π back bonding yields a larger net negative charge on the terminal carbon atoms than the central carbon atom as a result of metal → ligand π donation into the ligand $5e$ MO. Interestingly, for all three iron tricarbonyl compounds the charge on the CO ligands is very similar to that in $Fe(CO)_5$; hence butadiene, cyclobutadiene, and trimethylenemethane appear to be better π acceptors than CO.

The higher lying MOs of the three iron tricarbonyl compounds are given in Table XXIV. ΔSCF calculations have also been performed on the first few ionic states of all these molecules. As with many other transition metal systems these calculations indicate that large relaxation energies are associated with the photo-ionization of metal-rich MOs. The UPS assignments indicated in Table XXIV were arrived at using relative intensity criteria and by considering the effects of He(II) irradiation. Note that in some of the ionizations of (butadiene)$Fe(CO)_3$ and (trimethylenemethane)$Fe(CO)_3$ the ΔSCF calculations seem to underestimate the relaxation energies for the ionization of some of the metal-centered MOs.

Band I in the UPS of (butadiene)$Fe(CO)_3$ is ascribed to the ionization of four nearly degenerate MOs that are essentially metal localized. The second band is due to the ionization of what would be the HOMO ($1a_2$) of uncomplexed *cis*-butadiene, and band III is a result of electron loss from the lower π level ($1b_1$) of the free ligand. The UPS of several cyclic diene complexes of $Fe(CO)_3$ and $Ru(CO)_3$ have been assigned (153) by analogy with that of (butadiene)-$Fe(CO)_3$ and are presented in Table XXV. All these complexes show a decrease in IE for all bands as the ring size increases — a feature that is not observed in the UPS of the parent dienes (24, 29, 273). It is also interesting that the separation between bands II and III of the Fe and Ru diene complexes is almost constant, thus suggesting that the diene ligands are constrained to an essentially cis-planar conformation upon coordination.

When additional formally noncoordinating double bonds are added to the ligands as in (cycloheptatriene)$Fe(CO)_3$ and (cyclooctatetraene)$Fe(CO)_3$, two further peaks are apparent in the UPS at 10.23 and 10.61 eV, respectively. These are attributable to the ionization of the π-type MOs of the uncoordinated olefin

TABLE XXIV

Experimental and Theoretical Ionization Energies (eV) for Iron Tricarbonyl Complexes[a]

MO	%Fe(3d)	Ab initio (Koopmans' theorem)	ΔSCF	Relaxation energy	Experimental
		(Butadiene)Fe(CO)₃			
31a'	33	8.1	6.5	1.6	8.82[b]
18a''	10	10.8	9.9	0.9	9.93
30a'	51	12.6	7.7	4.9	8.82[b]
17a''	66	13.8	10.8	3.0	8.82[b]
29a'	34	14.1	13.4	0.7	11.52
28a'	65	14.7	9.0	5.7	8.82[b]
16a''	13	15.6	–	–	12.94
		(Cyclobutadiene)Fe(CO)₃			
17a''	21	9.2	8.1	1.1	8.17[c]
31a'	22	9.3	8.2	1.1	8.45
31a'	50	13.8	8.4	5.4	9.21
16a''	65	13.9	8.5	5.4	9.21
		(Trimethylenemethane)Fe(CO)₃			
15e	11	9.2	8.4	0.8	9.26
17a₁	37	13.0	8.2	4.8	8.62
14e	–	14.2	13.3	1.5	12.11
2a₂	–	14.3	13.6	0.7	12.57
13e	80	14.7	8.5	6.2	8.62
16a₁	43	15.0	13.3	1.7	11.07

[a] Data and assignments from Refs. 65, 66, and 164.
[b] Shoulder also observed at 8.23 eV.
[c] Shoulder.

moieties. In (cyclooctatetraene)Fe(CO)₃ there is structural and theoretical evidence (101) of some interaction between the free and complexed double bonds. The larger separation between UPS bands II and III for (cyclo-heptatriene)Fe(CO)₃ and (cyclooctatetraene)Fe(CO)₃ than for the diene complexes may be due to these interactions.

A somewhat similar UPS study of the coordination of pentahapto cyclic olefins toward the Mn(CO)₃, Re(CO)₃ and Fe(CO)₂X (X = Cl, Br, I and Me) moieties has been made by Lichtenberger et al. (202a, 279). Some data are presented in Table XXV. The study of the related molecule, (π-C₅H₅)Mn(CO)₂N₂ (203a) is of special interest since this represents the first UPS investigation of a coordinated N₂ molecule.

Very recently the UPS of the (cycloheptatriene)M(CO)₃ complexes of Group VIA were reported along with those of some substituted arene derivatives (145).

TABLE XXV
Ionization Energies (eV) for $M(CO)_3$ Complexes[a]

Compound	Band I [metal MOs]	Band II [ligand π]	Band III [ligand π]
a. Fe(CO)$_3$ and Ru(CO)$_3$ Complexes[a]			
$(C_6H_8)Fe(CO)_3$	7.98(sh), 8.56	9.33	11.04
$(C_7H_{10})Fe(CO)_3$	7.78(sh), 8.46	9.12	10.86
$(C_8H_{12})Fe(CO)_3$	7.45(sh), 8.27	8.87	10.44
$(C_6H_8)Ru(CO)_3$	8.01, 8.91	9.39(sh)	11.01
$(C_7H_{10})Ru(CO)_3$	7.96, 8.94	9.40(sh)	10.84
$(C_7H_8)Fe(CO)_3$	7.76(sh), 8.39	8.78	11.10
$(C_8H_8)Fe(CO)_3$	7.84	8.74	11.63
b. Mn(CO)$_3$ Complexes[b]			
$(C_6H_7)Mn(CO)_3$	8.06	8.59	10.25
$(C_7H_9)Mn(CO)_3$	7.86, 8.10	8.67	9.97
$(C_7H_7)Mn(CO)_3$	7.66, 7.86	8.33	10.33

[a] Data and interpretations from Ref. 153.
[b] Data and interpretations from Ref. 279.

V. TRIFLUOROPHOSPHINE COMPLEXES

There are many similarities in the behavior of PF_3 and CO as ligands. This is manifested by the existence of a large number of PF_3 complexes of the transition metals (197, 226). Several UPS studies of such complexes have now appeared in the literature (20, 152, 169, 182, 221, 227). As pointed out in Section II, it is difficult to arrive at a completely unequivocal set of assignments for these compounds other than for the predominantly metal MOs. It is seen below, however, that one of the advantages of studying the UPS of the PF_3 complexes is that many more ionizations pertinent to bonding discussions are discernible. Furthermore, trifluorophosphine complexes tend to be more thermally stable than their carbonyl counterparts.

Initially, let us consider the free ligand PF_3. In common with other phosphorus trihalides of C_{3v} symmetry, group theoretical considerations indicate that the P–X σ-bonding MOs span the irreducible representations, $a_1(2) + e$. One of the a_1 MOs [hereafter designated $a_1(n)$] involves principally the $P(3p_z)$ AO and, consequently, occurs at lower binding energy. Usually it is referred to as the phosphorus "lone pair." The other a_1 MO [hereafter designated $a_1(s)$] involves appreciable $P(3s)$ character and is therefore at much higher binding energy. The remaining valence electrons are distributed in halogen

TABLE XXVI

CNDO/2 MO Calculations on the Molecular Ground States of $Ni(PF_3)_4$, $Fe(PF_3)_5$, and $Cr(PF_3)_6$[a]

$Ni(PF_3)_4$		$Fe(PF_3)_5$		$Cr(PF_3)_6$	
MO	Eigenvalue, eV	MO	Eigenvalue, eV	MO	Eigenvalue, eV[b]
t_2	−14.33	e'	−14.40	t_{2g}	−14.81
				t_{2g}	−14.91
e	−17.10	e''	−15.87	e_g	−16.81
				e_g	−16.82
t_2	−19.06	a_1'	−16.08		
		a_2''	−16.96	t_{1u}	−17.08
		e'	−18.98	t_{1u}	−17.10

[a] Data from Ref. 251.

[b] Splitting due to C_{3v} symmetry of PF_3 ligand.

nonbonding MOs of symmetries a_2, a_1, and $e(2)$. The UPS data and assignments for PF_3 are shown in Table XXVII along with the assignments for the trifluorophosphine complexes (20–22, 152, 182, 210, 213, 221, 227, 237).

Structural and/or spectroscopic investigations have shown that the local symmetries around the metal atoms in the $M(PF_3)_4$ (189, 215), $M(PF_3)_5$ (2), and $M(PF_3)_6$ compounds are T_d, D_{3h}, and O_h, respectively. Unfortunately theoretical studies on trifluorophosphine complexes are confined to CNDO/2 calculations on $Ni(PF_3)_4$, $Fe(PF_3)_5$, and $Cr(PF_3)_6$ (251). In view of the discussion of relaxation effects and basis set dependences presented in Section II the CNDO/2 results (Table XXVI) should be used only as a qualitative guide to spectral assignment.

The first two bands in the UPS of $Ni(PF_3)_4$ (Fig. 26) are remarkably similar to those of $Ni(CO)_4$ and hence can be assigned to the $^2T_2$ and ^2E ionic states, which are produced by electron elimination from the t_2 and e MOs of predominantly $Ni(3d)$ character (20). The UPS of $Pd(PF_3)_4$ and $Pt(PF_3)_4$ can be assigned analogously (Table XXVII and Fig. 26) (20). The "heavy atom effect" (172) is clearly operative in this triad in the sense that the relative intensities of the spectral peaks are in the order $Pt > Pd > Ni$. The next two bands in the UPS of $Ni(PF_3)_4$ correspond to the ionization of the metal—phosphorus σ bonds of symmetries t_2 and a_1. The latter ionization is not detectable in the UPS of the heavier metal compounds and, presumably, is obscured by peaks of higher intensity. [Note that there are some differences between the preliminary reports (152, 182) and a subsequent full paper (20) regarding the spectra and assignments. For example, the weak 14.7 eV band of $Ni(PF_3)_4$ was not detected in one report (152). This band was detected in

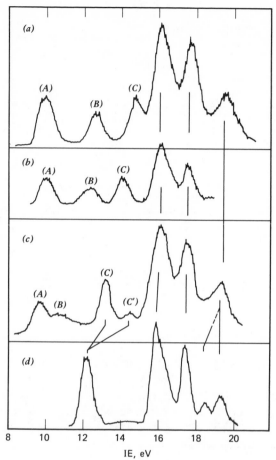

Fig. 26. He(I) UPS of (a) [Pt(PF$_3$)$_4$], (b) [Pd(PF$_3$)$_4$], (c) [Ni(PF$_3$)$_4$], and (d) PF$_3$. (From. Ref. 20.)

another report (182) but was given an assignment different than the one presented above.]

Like the corresponding hexacarbonyls the Group VIB triad M(PF$_3$)$_6$, M = Cr, Mo, and W (169, 227), exhibits only one UPS band below 12 eV that is attributable to the production of the $^2T_{2g}$ ionic state by way of electron ejection from the occupied metal-centered t_{2g} MO (Fig. 27). [However, in contrast to W(CO)$_6$ no spin—orbit coupling effects are discernible in the UPS of W(PF$_3$)$_6$.] Group theoretical considerations indicate that in O_h symmetry the six metal—phosphorus σ bonds span the irreducible representations a_{1g}, e_g, and t_{1u}. In the case of Cr(PF$_3$)$_6$ these ionizations are apparently degenerate, possibly implying that metal—phosphorus bonding is weaker in the Cr complex

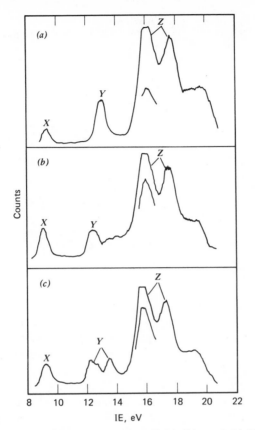

Fig. 27.   He(I) UPS of $[M(PF_3)_6]$ (a) M = Cr, (b) M = Mo, and (c) M = W. (From Ref. 169.)

than in those of the heavier elements. While the Mo and W complexes both display three peaks in the metal–phosphorus σ-bonding region (Table XXVII and Fig. 27), it is difficult to arrive at a persuasive set of assignments without *ab initio* calculations.

As in the case of $Fe(CO)_5$ the metal d orbitals of $Fe(PF_3)_5$ and $Ru(PF_3)_5$ split into a'_1, e', and e'' levels in D_{3h} symmetry, and both doubly degenerate MOs are fully occupied (Table XXVII). The first two UPS peaks of these compounds therefore correspond to the production of the $^2E'$ and $^2E''$ Fe 3d-like ionic states. Group theoretical considerations indicate that $^2E'$, $^2A'_1$ (2), and $^2A''_2$ ionic states should arise from ionization of the metal–phosphorus σ-bonding MOs. These ionizations are evidently degenerate in $Fe(PF_3)_5$, and only three of the four ionizations are detectable in $Ru(PF_3)_5$ (169). The assignments in Table XXVIII should therefore be viewed with caution.

TABLE XXVII
Ionization Energies (eV) and Assignments for PF_3 and $M(PF_3)_4$ Compounds
M = Ni, Pd, Pt

PF_3[a]		$M(PF_3)_4$[b]			
			M = Ni	M = Pd	M = Pt
MO	IE	MO	IE	IE	IE
$8a_1$ (n)	12.27	t_2	9.69	9.9	9.83
$6e$	15.88	e	10.74	12.2	12.45
$1a_2$	16.30	t_2	13.17	13.7	14.54
$5e$	17.46	a_1	14.65	—	—
$7a_1$ (s)	18.60		15.97	15.84	15.87
$4e$	19.50		17.48	17.4	17.53
			19.42		19.40

[a] Reference 213. See also Refs. 20–22, 152, 182, 210, 221, 227, and 237.
[b] Reference 20.
[c] Obscured beneath other peaks; see text.

The hydrides $HM(PF_3)_4$, M = Co, Rh, Ir, possess a structure similar to that of $HCo(CO)_4$. In C_{3v} skeletal symmetry the filled metal orbitals are of symmetry $e(2)$, the Rh–H σ bond transforms as a_1, and the metal–phosphorus σ bonds span the irreducible representations $a_1(2) + e$. Three low-energy peaks (Table XXIX) (169, 227) have been detected in the UPS of $HCo(PF_3)_4$, and overlapping of ionization occurs with the Rh and Ir compounds (Fig. 28). While the assignments cannot be regarded as definitive at the present time, the first two peaks in the UPS of $HCo(PF_3)_4$ probably correspond to the two ^2E ionic states of predominant metal character.

The hydrides $HMn(PF_3)_5$ and $HMn(CO)_5$ are expected to be similar with respect to their molecular and electronic structures. Since the sequence of the first three valence MOs in $HMn(CO)_5$ is $a_1(\sigma_{MnH}) < b_2(3d_{xy}) < e(3d_{xy}, 3d_{yz})$, a comparable ordering is anticipated for $HMn(PF_3)_5$. However, in the UPS of the latter (169) only two low IE peaks are apparent, consequently it has been assumed that the e and b_2 metal-localized MOs are degenerate (Table XXIX).

For all the trifluorophosphine complexes the IEs beyond 15 eV correspond very closely to those of uncomplexed PF_3 (20–22, 152, 182, 210, 213, 221, 227, 237), the most significant difference being the disappearance of the 18.6 eV peak upon coordination. This peak is assigned to the $a_1(s)$ MO of free PF_3 and features a modicum of phosphorus lone pair character; presumably it is stabilized slightly when coordination takes place, moving under the ~19.4 eV band. There is a noticeable constancy in the IEs of the F lone pair MOs among all the compounds studied (Tables XXVII–XXIX) and remarkably little difference between the IEs of free and complexed PF_3. Generally the IEs of PF_3 are

TABLE XXVIII

Ionization Energies (eV) and Assignments for PF_3, $M(PF_3)_6$, and $M(PF_3)_5$ Compounds

PF₃[a]			M(PF₃)₆[b]				M(PF₃)₅[b]	
MO	IE	MO	M = Cr	M = Mo	M = W	MO	M = Fe	M = Ru
$8a_1$ (n)	12.27	t_{2g}	9.29	9.17	9.30	e'	9.15	9.17
$6e$	15.88	e_g	12.84	12.94[c]	12.26[c]	e''	10.43	11.07
$1a_2$	16.30	a_{1g}	12.84	13.48[c]	12.64[c]	e'	13.08	12.8[c]
$5e$	17.46	t_{1u}	12.84	13.93[c]	13.52[c]	a_1'	13.08	12.8[c]
$7a_1$ (s)	18.60		15.80	15.80	15.85	a_1'	13.08	13.25[c]
$4e$	19.50		17.36	17.36	17.44	a_2''	13.08	13.65[c]
			19.3	19.1	18.7		15.83	15.75
							17.24	17.18
							19.1	18.9

[a] Data from Ref. 213. See also Refs. 20–22, 152, 182, 210, 221, 227, and 237.
[b] Data from Ref. 169.
[c] Assignments uncertain; see text.

111

TABLE XXIX
Ionization Energies (eV) and Assignments for PF_3, and Hydridotrifluorophosphine Complexes

PF_3 [a]		$HM(PF_3)_4$ [b]				$HMn(PF_3)_5$ [b]	
MO	IE	MO	M = Co	M = Rh	M = Ir	MO	IE
$8a_1$ (n)	12.27	e	9.58	9.70	9.82	e	9.47
$6e$	15.88	e	10.56	11.79	11.95	b_2	9.47
$1a_2$	16.30	a_1	12.12	13.83	14.18	a_1	11.30
$5e$	17.46	a_1	13.25	13.83	14.18	a_1	12.93
$7a_1$ (s)	18.60	a_1	13.25	13.83	14.18	e	12.93
$4e$	19.50	e	13.25	15.90	16.01	e	12.93
			16.00	17.42	17.42		15.85
			17.46	19.3	19.4		17.43
			19.4				19.4

[a] Data from Ref. 213. See also Refs. 20–22, 152, 182, 210, 221, 227, and 237.
[b] Data and assignments from Ref. 169.

112

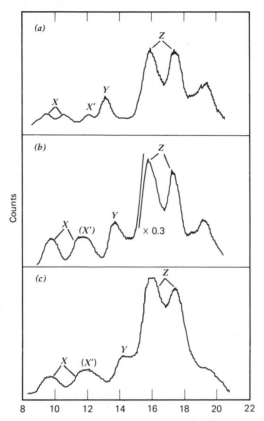

Fig. 28. He(I) UPS of $[HM(PF_3)_4]$ (a) M = Co, (b) M = Rh, and (c) M = Ir. (From Ref. 169.)

shifted by ~ 1 eV upon coordination to main group acceptors such as BF_3, BH_3, O, and S (21, 32, 113, 178, 210). Since the latter donor–acceptor interactions are anticipated to involve primarily σ bonding, the constancy of the IEs in the organometallic compounds has been attributed to back donation from the metal into PF_3 virtual orbitals of π symmetry.

For each of the PF_3 complexes the average IEs (weighted for degeneracies as appropriate) for the predominantly metal nd MOs are larger for the PF_3 complexes than for the corresponding carbonyls (Table XXX), thus suggesting higher positive charges on the metals in the PF_3 complexes. This is anticipated since PF_3 is a more electron-withdrawing ligand than CO.

The IE differences between the $^2T_2$ and ^2E states of the $M(PF_3)_4$ compounds, $\Delta_{T_2,E}$ (Table XXXI), correspond, within the limitations of Koopmans' theorem, to the ligand field splitting. The observation that $\Delta_{T_2,E}$

TABLE XXX

Average IEs for Metal nd MOs in PF_3 and CO Complexes, IEs for Lowest Energy
Metal–Phosphorus σ-Bonding MOs of PF_3 Complexes, and $T_2 - E$ Energy Gaps for
$M(PF_3)_4$ Compounds

	PF_3 complexes[a,b]			Carbonyls	
Compound	Average IE for nd MOs	$\Delta T_2, E$	$^\sigma(MP)^c$	Compound	Average IE for nd MOs
$Ni(PF_3)_4$	10.11	1.05	13.17(t_2)	$Ni(CO)_4$	8.49[d]
$Pd(PF_3)_4$	10.82	2.3	13.7(t_2)	–	
$Pt(PF_3)_4$	10.88	2.62	14.54(t_2)	–	
$Fe(PF_3)_5$	9.6	–	13.0(e')	$Fe(CO)_5$	8.28[e]
$Cr(PF_3)_6$	9.0	–	12.7(a_1g)	$Cr(CO)_6$	8.40[f]

[a] Data for Ni, Pd, and Pt compounds from Ref. 20.
[b] Data for Fe and Cr compounds from Ref. 227.
[c] Symmetry of MO indicated in parentheses.
[d] Reference 177.
[e] Reference 211.
[f] Reference 172.

TABLE XXXI

Average IEs (eV) for Metal nd MOs in PF_3 and CO Complexes and $T_2 - E$ or $E'' - E'$ Energy
Gaps for $M(PF_3)_4$ and $M(PF_3)_5$ Compounds

	PF_3 complexes[a,b]		Carbonyls	
Compound	Average IE for nd MOs	$\Delta T_2, E$ or $\Delta E'', E'$	Compound	Average IE for nd MOs
$Ni(PF_3)_4$	10.11	1.05	$Ni(CO)_4$	8.49[c]
$Pd(PF_3)_4$	10.82	2.3	–	
$Pt(PF_3)_4$	10.88	2.62	–	
$Fe(PF_3)_5$	9.79	1.28	$Fe(CO)_5$	9.25[d]
$Ru(PF_3)_5$	10.12	1.90	–	–
$Cr(PF_3)_6$	9.29	–	$Cr(CO)_6$	8.40[e]
$Mo(PF_3)_6$	9.17	–	$Mo(CO)_6$	8.50[e]
$W(PF_3)_6$	9.30	–	$W(CO)_6$	8.56[e]

[a] Data for Ni, Pd, and Pt compounds from Ref. 20.
[b] Data for other compounds from Ref. 169.
[c] Reference 177.
[d] Reference 10.
[e] Reference 172.

increases on proceeding down the group is in qualitative accord with the findings of optical spectroscopy. Note, however, that the UPS splittings increase in a nonuniform manner. Comparison of the T_2/E data (Table XXX) with the IE data for the $^2T_2$ and ^2E states (Table XXVII) reveals that the energy of the $^2T_2$ state remains relatively constant and that the changes in T_2/E are caused by the variations in the energy of the ^2E state. Since the e-type MO does not participate in σ bonding, the energy of the ^2E state is responding mainly to changes in the energy of the metal nd AOs. The insensitivity of the $^2T_2$ state to changes in the metal is caused by a near cancellation of stabilizing effects due to increasing the principal quantum number and dative π bonding to PF_3 virtual orbitals, and the destabilizing interaction between the metal MOs and metal–phosphorus MOs of t_2 symmetry.

The pentacoordinate trifluorophosphine complexes behave similarly. Thus the difference in energy between the $^2E'$ and $^2E''$ states, $\Delta_{E',E''}$, increases markedly in going from $Fe(PF_3)_5$ to $Ru(PF_3)_5$. This increase is caused by the change in energy of the $^2E''$ state, which, in turn, results from changing from a $3d$ to a $4d$ metal orbital.

VI. METAL NITROSYLS

An attempt has been made to study the relative ligand properties of CO and NO by examining the UPS of the isoelectric molecules $Ni(CO)_4$, $Co(CO)_3NO$, and $Fe(CO)_2(NO)_2$ (177, 211). The following abbreviated correlation table indicates qualitatively how the number of predominantly metal $3d$ MO ionizations should increase as the molecular symmetry descends from T_d to C_{2v}.

$T_d[Ni(CO)_4]$	$C_{3v}[Co(CO)_3NO]$	$C_{2v}[Fe(CO)_2(NO)_2]$
		$b_1(d_{xz})$
$t_2(d_{xz}, d_{yz}, d_{xy})$	$e(d_{xz}, d_{yz})$	$b_2(d_{yz})$
	$a_1(d_{z^2})$	$a_1(d_{z^2})$
$e(d_{x^2-y^2}, d_{xy})$	$e(d_{x^2-y^2}, d_{xy})$	$a_1(d_{x^2-y^2})$
		$a_2(d_{xy})$

However, it is clear from the UPS of $Co(CO)_3NO$ and $Fe(CO)_2(NO)_2$ that some overlapping of bands must occur because fewer ionizations are detected than predicted. In an effort to provide a more definitive basis for spectral assignment, *ab initio* MO carried out on the ground states of these molecules (177, 180). The energies of the lower lying ionic states have also been computed within the restricted Hartree–Fock formalism (177), thus affording direct calculations of the pertinent ionization energies. These data are summarized in Table XXXII, along with the computed relaxation energies and experimental ionization energies. Two general considerations emerge. First, while the first two MOs of $Ni(CO)_4$ are of predominant metal $3d$ character, the HOMOs of $Co(CO)_3NO$ and $Fe(CO)_2(NO)_2$ involve significantly less metal character and substantial amounts of nitrosyl ligand participation. Thus, while $Ni(CO)_4$ can be described as possessing a $3d^{10}$ electronic configuration, the traditional EAN rule view of the metal nitrosyls is not accurate. Rather than acting as a three-electron donor the NO

TABLE XXXII

Computed Eigenvalues, Ground State–Ionic State Energy Differences (ΔSCF), Relaxation Energies and Experimental Ionization Energies for $Ni(CO)_4$, $Co(CO)_3NO$, $Fe(CO)_2(NO)_2$, and $(\pi\text{-}C_5H_5)NiNO$ [a]

MO	% metal 3d character	92% of computed eigenvalue, eV	ΔSCF	Computed relaxation energy, eV	Experimental ionization energy, eV
		$Ni(CO)_4$			
$5t_2$	74	11.7	$7.0(^2T_2)$	4.7	8.90
$2e$	90	13.5	$7.8(^2E)$	5.7	9.77
		$Co(CO)_3(NO)$			
$8e$	57	8.7	$6.7(^2E)$	2.0	8.90
$8a_1$	78	12.9	$6.7(^2A_1)$	6.2	8.90
$7e$	86	14.1	$7.8(^2E)$	6.3	9.82
		$Fe(CO)_2(NO)_2$			
$6b_1$	15	8.7	$7.9(^2B_1)$	0.8	8.97
$10a_1$	19	9.1	$8.1(^2A_1)$	1.0	8.97
$6b_2$	65	12.6	$7.5(^2B_2)$	5.1	8.56
$9a_1$	83	15.1	$9.4^b(^2A_1)$	5.7	9.74
$3a_2$	80	16.0	$9.6(^2A_2)$	6.4	9.74
		$(\pi\text{-}C_5H_5)NiNO$			
$8e_1$	$-c$	9.4	8.9	0.5	$8.29(8.09)^a$
$7e_1$	60	13.5	11.6	1.9	$10.27(10.15)$
$15a_1$	72	15.2	8.6	6.6	$8.48(8.32)^a$
$4e_2$	74	15.2	8.4	6.8	$9.30(9.23)^c$

[a] Data for $(\pi\text{-}C_5H_5)NiNO$ from Ref. 124. Other data from Ref. 177.
[b] Estimated value.
[c] 63% carbon $2p\pi$ ring MO.
[d] Values in parentheses refer to $(\pi\text{-}C_5H_4Me)NiNO$.

TABLE XXXIII

Computed Second Moment Operators and Relaxation Energies for the First Four MOs of $(\pi\text{-}C_5H_5)NiNO^a$

MO	$\langle r^2 \rangle$	Computed relaxation energy, eV
$8e_1$	16.6	0.5
$7e_1$	9.7	1.9
$15a_1$	5.5	6.6
$4e_2$	6.0	6,8

^{a} Data from Ref. 124.

ligand seems to function principally as a π acceptor (181). The second significant feature concerns the large relaxation energies associated with metal-rich MOs. For each of the three compounds there is an approximate correlation between the percent metal $3d$ character and the relaxation energy. The very large ($\sim$6 eV) relaxation energies can cause a reshuffling of the MOs during the photoionization process. For example, on the basis of Koopmans' theorem the sequence of the first three MOs of $Fe(CO)_2(NO)_2$ is $6b_2 < 10a_1 < 6b_1$, yet because of the large relaxation energy accompanying the ionization of the predominantly metal $6b_2$ MO, the computed sequence of ionic-state energies is $^2B_2 < ^2B_1 < {}^2A_1$. A rather similar situation prevails in the complex $(\pi\text{-}C_5H_5)NiNO$ (124, 181). In the requisite C_{5v} symmetry the Ni($3d$) orbitals transform as $a_1(d_{z^2})$, $e_1(d_{xz}, d_{yz})$, and $e_2(d_{x^2-y^2}, d_{xy})$. (UPS studies with other metal arenes indicate that a C_5H_5 ring π-type MO of e_1 symmetry should be close in energy.) Relaxation energy calculations (Table XXXII), together with a study of the effect of ring methylation on the UPS, indicate that Koopmans' theorem fails to provide a satisfactory basis for spectral interpretation. From difference maps in which the electron densities of the Koopmans' theorem configurations are subtracted from the restricted Hartree–Fock configurations it is clear that large relaxation energies are dominated by the localization of electron density around the metal center during ionization (177). From Table XXXIII (124) it is clear that the calculated ionization energies of, for example, $(\pi\text{-}C_5H_5)NiNO$ correlate with the degree of localization of a molecular orbital, which can be measured by the magnitude of the second-moment operator $\langle r^2 \rangle$. A similar result has been found in the study of core holes in the sense that localized core holes involve larger relaxation energies (1, 13).

VII. TRANSITION METAL AMIDES

The transition metal amides constitute an extensive series of compounds with widely varying coordination numbers (37, 38). These factors, coupled with the observation that the ionizations from the metal nd MOs, nitrogen "lone pair" MOs, and metal–nitrogen σ bonds often fall into distinct regions, render these compounds ideal for photoelectronic spectroscopic study.

The majority of the UPS data reported so far (58, 144) pertain to the tetracoordinate dialkylamides. If the nitrogen geometry is taken to be trigonal planar (a common stereochemical facet of dialkyl- and disilylamides) and metal–nitrogen bond rotation is assumed to be slow on the UPS time scale, the maximum skeletal symmetry of a $M(NR_2)_4$ species is D_{2d}. There are, in fact, two possible structures that possess this symmetry; they differ by 90° rotations about the metal–nitrogen bonds. Very recently it was established that the Group VIB amide $Mo(NMe_2)_4$ adopts the D_{2d} geometry in which the C–N–C planes are perpendicular to the σ_d planes (55). This structure is presumed to be the preferred one for the other $M(NMe_2)_4$ molecules of concern here. The arrangement can be regarded as being derived from a tetrahedral array of nitrogen atoms around Mo, the descent to D_{2d} symmetry resulting from the spatial requirements of the NR_2 groups. In T_d symmetry the metal d orbitals split into the familiar e and t_2 sets; reduction of symmetry to D_{2d} results in these orbitals transforming as b_1, a_1, b_2, and e. In this particular D_{2d} geometry the nitrogen lone pair MOs, n_N, and metal–nitrogen σ-bonding MOs span the irreducible representations a_1, b_2, and e. Dative π bonding interactions are permitted between the nitrogen lone pair and metal d orbitals of symmetries a_1, b_2, and e as shown in Fig. 29. Note that since no interactions are possible with the $b_1(d_{x^2-y^2})$ orbital, this metal orbital is not perturbed and becomes the HOMO. The general correctness of this sequence of MOs has been confirmed by Fenske–Hall calculations on $Mo(NMe_2)_4$, which, in addition, revealed that the b_1 MO is 97% $4d_{x^2-y^2}$ in character and that the e, b_2, and a_1 MOs consist of 91, 94, and 71% ligand π character, respectively. From the UPS data for $Mo(NR_2)_4$, R = Me, Et and from the sequence of MOs indicated in Fig. 29, it is clear that the first peak (Table XXXIV and Fig. 30) corresponds to ionization of the b_1 MO producing the $^2B_1$ ionic state. The second, third, and fourth peaks correspond to electron ejection from the predominantly nitrogen lone pair MOs of symmetries e, b_2, and a_1, giving rise to the ^2E, $^2B_2$, and $^2A_1$ ionic states, respectively.

Interestingly $Cr(NEt_2)_4$ is paramagnetic, although the reasons for this are not clear at the present time. Possibly it is a reflection of the spin-pairing energies being in the order Cr > Mo and the ligand field energies being in the order Mo > Cr; alternatively, $Cr(NEt_2)_4$ could adopt the other type of D_{2d} structure. If this were so, the nitrogen lone pairs would transform as a_2, e, and b_2, and no conjugative interactions would be possible with the metal a_1 and b_1 MOs, thereby making them approximately degenerate. Ionization of the $^3B_1$ [$\ldots b_1^1 a_1^1$] ground state affords several ionic states. These are given in Table XXXV along with our tentative assignments.

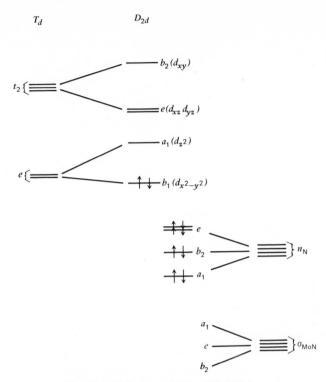

Fig. 29. Molecular orbitals of $Mo(NR_2)_4$.

The UPS of the dimagnetic tetra-amides of Ti, Zr, and Hf can also be interpreted according to Fig. 29. However, there are two less valence electrons for the Group IVB amides than for $Mo(NR_2)_4$; hence the b_1 metal orbital is unoccupied. All the Group IVB amide spectra exhibit splitting of the ^2E ionic state (Table XXXIV), presumably owing to the relief of Jahn–Teller degeneracy.

As expected the Group VB amide $V(NMe_2)_4$ possesses a paramagnetic ground state. On the basis of single occupancy of the b_1 metal orbital the molecular ground state is expected to be $^2B_1$, a deduction that has been confirmed by ESR studies of this compound (42, 184). Tentative assignments for the various predicted ionic states are indicated in Table XXXV.

Curiously the UPS of the main group tetraamides of Si, Ge, and Sn do not exhibit any separation of the ^2E, $^2B_2$, and $^2A_1$ peaks. This apparent degeneracy of ionic states has been attributed (144) to rapid rotation around M–N bonds and/or decreased interactions among the nitrogen lone pairs.

The use of the sterically demanding amido ligands such as $N(SiMe_3)_2$ and $N(^iPr)_2$ has given rise to an interesting variety of compounds with low coordination numbers (37, 38). For example, the tricoordinate silylamides,

TABLE XXXIV
Ionization Energy Data (eV) and Assignments for Tetrakis(dialkylamides)[a]

Compound	$^2B_1\,(d_{x^2-y^2})$	$^2E(n_N)$	$^2B_2\,(n_N)$	$^2A_1\,(n_N)$	M–N ionizations
		Ionic state			
Mo(NMe$_2$)$_4$	5.30	7.34	7.70	9.01	10.7
Mo(NEt$_2$)$_4$	5.3	7.0[b] 7.3[b]	7.56	8.7	–
Ti(NMe$_2$)$_4$	–	7.13[b] 7.36[b]	7.75	8.00	10.32
Ti(NEt$_2$)$_4$	–	6.83[b] 7.10[b]	7.47	7.75	9.78
Zr(NMe$_2$)$_4$	–	7.23[b] 7.54[b]	7.92	8.14	10.44
Zr(NEt$_2$)$_4$	–	6.76[b] 6.98[b]	7.35	7.54	9.55
Hf(NMe$_2$)$_4$	–	7.50[b] 7.82[b]	8.05	8.34	10.56
Hf(NEt$_2$)$_4$	–	7.15[b] 7.35[b]	7.68	7.91	9.97

[a] Data and assignments from Refs. 58 and 144.
[b] Splitting due to the Jahn-Teller effect. See text.

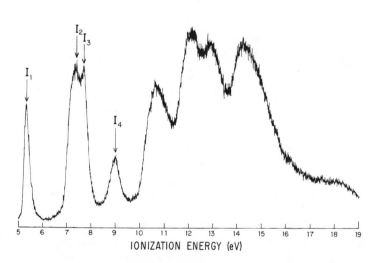

Fig. 30. He(I) UPS of Mo(NMe$_2$)$_4$. (From Ref. 58.)

TABLE XXXV
Ionic States Resulting from Lower Energy Ionizations of Open-Shell Metal Dialkylamides

Ground state	Ion configuration	Ionic states produced	Experimental ionization energy
	$Cr(NEt_2)_4$[a]		
$^3B_1\,(a_1^2 b_2^2 e^4 b_1^1 a_1^1)$	$a_1^2 b_2^2 e^4 b_1^1$	$^2B_1$	5.9
	$a_1^2 b_2^2 e^4 a_1^1$	$^2A_1$	6.3
	$a_1^2 b_2^2 e^3 b_1^1 a_1^1$	^2E	7.0
	$a_1^2 b_2^2 e^3 b_1^1 a_1^1$	^4E	7.0
	$a_1^2 b_2^1 e^4 b_1^1 a_1^1$	$^2A_2$	7.2
	$a_1^2 b_2^1 e^4 b_1^1 a_1^1$	$^4A_2$	7.2
	$a_1^1 b_2^2 e^4 b_1^1 a_1^1$	$^2B_1$	7.9
	$a_1^1 b_2^2 e^4 b_1^1 a_1^1$	$^4B_1$	7.9
	$V(NMe_2)_4$[b]		
$^2B_1\,(a_1^2 b_2^2 e^4 b_1^1)$	$a_1^2 b_2^2 e^4$	$^1A_1$	6.2
	$a_1^2 b_2^2 e^3 b_1^1$	^1E	7.08
	$a_1^2 b_2^2 e^3 b_1^1$	^3E	7.08
	$a_1^2 b_2^1 e^4 b_1^1$	$^1A_2$	7.60
	$a_1^2 b_2^1 e^4 b_1^1$	$^3A_2$	7.60
	$a_1^1 b_2^2 e^4 b_1^1$	$^1B_1$	8.28
	$a_1^1 b_2^2 e^4 b_1^1$	$^3B_1$	8.28
	$Cr(N^iPr_2)_3$[a]		
$^4A_2\,(e^4 a_2^2 a_1^1 e^2)$	$e^4 a_2^2 a_1^1 e^1$	^3E	6.53
	$e^4 a_2^2 e^2$	$^3A_2$	6.3
	$e^4 a_2^2 a_1^1 e^2$	$^3A_1$	7.38
	$e^4 a_2^1 a_1^1 e^2$	$^5A_1$	7.38
	$e^3 a_2^2 a_1^1 e^2$	^3E	7.9
	$e^3 a_2^2 a_1^1 e^2$	^5E	7.9

[a] Data and assignments from Ref. 58.
[b] Data and assignments from Ref. 144.

$M[N(SiMe_3)_2]_3$, $M = Sc$, Ti, Cr, and Fe, are stable, relatively volatile compounds that proved suitable for UPS measurements (200). With the exception of $Sc[N(SiMe_3)_2]_3$, which possesses a pyramidal (C_3) structure (143), these compounds are expected to adopt D_3 MN_3 skeletal geometries. The dihedral angle ϕ between the NSi_2 and MN_3 planes is not generally known for all the compounds; however, ϕ is 49° for $Fe[N(SiMe_3)_2]_3$ (41). Crystal field calculations

(4) (using the D_{3h} geometry), together with ESR (40) and magnetic data (4), indicate the following ground-state electronic configurations for the tricoordinate silylamides:

$$Ti[N(SiMe_3)_2]_3 [\ldots\ldots a_1{}^1]   {}^2A$$
$$Cr[N(SiMe_3)_2]_3 [\ldots\ldots a_1{}^1e^2]   {}^4A$$
$$Fe[N(SiMe_3)_2]_3 [\ldots\ldots e^2a_1{}^1e^2]   {}^6A$$

Curiously no ionizations attributable to the metal localized MOs were detected in the UPS of these open-shell compounds. The reason given (200) for this phenomenon is that the metal MOs are stabilized greatly by virtue of the strongly electron-withdrawing $N(SiMe_3)_2$ substituents, possibly causing them to become obscured by the ionizations of the nitrogen "lone pair" MOs. Interestingly, however, the ionizations from the metal orbitals are detectable (Fig. 31) for the dialkylamide $Cr(N^iPr_2)_3$ (58). The predicted ionic states and our tentative assignments are shown in Table XXXV.

In threefold symmetry the nitrogen "lone pair" MOs span the a and e irreducible representations. However, as is clear from the nodal properties of such MOs (Fig. 32) their relative energies are sensitive to the dihedral angle (82). Obviously, at intermediate values of ϕ, a crossover occurs, and at some particular value of ϕ the a and e MOs must be degenerate. Since the dihedral angles of the tricoordinate metal silylamides are ~50° it is difficult to be definitive about the assignments; accordingly only the nitrogen "lone pair" IE data are given in Table XXXVI.

As in the case of the main group tetrakis(amides), the tricoordinate

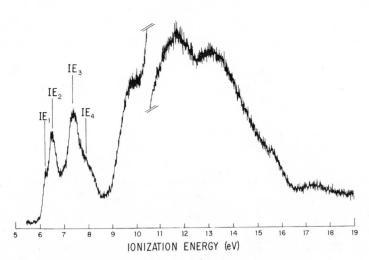

Fig. 31.   He(I) UPS of $Cr(N^iPr_2)_3$. (From Ref. 58.)

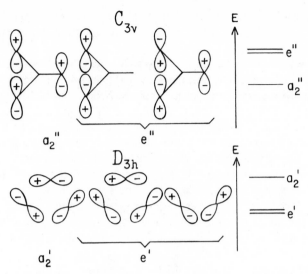

Fig. 32. Vertical ($\phi = 0°$) and horizontal ($\phi = 90°$) arrangements of N(2p) AOs in threefold symmetry.

TABLE XXXVI

Ionization Energies (eV) for Predominantly Nitrogen Lone Pair MOs of Dialkylamides and Disilylamides

$M[N(SiMe_3)_2]_3$ [a]	
M = Sc	8.18, 8.62
M = Ti	8.14, 8.70
M = Cr	8.07, 8.76
M = Fe	7.88, 8.74

$M[N(SiMe_3)_2]_2$ [b]	
M = Zn	8.50
M = Hg	8.33

$M(NMe_2)_5$ [c]	
M = Nb	6.77, 6.9, 7.63, 8.02, 8.21
M = Ta	6.89, 7.1, 7.78, 8.15, 8.35

[a] Data from Ref. 200.
[b] Data from Ref. 168.
[c] Data from Ref. 58.

bis(trimethylsilyl)amides of Ga and In exhibit no perceptible splittings in the nitrogen "lone pair" region (200). Possible reasons for this are advanced earlier in this section.

Two general classes of bicoordinate amides have been investigated by UPS (168), the d^{10} bis(trimethylsilyl)amides, $Zn[N(SiMe_3)_2]_2$ and $Hg[N(SiMe_3)_2]_2$, and main group species of the general types $M(NR_2)_2$ and $M(NRR')_2$, M = Ge, Sn, and Pb; R = $SiMe_3$, R' = CMe_3. The Zn and Hg compounds are presumed to be isostructural with $Be[N(SiMe_3)_2]_2$, which was found to adopt a linear (D_{2d}) geometry by electron diffraction (60), while the main group compounds are carbenoids and, as such, possess a lone pair of electrons and a C_{2v} framework. As anticipated for a linear (D_{2d}) MN_2 framework, no splitting of the nitrogen "lone pair" peaks is observed, since these orbitals are orthogonal.

High coordination numbers also exist with amido ligands. However, because of the steric requirements of even the smaller ligands, such as NMe_2, examples of five- and six-coordination are encountered only with the heavier elements.

One of the intriguing aspects of the pentacoordinate amides concerns the apparently close energies of the square pyramidal and trigonal bipyramidal MN_5 geometries. Thus it has been demonstrated by x-ray crystallography that the NbN_5 moiety of $Nb(NMe_2)_5$ approaches a square pyramidal structure (170), and $Ta(NEt_2)_5$ has been found (266) to adopt a trigonal bipyramidal geometry for the TaN_5 skeleton. My colleagues and I have found (58) that five ionizations are detectable in the nitrogen "lone pair" region of $Nb(NMe_2)_5$ and $Ta(NMe_2)_5$ (Fig. 33). In a C_2 arrangement of MNC_2 moieties these would correspond to the ionization of two a and three b MOs. However, it is difficult to be more specific

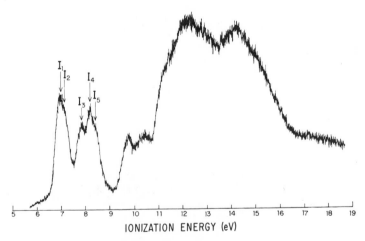

IONIZATION ENERGY (eV)

Fig. 33.   He(I) UPS of $Ta(NMe_2)_5$. (From Ref. 58.)

without MO calculations. The higher IEs are therefore given without assignments in Table XXXVI.

The six-coordinate tungsten compound $W(NMe_2)_6$ adopts a T_h structure. The $W-NC_2$ units are arranged about a locally octahedral WN_6 geometry in three mutually perpendicular pairs (39). A qualitative MO diagram appropriate to this symmetry has been assembled in Fig. 34 in the usual manner. The first two intense peaks in the UPS of $W(NMe_2)_6$ at 6.74 and 7.92 eV can be attributed to the ionization of the mainly nitrogen "lone pair" MOs of symmetries t_u and t_g, respectively. Ionization of the tungsten–nitrogen σ-bonding MOs of t_u and e_g symmetry leads to the production of the ^2T_u and ^2E_g states at 9.55 and 9.94 eV, respectively (144).

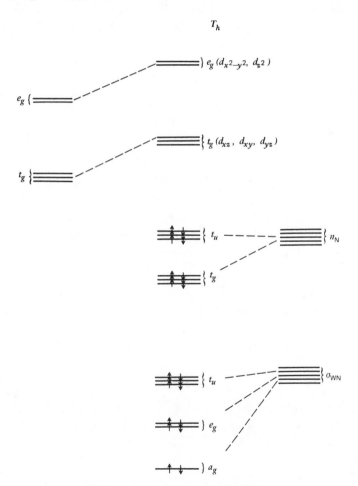

Fig. 34. Molecular orbitals of $W(NMe_2)_6$.

VIII. COMPOUNDS WITH METAL–METAL MULTIPLE BONDS

The development of systematic synthetic methods for compounds with triple and quadruple bonds between transition metals, together with the structural characterization of such species, has turned out to be a highly significant chemical development (54, 72). All the $M_2 X_6$ triply bonded structures possess basic ethane-like (D_{3d}) skeletal geometries (54). In the case of X = dialkylamino the nitrogen geometries are essentially trigonal planar with two distinct sets of alkyl environments, those pointing toward the M≡M bond (proximal) and those directed away (distal). The quadruply bonded compounds generally feature D_{4h} skeletal geometries as a consequence of an eclipsed arrangement of the two locally square planar subunits.

Initially the view was proposed (71) that metal–metal triple and quadruple bonds are, in qualitative terms, describable as $\sigma^2 \pi^4$ and $\sigma^2 \pi^4 \delta^2$, respectively, by effecting the appropriate combinations of valence metal d orbitals. Several SCF–X_α calculations have now been performed on a variety of compounds with triple (80) and quadruple (78, 79a, 219, 228–230) metal–metal bonds. Collectively these calculations have confirmed the essential correctness of the earlier, qualitative view of the bonding in compounds with multiple metal–metal bonds. However, as expected, mixing of orbitals of metal and ligand character takes place, thereby complicating the purely metal nd orbital description of the metal–metal binding. In fact, orbital mixing has been found (78) to be quite substantial in $Tc_2 Cl_8^{3-}$ and $Re_2 Cl_8^{2-}$; hence in these anions the various contributions to the quadruple bonds are not as simply described. *Ab initio* calculations (27a, 159a) reveal that, while a single determinental description is adequate for Mo≡Mo complexes, correlation effects are very important for the analogous dichromium complexes.

A number of UPS investigations of metal–metal multiple bonding have been carried out (73, 79, 83, 142a, 149). SCF–X_α calculations on the model compound $(HO)_3 Mo≡Mo(OH)_3$ reveal that the $5e_u(\pi)$ and $4a_{1g}(\sigma)$ MOs (Fig. 35), which involve large amounts of metal character, are principally responsible for the bonding between the molybdenum atoms (80). The orbitals $1a_{2g}$ through $3e_g$ correspond to those anticipated by group theory for the oxygen lone pairs. The observed UPS of $Mo_2 (OCH_2 CMe_3)_6$ in the low-energy region can be predicted very satisfactorily by the IEs calculated by the X_α transition state method if account is taken of the inductive shifts of replacing H atoms of the model compound by neopentyl groups (Fig. 36). Note that the first two peaks are associated with the ionization of the π and σ components of the molybdenum–molybdenum triple bond.

The situation is somewhat more complicated for the amido derivatives, $Mo_2 (NR_2)_6$ and $W_2 (NR_2)_6$. In D_{3d} symmetry the nitrogen lone pair MOs, n_N, span the irreducible representations a_{2g}, a_{1u}, and e_g. Rather close in energy are

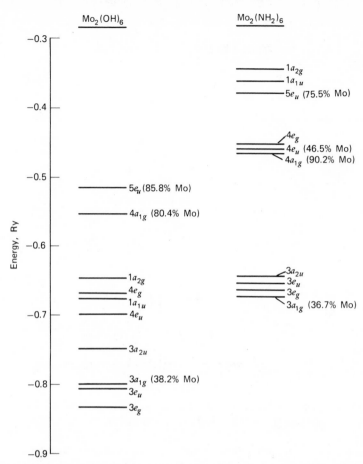

Fig. 35. Higher lying MOs for $Mo_2(NH_2)_6$ and $Mo_2(OH)_6$ computed by the SCF–X$_\alpha$ method. (From Ref. 80.)

the $\pi(5e_u)$ and $\sigma(4a_{1g})$ components of the metal–metal triple bond. Since the π bond is of the same symmetry (e_u) as one of the nitrogen lone pair MOs, they mix extensively. The extent of mixing is quite substituent dependent, as evidenced by the SCF–X$_\alpha$ calculations (80) on $Mo_2(NH_2)_6$ and $Mo_2(NMe_2)_6$. The tungsten–nitrogen σ bonds, σ_{MN}, which transform as a_{2u}, e_u, e_g, and a_{1g}, are found at higher binding energies. Note that the σ component of the metal–metal triple bond transforms as a_{1g}; consequently this MO undergoes mixing with the $\sigma_{MN}(a_{1g})$ MO. The overall energy level scheme is shown in Fig. 35. The UPS of $Mo_2(NMe_2)_6$, $W_2(NMe_2)_6$, and $W_2(NEt_2)_6$ have been assigned on the basis of this scheme and the data are presented in Table XXXVII.

X-ray crystal structure studies indicate that $(Et_2N)_4W_2Me_2$ (56) and

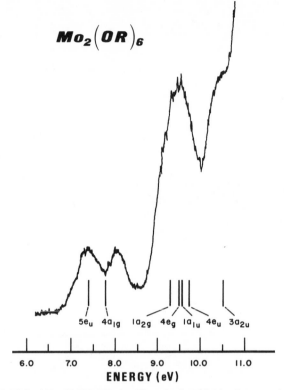

$Mo_2(OR)_6$

5e$_u$ 4a$_{1g}$ 1a$_{2g}$ 4e$_g$ 1a$_{1u}$ 4e$_u$ 3a$_{2u}$

ENERGY (eV)

Fig. 36. He(I) UPS of $Mo_2(OCH_2CMe_3)_6$. The calculated ionization energies for $Mo_2(OH)_6$ using the transition-state method are also shown. (From Ref. 80.)

$(Et_2N)_4W_2Cl_2$ (57) both possess trans stereochemistry in the solid state. In solution, however, $(Et_2N)_4W_2Me_2$ exists as an approximately 3:2 mixture of gauche (C_2) and trans (C_i) conformers with a barrier to interconversion of ~21 kcal mole^{-1} (56). There is considerable reduction in symmetry in proceeding from $W_2(NR_2)_6$ to either of the conformers of $(R_2N)_4W_2X_2$. As shown in the following correlation table many of the MOs are of the same symmetry; consequently orbital interactions become very pronounced, and only qualitative interpretations can be attempted (Table XXXVII).

	D_{3d}	C_2	C_i
n_N	$a_{2g} + e_g + a_{1u} + e_u$	$2a + 2b$	$2a_g + 2a_u$
π_{WW}	e_u	$a + b$	$2a_u$
σ_{WW}	a_{1g}	a	a_g
σ_{WN}	$a_{1g} + a_{2u} + e_g + e_u$	$2a + 2b$	$2a_g + 2a_u$
σ_{WX}	$-$	$a + b$	$a_g + a_u$

TABLE XXXVII

Higher Lying Molecular Orbitals and UPS Data (eV) for $Mo_2(NR_2)_6$, $W_2(NR_2)_6$, and Related Compounds

MO[a]	Description	Experimental[b]		
		$Mo_2(NMe_2)_6$	$W_2(NMe_2)_6$	$W_2(NEt_2)_6$
$1a_{2g}$	n_N	6.6	6.5	6.4
$1a_{1u}$	n_N	6.6	6.5	6.4
$5e_u$	π_{MM}/n_N	7.01	6.90	6.79
$4e_g$	n_N	7.5	7.95	7.77
$4e_u$	n_N/π_{MM}	8.1	8.2	8.02
$4a_{1g}$	σ_{MM}	8.1	8.2	8.02
$3a_{2u}$	σ_{MN}	10.8	10.7	10.3
$3e_u$	σ_{MN}			
$3e_g$	σ_{MN}			
$3a_{1g}$	σ_{MN}/σ_{MM}			

$(R_2N)_4W_2X_2$ Compounds[c]

	R = Me; X = Me	R = Et; X = Me
n_N	6.71; 6.94	6.5; 6.7
π_{WW}/n_N	7.75; 8.13	7.50; 7.95
π_{WW}	9.34	9.3
σ_{WN}	10.81	10.26

[a] SCF$-X_\alpha$ calculations from Ref. 80.
[b] Data from Refs. 80 and 83.
[c] Data from Ref. 83.

By analogy with $W(NMe_2)_6$, and other metal dialkylamides the first two peaks in the UPS of $W_2(NMe_2)_4Me_2$ and $W_2(NEt_2)_4Me_2$ are probably associated with the removal of electrons from MOs that are primarily nitrogen lone pair in character. The third and fourth correspond to an ionization of MOs that are admixtures of the tungsten–tungsten π bonds and n_N MOs. Both $W_2(NMe_2)_4Me_2$ and $W_2(NEt_2)_4Me_2$ feature ionizations at ~9.3 eV. This is too high an IE for a pure nitrogen lone pair MO, so these peaks probably correspond to the ionization of an MO that is primarily tungsten–tungsten σ bonding.

In the requisite D_{4h} geometry the constituents of the metal–metal quadruple bond comprise a σ bond of a_{1g} symmetry, formed by the linear union of d_{z^2} AOs, a π bond of e_u symmetry, which results from the end-to-end overlap of d_{xz} and d_{yz} AOs, and the δ bond of b_{2g} symmetry, which arises from the face-to-face interaction of eclipsed d_{xy} AOs. When the ligands are considered many additional interactions take place and MO calculations must be used to delineate a cogent MO scheme. For the parent tetracarboxylate, $Mo_2(O_2CH)_4$,

X_α calculations (230) indicate that while the δ bond ($2b_{2g}$) is very high in metal character, the Mo–Mo π ($6e_u$) and σ ($4a_{1g}$) bonding orbitals are more extensively mixed with Mo–O orbitals. The IEs computed by the transition state method agree very well with those observed experimentally for $Mo_2(O_2CH)_4$ (Table XXXVIII). Related tetracarboxylates can be assigned in an analogous manner. In the UPS of each of the dimolybdenum tetracarboxylates (79, 149), the first two peaks are due to $^2B_{2g}$ and ^2E_u ionic states, which originate in the ionization of the $\delta(2b_{2g})$ and $\pi(6e_u)$ components of the quadruple metal–metal bond. It is difficult to locate the $^2A_{1g}$ ionic state arising from ionization of the metal–metal σ bond ($4a_{1g}$) because of the close spacing of ionizations in this region. However, this $^2A_{1g}$ state probably occurs at 12.7 and 11.95 eV in the formate and acetate, respectively. The UPS of $Cr_2(O_2CMe)_4$ (142a) differs substantially from that of the molybdenum analogue and poses considerable assignment problems.

Preliminary UPS studies have been made (73) on the new (75) air-stable complexes of general formula $M_2(mhp)_4$, M = Cr, Mo, where mhp represents the anion of 2-hydroxy-6-methylpyridine. These complexes, which involve quadruple metal–metal bonds, adopt structure **1**, which approximates D_{2d} symmetry rather closely. Both the Cr and Mo compounds exhibit three distinct

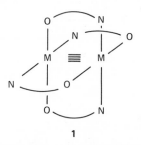

1

symmetry rather closely. Both the Cr and Mo compounds exhibit three distinct peaks at low energy (Fig. 37), the first of which derives presumably from ionization of the δ bond. However, at this point it is not clear whether the second and third peaks are due to Jahn–Teller splitting of the ^2E state (from the π component of the M≡M bond) or to the presence of an ionization of an MO with ligand character. *Ab initio* MO calculations are necessary for further discussion on this point.

A rather interesting case of metal–metal multiple bonding occurs in the molecule $(allyl)_4Re_2$. A very recent x-ray crystallographic study (74) has established that this species possesses virtual D_{2d} symmetry and that no bridging allyl groups are present. If the terminal allyl group is considered to be a uninegative, bidentate ligand the compound contains two Re(II) centers, thereby implying a total of 10 electrons in the metal–metal bonding region, and hence the electronic configuration $(\sigma)^2(\pi)^4(\delta)^2(\delta^*)^2$. The electronic structure of $(allyl)_4Re_2$

TABLE XXXVIII

Calculated and Experimental Ionization Energies (eV) for Dimolybdenum Tetracarboxylates

MO	Description	Calculated IE for Mo$_2$(O$_2$CH)$_4$ [a]	Experimental [b]			
			Mo$_2$(O$_2$CH)$_4$	Mo$_2$(O$_2$CMe)$_4$	Mo$_2$(O$_2$CCMe$_3$)$_4$	Mo$_2$(O$_2$CCF$_3$)$_4$
$2b_2g$	Mo—Moδ(89% Mo)	8.00	7.7	6.8	6.7	8.7
$6e_u$	Mo—Moπ, Mo—Oπ(65% Mo)	9.68	9.5	8.7	8.5	10.4
$4e_g$	n_0	(10.2)	10.0(sh)			
$1a_1u$	n_0	10.23	10.0(sh)			
$5e_u$	Mo—Oπ, Mo—Moπ(38% Mo)	(10.6)	11.0(sh)			
$5a_1g$	Mo—Oσ, Mo—Moσ(48% Mo)	(10.9)	11.0(sh)	10.4	10.1	12.5
$3e_g$	n_0	(10.7)	11.0(sh)			
$3a_2u$	Mo—Oσ	(11.0)	11.0(sh)			
$3b_2u$	Mo—Oσ	11.29	11.6	11.1	10.8	
$1b_1u$	Mo—Oπ	(11.6)	11.6			
$4b_1g$	Mo—Oσ	(12.0)	12.0(sh)			
$4a_1g$	Mo—Moσ, Mo—Oσ(75% Mo)	13.06	12.7	11.95		

[a] X$_\alpha$ transition state method. See Ref. 230.
[b] Data from Refs. 79 and 149.

131

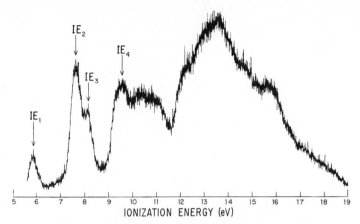

Fig. 37. He(I) UPS of $Mo_2(mhp)_4$. (From Ref. 73.)

has been investigated by UPS (73) and by means of the Fenske—Hall MO calculations (74) on the hypothetical technetium analogue $(allyl)_4Tc_2$ (Table XXXIX). The calculations do, indeed, confirm that the HOMO ($3a_1$) has δ-antibonding character. The δ, π, and σ components of the metal—metal bonding are represented by the third, fourth, and seventh MOs. The agreement with the UPS of $(allyl)_4Re_2$ is, however, not satisfactory. Two comments should be made on this point. First, the computations were performed on the technetium analogue, and, second, Koopmans' theorem has been shown to be inadequate for even the simple π-allyl system $(\pi\text{-}C_3H_5)_2Ni$, as discussed else-

TABLE XXXIX
Eigenvalues for the Molecular Ground State of $(allyl)_4Tc_2$ and UPS Data for $(allyl)_4Re_2$

| MO | $(allyl)_4Tc_2$[a] | | | $(allyl)_4Re_2$[b] |
	Energy, eV	% Tc(4d)	Description	Experimental IEs, eV
$3a_1$	−10.39	70	M−Mδ*	6.73
$3e$	−11.32	12	Ligand	7.77
$2b_2$	−12.09	81	M−Mδ	7.98
$2e$	−13.90	86	M−Mπ	8.94
$1b_1$	−14.26	40	M−ligand	9.33
$1a_2$	−14.71	43	M−ligand	
$2a_1$	−15.18	58	M−Mσ	
$1e$	−16.84	1	Ligand	
$1b_2$	−18.05	10	Ligand	
$1a_1$	−18.33	29	M−ligand	

[a] Data from Ref. 74.
[b] Data from Ref. 73.

where in this chapter. For these reasons Cotton et al. are undertaking X_α calculations on $(allyl)_4 Re_2$.

IX. TRANSITION METAL ALKYLS

The discovery of hexamethyltungsten by Shortland and Wilkinson (259, 260) has prompted an interest in the synthesis and physical properties of alkylated transition metal compounds. The relatively high thermal stability of these derivatives is plausibly ascribed to the lack of a low-energy decomposition pathway. This is particularly true in the case of ligands such as $Me_3 SiCH_2-$ and $Me_3 CCH_2-$, for which β-hydrogen transfer is precluded (43, 282).

From the UPS standpoint, by far the most work has been done with the tetraalkyls, MR_4 (35, 118, 120, 191, 199). Let us start with a simple MO diagram for the tetramethyl compounds that is based on the localized bond orbital concept (Fig. 38). In T_d symmetry the four metal–carbon σ-bonding orbitals transform as $t_2 + a_1$ and the corresponding $^2 T_2$ and $^2 A_1$ ionic states are recognizable in the UPS of the Group IVA tetramethyls (Table XL and Fig. 39). The 12 C–H σ-bonding orbitals span the irreducible representations $a_1 + e + t_1 + 2t_2$. Unless d-orbital effects are important, the separation of the t_1, e, and t_2 symmetry-adapted combinations originating from the methyl group e MOs is expected to diminish steadily with increasing size of the central metal. As pointed out by Lappert et al. (35), of the $1e$ and $1t_1$ levels (Fig. 38) only the former can interact with vacant d orbitals in T_d symmetry, and the near-zero splitting of the ionic states stemming from the ionization of these MOs in $Me_4 Si$, $Me_4 Ge$, $Me_4 Sn$, and $Me_4 Pb$ (Table XL and Fig. 39) has been taken as evidence against d-orbital participation in the bonding in these compounds.

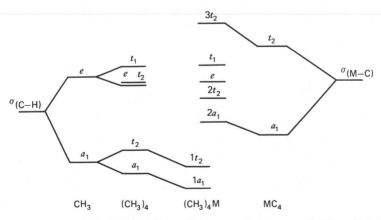

Fig. 38. Qualitative MO scheme for Group IVA tetramethyl compounds. (From Ref. 120.)

TABLE XL

Ionization Energy Data (eV) and Assignments for Metal Tetraalkyls

Compound	d electrons	Metal–carbon σ		C–H σ	Ref.
		$3t_2$	$2a_1$	$t_1 + e + 2t_2$	
$Me_4 Si$	–	10.57	15.58	13.06, 14.08	120
$Me_4 Ge$	–	10.23	~ 15.9	~ 13.0(sh), 13.85	120
$Me_4 Sn$	–	9.70	–	~ 13.4	120
$Me_4 Pb$	–	8.81 U′	~ 15.3	~ 13.3	120
		9.09 U′	~ 15.3	~ 13.3	120
		9.86 E″			
$(Me_3 CCH_2)_4 Ge$	–	9.0_1			199
$(Me_3 CCH_2)_4 Sn$	–	8.5_8			199
$(Me_3 SiCH_2)_4 Sn$	–	8.71			118
$(Me_3 SiCH_2)_4 Pb$	–	8.14			118
		8.86			118
$(Me_3 CCH_2)_4 Ti$	–	8.3_3			199
$(Me_3 CCH_2)_4 Zr$	–	8.3_3			199
$(Me_3 CCH_2)_4 Hf$	–	8.5_1			199
$(Me_3 CCH_2)_4 Cr$	7.25	8.37			120
$(Me_3 SiCH_2)_4 Cr$	7.26	8.69			120

The UPS of the $(Me_3 SiCH_2)_4 M$ and $(Me_3 CCH_2)_4 M$ compounds resemble those of $Me_4 C$ and $Me_4 Si$ above 10 eV. Below 10 eV only the $^2 T_2$ ionic state arising from ionization of the t_2 metal–carbon σ-bond is discernible for the diamagnetic compounds. The Cr complexes possess two unpaired electrons. In a crystal field of T_d symmetry constituted by the four R^- groups the $Cr(3d)$ electrons are expected to occupy an orbital of e symmetry; consequently the

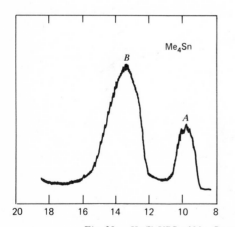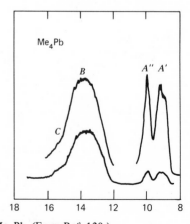

Fig. 39.   He(I) UPS of $Me_4 Sn$ and $Me_4 Pb$. (From Ref. 120.)

ground state is ^3E. Electron ejection from the ^3E ground state affords the ^2E ionic state, which occurs at 7.25 and 7.26 eV in $(Me_3CCH_2)_4Cr$ and $[Me_3SiCH_2]_4Cr$, respectively. The similarities of these IEs suggest (1) that the electronegativities of the Me_3CCH_2 and Me_3SiCH_2 ligands are close and (2) there is no significant interaction between the $Cr(3d)$ orbitals and vacant receptor orbitals on silicon.

It is evident from the trends pertaining to the ionization of the metal—carbon σ bonds of t_2 symmetry that the IEs are less for the Me_3XCH_2-substituted compounds than for the methylated compounds. On the basis of Koopmans' theorem this would imply that the metal—carbon bonding is weaker in the Me_3XCH_2 derivatives.

Some confusion has developed regarding the UPS of hexamethyltungsten. The spectrum reported originally (89) was apparently not that of WMe_6, as demonstrated in subsequent work (142) on an authentic sample prepared by a new route. On the basis of He(II) spectra it was suggested (142) that, as expected on group theoretical grounds, the first three spectral bands at 8.61, 9.33, and 10.17 eV are associated with the ionization of the t_{1u}, a_{1g}, and e_g tungsten—carbon σ-bonding MOs, respectively. It was also suggested that the considerable splitting of the C—H σ-bonding ionizations in the $11-16$ eV range may be due to steric compression of the Me groups as in neopentane.

Wilkinson et al. (142) also reported the He(I) and He(II) UPS of Me_5Ta (Fig. 40). Four low-energy ionizations (8-11 eV) presumably correspond to electron ejection from the $a_1'(2), a_2''$, and e' tantalum—carbon σ-bonding MOs.

The dialkyls of the d^{10} systems, Zn, Cd, and Hg, possess linear skeletons, and if rotation around the metal—carbon bonds is rapid (9) the effective symmetry is $D_{\infty h}$. Under these conditions two metal—carbon σ-bonding MOs are anticipated: the HOMO, which consists of the unsymmetrical combination of alkyl group fragment MOs interacting with a metal p AO, and, at higher binding energy, an orbital formed by overlap of the symmetrical combination of alkyl group fragment MOs with the metal s AO. A series of C—H σ bonding is expected to follow the two metal—carbon σ-bonding MOs. Recent *ab initio* MO calculations on Me_2Cd (18) (Table XLI) confirm the essential correctness of this picture. (For convenience these computations were performed by assuming a C_{2v} molecular symmetry.) Thus the symmetric MC_2 σ-bonding MO ($15a_1$) is found to be more stable than its antisymmetric counterpart ($9b_1$), which, in fact, is only slightly bonding. The UPS of Me_2Cd (17, 18) (Fig. 41), and the spectra of Me_2Zn (112), Me_2Hg (48, 112), and higher alkyls (131, 168) can be assigned on this basis. In a study of the first ionization energy of a large number of alkyl mercurials, Fehlner et al. (131) noted a distinct "saturation effect" for the various alkyl groups, rather than the "additive" substituent relationships frequently observed in UPS experiments. These authors attributed the saturation effect to the bonding character of the HOMO.

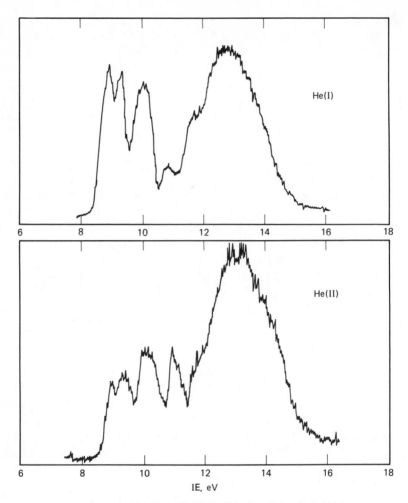

Fig. 40. He(I) and He(II) UPS of Me₅Ta. (From Ref. 142.)

Without doubt the most fascinating aspect of the UPS of the d^{10} dialkyls is that it is possible to detect the effects of the ligand field on the d^9 hole states of the cation. Since the major part of the splitting is due to the C_2^0 asymmetric crystal field term, which transforms in the same manner as the electric field gradient, measurement of the d orbital splittings in the UPS experiment might prove to be a complementary method to Mössbauer spectroscopy and NQR for estimating electric field gradients (18). Spectroscopically the crystal field effects are apparent in the 15–18 eV region of the UPS of Me_2M, M = Zn, Cd, Hg. The dialkylcadmium compounds Me_2Cd and Et_2Cd (Fig. 41) have been examined in

TABLE XLI

Ab initio Eigenvalues for Me_2Cd^a

MO	Eigenvalue, eV	Nature
$9b_1$	−9.13	Cd−C
$15a_1$	−10.37	Cd−C
$3a_2$	−13.98	C−H
$8b_1$	−13.98	C−H
$6b_2$	−14.24	C−H
$14a_1$	−14.24	C−H
$13a_1$	−20.18	$Cd(4d)\ (\sigma)$
$7a_2$	−20.54	$Cd(4d)\ (\pi)$
$7b_1$	−20.54	$Cd(4d)\ (\pi)$
$12a_1$	−20.71	$Cd(4d)\ (\delta)$
$5b_2$	−20.71	$Cd(4d)\ (\delta)$

^{a} Data from Ref. 18.

particular detail (18). From a purely crystal field point of view the sequence of d orbitals should be $\sigma(d_{z^2}) > \pi(d_{xz}, d_{yz}) > \Delta(d_{x^2-y^2}, d_{xy})$. Interestingly this order was revealed in the *ab initio* calculation on Me_2Cd (Table XLI), which also indicated that there was little or no involvement of the d electrons in the

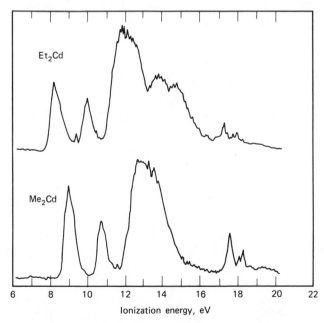

Fig. 41.   He(I) UPS of Me_2Cd and Et_2Cd. (From Ref. 18.)

TABLE XLII

Calculated and Observed Ionization Energies (eV) for the Cd(4d) Ionic States of
Me$_2$Cd and Et$_2$Cd[a]

Ionic state	Me$_2$Cd		Et$_2$Cd	
	Calculated	Experimental	Calculated	Experimental
$^2\Sigma_{1/2}$	17.344	17.349	17.075	17.074
$^2\pi_{3/2}$	17.475	17.483	17.244	17.251
$^2\Delta_{5/2}$	17.601	17.589	17.351	17.341
$^2\pi_{1/2}$	18.064	18.053	17.815	17.805
$^2\Delta_{3/2}$	18.252	18.262	18.000	18.008

[a] Data from Ref. 18.

bonding. The latter point is consistent with the sharpness of the UPS bands in
the d-orbital region. When spin—orbit coupling is introduced in addition to the
crystal field perturbation, the ionic states shown in Table XLII are produced
(18). Exceptionally good agreement is found between the computed and experimental energies. In the case of Me$_2$Hg, however, analysis of the 5d structure
suggests some involvement of these orbitals in covalent bonding (48).

The UPS of other alkylated transition metal species, namely, cyclopentadienyl metal alkyls and carbonyl metal alkyls, are discussed in other sections of
this chapter. A summary of the IE data pertaining to the ionization of the

TABLE XLIII

Ionization Energies Relating to Ionization of Metal—Carbon Bonds of
Transition Metal Alkyls[a]

Compound	IE, eV			Weighted average IE, eV
WMe$_6$	8.61	9.33	10.17	9.11
ReMe$_6$	8.47	9.77	10.48	9.24
(Me$_3$CCH$_2$)$_4$Cr	8.37			
(Me$_3$SiCH$_2$)$_4$Cr	8.69			
MeMn(CO)$_5$	9.49			
MeRe(CO)$_5$	9.51			
(C$_5$H$_5$)$_2$MoMe$_2$	8.3, 9.6			8.95
(C$_5$H$_5$)$_2$WMe$_2$	8.3, 9.6			8.95
(C$_5$H$_5$)Mo(CO)$_3$Me	9.07			
(C$_5$H$_5$)W(CO)$_3$Me	9.21			
(C$_5$H$_5$)Fe(CO)$_2$Me	9.15			
(C$_5$H$_5$)Ru(CO)$_2$Me	9.48			

[a] Data from Ref. 150 and references cited therein.

metal–carbon σ bonds is presented in Table XLIII, along with some recent data (150) for cyclopentadienyl metal carbonyl alkyls. It is clear that the higher IEs (or average IEs) arise from compounds of the metals to the right of the transition series or from those with a substantial number of carbonyl ligands.

X. CHELATED TRANSITION METAL COMPOUNDS

Most of the coordination compounds discussed in this chapter involve "soft" ligands such as CO, PF_3, arenes, and olefins. Part of the motivation for examining the photoelectron spectra of chelated compounds is that they provide an opportunity for studying the coordination behavior of "hard" ligands such as ketones. An additional advantage of chelated compounds is that they permit the study of a wide variety of d electron configurations within an isostructural framework.

Since most of the UPS work on chelated compounds has been concerned with β-diketone ligands (45, 51, 125–127, 207, 208), an interpretational framework can be developed by allowing the MOs of, for example, the acetylacetonate anions (acac), to interact with metal d orbitals in the appropriate symmetry. For the tris-chelates $M(acac)_3$, the prevalent geometry is D_3, the oxygen atom array around the metal being approximately O_h symmetry (204). Even when Jahn–Teller distortions are anticipated as in $Mn(acac)_3$, the oxygen atoms still preserve their approximately octahedral geometry (96). The π system for the acetylacetonate anion comprises five MOs; however, only $\pi_3(b_2)$, the highest lying MO, is pertinent to the present orbital syntheses. The other important ligand orbitals are the symmetric $(n+)$ and antisymmetric $(n-)$ combinations of ketonic oxygen lone pairs, which transform as b_1 and a_1, respectively, in C_{2v} symmetry. For the tris-chelates, the inter-ligand interactions produce the symmetry-adapted combinations indicated in Fig. 42, the ordering of the $n+$ and $n-$ combinations being predicated on the assumption of a through space interaction. The scheme is completed by introducing the metal nd AOs, which span the irreducible representations $a_1(d_{z^2})$, $e(d_{xz}, d_{yz})$, and $e(d_{x^2-y^2}, d_{xy})$ in D_3 symmetry. Obviously interaction between the ligand a_2 MOs and metal orbitals is precluded on symmetry grounds. Of the remaining possible interactions, overlap criteria indicate that the major interaction is between the metal d_{xy} and $d_{x^2-y^2}$ AOs and the $e(n-)$ antisymmetric combination of oxygen lone pairs. Only a small interaction is expected between the $e(\pi_3)$ ring MO and the metal d orbitals. The same scheme can be employed for hexafluoroacetoacetonate (hfa) and trifluoroacetoacetonate (tfa) complexes if allowance is made for the inductive effects of introducing fluorine atoms into the ligands.

The UPS of the d^0 complex $Sc(hfa)_3$ (Fig. 43) can be assigned fairly straightforwardly (Table XLIV) using the qualitative orbital sequence in Fig. 42

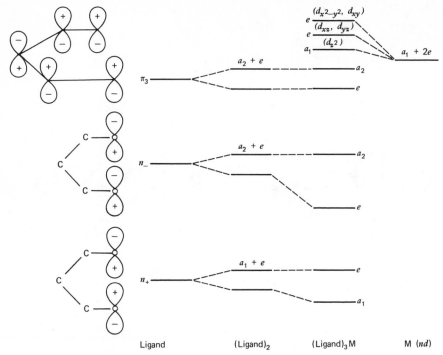

Fig. 42. A qualitative MO diagram for the interaction of a tris-chelate ligand system with metal d orbitals. (From Ref. 127.)

and, of course, assuming the validity of Koopmans' theorem. When the metal d orbitals become occupied, as in the remaining tris-chelates, the important question of the electronic ground state is raised. The following probable ground states for some relevant tris(β-diketonato) complexes, ML_3, have been deduced by magnetic measurements or ESR (appropriate literature cited in Ref. 127):

$$
\begin{array}{ll}
TiL_3 & {}^2A_1(a_1) \\
VL_3 & {}^3A_2(e^2) \\
CrL_3 & {}^4A_2(a_1e^2) \\
MnL_3 & {}^5E(a_1e^2e) \\
FeL_3 & {}^6A_1(a_1e^2e^2) \\
RuL_3 & {}^2E(a_1^2e^3) \\
CoL_3 & {}^1A_1(a_1^2e^4)
\end{array}
$$

Thus all the tris-chelates studied except those of Co(III) are paramagnetic. As is discussed earlier, spectral complexities occur when open-shell systems are ionized. For instance, ionization of the d shell of the tris-chelates of Cr(III)

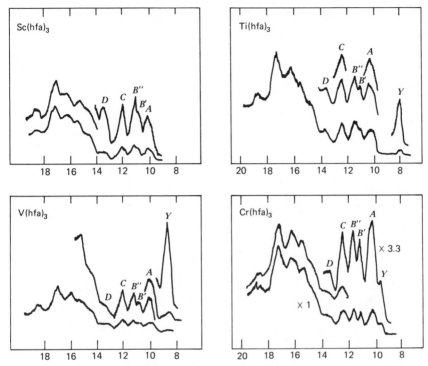

Fig. 43.   He(I) UPS of Sc(hfa)$_3$, Ti(hfa)$_3$, V(hfa)$_3$, and Cr(hfa)$_3$. (From Ref. 127.)

should produce the ionic states $^3A_2(e^2)$ and $^3E(a_1 e)$. However, the anticipated splittings were not discernible presumably because of the closeness in energy of the ionic states.

Regarding the trends in the IEs (Table XLIV), it is clear that within, for example, the M(hfa)$_3$ series there is a rapid increase in the IEs associated with the ionization of the predominantly d-electron MOs on traversing the transition series such that at Co(hfa)$_3$ the metal-localized MO is no longer the HOMO. This causes assignment difficulties for the tris chelates of Co(III), Mn(III), and Fe(III). The assignments shown in Table XLIV for the Co(III) complexes are considered to be the most reasonable of four possible schemes. No assignments are shown for the high-spin Mn(III) and Fe(III) complexes because of the amorphous nature of the spectra. Interestingly, however, the UPS of the low-spin hfa chelate of Ru(III) is well defined and easily assigned, as is that of (hfa)Mn(CO)$_4$.

Another noteworthy trend is that on crossing the transition series the difference in the IEs associated with the $a_2(n-)$ and $e(n-)$ MOs increases. This occurs because, as pointed out above, the metal d orbitals are progessively

TABLE XLIV

Ionization Energies (eV) and Proposed Assignments for Some Tris-Chelate Complexes[a]

Complex	d^n	$a_2 + e(\pi_3)$	$a_2(n-)$	$e(n-)$	$a_1 + e(n+)$
Sc(hfa)$_3$		10.13	10.82	11.12	12.05
Ti(hfa)$_3$	7.94	10.24	10.87	11.28	12.24
V(hfa)$_3$	8.68	10.10	10.96	11.39	12.20
Cr(hfa)$_3$	9.57	10.18	11.10	11.61	12.44
Cr(tfa)$_3$	8.58	9.12	10.01	10.54	11.40
Cr(acac)$_3$	7.46	8.06	8.96	9.48	10.26
Ru(hfa)$_3$	8.85, 9.07	10.30	11.06	11.65	12.50
Co(hfa)$_3$	10.13(a_1), 10.73(e)	9.73(e), 10.13(a_2)	11.15	11.75	12.56
Co(acac)$_3$	8.03(a_1), 8.49(e)	7.52(e), 8.03(a_2)	8.99	9.54	10.36

[a] Data and assignments from Ref. 127.

stabilized, thus permitting an increasingly strong interaction with the $e(n-)$ oxygen lone pair MOs. Note also that, consistent with their nonbonding nature, the a_2 MOs exhibit relatively constant ionization energies.

A very similar approach can be used to interpret the UPS of bis-chelate complexes of acac, hfa, and related monothio and bisthio analogues (45, 51). Monomeric Ni(acac)$_2$ has been found to possess D_{2h} symmetry in the vapor phase (258), and this geometry is assumed here for the other bis-chelates. In D_{2d} symmetry the π_3, $n+$, and $n-$ MOs of each β-diketonate ligand combine to afford the symmetry-adapted combinations shown in Fig. 44. Introduction of the metal d orbitals, which transform as $a_g(d_{z^2})$, $a_g(d_{x^2-y^2})$, $b_{1g}(d_{xy})$, $b_{2g}(d_{xz})$, and $b_{3g}(d_{yz})$, completes the scheme. By analogy with the tris-chelates, the strongest interaction anticipated is that between the d_{xy} orbital and the antisymmetric lone pair combination of b_{1g} symmetry.

The assignments for various bis-chelates of Ni(II), Cu(II), and Zn(II) are presented in Table XLV. There are considerable differences in the spectra reported (45, 51) for Ni(acac)$_2$. Furthermore, the interpretations in these publications also differ. The interpretation presented in Ref. 51 appears to be the more reasonable. A number of points merit comment. First, in contrast to

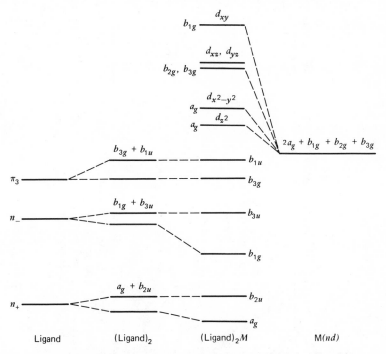

Fig. 44. A qualitative MO diagram for the interaction of a bis-chelate ligand system with metal d orbitals. (Scheme from Ref. 51.)

TABLE XLV

Ionization Energies (eV) and Proposed Assignments for Some Bis-Chelate Complexes[a,b,c]

Complex	d^n	$b_{1u} + b_{3g}(\pi_3)$	$b_{3u}(n-)$	$b_{1g}(n-)$	$b_{2u} + a_g(n+)$
Ni(acac)$_2$[c]	7.41, 7.89, 8.15	8.38	8.75	9.26	10.06
Ni(Sacac)$_2$[c]	6.99, (7.54), 7.63	8.44	8.84	9.46	9.84
Ni(S$_2$acac)$_2$[c]	6.92, (7.63), 7.73	8.31	8.58	9.26	9.68, 10.06
Ni(tfa)$_2$	8.25, (8.75), 8.92	9.30	9.65	10.05	10.98
Ni(Stfa)$_2$[c]	7.80, 8.51	9.28	9.58	10.33	10.88
Ni(S$_2$tfa)$_2$[c]	7.65, (8.38), 8.58	9.08	9.38	9.98	10.36, 10.68
Ni(hfa)$_2$	9.35	9.84	10.67	11.11	12.01
Cu(acac)$_2$	—	8.20	8.53	9.13	9.75, 10.23
Cu(hfa)$_2$	—	9.92	10.20	10.88	11.93
Zn(acac)$_2$	—	8.46	9.22	10.39	10.72
Zn(hfa)$_2$	—	10.25	11.17	12.53	12.78

[a] Data for Ni(hfa)$_2$, Cu(II), and Zn(II) complexes from Ref. 45. All other data from Ref. 51.

[b] Interpretation from Ref. 51.

[c] Abbreviations: Sacac = monothioacetylacetonato; S$_2$acac = dithioacetylacetonato; Stfa = monothiotrifluoromethylacetylacetonato; S$_2$tfa = dithiotrifluoromethylacetylacetonato.

the trend of increasing d-orbital ionization energy in crossing the transition series, the first ionizations of the Ni(II) chelates are attributed to the ionization of predominantly metal $3d$ MOs. The fact that such ionizations are not detectable for the d^9 chelates of Cu(II) may imply that they are obscured beneath the ligand-type MOs. Second, note that for a given ligand type the mainly metal orbital IEs are lower for the thio analogues than for the β-diketonates, thus implying a higher electron density on Ni(II) in the former. It is interesting, however, that the IE associated with the ionization of the $b_{1g}(n-)$ MO is approximately the same in the oxo and thio ligands. This would imply that the effect of the decreased electronegativity of sulfur is compensated by a stronger interaction between the antisymmetric combination of sulfur lone pairs and the metal d_{xy} orbital.

UPS studies of some dithiocarbamates (50) and acetylacetoneiminato (64, 130a) complexes have also been reported. Precise assignments have not been made; however, the ionization of the metal-localized MOs appears at higher energy than those of the ligand-type MOs. Comparable studies of diethyldithiophosphate (212) and difluorodithiophosphate (6) complexes of several metal ions have also been reported.

XI. TRANSITION METAL HALIDES AND OXIDES

The wealth of information that has accumulated regarding the UPS of the main group halides is not considered here, since it falls outside the purview of this Chapter. Of the transition metal halides the largest amount of work has been done with the halides of the Group 2B metals and their monomethyl derivatives (28, 33, 63, 108, 112, 129, 233). In these studies it was found that the zinc and cadmium dihalides require significantly higher vaporization temperatures than the mercury compounds; hence specialized high-temperature techniques were necessary. A summary of these and related methods may be found in a recent review article by Schweitzer (255).

The upper valence orbitals of the linear halides are easily understood in qualitative terms. Thus, starting with the parent halogen molecules, X_2, the valence p orbitals combine to form MOs of symmetries σ_g, π_u, π_g^*, and σ_u^* in order of increasing energy. Combination of these MOs with metal s AOs stabilizes the σ_g level, and interaction with the metal p AOs stabilizes the σ_u^* level, resulting in the sequence $\sigma_g^2 \sigma_u^2 \pi_u^4 \pi_g^4$. This qualitative MO synthesis has been corroborated by *ab initio* MO calculations (283) on ZnF_2 and by semi-empirical MO calculations (92) on several Group 2B halides. Of course, when an electron is ejected from one of the π levels, the remaining electrons possess both spin and orbital momentum and, as a consequence, each $^2\pi$ state splits into two components, $^2\pi_{3/2}$ and $^2\pi_{1/2}$, under the influence of spin–orbit coupling.

The $^2\pi_g$ states exhibit spin–orbit couplings typical of the halogen ligands because of the symmetry exclusion of metal p AOs. On the other hand, the $^2\pi_u$ states can mix with metal p AOs, and the magnitudes of the spin–orbit splittings are more difficult to predict.

As in the case of the Group 2B dialkyls, the most interesting aspect of the UPS of the dihalides is the rich spectral detail associated with the ionization of the inner nd electrons. This is particularly true for the zinc dihalides. Thus $ZnCl_2$ and $ZnBr_2$ exhibit three bands and ZnI_2 has five bands in the $3d$ region of the spectra. For ZnI_2, as for the cadmium dialkyls, the ionic states are believed (233) to be in the order $^2\Sigma_{1/2} < {}^2\Pi_{3/2} < {}^2\Delta_{5/2} < {}^2\Pi_{1/2} < {}^2\Delta_{3/2}$. For the mercuric halides, however, the spin–orbit coupling is larger than the crystal field splitting, hence the foregoing labeling is inappropriate and the symbols $^2D_{3/2}$ and $^2D_{1/2}$ should be employed. These states are, in turn, split into three and two components, respectively, by means of the crystal field.

Even though the d orbitals are virtually uninvolved in the chemical bonding, the IEs corresponding to the ionizations of the nd orbitals vary substantially with respect to both the metal and the electronegativity of the ligand. These shifts are electrostatic in origin and can be treated by a potential model similar to that developed to account for ESCA shifts. In this manner various estimates have been made of metal and ligand electronegativities (28, 63, 112). There is, however, some disagreement (28, 63) regarding the electronegativities of Zn, Cd, and Hg.

The electronic structures of the titanium and vanadium tetrahalides have received considerable attention because of their importance in transition metal chemistry. For a qualitative discussion of the bonding in these compounds, one can consider only the valence p orbitals on the halogens and the d orbitals on the metal atoms. In the requisite T_d symmetry the metal–halogen σ bonds transform as $a_1 + t_2$ and the halogen "lone pair" orbitals span the irreducible representations $t_1 + t_2 + e$. Generally one might expect the halogen lone pair combinations to be higher in energy than the metal–halogen σ-bonding MOs; however, without recourse to MO calculations it would be difficult to predict a reliable sequence of MOs, not only because of the possibility of $p_\pi - d_\pi$ bonding from the halogen lone pair MOs of symmetries e and t_2, but also because of $\sigma-\pi$ mixing of the t_2 metal–halogen σ-bonding MOs and halogen lone pair MOs. From a purely electrostatic standpoint the metal $3d$ orbitals split in the T_d crystal field into the familiar t_2 and e levels of relative energy $t_2 > e$. The ground state of paramagnetic VCl_4 should therefore be ^2E.

Several MO calculations have been performed on the Ti(IV) and V(IV) halides (26, 27, 49, 69, 135, 177a, 234, 270, 271). Of these, the INDO (270, 271), SCF–X$_\alpha$ (49, 234) and *ab initio* (177a) are probably the most reliable and there is reasonable agreement between these methods regarding the sequences of computed eigenvalues for both $TiCl_4$ (Table XLVI) and VCl_4 (Table XLVII). All

TABLE XLVI
Computed Eigenvalues, Computed Ionization Energies, and Experimental Ionization Energies
for $TiCl_4$ and $TiBr_4$

		Eigenvalues, eV			Computed IEs, eV (INDO)[a]	Experimental IEs, eV[d]
MO	Description	INDO[a]	SCF–X$_\alpha$[b]	SCF–X$_\alpha$[c]		
			$TiCl_4$			
$1t_1$	Ligand π	12.54	7.52	12.2	12.55	11.76
$4t_2$	Ligand σ/π	12.83	7.53	12.7	12.83	12.77
$1e$	Ligand π	13.31	8.16	13.0	13.08	13.25, 13.96
$3a_1$	Ligand σ	13.77	8.23	14.1	15.42	13.25, 13.96
$3t_2$	Ligand σ/π	17.74	8.37	13.8	16.87	13.25, 13.96
			$TiBr_4$			
$1t_1$	Ligand π	10.76				10.55[e] U'
						10.63[e] U'
						10.86 E'
$4t_2$	Ligand σ/π	11.31 ⎫				(11.56)
						11.68
						12.00
$1e$	Ligand π	11.80 ⎬				12.31
						12.42
						(12.55)
$3a_1$	Ligand σ	12.89 ⎭				13.04
$3t_2$	Ligand σ/π	16.01				

[a] Reference 270.
[b] Reference 49.
[c] Reference 234.
[d] Data from Refs. 49 and 108.
[e] Splitting of $^2T_1$ ionic state as a result of spin–orbit coupling.

three methods predict a ^2E ground state for VCl_4, consistent with the simple crystal field view. By contrast, using the CNDO method, Becker and Dahl (27) found a $^2T_2$ ground state for VCl_4, and in the unrestricted Hartree-Fock CNDO calculation of Copeland and Ballhausen (69) a $^2A_1$ state emerged as the most stable. In turn, these results have raised serious questions regarding the applicability of CNDO methods to transition metal systems (234). The difficulties arise from both the approximations inherent in the method and the choice of basis set. Configuration interaction calculations (177a) on the singlet and triplet states of VCl_4^+ indicate that a simple orbital interpretation of the UPS of VCl_4 is not possible.

TABLE XLVII

Computed Eigenvalues, Computed Ionization Energies, and Experimental Ionization
Energies for VCl_4, and Experimental Ionization Energies for WCl_5

| | | Eigenvalues, eV | | Computed IEs, eV | Experimental |
MO	Description	INDO[a]	SCF–X$_\alpha$[b]	(INDO)[a]	IE, eV
			VCl_4		
$2e$	Metal $3d$	8.55	7.4	8.55	9.36
$1t_1$	Ligand π	12.37	12.0	12.37	11.77
$4t_2$	Ligand σ/π	12.74	12.6	12.74	12.70, 12.90
$1e$	Ligand π	13.13	12.8	13.13 ⎫	13.53
$3t_2$	Ligand σ/π	17.10	13.7	17.10 ⎬	13.64
$3a_1$	Ligand σ	17.53	14.1	17.54 ⎭	14.27
			WCl_5[c]		
W(5d)					8.82
$2a_2'' + 1a_2' + 3e' + 4e' + 2e''$					11.44, 12.13
$1e'' + 1a_2''$					13.08, 13.32
$3a_1' + 4a_1' + 2e'$					14.31

[a] Reference 271.
[b] Reference 234.
[c] Data from Ref. 108.

Four papers have now appeared (49, 84, 108, 147) in which the UPS of the Ti(IV) and V(IV) halides are discussed. The spectra presented in the later papers are of higher resolution; furthermore, some spurious features are apparent in the earlier data. There is little doubt that the first UPS band of $TiCl_4$ (Fig. 45) and $TiBr_4$ corresponds to photoionization of the $1t_1$ MO, resulting in the $^2T_1$ ionic state. This is in accord with the INDO and X$_\alpha$ calculations (Table XLVI). Furthermore, the intensity of this band is not increased in changing from He(I) to He(II) irradiation (108), which is consistent with the prelusion of d orbital participation in the $1t_1$ MO. For $TiBr_4$ the effects of bromine spin–orbit coupling are clearly apparent in the $^2T_1$ ionic state (49). For $TiCl_4$ the INDO and X$_\alpha$ calculations indicate that the second UPS band is due to photoionization of the $4t_2$ ligand σ/π MO. However, the remaining bands in the TiX_4 spectra cannot be assigned unequivocally because (1) the X$_\alpha$ calculations (49, 234) differ regarding the sequencing of the $3a_1$ and $3t_2$ MOs and (2) the UPS bands are quite closely spaced. Unfortunately He(II) spectra did not provide unequivocal information on these assignments (108).

The UPS of VCl_4 exhibits a low-energy ionization not apparent in the

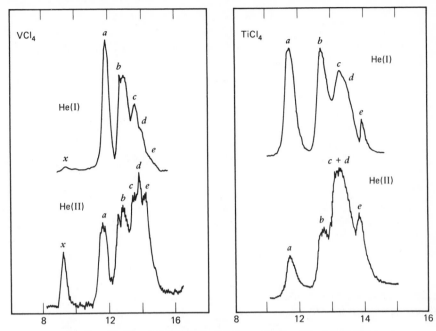

Fig. 45. He(I) and He(II) UPS of VCl_4 and $TiCl_4$. (From Ref. 108.)

spectrum of $TiCl_4$ (Fig. 45). In consonance with the MO calculations and qualitative crystal field considerations this peak must be due to ionization of the unpaired d electron, thus producing the $^1A_1$ state of the VCl_4^+ ion. The ^2E state of neutral VCl_4 should, of course, exhibit a static Jahn-Teller distortion. However, there is no evidence for such a distortion in the UPS. This is perhaps not surprising, since Johannesen et al. (186) have found only very small static distortion in EPR measurements. There is, on the other hand, evidence of a dynamic Jahn-Teller effect in the optical spectrum (30). The fact that VCl_4 a paramagnetic molecule means that spectral complications arise because each subsequent ionization is capable of producing more than one ionic state. Specifically ionization of t_1 or t_2 MOs produces $^{3,1}T_1$ or $^{3,1}T_2$ ionic states, respectively, while ionization of an e MO results in $^{3,1}A_1 + ^{3,1}A_1 + ^{3,1}E$ ionic states and ionization of the a_1 MO produces the $^{3,1}E$ ionic states. Thus VCl_4 should show a total of 20 UPS peaks for ionization of the first five MOs compared to only five peaks for the diamagnetic molecule $TiCl_4$. Inferentially, however, the separation of these ionic states is not very large, since, for example, only a 0.2 eV splitting is observed, corresponding to the $^1T_2$ and $^3T_2$ states that arise from ionization of the $4t_2$ MO.

Finally, some UPS data and tentative assignments pertaining to the paramagnetic, trigonal bipyramidal molecule WCl_5 are included in Table XLVII.

There is widespread interest (220) in the electronic structures of tetra-hedral oxyanions of the type MO_4^-. Being quasi-isoelectronic the neutral tetroxides, MO_4, M = Ru, Os, are well suited for vapor-phase UPS studies (49, 102, 140) of the bonding in this type of system. Unfortunately impurities are apparent in two of these spectral studies (102, 140), and in one of them (140) doubts have been raised (49) about the validity of the reported vibrational fine structure. Furthermore, the interpretations presented in the three papers are somewhat different. Overall, the interpretation of Orchard et al. (49) appears to be the most reasonable.

Overlap of oxygen $2p_\sigma$ orbitals with the appropriate metal orbitals in T_d symmetry affords the familiar t_2 and a_1 σ-bonding orbitals, and the orthogonal oxygen $2p_\pi$ orbitals yield the t_2, t_1, and e symmetry-adapted combinations. Clearly σ–π mixing of the t_2 MOs is symmetry allowed, causing some un-certainty regarding their relative energies. Normally, however, the most plausible ground-state electronic configuration would be $[a_1(\sigma)]^2 [t_2(\sigma/\pi)]^6 [e(\pi)]^4$

TABLE XLVIII

Ionization Energy Data (eV) and Tentative Assignments for OsO_4 and RuO_4[a]

Band system	OsO_4	RuO_4	Assignment
1	12.35		$1t_1$
	12.38		$1t_1$
	(12.425)		$1t_1$
	(12.465)	12.15	$1t_1$
	12.48	12.25	$1t_1$
	12.51	(12.35)	$1t_1$
	(12.545)	(12.45)	$1t_1$
	12.57		$1t_1$
	(12.68)		$1t_1$
	(12.61)		$1t_1$
2	13.14		$3t_2$
	13.26		$3t_2$
	13.36	12.92	$3t_2$
		13.01	
3	13.54	(13.10)	$1e$
	13.63	(13.20)	$1e$
	13.74		$1e$
4	14.66	13.93	$2a_1$
	14.78	13.93	$2a_1$
5	16.4	16.1	$2t_2$
	16.8	16.1	$2t_2$

[a] Data and assignments from Ref. 49.

$[t_2(\sigma/\pi)]^6 [t_1(\pi)]^6$. Thus, as in the case of the titanium tetrahalides, the first UPS band should be ascribable to ionization of the oxygen lone pair combination of t_1 symmetry (Table XLVIII). Vibrational fine structure corresponding to v_1, the totally symmetric stretching mode of a_1 symmetry, is apparent on this band for both OsO_4 and RuO_4. The fifth band at ~16.0 eV is assigned to the production of the $^2T_2$ state by ionization of the metal–oxygen σ/π-bonding MO of t_2 symmetry. This assignment is based on the great breadth and splitting of this ionization, which is probably due to the Jahn-Teller effect. This, in turn, would suggest that, somewhat unusually, the t_2 σ/π-bonding MO is of lower energy than the a_1 σ-bonding MO. The greater breadth of the $t_2(\sigma/\pi)$ ionization in OsO_4 than in RuO_4 is consistent with the greater thermal stability of the former.

A number of oxyhalides, such as MnO_3X, X = F, Cl (103), $VOCl_3$ (49), CrO_2Cl_2 (49, 200a), MoO_2Cl_2 (49) have also been studied by UPS. Finally, UPS data have been reported for the carbonyl halides, $Fe(CO)_4X_2$, X = Br, I (201).

Acknowledgements

The author is grateful to the National Science Foundation for the purchase of a UV photoelectron spectrometer and for their continued support of our endeavors in photoelectron spectroscopy through NSF Grant CHE 76-10331. The author is deeply appreciative of a Guggenheim Fellowship, during the tenure of which most of this chapter was written. Finally, he would like to express his gratitude to the Department of Chemistry, Princeton University, and to Professors K. Mislow and Malcolm H. Chisholm, in particular, for providing a convivial and intellectually stimulating atmosphere.

ADDENDUM

Some significant developments have come to the author's attention since the manuscript was completed.

The bonding descriptions and UPS assignments for compounds with quadruple metal–metal bonds continue to be topics of lively discussion. An *ab initio* calculation of $Cr_2(O_2CH)_4 2H_2O$ (142a) produced the curious result that the ground state of this molecule involves the configuration $(\sigma)^2(\delta)^2(\sigma^*)^2(\delta^*)^2$ at a Cr–Cr separation of 2.362(1) Å, thus leading to the conclusion that there is no net bonding between the chromium atoms! However, more recent *ab initio* calculations (27a, 159a, 284, 285) in which limited configuration interaction (CI) was included have shown that this result was artifactitious and due to the inability of a single-configuration calculation to take proper account of electron correlation. Even these limited CI treatments lowered the energy of the $(\sigma)^2 (\pi)^4 (\delta)^2$ configuration below that of the nonbonded state.

Considerable doubt persists about the UPS assignments for dichromium and dimolybdenum tetracarboxylates. It has been suggested (285) that the δ, π, and σ bond ionizations of $Cr_2(O_2CMe)_4$ all occur in one (albeit broad) UPS band with a maximum at 9.65 eV. While both the *ab initio* (27a, 159a, 284, 285) and SCF–X_α (79a) calculations agree in placing the $^2B_{2g}$, ^2E_u, and $^2A_{1g}$ ion states closer in energy for $Cr_2(O_2CH)_4$ than for $MO_2(O_2CH)_4$, this assignment definitely warrants further investigation.

Recently, it has been suggested (285) on the basis of *ab initio* calculations that the second ionization band of dimolybdenum tetracarboxylates contains *both* the π and σ ionizations as opposed to the assignment presented in the text. This interpretation must be viewed with caution because *inter alia* it requires the degeneracy of the ^2E_u and $^2A_{1g}$ ion states for a wide range of substituent electronegativity.

A fairly simple "pseudo-atomic" method has been developed (286) for calculating the ligand field splittings in the linear molecules Me_2Cd and Me_2Zn. The paper also contains a useful list of references to previous work in this area.

Recently a series of 15 metal carbonyls has been studied by UPS (and XPS) in both the solid and vapor phases (287). The observations on these carbonyls were compared to the photoelectron spectra of CO adsorbed on transition metal surfaces.

Interest has been sustained in the UPS of the $LM(CO)_5$ complexes of the group VI A transition elements. Both nitrogen (288) and phosphorus containing (289, 290) ligands have been studied.

Finally a series of (alkene)$Fe(CO)_4$ complexes has been studied by Cardaci et al. (291).

References

1. L. J. Aarons, M. F. Guest, W. B. Hall, and I. H. Hillier, *J. Chem. Soc. Faraday Trans. 2*, *69*, 563 (1973).
2. A. Almenningen, B. Andersen, and E. E. Astrup, *Acta Chem. Scand.*, *24*, 1579 (1970).
3. A. Almenningen, A. Haaland, and K. Wahl, *Chem. Commun.*, *1968*, 1027.
4. E. C. Alyea, D. C. Bradley, R. C. Copperthwaite, and K. D. Sales, *J. Chem. Soc. Dalton Trans.*, *1973*, 185.
5. S. E. Anderson and R. S. Drago, *Inorg. Chem.*, *11*, 1564 (1972).
6. M. V. Andreocci, P. Dragoni, A. Flamini, and C. Furlani, *Inorg. Chem.*, *17*, 291 (1978).
7. A. T. Armstrong, D. G. Carroll, and S. P. McGlynn, *J. Chem. Phys.*, *47*, 1104 (1967).
8. H. Aulich, L. Nemec, and P. Delahay, *J. Chem. Phys.*, *61*, 4235 (1974).
9. A. M. W. Baake, *J. Mol. Spectrosc.*, *41*, 1 (1972).
10. E. J. Baerends, C. Oudshoorn, and A. Oskam, *J. Electron Spectrosc. Relat. Phenom.*, *6*, 259 (1975).
11. E. J. Baerends and P. Ros, *Chem. Phys. Lett.*, *23*, 391 (1973).

12. E. J. Baerends and P. Ros, *Mol. Phys.*, 1735 (1975).
13. P. S. Bagus and H. F. Schaefer, *J. Chem. Phys.*, *56*, 224 (1972).
14. P. S. Bagus, U. I. Walgren, and J. Almlof, *J. Chem. Phys.*, *64*, 2324 (1976).
15. E. C. Baker, G. W. Halstead, and K. N. Raymond, *Struct. Bonding*, *25*, 23 (1976).
16. C. J. Ballhausen and J. P. Dahl, *Acta Chem. Scand.*, *15*, 1333 (1961).
17. G. M. Bancroft, L. Adams, D. K. Creber, D. E. Eastman, and W. Gudat, *Chem. Phys. Lett.*, *38*, 83 (1976).
18. G. M. Bancroft, D. K. Creber, and H. Basch, *J. Chem. Phys.*, *67*, 4891 (1977).
19. M. Barber, J. A. Connor, I. H. Hillier, and W. N. E. Meredith, *J. Electron Spectrosc. Relat. Phenom.*, *1*, 110 (1973).
20. P. J. Bassett, B. R. Higginson, D. R. Lloyd, N. Lynaugh, and P. J. Roberts, *J. Chem. Soc. Dalton Trans., 1974*, 2316.
21. P. J. Bassett and D. R. Lloyd, *J. Chem. Soc. Dalton Trans., 1972*, 248.
22. P. J. Bassett, D. R. Lloyd, I. H. Hillier, and V. R. Saunders, *Chem. Phys. Lett.*, *6*, 253 (1970).
23. C. Batich, *J. Am. Chem. Soc., 99*, 7585 (1977).
24. C. Batich, P. Bischof, and E. Heilbronner, *J. Electron Spectrosc. Relat. Phenom.*, *1*, 333 (1973).
25. N. A. Beach and H. B. Gray, *J. Am. Chem. Soc., 90*, 5713 (1968).
26. C. A. L. Becker and J. P. Dahl, *Theor. Chim. Acta*, *14*, 26 (1969).
27. C. A. L. Becker and J. P. Dahl, *Theor. Chim. Acta*, *19*, 135 (1970).
27a. M. Benard and A. Veillard, *Nouv. J. Chim.*, 1, 97 (1977).
28. J. Berkowitz, *J. Chem. Phys., 61*, 407 (1974).
29. P. Bischof and E. Heilbronner, *Helv. Chim. Acta, 53*, 1677 (1970).
30. F. A. Blankenship and R. T. Bedford, *J. Chem. Phys., 36*, 633 (1962).
31. T. F. Block and R. F. Fenske, *J. Am. Chem. Soc., 99*, 4321 (1977).
32. H. Bock, *Rev. Pure Appl. Chem.*, *44*, 343 (1975).
33. G. W. Boggess, J. D. Allen, Jr., and G. K. Schweitzer, *J. Electron Spectrosc. Relat. Phenom.*, *6*, 467 (1973).
34. R. K. Bohn and A. Haaland, *J. Organomet. Chem.*, *5*, 470 (1966).
35. R. Boschi, M. F. Lappert, J. B. Pedley, W. Schmidt, and B. T. Wilkins, *J. Organomet. Chem.*, *50*, 69 (1973).
36. A. Botrel, P. Dibout, and R. Lissillour, *Theor. Chim. Acta, 37*, 37 (1975).
37. D. C. Bradley, *Adv. Inorg. Chem. Radiochem.*, *15*, 259 (1972).
38. D. C. Bradley and M. H. Chisholm, *Acc. Chem. Res.*, *9*, 273 (1976).
39. D. C. Bradley, M. H. Chisholm, C. E. Heath, and M. B. Hursthouse, *Chem. Commun., 1969*, 1261.
40. D. C. Bradley, R. G. Copperthwaite, S. A. Cotton, K. D. Sales, and J. F. Gibson, *J. Chem. Soc. Dalton Trans., 1973*, 191.
41. D. C. Bradley, M. G. Hursthouse, and P. F. Rodesiler, *Chem. Commun., 1969*, 14.
42. D. C. Bradley, R. H. Moss, and K. D. Sales, *Chem. Commun., 1969*, 1255.
43. P. S. Braterman and R. J. Cross, *J. Chem. Soc. Dalton Trans., 1972*, 657.
44. D. M. Brink and G. R. Satchler, *Angular Momentum*, Oxford University Press, London, 1968.
45. H. G. Brittain and R. L. Disch, *J. Electron Spectrosc. Relat. Phenom.*, *7*, 475 (1975).
46. D. A. Brown and W. J. Chambers, *J. Chem. Soc. (A), 1971*, 2083.
47. D. A. Brown and A. Owens, *Inorg. Chim. Acta, 5*, 675 (1971).
48. P. Burroughs, S. Evans, A. Hamnett, A. F. Orchard, and N. V. Richardson, *Chem. Commun., 1974*, 921.
49. P. Burroughs, S. Evans, A. Hamnett, A. F. Orchard, and N. V. Richardson, *J. Chem. Soc. Faraday Trans., 2, 70*, 1895 (1974).

50. C. Cauletti, N. V. Duffy, and C. L. Furlani, *Inorg. Chim. Acta, 23,* 181 (1977).
51. C. Cauletti and C. Furlani, *J. Electron Spectrosc. Relat. Phenom., 6,* 465 (1975).
52. K. G. Caulton and R. F. Fenske, *Inorg. Chem., 7,* 1273 (1968).
53. G. P. Ceasar, P. Milazzo, J. L. Cihonski, and R. A. Levenson, *Inorg. Chem., 13,* 3035 (1974).
54. M. H. Chisholm and F. A. Cotton, *Acc. Chem. Res.,* 11, 356 (1978).
55. M. H. Chisholm, F. A. Cotton, and M. W. Extine, *Inorg. Chem.,* 17, 1329 (1978).
56. M. H. Chisholm, F. A. Cotton, M. Extine, M. Millar, and B. R. Stults, *Inorg. Chem., 15,* 2244 (1976).
57. M. H. Chisholm, F. A. Cotton, M. Extine, M. Millar, and B. R. Stults, *J. Am. Chem. Soc., 98,* 4486 (1976).
58. M. H. Chisholm, A. H. Cowley, and M. Lattman, to be published.
59. D. W. Clack, *Theor. Chim. Acta, 35,* 157 (1974).
60. A. H. Clark and A. Haaland, *Acta Chem. Scand., 24,* 3024 (1970).
61. J. P. Clark and J. C. Green, *J. Chem. Soc. Dalton Trans., 1977,* 505.
62. J. P. Clark and J. C. Green, *J. Organomet. Chem., 112,* C14 (1976).
63. B. G. Cocksey, J. H. D. Eland, and C. J. Danby, *J. Chem. Soc. Faraday Trans. 2, 69,* 1558 (1973).
64. G. Condorelli, I. Fragalà, G. Centineo, and E. Tondello, *Inorg. Chim. Acta, 7,* 725 (1973).
64a. G. Condorelli, I. Fragalà, A. Centineo, and E. Tondello, *J. Organomet. Chem., 87,* 311 (1975).
65. J. A. Connor, L. M. R. Derrick, M. B. Hall, I. H. Hillier, M. F. Guest, B. R. Higginson, and D. R. Lloyd, *Mol. Phys., 28,* 1193 (1974).
66. J. A. Connor, L. M. R. Derrick, I. H. Hillier, M. F. Guest, and D. R. Lloyd, *Mol. Phys., 31,* 23 (1976).
67. J. A. Connor and O. S. Mills, *J. Chem. Soc. A, 1969,* 334.
68. M. Considine, J. A. Connor, and I. H. Hillier, *Inorg. Chem., 16,* 1392 (1977).
69. D. A. Copeland and C. J. Ballhausen, *Theor. Chim. Acta, 20,* 317 (1971).
70. F. A. Cotton, *Chemical Applications of Group Theory,* Wiley, New York, 1971. (2nd edition).
71. F. A. Cotton, *Acc. Chem. Res., 2,* 240 (1969).
72. F. A. Cotton, *Chem. Soc. Rev., 4,* 27 (1975).
73. F. A. Cotton, A. H. Cowley, and M. Lattman, to be published.
74. F. A. Cotton and M. W. Extine, *J. Am. Chem. Soc.,* 100, 3788 (1978).
75. F. A. Cotton, P. E. Fanwick, R. H. Niswander, and J. C. Sekutowski, *J. Am. Chem. Soc.,* 100, 4725 (1978).
76. F. A. Cotton, A. K. Fischer, and G. Wilkinson, *J. Am. Chem. Soc., 78,* 5168 (1956).
77. F. A. Cotton, A. K. Fischer, and G. Wilkinson, *J. Am. Chem. Soc., 81,* 800 (1959).
78. F. A. Cotton and B. J. Kalbacher, *Inorg. Chem., 16,* 2386 (1977).
79. F. A. Cotton, J. G. Norman, B. R. Stults, and T. R. Webb, *J. Coord. Chem., 5,* 217 (1976).
79a. F. A. Cotton and G. G. Stanley, *Inorg. Chem.,* 16, 2668 (1977).
80. F. A. Cotton, G. G. Stanley, B. Kalbacher, J. C. Green. E. Seddon, and M. H. Chisholm, *Proc. Natl. Acad. Sci., 74,* 3109 (1977).
81. M. Coutière, J. Demuynck, and A. Veillard, *Theor. Chim. Acta, 27,* 281 (1972).
82. A. H. Cowley, *Phosphorus Sulfur, 2,* 283 (1976).
83. A. H. Cowley, M. Lattman, and M. H. Chisholm, to be published.
84. P. A. Cox, S. Evans, A. Hamnett, and A. F. Orchard, *Chem. Phys. Lett., 7,* 414 (1970).

85. P. A. Cox, S. Evans, and A. F. Orchard, *Chem. Phys. Lett., 13*, 386 (1972).
86. P. A. Cox and A. F. Orchard, *Chem. Phys. Lett., 7*, 273 (1970).
87. S. Cradock, E. A. V. Ebsworth, and A. Robertson, *J. Chem. Soc. Dalton Trans., 1973*, 22.
87a. S. Cradock, E. A. V. Ebsworth, and A. Robertson, *Chem. Phys. Lett.,* 30, 413 (1975).
88. S. Cradock, E. A. V. Ebsworth, W. J. Savage, and R. A. Whiteford, *J. Chem. Soc. Faraday Trans. 2, 68*, 934 (1972).
89. S. Cradock and W. Savage, *Inorg. Nucl. Chem. Lett., 8*, 753 (1972).
90. S. Cradock and R. W. Whiteford, *J. Chem. Soc. Faraday Trans. 2, 68*, 281 (1972).
91. S. Cradock and R. A. Whiteford, *Trans. Faraday Soc., 67*, 3425 (1971).
92. L. C. Cusachs, F. A. Grimm, and G. K. Schweitzer, *J. Electron Spectrosc. Relat. Phenom., 3*, 229 (1974).
93. J. P. Dahl and C. J. Ballhausen, *K. Dan. Vidensk. Selsk. Mat.-Fys. Medd., 33*, 5 (1961).
94. L. F. Dahl and R. E. Rundle, *Acta Crystallogr., 16*, 419 (1963).
95. M. I. Davis and C. S. Speed, *J. Organomet. Chem., 21*, 401 (1970).
96. T. S. Davis, J. P. Fackler, and M. J. Weeks, *Inorg. Chem., 7*, 1994 (1968).
97. J. Demuynck and A. Veillard, *Theor. Chim. Acta, 28* 241 (1973).
98. M. J. S. Dewar, A. Komornicki, W. Thiel, and A. Schweig, *Chem. Phys. Lett., 31*, 286 (1975).
99. M. J. S. Dewar and S. D. Worley, *J. Chem. Phys., 50*, 654 (1969).
100. M. J. S. Dewar and S. D. Worley, *J. Chem. Phys., 51*, 1672 (1969).
101. B. Dickens and W. N. Lipscomb, *J. Chem. Phys., 37*, 2084 (1962).
102. E. Diemann and A. Müller, *Chem. Phys. Lett., 19*, 538 (1973).
103. E. Diemann, E. L. Varetti, and A. Müller, *Chem. Phys. Lett., 51*, 460 (1977).
104. H. Dietrich and R. Uttech, *Naturwissenschaften, 50*, 613 (1963).
105. H. Dietrich and R. Uttech, *A. Kristallogr., Kristallgeom., Kristallphys., Kristallchem., 122*, 60 (1965).
106. J. D. Dunitz, L. E. Orgel, and R. Rich, *Acta Crystallogr., 9*, 373 (1956).
107. D. E. Eastman and M. Kuznietz, *J. Appl. Phys., 42*, 1396 (1971).
108. R. G. Edgell, A. F. Orchard, D. R. Lloyd, and N. V. Richardson, *J. Electron Spectrosc. Relat. Phenom., 12*, 415 (1977).
109. W. F. Edgell, W. E. Wilson, and R. Summit, *Spectrochem. Acta, 19*, 863 (1963).
110. B. Eischner and S. Herzog, *Z. Naturforsch., 12a*, 860 (1957).
111. J. H. D. Eland, *Photoelectron Spectroscopy*, Butterworths, London, 1974.
112. J. H. D. Eland, *Int. J. Mass Spectrom. Ion Phys., 4*, 37 (1970).
113. S. Elbel and H. tom Dieck, *J. Chem. Soc. Dalton, 1976*, 1757.
114. D. E. Ellis and G. H. Painter, *Phys. Rev. (B), 2*, 2887 (1970).
115. F. O. Ellison, *J. Chem. Phys., 61*, 507 (1974).
116. S. Evans, J. C. Green, M. L. H. Green, A. F. Orchard, and D. W. Turner, *Discuss. Faraday Soc., 54*, 112 (1969).
117. S. Evans, J. C. Green, and S. E. Jackson, *J. Chem. Soc. Faraday Trans. 2, 68*, 249 (1972).
118. S. Evans, J. C. Green, and S. E. Jackson, *J. Chem. Soc. Faraday Trans. 2, 69*, 191 (1973).
119. S. Evans, J. C. Green, S. E. Jackson, and B. Higginson, *J. Chem. Soc. Dalton Trans., 1974*, 304.
120. S. Evans, J. C. Green, P. J. Joachim, A. F. Orchard, D. W. Turner, and J. P. Maier, *J. Chem. Soc. Faraday Trans. 2, 68*, 905 (1972).
121. S. Evans, J. C. Green, A. F. Orchard, T. Saito, and D. W. Turner, *Chem. Phys. Lett., 4*, 361 (1969).

122. S. Evans, M. L. H. Green, B. Jewitt, G. H. King, and A. F. Orchard, *J. Chem. Soc. Faraday Trans. 2*, **70**, 356 (1974).

123. S. Evans, M. L. H. Green, B. Jewitt, A. F. Orchard, and C. F. Pygall, *J. Chem. Soc. Faraday Trans. 2*, *68*, 1847 (1972).

124. S. Evans, M. F. Guest, I. H. Hillier, and A. F. Orchard, *J. Chem. Soc. Faraday Trans. 2*, *70*, 417 (1974).

125. S. Evans, A. Hamnett, and A. F. Orchard, *Chem. Commun.*, *1970*, 1282.

126. S. Evans, A. Hamnett, and A. F. Orchard, *J. Coord. Chem.*, *2*, 57 (1972).

127. S. Evans, A. Hamnett, A. F. Orchard, and D. R. Lloyd, *Discuss. Faraday Soc.*, *54*, 227 (1972).

128. S. Evans and A. F. Orchard, *Electronic Structure and Magnetism of Inorganic Compounds*, Vol. 2, The Chemical Society, London, 1973.

129. S. Evans and A. F. Orchard, *J. Electron Spectrosc. Relat. Phenom.*, *6*, 207 (1975).

130. S. Evans, A. F. Orchard, and D. W. Turner, *Int. J. Mass Spectrom. Ion Phys.*, *7*, 261 (1971).

130a. F. Fantucci, V. Valenti, F. Cariati, and I. Fragalà, *Inorg. Nucl. Chem. Lett.*, **11**, 585 (1975).

131. T. P. Fehlner, J. Ulman, W. A. Nugent, and J. K. Kochi, *Inorg. Chem.*, *15*, 2544 (1976).

132. R. D. Feltham. P. Sogo, and M. Calvin, *J. Chem., Phys.*, *26* 1354 (1957).

133. R. F. Fenske, *Prog. Inorg. Chem.*, *21*, 179 (1976).

134. R. F. Fenske and R. L. Dekock, *Inorg. Chem.*, *9*, 1053 (1970).

135. R. F. Fenske and D. D. Radtke, *Inorg. Chem.*, *7* 479 (1968).

136. B. N. Figgis, *Introduction to Ligand Fields*, Interscience, New York, 1966.

137. A. K. Fischer, F. A. Cotton, and G. Wilkinson, *J. Am. Chem. Soc.*, *79*, 2044 (1957).

138. E. O. Fischer and H. P. Kogler, *Z. Naturforsch.*, *13b*, 197 (1958).

139. R. D. Fischer, *Theor. Chim. Acta, 1*, 418 (1963).

140. S. Foster, S. Felps, L. C. Cusachs, and S. P. McGlynn, *J. Am. Chem. Soc.*, *95*, 5521 (1973).

141. D. C. Frost, F. G. Herring, A. Katrib, R. A. N. McLean, J. E. Drake, and N. P. C. Westwood, *Can. J. Chem.*, *49*, 4033 (1971).

142. L. Galyer, G. Wilkinson, and D. R. Lloyd, *Chem. Commun.*, *1975*, 497.

142a. C. D. Garner, I. H. Hillier, M. F. Guest, J. C. Green, and A. W. Coleman, *Chem. Phys. Lett.*, *41*, 91 (1976).

143. J. S. Ghotra, M. G. Hursthouse, and A. J. Welch, *J. Chem. Soc. Chem. Commun.*, *1973*, 669.

144. S. G. Gibbins, M. F. Lappert, J. B. Pedley, and G. J. Sharp, *J. Chem. Soc. Dalton Trans. 1975*, 72.

145. M. Gower, L. A. P. Kane-Maguire, J. P. Maier, and D. A. Sweigart, *J. Chem. Soc. Dalton Trans., 1977*, 316.

146. W. A. G. Graham, *Inorg. Chem.*, *7*, 315 (1968).

147. J. C. Green, M. L. H. Green, P. J. Joachim, A. F. Orchard, and D. W. Turner, *Philos. Trans., A268*, 111 (1970).

148. J. C. Green, M. L. H. Green, and C. K. Prout, *J. Chem. Soc. Chem. Commun.*, *1972*, 421.

149. J. C. Green and A. J. Hayes, *Chem. Phys. Lett. Trans.*, *31*, 306 (1975).

150. J. C. Green and S. E. Jackson, *J. Chem. Soc. Dalton Trans.*, *1976*, 1698.

151. J. C. Green, S. E. Jackson, and B. Higginson, *J. Chem. Soc. Dalton Trans.*, *1975*, 403.

152. J. C. Green, D. I. King, and J. H. D. Eland, *Chem. Commun.*, *1970*, 1121.

153. J. C. Green, P. Powell, and J. van Tiborg, *J. Chem. Soc. Dalton Trans.*, *1976*, 1974.

154. M. L. H. Green, *Organometallic Compounds*, Vol. 2, Methuen, London, 1967.
155. J. S. Griffith, *The Irreducible Tensor Method for Molecular Symmetry Groups*, Prentice-Hall, London, 1962.
156. J. S. Griffith, *Theory of Transition Metal Ions*, Cambridge University Press, London, 1961.
156a. C. J. Groenenboom, H. J. de Liefde Meijer, F. Jellinek, and A. Oskam, *J. Organomet. Chem.*, *97*, 73 (1975).
157. M. F. Guest, M. B. Hall, and I. H. Hillier, *Chem. Phys. Lett.*, *15*, 592 (1972).
158. M. F. Guest, M. B. Hall, and I. H. Hillier, *Mol. Phys.*, *25*, 629 (1973).
159. M. F. Guest, B. R. Higginson, D. R. Lloyd, and I. H. Hillier, *J. Chem. Soc. Faraday Trans. 2*, *71*, 902 (1975).
159a. M. F. Guest, I. H. Hillier, and C. D. Garner, *Chem. Phys Lett.*, *48*, 587 (1977).
160. M. F. Guest, I. H. Hillier, B. R. Higginson, and D. R. Lloyd, *Mol. Phys.*, *29*, 113 (1975).
161. M. B. Hall, *J. Am. Chem. Soc.*, *97*, 2057 (1975).
162. M. B. Hall and R. F. Fenske, *Inorg. Chem.*, *11*, 768 (1972).
163. M. B. Hall and R. F. Fenske, unpublished results.
164. M. B. Hall, I. H. Hillier, J. A. Connor, M. F. Guest, and D. R. Lloyd, *Mol. Phys.*, *31*, 839 (1975).
165. A. Hamnett and A. F. Orchard, *Electronic Structure and Magnetism of Inorganic Compounds*, Vol. 1, The Chemical Society, London, 1972.
166. A. Hamnett and A. F. Orchard, *Electronic Structure and Magnetism of Inorganic Compounds*, Vol. 3, The Chemical Society, London, 1974.
167. G. C. Hardgrove and D. H. Templeton, *Acta Crystallogr.*, *12*, 28 (1959).
168. D. H. Harris, M. F. Lappert, J. B. Pedley, and G. J. Sharp, *J. Chem. Soc. Dalton Trans.*, *1976*, 945.
169. R. A. Head, J. F. Nixon, G. J. Sharp, and R. J. Clark, *J. Chem. Soc. Dalton Trans.*, *1975*, 2054.
170. C. E. Heath and M. B. Hursthouse, *Chem. Commun.*, *1971*, 143.
171. G. Henrici-Olivé and S. Olivé, *Z. Phys. Chem. (Frankfurt)*, *56*, 223 (1967).
172. B. R. Higginson, D. R. Lloyd, P. Burroughs, D. M. Gibson, and A. F. Orchard, *J. Chem. Soc. Faraday Trans. 2*, *69*, 1659 (1973).
173. B. R. Higginson, D. R. Lloyd, J. A. Connor, and I. H. Hillier, *J. Chem. Soc. Faraday Trans. 2*, *70*, 1418 (1974).
174. B. R. Higginson, D. R. Lloyd, S. Evans, and A. F. Orchard, *J. Chem. Soc. Faraday Trans. 2*, *71*, 1913 (1975).
175. I. H. Hillier, *J. Chem. Phys.*, *52*, 1948 (1970).
176. I. H. Hillier and R. M. Canadine, *Discuss. Faraday Soc.*, *47*, 27 (1969).
177. I. H. Hillier, M. F. Guest, B. R. Higginson, and D. R. Lloyd, *Mol. Phys.*, *27*, 215 (1974).
177a. I. H. Hillier and J. Kendrick, *Inorg. Chem.*, *15*, 520 (1976).
178. I. H. Hillier, J. C. Marriott, V. R. Saunders, M. J. Ware, D. R. Lloyd, and N. Lynaugh, *Chem. Commun.*, *1970*, 1586.
179. I. H. Hillier and V. R. Saunders, *Chem. Commun.*, *1971*, 642.
180. I. H. Hillier and V. R. Saunders, *Mol. Phys.*, *22*, 1025 (1971).
181. I. H. Hillier and V. R. Saunders, *Mol. Phys.*, *23*, 449 (1972).
182. I. H. Hillier, V. R. Saunders, M. J. Ware, P. J. Bassett, D. R. Lloyd, and N. Lynaugh, *Chem. Commun.*, *1970*, 1316.
183. R. J. Hoare and O. S. Mills, *J. Chem. Soc. Dalton Trans.*, *1972*, 653.
184. C. E. Holloway, F. E. Mabbs, and W. R. Smail, *J. Chem. Soc. A, 1968*, 2980.

185. H. B. Jansen and P. Ros, *Theor. Chim. Acta, 34*, 85 (1974).

186. R. B. Johannesen, G. Candela, and R. J. Tsang, *J. Chem. Phys., 48*, 5544 (1968).

187. J. B. Johnson and W. G. Klemperer, *J. Am. Chem. Soc., 99*, 7132 (1977).

188. K. H. Johnson, *Advances in Quantum Chemistry, Vol. 7*, P. O. Lowdin, Ed., p. 143, Academic Press, New York, N.Y.

189. K. H. Johnson and F. C. Smith, Jr., *Phys. Rev. B, 5*, 831 (1972).

190. K. H. Johnson and U. Wahlgren, *Int. J. Quantum Chem.*, Symp. 6, 243 (1972).

191. A. E. Jonas, G. K. Schweitzer, F. A. Grimm, and T. A. Carlson, *J. Electron Spectrosc. Relat. Phenom., 1*, 29 (1972).

192. L. H. Jones, *J. Chem. Phys., 28*, 1215 (1958).

193. L. H. Jones, *Spectrochim. Acta, 19*, 329 (1963).

194. B.-I. Kim, H. Adachi, and S. Imoto, *J. Electron Spectrosc. Relat. Phenom., 11*, 349 (1977).

195. R. F. Kirschner, G. H. Loew, and V. T. Mueller-Westerhuff, *Theor. Chim. Acta, 41*, 1 (1970).

196. T. Koopmans, *Physica, 1*, 104 (1934).

197. Th. Kruck, *Angew. Chem. Int. Ed., 6*, 53 (1967).

198. Th. Kruck and A. Prasch, *Z. Anorg. Allg. Chem., 356*, 118 (1968).

199. M. F. Lappert, J. B. Pedley, and G. Sharp, *J. Organomet. Chem., 66*, 271 (1974).

200. M. F. Lappert, J. B. Pedley, G. J. Sharp, and D. C. Bradley, *J. Chem. Soc. Dalton Trans., 1976*, 1737.

200a. T. H. Lee and J. W. Rabalais, *Chem. Phys. Lett., 34*, 135 (1975).

201. R. A. Levenson, J. L. Cihonski, P. Milazzo, and G. P. Ceasar, *Inorg. Chem., 14*, 2578 (1975).

202. D. L. Lichtenberger and R. F. Fenske, *Inorg. Chem., 13*, 486 (1974).

202a. D. L. Lichtenberger and R. F. Fenske, *J. Am. Chem. Soc., 98*, 50 (1976).

202b. D. L. Lichtenberger and R. F. Fenske, *Inorg. Chem., 15*, 2015 (1976).

203. D. L. Lichtenberger, A. C. Sarapu, and R. F. Fenske, *Inorg. Chem., 12*, 702 (1973).

203a. D. L. Lichtenberger, D. Sellman, and R. F. Fenske, *J. Organomet. Chem., 117*, 253 (1976).

204. E. C. Lingafelter, *Coord. Chem. Rev., 1*, 151 (1966).

205. E. R. Lippincott and R. D. Nelson, *Spectrochim. Acta, 10*, 307 (1958).

206. E. R. Lippincott, J. Xavier, and D. Steele, *J. Am. Chem. Soc., 83*, 2262 (1961).

207. D. R. Lloyd, *Chem. Commun., 1970*, 868.

208. D. R. Lloyd, *Int. J. Mass Spectrum Ion Phys., 4*, 500 (1970).

209. D. R. Lloyd and N. Lynaugh, *Electron Spectroscopy*, D. E. Shirley, Ed., North-Holland, Amsterdam, 1972.

210. D. R. Lloyd and N. Lynaugh, *J. Chem. Soc. Faraday Trans. 2, 68*, 947 (1972).

211. D. R. Lloyd and E. W. Schlag, *Inorg. Chem., 8*, 2544 (1969).

212. J. P. Maier and D. A. Sweigart, *Inorg. Chem., 15*, 1989 (1976).

213. J. P. Maier and D. W. Turner, *J. Chem. Soc. Faraday Trans. 2*, 711 (1972).

214. A. H. Maki and T. E. Berry, *J. Am. Chem. Soc., 87*, 4437 (1965).

215. J. C. Marriott, J. A. Salthouse, M. J. Ware, and J. M. Freeman, *Chem. Commun., 1970*, 595.

216. O. S. Mills and A. D. Redhouse, *J. Chem. Soc. A*, 642 (1968).

217. O. S. Mills and A. D. Redhouse, *J. Chem. Soc. A*, 1274 (1969).

218. O. S. Mills and G. Robinson, *Acta Crystallogr., 16*, 758 (1973).

219. A. P. Mortola, J. W. Moskowitz, N. Rösch, C. D. Cowman, and H. B. Gray, *Chem. Phys. Lett., 32*, 283 (1975).

220. A. Müller, E. Diemann, and C. K. Jørgensen, *Struct. Bonding, 14*, 23 (1973), and references therein.

221. J. Müller, K. Fenderl, and B. Mertschenk, *Chem. Ber., 104,* 700 (1971).
222. L. Nemec, L. Chia, and P. Delahay, *J. Phys. Chem., 79,* 2935 (1975).
223. L. Nemec, H. J. Gaehrs, L. Chia, and P. Delahay, *J. Chem. Phys., 66,* 4450 (1977).
224. L. Ngai, F. Stafford, and L. Schafer, *J. Am. Chem. Soc., 91,* 48 (1969).
225. W. C. Nieuwpoort, *Philips Res. Rep. Suppl., 6,* 1 (1965).
226. J. F. Nixon, *Adv. Inorg. Chem. Radiochem., 13,* 363 (1970).
227. J. F. Nixon, *J. Chem. Soc. Dalton Trans., 1973,* 2226.
228. J. G. Norman and H. J. Kolari, *J. Am. Chem. Soc., 97,* 33 (1975).
229. J. G. Norman and H. J. Kolari, *Chem. Commun., 1974,* 303.
230. J. G. Norman, H. J. Kolari, H. B. Gray, and W. C. Trogler, *Inorg. Chem., 16,* 987 (1977).
231. A. F. Orchard, *Electronic States of Inorganic Compounds, New Experimental Techniques,* P. Day, Ed., D. Reidel Publ. Co., Dordrecht, Holland, and Boston, Mass., 1975.
232. A. F. Orchard, *Discuss. Faraday Soc., 54,* 225 (1972).
233. A. F. Orchard and N. V. Richardson, *J. Electron Spectrosc. Relat. Phenom., 6,* 61 (1975).
234. T. Parameswaran and D. E. Ellis, *J. Chem. Phys., 58,* 2088 (1973).
235. J. L. Petersen, D. L. Lichtenberger, R. R. Fenske, and L. F. Dahl, *J. Amer. Chem. Soc., 97,* 6433 (1975).
236. C. G. Pitt and H. Bock, *Chem. Commun., 1972,* 28.
237. A. W. Potts, H. J. Lempka, D. G. Streets, and W. C. Price, *Philos. Trans., A268,* 59 (1970).
238. R. Prins, *J. Chem. Phys., 50,* 4804 (1969).
239. R. Prins, *Mol. Phys., 19,* 603 (1970).
240. R. Prins and F. J. Reinders, *Chem. Phys. Lett., 3,* 45 (1969).
241. R. Prins and F. J. Reinders, *J. Am. Chem. Soc., 91,* 4929 (1969).
242. J. W. Rabalais, *Principles of Ultraviolet Photoelectron Spectroscopy,* Wiley – Interscience, New York, 1977.
243. J. W. Rabalais, L. O. Werme, T. Bergmark, L. Karlsson, M. Hussain, and K. Siegbahn, *J. Chem. Phys., 57,* 1185 (1972).
244. M. F. Rettig, C. D. Stout, A. Klug, and P. Farnham, *J. Am. Chem. Soc., 92,* 5100 (1970).
245. A. G. Robiette, G. M. Sheldrick, R. N. F. Simpson, B. J. Aylett, and J. M. Campbell, *J. Organomet. Chem., 14,* 279 (1968).
246. N. Rösch and K. H. Johnson, *Chem. Phys. Lett., 24,* 179 (1974).
247. M.-M. Rohmer, J. Demuynck, and A. Veillard, *Theor. Chim. Acta, 36,* 93 (1974).
248. M.-M. Rohmer and A. Veillard, *Chem. Commun., 1973,* 250.
249. M. Rosenblum, *Chemistry of the Iron Group Metallocenes,* Interscience, New York, 1965.
250. W. R. Salanek, N. D. Lipari, A. Paton, R. Zalten, and K. S. Liang, *Phys. Rev. B, 12,* 1493 (1975).
251. J.-M. Savariault, A. Serafini, M. Pelissier, and P. Cassoux, *Theor. Chim. Acta, 42,* 1555 (1976).
252. J. H. Schachtschneider, R. Prins, and P. Ros, *Inorg. Chim. Acta, 1,* 462 (1967).
253. A. F. Schreiner and T. L. Brown, *J. Am. Chem. Soc., 90,* 3366 (1968).
254. A. Schweig and W. Thiel, *J. Chem. Phys., 60,* 951 (1974).
255. G. K. Schweitzer, *Applied Spectroscopy,* E. G. Brame, Jr., Ed., Dekker, New York, 1976.
256. A. Serafini, M. Pelissier, J.-M. Pelissier, M.-M. Savariault, P. Cassoux, and J.-F. Labarre, *Theor. Chim. Acta, 39,* 229 (1975).

160

257. A. Serafini, J.-M. Savariault, P. Cassoux, and J.-F. Labarre, *Theor. Chim. Acta, 36,* 241 (1975).
258. S. Shibata, *Bull. Chem. Soc. Jap., 30,* 753 (1957).
259. A. Shortland and G. Wilkinson, *Chem Commun., 1972,* 318.
260. A. Shortland and G. Wilkinson, *J. Chem. Soc. Dalton Trans., 1973,* 872.
261. E. M. Shustorovich and M. E. Dyatkina, *Dokl. Chem., 128,* 1234 (1959).
262. E. M. Shustorovich and M. E. Dyatkina, *Dokl. Chem., 131,* 113 (1960).
263. E. M. Shustorovich and M. E. Dyatkina, *Zh. Strukt. Khim., 1,* 58 (1960).
264. E. M. Shustorovich and M. E. Dyatkina, *Zh. Strukt. Khim., 2,* 49 (1961).
265. J. C. Slater and K. H. Johnson, *Phys. Rev. B, 5,* 844 (1972).
266. R. J. Smallwood, Ph.D. Thesis, University of London, London, 1975.
267. Y. S. Sohn, D. N. Hendrickson, and H. B. Gray, *J. Am. Chem. Soc., 93,* 3603 (1971).
268. D. A. Symon and T. C. Waddington, *J. Chem. Soc. Dalton Trans., 1975,* 2141.
269. W. Thiel and A. Schweig, *Chem. Phys. Lett., 12,* 49 (1971).
270. D. R. Truax, J. A. Geer, and T. Ziegler, *J. Chem. Phys., 59,* 6662 (1973).
271. D. R. Truax, J. A. Geer, and T. Ziegler, *Theor. Chim. Acta, 33,* 299 (1974).
272. D. W. Turner, *Physical Methods in Advanced Inorganic Chemistry,* H. A. O. Hill and P. Day, Eds., Interscience, New York, 1968.
273. D. W. Turner, A. D. Baker, C. Baker, and C. R. Brundle, *Molecular Photoelectron Spectroscopy,* Wiley, London, 1970.
274. A. Veillard, *Chem. Commun., 1969,* 1022.
275. A. Veillard, *Chem. Commun., 1969,* 1427.
276. K. D. Warren, *Inorg. Chem., 13,* 1243 (1974).
277. M. A. Weiner, A. Gin, and M. Lattman, *Inorg. Chim. Acta, 24,* 235 (1977).
278. M. A. Weiner and M. Lattman, *Inorg. Chem., 17,* 1084 (1978).
279. T. H. Whitesides, D. L. Lichtenberger, and R. A. Budnik, *Inorg. Chem., 14,* 67 (1975).
280. G. Wilkinson and F. A. Cotton, *Prog. Inorg. Chem., 1,* 1 (1959).
281. S. D. Worley, *Chem. Commun., 1970,* 980.
282. G. Yagupsky, W. Mowat, A. Shortland, and G. Wilkinson, *J. Chem. Soc. Dalton Trans., 1972,* 533.
283. D. R. Yarkony and H. F. Schaefer, *Chem. Phys. Lett., 15,* 514 (1972).
284. M. Benard, *J. Am. Chem. Soc., 100,* 2354 (1978).
285. M. F. Guest, C. D. Garner, I. H. Hillier, and I. B. Walton, *J. Chem. Soc. Faraday Trans. II, 74,* 2092 (1978).
286. G. M. Bancroft and R. P. Gupta, *Chem. Phys. Lett., 54,* 226 (1978).
287. E. W. Plummer, W. R. Salaneck, and J. S. Miller, *Phys. Rev., B, 18,* 1673 (1978).
288. H. Daamen and A. Oskam, *Inorg. Chim. Acta, 26,* 81 (1978).
289. L. W. Yarbrough, II and M. B. Hall, *Inorg. Chem., 17,* 2269 (1978).
290. H. Daamen, G. Boxhoorn, and A. Oskam, *Inorg. Chim. Acta, 28,* 263 (1978).
291. A. Flamini, E. Semprini, F. Stefani, G. Cardaci, G. Bellachioma, and M. Andreocci, *J. Chem. Soc. Dalton Trans.* 698 (1978).

Direct Fluorination: A "New" Approach to Fluorine Chemistry

RICHARD J. LAGOW

Department of Chemistry, The University of Texas at Austin, Austin, Texas

and

JOHN L. MARGRAVE

Department of Chemistry, Rice University Houston, Texas

CONTENTS

I. INTRODUCTION

New concepts and experimental methods for controlling the reaction of elemental fluorine with organic and inorganic compounds and polymers have recently been developed in our laboratories. The rate of these reactions has been so successfully controlled that direct fluorination shows promise of becoming the method of choice for preparing many organic and inorganic fluorine compounds. These techniques have provided and will continue to provide a source of many novel and potentially valuable fluorocarbons and fluorine compounds. Capabilities have been developed for handling gaseous, liquid, and solid starting materials, and reactions of elemental fluorine have been successful with a variety of compounds ranging from extremely reactive inorganic hydrides to functional oxygen-containing hydrocarbons. Recently very significant advances have been made by a number of other academic and industrial laboratories.

II. EARLY WORK IN DIRECT FLUORINATION

The difficulty in controlling the reactions of elemental fluorine is so widely recognized that, it has been a general practice to avoid the use of elemental fluorine as a fluorinating agent in the synthesis of fluorocarbons and other fluorine-containing compounds. While many significant developments in fluorine chemistry have been reported in the last 30 years, apparently no great amount of effort has been directed toward solving this basic problem, because of early difficulties and the corresponding prejudices that developed due to these difficulties. Most of the modern approaches for controlling elemental fluorine are really attempts to circumvent the problem of the great reactivity of F_2 rather then to solve the basic kinetic and thermodynamic problems.

Since the time of the earliest work concerned with the reaction of hydrocarbons and fluorine in 1890 by Moissan (who isolated fluorine in 1886), numerous difficulties have been reported. According to Lovelace et al; "the action of fluorine on a carbon compound can be likened to a combustion process where the products are carbon tetrafluoride and hydrogen fluoride" (1).

The reaction proceeds as follows:

$$C_xH_y + \frac{4x + y}{2} F_2 \longrightarrow xCF_4 + yHF + \text{large amount of energy}$$

This reaction has often reached explosive proportions in the laboratory. In the period 1940–1965 methods were developed for controlling the reactions to the extent that, in the case of a hydrocarbon of low molecular weight reacting with F_2, as much as an 80% yield of the perfluorinated product was reported in one case (2). However, in a typical fluorination, appreciable fragmentation of the carbon–carbon bonds may occur and often the yield is in the range of a few percent (3).

Until 1970 the use of elemental fluorine was usually considered a "classical" method of fluorination, while other approaches were regarded as "modern" methods; for example, according to Sharts: "use of fluorine for these reactions (addition to carbon-carbon double bonds and for substitution for hydrogen) cannot be considered to be a useful general method. Fluorination by fluorine is unlikely to be used in normal organic syntheses (4).

Earlier Banks had commented: "The yield of required fluorocarbon decreases as the molecular complexity of its hydrocarbon precursor increases, and it is difficult to fluorinate hydrocarbons above C_{10} without extensive decomposition occurring (5).

A very diversified art has been developed utilizing certain metal fluorides, inorganic fluorides, halogen fluoride, or electrochemical cells as media for fluorination. The more important approaches which have been developed to produce fluorocarbons and yet avoid the great reactivity of elemental fluorine are:

1. The use of inorganic metal fluorides (BF_3, CoF_3, AgF_2, AsF_5, SbF_5, etc.) or halogen fluorides (ClF_3, BrF_3, $KBrF_4$, etc.) as fluorinating agents.

2. The electrolysis of solutions of organic compounds in anhydrous hydrogen fluoride under conditions where no free fluorine is produced.

3. Addition of gaseous HF to olefins or acetylenes.

4. Fluorination of slurries of organic compounds in nonreactive liquids.

5. Use of SF_4, $FClO_3$, and the use of alkali halides in polar solvents.

The basic problem of direct fluorination involves both kinetics and thermodynamics. The rate of the reaction must be slowed down so that the energy liberated from the reaction may be absorbed or carried away. The most significant and crucial innovations in the evolution of the direct fluorination process recently developed by Lagow and Margrave (6) have been kinetic considerations. Their technique has been named the "La-Mar" direct fluorination process (7). Most of the kinetic considerations involve fluorine dilution schemes and probability considerations.

III. A "NEW" APPROACH TO DIRECT FLUORINATION

A. Thermochemistry of Direct Fluorination

Many current fluorine texts contain sections on the "unfavorable thermochemistry" of elemental fluorine reactions. A discussion of this thermochemistry, therefore, is helpful to establish the feasibility of direct fluorination and gain insight into reaction conditions, to select, for example, the reaction temperatures that are most likely to lead to successful results. The tables of data presented and discussed deal with the thermochemistry of fluorination of hydrocarbons. This choice was made because such data were most readily available. Similar arguments may certainly be made for inorganic compounds or polymers.

Table I is based on data from the JANAF Tables for CH_4, which yield an average carbon–hydrogen bond strength of 98 kcal mole^{-1} based on the atomization energy of CH_4. Table II is based on C_2F_4 data from the JANAF Tables. This is, perhaps, a poor choice, since the double bond in C_2F_4 is extremely weak. Reliable enthalpies and free energy functions for the "average" double bond are not available at temperatures other than 298°K. The reaction of hexafluorobenzene with fluorine to form perfluorocyclohexane C_6F_{12} might be a better choice. However, ideal gas free energy calculations on perfluorocyclohexane would probably be no more accurate than reasonable estimates for these values because of boat–chair isomerization and many low-lying molecular energy levels. The $\Delta H_{298°K}$ included in the table is based on bond strengths for average bonds rather than on some other approximations. The limiting parameter to be considered in attempting to develop a satisfactory method for controlling reactions of elemental fluorine is the weakest bond in the reactant compound. For hydrocarbons presently under consideration the example, the average carbon–carbon single bond strength is 84–88 kcal mole^{-1}.

TABLE I

Thermodynamic Data for Steps in Fluorination of CH_4[a]

Step		Reaction	$\Delta H_{298°K}$ kcal mole^{-1}	$\Delta H_{598°K}$ kcal mole^{-1}	$\Delta G_{298°K}$ kcal mole^{-1}	$\Delta G_{598°K}$ kcal mole^{-1}
Initiation	1a	$F_2 \rightarrow 2F\cdot$	+ 37.7	+ 38.5	+ 29.55	+ 20.9
	1b	$F_2 + RH \rightarrow R\cdot + HF + F\cdot$	+ 3.9	+ 5.1	− 5.84	− 18.904
Propagation	2a	$RH + F\cdot \rightarrow R\cdot + HF$	− 33.8	− 33.4	− 36.215	− 37.51
	2b	$R\cdot + F_2 \rightarrow RF + F\cdot$	− 69.1	− 69.5	− 68.1	− 64.15
Termination	3a	$R\cdot + F\cdot \rightarrow RF$	− 106.8	− 108.0	− 97.5	− 85.091
	3b	$R\cdot + R\cdot \rightarrow R - R$	− 83.8	− 83.06	− 70.3	− 57.5
Overall reaction		$R - H + F_2 \rightarrow RF + HF$	− 102.9	− 102.9	− 103.4	− 103.9

[a] Based on JANAF Table data for CH_4.

TABLE II

Thermodynamic Data for Steps in Fluorination of Double Bonds

Step		Reaction	$\Delta H_{298^\circ K}$[a]	$\Delta H_{598^\circ K}$[a]	$\Delta G_{298^\circ K}$[a]	$\Delta G_{598^\circ K}$[a]
Initiation	1a	$F_2 \rightarrow 2F\cdot$	+37.7	+38.5	+29.53	+20.9
	1b	$R_2C{=}CR_2 + F_2 \rightarrow \cdot R_2C{-}FCR_2 + F\cdot$	−37.5	−35.9	−36.1	−34.8
Initiation and termination	2	$R_2C{=}CR_2 + F_2 \rightarrow R_2CF{-}FCR_2$	−161.3	−161.2	−149.2	−136.8
Propagation	3a	$R_2C{=}CR_2 + F\cdot \rightarrow \cdot R_2C{-}FCR_2$	−75.2	−74.4	−65.6	−56.3
	3b	$\cdot CR_2{-}CFR_2 + F_2 \rightarrow R_2CF{-}FCR_2 + F\cdot$	−86.0	−86.7	−83.5	−80.4
Termination	4a	$F\cdot + \cdot CR_2{-}FCR_2 \rightarrow R_2CF{-}FCR_2$	−163.7	−125.3	−113.1	−101.4
	4b	$F\cdot + F\cdot \rightarrow F_2$	−37.7	−38.5	−29.53	−20.9
	4c	$\cdot CR_2{-}FCR_2 + \cdot CR_2FCR_2 \rightarrow FCR_2CR_2{-}CR_2FCR_2$	—	—	—	—
Overall reaction		$R_2C{=}CR_2 + F_2 \rightarrow R_2CF{-}FCR_2$	−161.3	−161.2	−149.2	−136.8

Step		$\Delta H_{298^\circ K}$[b,c]	$\Delta H_{298^\circ K}$[b,d]	$\Delta H_{598^\circ K}$[b,d]	$\Delta G_{298^\circ K}$[b,d]	$\Delta G_{598^\circ K}$[b,d]	$\Delta H_{298^\circ K}$ kcal mole⁻¹ carbon[b,e]
Initiation	1a	−37.7	—	—	—	—	—
	1b	+8.5	−18.77	−17.95	−18.0	−17.4	+4.25
Initiation and termination	2	−123.3	−80.65	−80.5	−74.6	−68.4	−61.65
Propagation	3a	−46.5	−37.6	−37.2	−32.3	−28.2	−23.25
	3b	−69.3	−43.0	−43.4	−41.8	−40.2	−34.65
Termination	4a	−107	−61.9	−62.6	−56.5	−50.7	−53.5
	4b	−37.7	—	—	—	—	—
	4c	−104	—	—	−52	—	—
Overall reaction		−123.3	−80.65	−80.5	−74.6	−68.4	−61.65

[a] Based on JANAF Table data for C_2F_4. Values given in units of kcal mole⁻¹.

[b] Values given are in units of kcal mole⁻¹ per carbon atom.

[c] Based on bond strengths for average carbon–carbon double bond.

[d] Energy per carbon atom based on JANAF Table data for C_2F_4.

[e] Energy per carbon atom based on bond strengths for average carbon–carbon double bond.

165

Any successful fluorination process must minimize the chances of this amount of energy being appropriately localized and available per carbon–carbon bond to preserve the carbon–carbon molecular skeleton and avoid fragmentation (Table III).

The overall reaction in the replacement of hydrogen is exothermic enough $(\Delta G_{298} = -103.4 \text{ kcal mole}^{-1})$ to fracture carbon–carbon bonds if it were to occur by a concerted mechanism or if it were to occur on several adjacent carbon atoms simultaneously. The comparison of 86 versus 103 kcal mole^{-1} has been cited in many previous discussions as an obvious basis for the prediction of the failure of direct fluorination methods. For the rapid reaction rates employed in most previous experiments, this is a valid argument.

The most fundamental and obvious observation to be made concerning the thermochemistry in Table I is that no individual step in this reaction sequence is exothermic enough to break carbon–carbon bonds except the termination step 3a of -97.5 kcal mole^{-1}. Consequently procedures or conditions that minimize the atomic fluorine population or decrease the mobility of hydrocarbon radical intermediates, such as keeping them in the solid state during reaction, are desirable. It is necessary to decrease the reaction rate to the extent that these hydrocarbon radical intermediates have finite lifetimes so that the advantages of fluorination in individual steps may be realized experimentally. It has been demonstrated by EPR (8) methods that under reaction conditions with high fluorine dilution, the various radicals do indeed have appreciable lifetimes.

There are two possible initiation steps for the free radical reaction: step 1b and the combination of steps 1a and 2a. The role of step 1b in the reaction scheme is an important consideration in minimizing the population of atomic fluorine. This step was first postulated by Miller et al. (9) on the basis of reaction products. As indicated in Table I, this process is exothermic at room temperature $(\Delta G_{298^\circ \text{K}} = -5.84 \text{ k cal mole}^{-1})$ although the enthalpy is slightly positive. The validity of this step has not yet been conclusively established by spectroscopic methods, and this remains an unsolved problem of prime importance in fluorine chemistry. The fact that fluorine reacts at a significant rate with hydrocarbons in the dark at temperatures lower than -78° C is an indication that step 1b is a significant step that may have very little or no activation energy at room temperature. At very low temperatures there is no reaction between molecular F_2 and CH_4 or C_2H_6 when isolated in $\sim 10^\circ$ K matrices (10).

TABLE III

Thermodynamic Data for Fragmentation of C_2H_6 [a]

Step	Reaction	$\Delta H_{298^\circ \text{K}}$	$\Delta H_{598^\circ \text{K}}$	$\Delta G_{298^\circ \text{K}}$	$\Delta G_{598^\circ \text{K}}$
1	$R_3C - CR_3 + F_2 \rightarrow 2R_3CF$	-63.6	-64.1	-63.2	-62.5
2	$R_3C - CR_3 + 2F\cdot \rightarrow 2R_3CF$	-82.4	-83.3	-77.9	-73.0

[a] Based on JANAF Table data for C_2H_6. Values given in kcal mole^{-1} per carbon atom.

A simple equilibrium calculation reveals that at $298°K$ and atmospheric pressure, fluorine is less than 1% dissociated, and at $598°K$, a value of 4.6% dissociation of molecular fluorine is obtained from this calculation. It is obvious, therefore, that less than 1% of the collisions occurring at room temperature would result in reaction if step 1a were the only important initiation step. By $598°K$ the free radical initiation should become more important. From an energy control viewpoint, as seen in Table I, it would be advantageous to have step 1b predominate over step 2a and promote attack by molecular, rather than atomic, fluorine. Ambient or lower temperatures would lower the atomic fluorine population.

In the addition of fluorine to double bonds (Table II), the energetic situation is less severe. Note that the addition of fluorine to double bonds releases only 60–70 kcal per carbon–carbon bond. This is not enough energy to fracture the carbon skeleton if care is taken to keep addition from occurring on several adjacent carbon atoms simultaneously. Here, as in the case of hydrogen removal, the individual steps are less exothermic than the overall reaction. It has been established experimentally that less fragmentation occurs and, correspondingly, a higher yield is obtained, with most conventional fluorination processes when the starting material is an aromatic hydrocarbon or one containing double bonds rather than the corresponding aliphatic or alicyclic hydrocarbons. This is due to the greater exothermicity of the reaction with hydrogen (-103.9 kcal mole^{-1} per carbon atom as compared with 50–70 kcal mole^{-1} per unsaturated carbon atom).

In the case of addition of fluorine to double bonds (Table II), note that the corresponding initiation step (1b) is probably exothermic by 5–36 kcal and thus plays an important role. A second important possibility is the concerted reaction mechanism 2, which is exothermic by 50–68.4 kcal per carbon atom.

B. Steric Factors in Direct Fluorination

Another factor that works to the advantage of the successful direct fluorination process is the stereochemistry of fluorocarbons. Initially most of the collisions of fluorine molecules with hydrocarbons or aromatic compounds are likely to occur at a hydrogen site or at a π-bond site. When collisions occur with the π-electron bond, the double bond is fractured, but the single bond remains because the energy released in step 1b (Table II) is not enough to fracture the carbon–carbon single bond. Once carbon–fluorine bonds have begun to form on the carbon skeleton of either an unsaturated or alkane system, this skeleton is somewhat sterically protected by the sheath of fluorine atoms. Figure 1, which shows the crowded helical arrangement of fluorine around the carbon backbone of polytetrafluoroethylene, with a distance between consecutive points of just over 16 Å, is an illustration of an extreme case of this "steric protection" of carbon–carbon bonds.

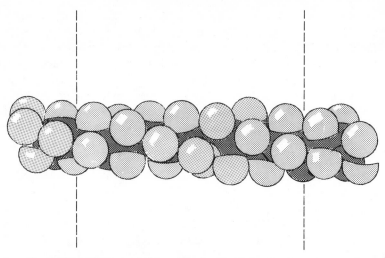

Fig. 1.   The steric protection of the carbon backbone by fluorine of a polytetrafluoro-ethylene chain. The helical configuration of fluorine with a repeat distance of 16.8 Å (– – – –) results from the steric crowding of adjacent fluorine.

The nonbonding electron cloud of the attached fluorine atoms would tend to repel some of the incident fluorine molecules as they approach the carbon skeleton. This reduces the number of effective collisions, making it possible to increase the total number of collisions and still not accelerate the reaction rate as the reaction proceeds toward completion. This sheath of fluorine atoms is one of the reasons for the inertness of Teflon and other fluorocarbons and also explains the greater success commonly reported in the literature when the hydrocarbon to be fluorinated is partially fluorinated in advance by some other process or is prechlorinated.

C. Kinetic Control of the Reactions of Elemental Fluorine

The most crucial element in the control of direct fluorination, a kinetic consideration, is the dilution technique. In most previous work on reactions of elemental fluorine, dilution with nitrogen or helium has been employed. However, the concentration of fluorine in the reactor has been kept at a constant value (usually 10% or greater) by introducing a specified mixture of fluorine and nitrogen, for example, a 10:1 nitrogen-to-fluorine ratio, relatively rapidly into a reactor. Such a dilution scheme corresponds to the horizontal straight line in Fig. 2. The rate of reaction between a hydrocarbon compound and a 10% fluorine mixture is relatively high and the very exothermic process leads to fragmentation (Table III) and in some cases, to combustion. The initial stages of reaction are most critical and nearly all the fragmentation occurs at this time. A

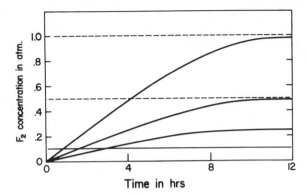

Fig. 2.   Fluorine dilution scheme direct fluorination.

10% initial concentration of fluorine is, for most compounds, much too high for smooth fluorination.

Molecular relaxation processes such as vibration or rotational relaxation and thermal conduction make it possible to dissipate energy released during fluorination. For an individual molecule or a polymer chain, such processes are capable of minimizing the chances of the activation energy required to break the weakest bond from being appropriately localized only if the reaction sites are widely distributed over the system. The probability of a molecule or polymer chain surviving a fluorine reaction on two adjacent carbon atoms at approximately the same time is very low. It is questionable whether a low-molecular-weight hydrocarbon such as a cyclohexane ring can survive simultaneous fluorine reactions on any two carbon—hydrogen adjacent sites without fragmentation. Therefore, in the initial stages of fluorination it is advantageous to reduce the probability of more than one reaction site on the same molecule or adjacent molecules in a crystal, such that relaxation processes distribute the energy more widely over the molecules and fragmentation is avoided.

A very low initial fluorine concentration is used in the La-Mar fluorination process. Initially a helium or nitrogen atmosphere is used in the reactor and fluorine is bled slowly into the system. If pure fluorine is used as the incoming gas, one may elect to asymptotically approach a concentration of 1 atm. of fluorine over any time period (see Fig. 2). It is also possible to approach asymptotically any fluorine partial pressure in the same manner. The very low initial concentrations of fluorine in the system greatly decreases the probability of simultaneous fluorine collisions on the same molecules or on adjacent reaction sites. As previously discussed, reactant molecules, as they become more highly fluorinated, are able to withstand more fluorine collisions without decomposition because sites are sterically protected by fluorine. Such collisions at carbon–fluorine sites are obviously nonreactive collisions. The fluorine concentration may therefore be increased as the reaction proceeds to obtain a practical reaction rate. Actual dilution schemes to obtain successful fluorination must be individually tailored for a specific reaction system. Sometimes stepwise procedures are used, giving rise to plateaus in the curves of Fig. 1 instead of the smooth concentration rise with time. Detailed knowledge of fluorine delivery procedures has been developed in our laboratory as an art but is rapidly approaching a more scientific plane. More information on dilution schemes may be obtained from the references given later in the discussion.

One may conclude, in summary, that a successful fluorination process with elemental fluorine should reduce the number of collisions initially and provide a heat sink to remove or absorb the resulting energy such that in the case of a hydrocarbon, the probability of 84—88 kcal per mole of carbon being available and properly localized to break carbon—carbon bonds is very low.

IV. DESIGN OF APPARATUS FOR ELEMENTAL FLUORINATION

A. Early Methods of Fluorination

Important pioneering contributions were made in the area of the fluorination of hydrocarbons with elemental fluorine, notably, the wartime efforts of Cady et al. (2) and the work of Haszeldine and Smith (11). Their reaction systems were of similar design, and both groups vaporized the hydrocarbon mixed with nitrogen into a reactor mixing the vapor with a nitrogen—fluorine mixture from another preheated source. Haszeldine and Smith used gold as a catalyst in their work. Another important design was the jet reactor of Bigelow and Tyczkowski (3), which produced low-molecular-weight fluorocarbons. In these early studies, the reaction temperatures were in the $200°-370°C$ range and fluorinations were conducted in the vapor phase.

At temperatures in the $200-300°C$ range, many organic and inorganic compounds are only marginally stable and often are not in their lowest vibrational states. Addition to the system of the quantities of energy produced by the reaction with fluorine is likely to produce substantial fragmentation. Vaporization of higher molecular weight hydrocarbons is difficult, and the fluorination of most high-molecular-weight hydrocarbons was not attempted in these pioneering studies. One reason the original workers chose these reaction conditions may have been that similar work had been performed previously with chlorine and higher temperatures were used. Chlorine has a dissociation energy of $+58$ kcal $mole^{-1}$ compared with only 37.7 kcal $mole^{-1}$ for fluorine, yet carbon—chlorine bonds are weaker so that an entirely different thermochemistry results. Comparing step 1a ($\Delta G_{298°K} = +56.8$ kcal, $\Delta H_{298°K} = 3.9$ kcal $mole^{-1}$) with the similar step for chlorination:

$$RH + Cl_2 \longrightarrow R\cdot + Cl\cdot + HCl$$

one finds that $\Delta F_{298°K} = 47.1$ kcal mole^{-1} and $\Delta H_{298°K} = 56.8$ kcal mole^{-1}. This indicates that, as is well documented, the initiation step in chlorination is largely free radical and use of high temperatures or U.-V. activated dissociation is necessary.

B. Experimental Techniques for Precision Control of Elemental Fluorine Reactions

Most of the apparatus used has been simple in design, although fluorine-resistant materials are required. Figure $3a$ shows typical fluorine-handling apparatus comprised of (1) a fluorine source and (2) a helium source, (3) a Hasting-Raydist model LF-50 mass flowmeter and model F-50 M transducer for measuring fluorine flow rates, (4) is a simple gas flowmeter, (5 and 6) are needle valves to control the gas flow rates, (7) a brass mixing chamber packed with fine copper turnings, (8) a reaction chamber, which consists of a 1 in. x 1.5 in. nickel tube containing a prefluorinated nickel boat, $7\frac{1}{2}$ in. long and $\frac{1}{2}$ in. wide, used to hold a solid sample, (9) an alumina-packed cylinder to dispose of unreacted fluorine, (10) the oxygen outlet.

To eliminate back diffusion of the oxygen produced when fluorine reacts with the alumina, as well as possible sources of air and moisture, a 1 in. by 6 in. long T-joint is placed between the reactor (8) and the alumina trap (9). The gases from (8) are exhausted into the side of the T-joint. A constant flow of nitrogen (100 ml min^{-1}) is used to constantly flush the alumina trap and is connected with copper tubing to the alumina trap (9) so that both the nitrogen and the waste gases from (8) are exhausted through the alumina trap. A standard nitrogen gas bubbler is placed in the line after the alumina trap (9) to prevent air and moisture from entering the alumina trap when the system is not in use. When the system is in use, the nitrogen flow also exhausts through the bubbler.

It is extremely important to remove all sources of oxygen and water from the fluorination system. When free oxygen is present, it can cross-link the material presumably with epoxy bridges, and form carbonyl groups such as acid fluorides and peroxides on contact with carbon radical sites. Cross-linking greatly decreases the yield of perfluorocarbon. Should this occur, it may be detected by infrared activity in the 1600–2000 cm^{-1} region and by noting the polymeric nature of the products.

All connections between the pieces of equipment are made with $\frac{1}{4}$ in. copper tubing. Whenever the fluorocarbon obtained in the reaction is volatile at room temperature, a cold finger-type trap is placed between 8 and 9 to catch the product. The temperature of the trap must be high enough for passage of

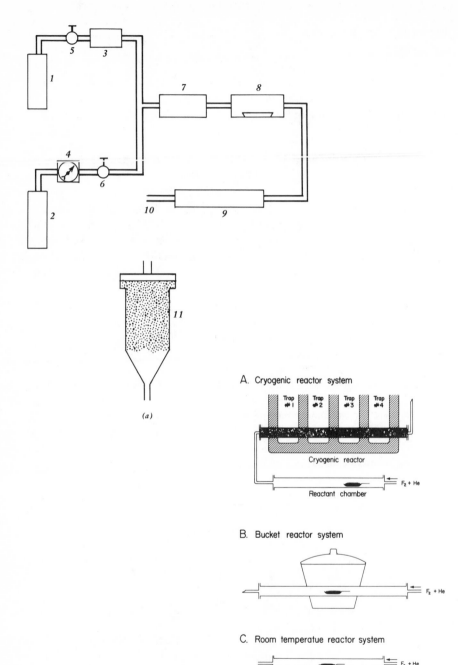

Fig. 3. (a) Diagram of fluorine-handling apparatus. (b) Fluorine reactor systems.

172

fluorine but cold enough to catch the product. In such a 1 in. diameter reactor, typical fluorine flow rates are $\frac{1}{2}$–4 ml/min into an atmosphere of pure helium or nitrogen. In the initial stage, the fluorine flow is often diluted with nitrogen. When solid samples are to be perfluorinated, it is necessary to grind the hydrocarbon starting material to a particle size of less than 100 mesh so that the fluorine is able to diffuse into the center of the particle to react. If larger particles are used, an unfluorinated core of hydrocarbon material may remain in the center of the particle.

A cryogenic reactor which has been used successfully to control reactions of fluorine with liquid and gaseous samples was developed by the R. J. Lagow research group while at Massachusetts Institute of Technology (12–14). The dimensions of the reactor are shown in Fig. 3b. Reactants are volatilized into the reaction zone of the cryogenic reactor from the heated oil evaporator prior to initiation of the reaction. The main reaction tube is a 1 in. nickel chamber packed with fluorinated copper filings. The individual compartments are constructed of stainless steel and are insulated with urethane foam. They each measure 4 in. x 4 in. x 8 in. All the parts are connected by 0.25-in. copper or aluminum tubing. A sodium fluoride trap is used to remove hydrogen fluoride from the reaction products and a glass trap is placed in the line to collect the reaction products. The separated compartments in the cryogenic reactor, aside from their use as a heat sink, are used to create a temperature gradient along the reaction tube by successively cooling and warming the various compartments. As the reactant becomes more highly fluorinated, its volatility increases and it moves through the reactor more rapidly. This process provides a continually renewed surface for fluorination. Combined with the large surface area provided

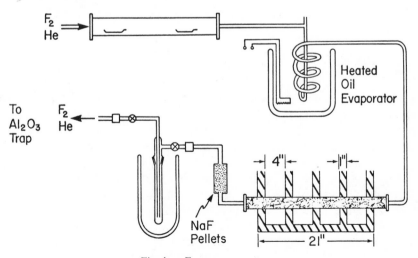

Fig. 4.   Four-zone reactor.

by the fluorinated copper turnings, a maximization of compound surface exposed to fluorine is obtained. The individual zones of the reactor may be cooled with solvent – solid CO_2 or solvent–liquid nitrogen slushes. Preferably the temperature is precisely regulated with an automatic liquid nitrogen temperature controller. The reactor shown in Fig. 4 is a four-zone reactor; recent designs in our laboratory which have been more successful involve multizones reactors (an eight-zone reactor has been found to be particularly efficient) and internal Freon cooling.

V. RECENT APPLICATIONS OF DIRECT FLUORINATION IN SYNTHETIC CHEMISTRY

The synthetic applications of direct fluorination are very broad and cover the range from polymer chemistry to inorganic and organic synthesis. Since the announcement of new procedures for direct fluorination by Lagow and Margrave (6, 15) considerable experimental experience has been obtained with these procedures. Control of the very exothermic reaction of elemental fluorine with the most delicate or the most reactive organic compound, reactive inorganic compound, polymer, or even an element in its most reactive form may be accomplished using these direct fluorination techniques. The degree of success obtained with such a fluorination varies greatly from system to system and according to the exact reaction conditions used. It is possible to safely control the reaction of fluorine with almost any substrate; however, more skill is required to obtain a desired reaction product.

A. Applications in Fluorocarbon Polymer Chemistry

1. General Applications

One of the most important applications of direct fluorination is the synthesis of fluorocarbon polymers from hydrocarbon polymers and in the conversion of the surface of hydrocarbon polymers to fluorocarbon polymer surfaces (16, 17). Idealized structures and reaction schemes for these fluorination studies of hydrocarbon polymers are shown in Figs. 5 and 6. It is possible to produce fluorocarbon polymers with chemical compositions and physical properties very similar to those of polytetrafluoroethylene from the reaction of fluorine with polyethylene and other perfluoropolymers from analogous hydrocarbon starting materials. The new fluorocarbon polymers differ from the idealized structures primarily because, in such-high molecular-weight species, carbon–carbon cross-linking occurs to a significant extent during fluorination. This cross-linking may be controlled, but in general the fluorocarbon polymer is

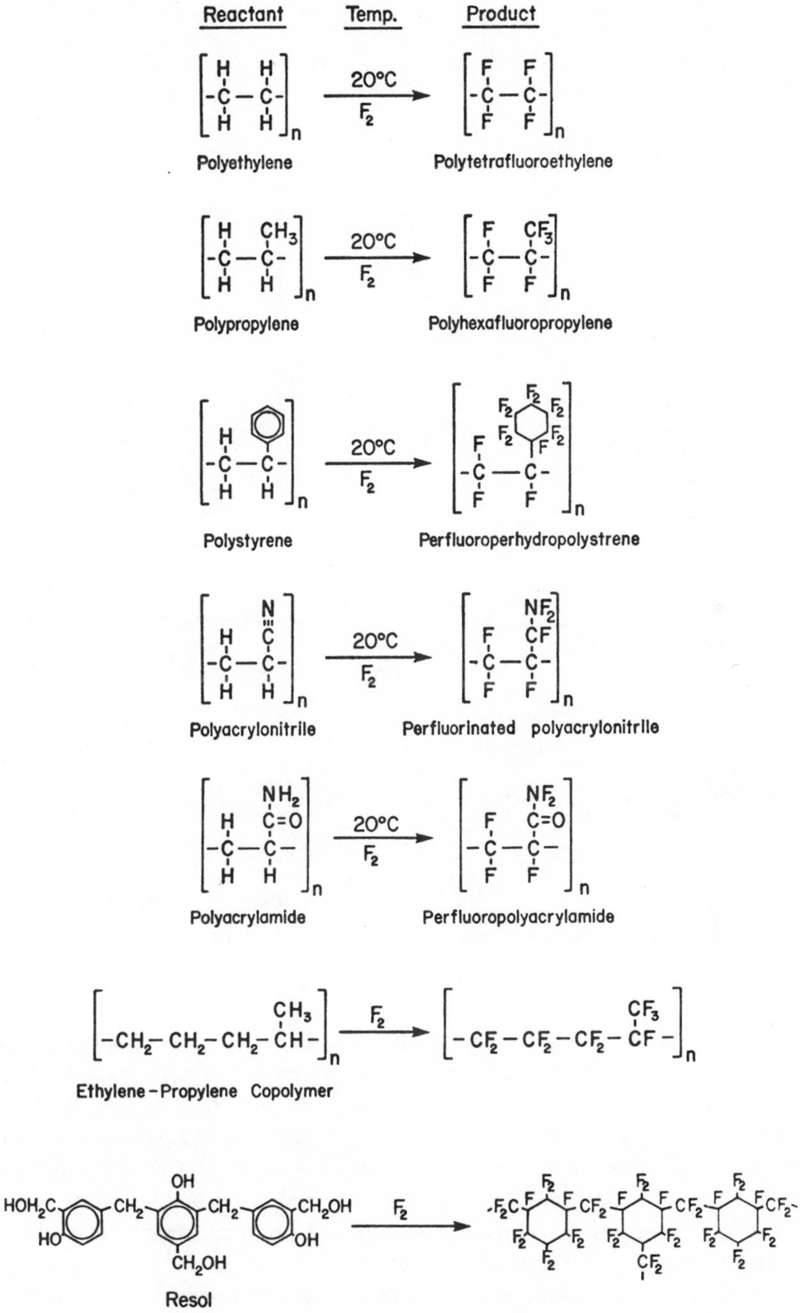

Reactant	Temp.	Product

$$\begin{bmatrix} \overset{H}{\underset{H}{-C}} - \overset{H}{\underset{H}{C}} - \end{bmatrix}_n \xrightarrow[F_2]{20°C} \begin{bmatrix} \overset{F}{\underset{F}{-C}} - \overset{F}{\underset{F}{C}} - \end{bmatrix}_n$$

Polyethylene   Polytetrafluoroethylene

$$\begin{bmatrix} \overset{H}{\underset{H}{-C}} - \overset{CH_3}{\underset{H}{C}} - \end{bmatrix}_n \xrightarrow[F_2]{20°C} \begin{bmatrix} \overset{F}{\underset{F}{-C}} - \overset{CF_3}{\underset{F}{C}} - \end{bmatrix}_n$$

Polypropylene   Polyhexafluoropropylene

$$\begin{bmatrix} \overset{H}{\underset{H}{-C}} - \overset{\bigcirc}{\underset{H}{C}} - \end{bmatrix}_n \xrightarrow[F_2]{20°C} \begin{bmatrix} \overset{F}{\underset{F}{-C}} - \overset{\bigcirc}{\underset{F}{C}} - \end{bmatrix}_n$$

Polystyrene   Perfluoroperhydropolystrene

$$\begin{bmatrix} \overset{H}{\underset{H}{-C}} - \overset{\overset{N}{\|}C}{\underset{H}{C}} - \end{bmatrix}_n \xrightarrow[F_2]{20°C} \begin{bmatrix} \overset{F}{\underset{F}{-C}} - \overset{NF_2 \atop CF}{\underset{F}{C}} - \end{bmatrix}_n$$

Polyacrylonitrile   Perfluorinated polyacrylonitrile

$$\begin{bmatrix} \overset{H}{\underset{H}{-C}} - \overset{NH_2 \atop C=O}{\underset{H}{C}} - \end{bmatrix}_n \xrightarrow[F_2]{20°C} \begin{bmatrix} \overset{F}{\underset{F}{-C}} - \overset{NF_2 \atop C=O}{\underset{F}{C}} - \end{bmatrix}_n$$

Polyacrylamide   Perfluoropolyacrylamide

$$\begin{bmatrix} -CH_2- CH_2- CH_2- \overset{CH_3}{\underset{}{CH}} - \end{bmatrix}_n \xrightarrow{F_2} \begin{bmatrix} -CF_2- CF_2- CF_2- \overset{CF_3}{\underset{}{CF}} - \end{bmatrix}_n$$

Ethylene–Propylene Copolymer

Resol $\xrightarrow{F_2}$

Fig. 5.   Fluorination of polymers.

175

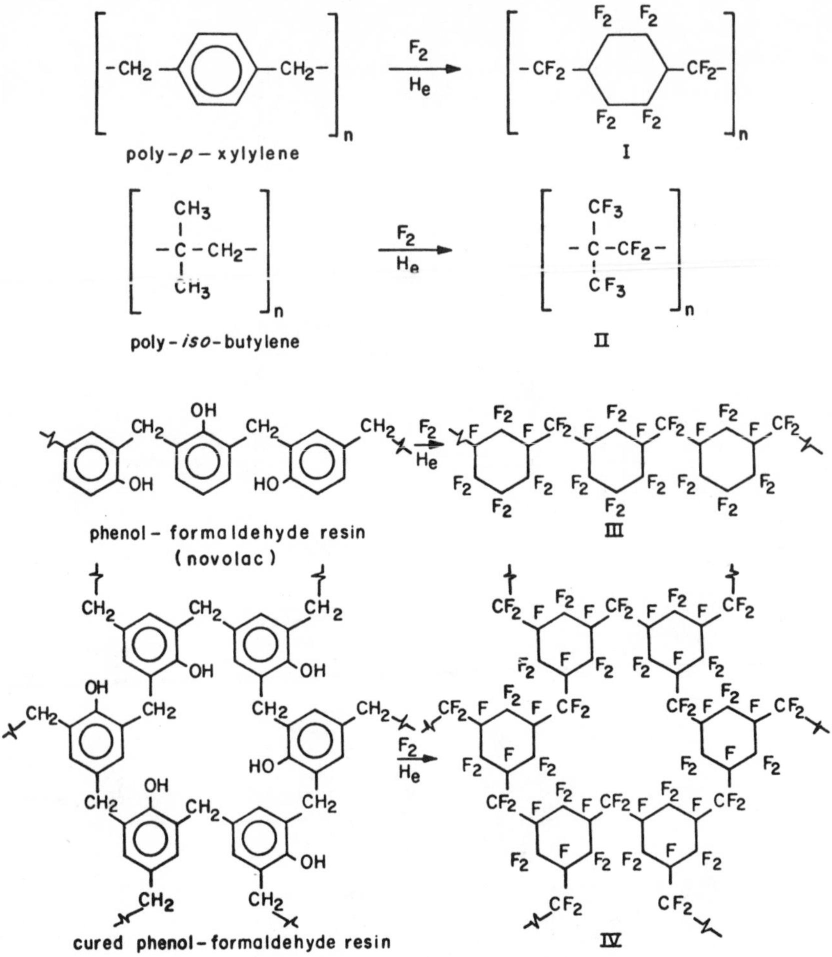

Fig. 6.   Fluorination of polymers.

of higher molecular weight than the hydrocarbon precursor. Polymer-chain fission can be almost entirely eliminated under proper conditions. Most of these fluorocarbon polymers are white solids with high thermal stability, and some can be heated as much as $300°C$ above the decomposition temperature of their hydrocarbon precursor before decomposition. The elemental analyses indicate complete conversion to fluorocarbon polymers and correspond closely to the idealized compositions. The infrared spectra and other spectroscopic evidence are consistent with perfluorocarbon materials. A number of very interesting ESCA studies by D. T. Clark and co-workers have confirmed that hydrocarbon polymers such as polyethylene (18) are converted to fluorocarbon polymers by direct fluorination with very little or no structural rearrangement. These studies, in addition to providing further information on the structure of the fluoropolymer, provide information on subtle surface effects, such as oxygen incorporation, and comment on the rate and depth of fluorination.

A primary advantage of this procedure for producing fluorocarbon polymers is that many monomers such as hexafluoropropylene, which would be polymerization precursors, are difficult to polymerize due to the steric repulsion of fluorine and lead to relatively low-molecular-weight species; also, many of the corresponding monomers are difficult or impossible to synthesize.

If solid polymer objects are fluorinated or polymer particles much larger than 100 mesh are used, only surface conversion to fluorocarbon results. Penetration of fluorine and conversion of the hydrocarbon to fluorocarbon to depths of at least 0.1 mm is a result routinely obtained and this assures nearly complete conversion of finely powdered polymers. These fluorocarbon coatings appear to have a number of potentially useful applications ranging from increasing the thermal stability of the surface and increasing the resistance of polymer surfaces to solvents and corrosive chemicals, to improving friction and wear properties of polymer surfaces. It is also possible to fluorinate polymers and polymer surfaces partially to produce a number of unusual surface effects. The fluorination process can be used for the fluorination of natural rubber and other elastomeric surfaces to improve frictional characteristics and increase resistance to chemical attack.

Interest in the direct fluorination of polymers in Japan was established about the same time the initial work on direct fluorination from Rice University was announced. The Japanese workers Okada and Makuuchi (19) studied the fluorination of polyethylene powder and described the reaction as "a fiery intense reaction," but were finally able to control it by dilution with nitrogen or helium and use of evacuated reaction vessels. The focus of their study was the rate of reaction.

More recently there has been interest in Japan in the surface fluorination of poly(vinyl fluoride) (2) and the surface fluorination of polycarbonates, polystyrene, and poly(methyl methacrylate) (21).

The surface fluorination of polymers has an interesting history. Aside from some early "paper" patents (i.e., not backed by experimental data), the first significant work reported on the surface fluorination of polymers was a patent issued to A. J. Rudge in 1954 (22) on the surface fluorination of polyethylene. This intriguing development certainly provided an early indication of the potential richness of the field of direct fluorination but was a sign of great potential largely ignored. From 1964 to 1970 direct fluorination work was conducted at Rice University and concurrently new directions were taken in Japan (19), and at the Bell Laboratories (23).

In the interim period there was direct fluorination activity, but it was focused largely on surface fluorination of polyethylene (24–28) In fact, in view of the early Rudge patent it is difficult to rationalize the issue of other patents (24, 25, 27) on substantially the same surface fluorination process. Perhaps this is an indication of the relative obscurity of the Rudge work, which was not mentioned in reviews or fluorine texts, even in 1970, and was perhaps unknown by other workers and patent examiners alike. Indeed, in 1964, when work on direct fluorination of organic compounds began at Rice University as a part of the long standing research effort of the J. L. Margrave research group in fluorine chemistry, this work was not recognized there, nor was there general knowledge among fluorine chemists of the work of A. J. Rudge.

It is difficult to establish the extent to which these "interim" workers explored the fluorination of species other than polyethylene and a few closely related polymers. Polyethylene is uniquely easy to fluorinate without visual degradation and therefore no sophisticated experimental strategy is needed to prevent combustion. Even the reaction of polypropylene, which has pendant methyl groups, is an order of magnitude more difficult. It appears likely that a number of workers who used simply a 10:1 nitrogen–fluorine dilution scheme for polyethylene were dissuaded from further pursuit of the general area after observing degradation, polymerization, and/or combustion and explosions with other classes of reactants. This is evident in at least one report (29), to which the author has called our attention.

D. Dixon and L. Hayes of Air Products Corporation have launched a significant effort directed toward commercial exploitation of direct fluorination. They have reported a study of partial fluorination (in the range from $\frac{1}{2}$ to 3%) of poly(ethylene terephtalate) (30) polyesters and have discovered that this treatment produces a high degree of polymer wetability and results in unusually good soil release properties in polyester garments. Similarly, they have reported a wetability increase in polyamides such as nylon 6,6 (31). They have also studied the contact angle and relation to surface energy of fluorinated polymer surfaces (32, 33).

Hayes and Dixon have also contributed toward the development of an industrial process for surface fluorination of polyethylene bottles to improve solvent resistance (32)

2. The Synthesis of Fluorocarbon Polyethers by Direct Fluorination

A promising and extremely interesting new area of research is the synthesis of perfluoropolyethers by the reaction of elemental fluorine with hydrocarbon polyethers (34, 35). An extensive recent research program at the University of Texas at Austin has succeeded in establishing the generality of this approach. In the initial study, polyethylene oxide was reacted with elemental fluorine under conditions carefully regulated to fragment and perfluorinate the polyether system and yield a number of new fluorocarbon ethers.

Saturated perfluoropolyethers are of current interest for new material applications due to their unusual properties. Lack of chemical reactivity and thermal stability ($>300°C$) is their outstanding feature. They have been described as being equally stable as perfluoroalkanes and unaffected by concentrated acids and bases at elevated temperatures over extended periods of time (36). The only reported reaction of saturated perfluoropolyethers is chain cleavage at the ether linkage by aluminum chloride (at elevated temperatures and autogenous pressure) to produce acyl chloride and trichloromethyl end groups (37). These remarkable stabilities, along with their interesting surface properties, viscosities, and the broad liquid ranges of the low-molecular-weight compounds, make saturated perfluoropolyethers attractive for numerous applications (38) as solvents, hydraulic fluids, heat-transfer agents, lubricants, greases, sealants, elastomers, and plastics.

Synthetic methods have limited the preparation of saturated perfluoropolyethers. The most successful perfluoropolyether synthetic chemistry has been DuPont's anionic polymerization of perfluoroepoxides, particularly hexafluoropropylene oxide and tetrafluoroethylene oxide (39). Their synthetic procedure is a three-step scheme for saturated perfluoropolyether production involving oxidation of perfluoroolefins to perfluoroepoxides, anionic polymerization to acyl fluoride terminated perfluoropolyethers, and conversion of acyl fluoride end groups to unreactive end groups by decarboxylation reactions or chain-coupling photolytic decarboxylate reactions.

A useful new synthetic method for the production of saturated perfluoropolyethylene glycol ethers has been developed using a one-step reaction, the direct fluorination of polyethylene oxide polymer with reaction conditions chosen to promote fragmentation of the polymer during the fluorination process. This method has made possible the synthesis of saturated perfluoropolyethers over a broad range of molecular weights. This is achieved by perfluorination of the polymer using normal procedures under ambient temperatures, followed by fluorination in pure fluorine at $110°C$ to promote fragmentation.

$$(CH_2-CH_2-O)_n \xrightarrow[\substack{1.\ ambient \\ 2.\ \Delta 110°C}]{F_2/He} R_f-(O-CF_2CF_2)_n-O-R_f$$
$$R_f = CF_3, C_2F_5$$

The low-molecular-weight compounds ($n = 1-6$) isolated are volatile liquids, the medium-molecular-weight compounds are nonvolatile oils, and the high-molecular-weight compounds are gel-like and powdery solids. Previously we found that milder fluorination conditions designed to prevent fragmentation lead to an extremely stable high-molecular-weight perfluoropolyether (40).

If the following scheme is used to classify the perfluoro ethers,

$$CF_3-O-(CF_2-CF_2-O)_a-CF_3 \qquad a = 1-6$$
$$CF_3-O-(CF_2-CF_2-O)_b-C_2F_5 \qquad b = 1-6$$
$$C_2F_5-O-(CF_2-CF_2-O)_c-C_2F_5 \qquad c = 1-3$$

the range of volatile compounds produced is that shown in Table IV along with their NMR data, which was one of the principal means of characterization.

Similar reactions have been studied with polymethylene oxide and poly-propylene oxide (41).

$$-(CH_2O)_{\overline{n}} \xrightarrow{\text{He/F}_2} -(CF_2O)_{\overline{n}} \qquad \text{(a)}$$

$$\left(\begin{matrix} CH_2-CH-O \\ | \\ CH_3 \end{matrix} \right)_n \xrightarrow[\substack{1.\ \text{ambient} \\ 2.\ \Delta}]{\text{He/F}_2} R_f-O(CF_2CF-O)_n-R_f \qquad \text{(b)}$$
$$\overset{|}{CF_3}$$

$$R_f = CF_3, C_2F_5, {}_nC_3F_7$$

An idea of the range of products obtained in the reaction and their ^{19}F NMR spectra are given in Table V and Fig. 7.

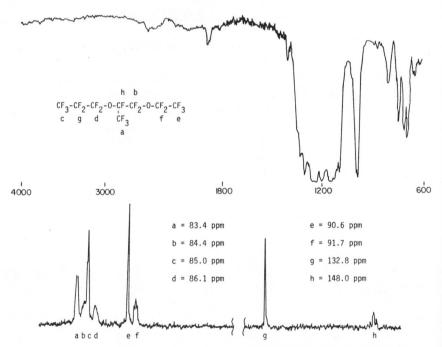

$$\underset{c\quad g\quad d\quad CF_3\quad\quad f\quad e}{\overset{h\quad b}{CF_3\text{-}CF_2\text{-}CF_2\text{-}O\text{-}CF\text{-}CF_2\text{-}O\text{-}CF_2\text{-}CF_3}}$$
$$a$$

a = 83.4 ppm	e = 90.6 ppm
b = 84.4 ppm	f = 91.7 ppm
c = 85.0 ppm	g = 132.8 ppm
d = 86.1 ppm	h = 148.0 ppm

Fig. 7.   Infrared and ^{19}F NMR spectra of perfluoro-4-methyl-3,6-dioxanonane.

TABLE IV

Properties of Volatile Perfluoroethyl Glycol Ethers

	Physical properties of perfluoroethylene glycol ethers		^{19}F NMR signals of perfluoroethylene glycol ethers shift in ppm vs. CFCl$_3$ external				
Compound	Melting point, °C	Boiling point, °C	CF_3-O (t)	CF_3-O-CF_2 (q)	Internal $-CF_2-O-$(s)	CF_3-CF_2-O (s)	$J_{CF_3-O-CF_2}$, Hz
$a = 1$	—	—	59.4	94.0	91.9 (4)	90.7	10
$b = 1$	—	—	59.4	93.9	91.9 (4)	90.7	10
$c = 1$	—	—	—	—	91.7 (4)	90.5	—
$a = 2$	—	66 to 66.5	59.3	93.8	91.8 (2)	—	9
$b = 2$	—	81.5 to 82 [a]	59.2	93.7	91.7 (8)	90.5	9
$c = 2$	−78.5 to −77	[a]	—	—	91.6 (6)	90.4	—
$a = 3$	−82 to −80.5	104.5 to 105	59.1	93.8	91.7 (4)	—	9
$b = 3$	−80 to −78	117.5 to 118.5 [a]	59.0	93.7	91.6 (12)	90.4	9
$c = 3$	[b]	[a]					
$a = 4$	−71 to −69.5	138 to 138.5	59.1	93.7	91.6 (6)	—	9
$b = 4$	−60.5 to −60	146.5 to 148	58.9	93.5	91.5 (16)	90.3	10
$a = 5$	−47 to −46	164 to 164.5	58.9	93.5	91.5 (8)	—	10
$b = 5$	−47 to −46	173.5 to 174	59.0	93.6	91.5 (20)	90.3	10
$a = 6$	−43.5 to −43	186 to 186.5	59.0	93.6	91.4 (10)	—	9
$b = 6$	−35 to −34	193 to 194	59.1	93.7	91.6 (24)	90.4	9

[a] Insufficient compound isolated to obtain a boiling point.
[b] Insufficient compound isolated to obtain a melting point.
[c] Insufficient compound isolated to obtain NMR spectrum.

t = triplet, q = quartet, s = singlet. Relative intensities: CF_3-O = 3; CF_3-O-CF_2 = 2; CF_3-CF_2-O = 3; internal $-CF_2-O-$ = as indicated in parentheses.

181

TABLE V

Identified Volatile Compounds from the Fluorination of Polypropylene Oxide with Assigned ^{19}F NMR Data

m/e in mass spec.	GLC 'cut'	Molecular weight	Identified compound assigned ^{19}F NMR data	Observed relative intensity	Theoretical relative intensity
No mass spec.	131 – 3	166	$\underset{a}{CF_3}-\underset{b}{CF_2}-\overset{\overset{O}{\parallel}}{C}-\underset{c}{F}$		
			$a = 87.1$ d of t ($J = 5.5, 1.5$)	45	3
			$b = 124.9$ d of quart. ($J = 9.0, 1.0$)	30	2
			$c = -19.2$ multiplet	15	1
285($C_5F_{11}O$), P–F	131 – 7, 78 – 2	304 $MW_{gas\ phase} = 307.5$	$\underset{a\quad e}{CF_3}-\underset{b\quad b}{CF_2}-\underset{d}{CF_2}-O-\underset{c}{CF_2}-CF_3,$		
			$a = 84.9$	28	3
			$b = 87.5$	18	2
			$c = 90.7$	27	3
			$d = 91.7$	15	2
			$e = 133.2$	16	2
351($C_6F_{13}O_2$), P–F	131 – 23, 78 – 8	370 $MW_{gas\ phase} = 374$	$\underset{f}{CF_3}-O-\underset{\underset{\underset{a}{CF_3}}{\mid}}{\underset{e}{CF}}-\underset{b}{CF_2}-O-\underset{d}{CF_2}-\underset{c}{CF_3}$		
			$a = 83.4$	11	3
			$b = 86.0$	10	2
			$c = 90.6$	10	3
			$d = 91.7$	7	2
			$e = 149.0$	Trace	1
			$f = 57.2$ d of quart. ($J = 15, 4$)	10	3

451($C_8F_{17}O$), P–F   78 – 38

470

$MW_{gas\ phase} = 466$

$$CF_3-CF_2-CF_2-O-\overset{\overset{h}{\underset{a}{|}}}{\underset{CF_3}{C}F}-CF_2-O-CF_2-CF_3$$

$a = 83.4$	58	10
$b = 84.4$	58	10
$c = 85.0$	58	10
$d = 86.1$	58	10
$e = 90.6$	15	3
$f = 91.7$	10	2
$g = 132.8$	10	2
$h = 148.0$	4	1

617($C_{11}F_{23}O_3$) P–F   30 – 35

636

$$CF_3-CF_2-CF_2-O-(\overset{\overset{h}{\underset{a}{|}}}{\underset{CF_3}{C}F}-CF_2-O)_2-CF_2-CF_3$$

$a = 83.2$	63	15
$b = 84.3$	63	15
$c = 84.9$	63	15
$d = 86.0$	63	15
$e = 90.5$	11	3
$f = 91.4$	7	2
$g = 132.7$	7	2
$h = 147.9$	6	2

183

TABLE VI

^{19}F NMR Data and Structures of Perfluoroethers from the Fluorination of Hexafluoroacetone–Ethylene Copolymer

Compound	Glc retention time, min	Highest m/e in mass spec.	Assigned ^{19}F NMR data, ppm	Relative intensities Obs.	Relative intensities Theor.
$(CF_3)_2CFOCF_2CF_2CF(CF_3)_2$ $\quad b \quad\; e \quad\; c \quad\; d \; f \quad a$	30	$435(C_8F_{17}O)$, P–F	$a = 75.2$ (octet) $(J = 8.5)$	16	6
			$b = 83.4$ (t) $(J = 5)$	16	6
			$c = 82.5$ (m)	6	2
			$d = 120.1$ (septet) $(J = 11.5)$	5	2
			$e = 147.0$ (t) $(J = 22)$	3	1
			$f = 187.9$ (m)	2.5	1
$(CF_3)_3COCF_2CF_2CF(CF_3)_2$ $\quad a \qquad\; c \quad\; d \quad e \;\; b$	56	$485(C_9F_{19}O)$, P–F	$a = 72.9$ (t) $(J = 9.5)$	165	9
			$b = 75.2$ (octet) $(J = 5.8)$	106	6
			$c = 81.8$ (m)	34	2
			$d = 119.7$ (septet) $(J = 11.5)$	34	2
			$e = 188.0$ (m)	17	1
$(CF_3)_2CFOCF_2CF_2C(CF_3)_2OCF_2CF_3$ $\quad b \quad\; g \quad\; d \;\; f \;\; a \qquad\; e \quad\; c$	64	$551(C_{10}F_{21}O_2)$, P–F	$a = 72.3$ (m)	10	6
			$b = 84.0$ (t) $(J = 5)$	9	6
			$c = 88.1$ (m)	4	3
			$d = 73.6$ (m)	3	2
			$e = 89.0$ (m)	2.5	2
			$f = 118.9$ (m)	3	2
			$g = 146.5$ (t) $(J = 20)$	1.5	1

(CF$_3$)$_2$CFOCF$_2$CF$_2$C(CF$_3$)$_2$OCF$_2$CF$_2$CF(CF$_3$)$_2$ 92 701(C$_{13}$F$_{27}$O$_2$), P–F
c h   d f   a   e   g i b

a = 69.7 (m)	7.5	6
b = 74.8 (octet) (J = 5.8)	8	6
c = 82.9 (t) (J = 5)	7	6
d = 71.9 (m)	3	2
e = 80.3 (m)	3	2
f = 116.3 (m)	3	2
g = 118.7 (septet) (J = 11.5)	2.5	2
h = 147.0 (t) (J = 20)	1	1
i = 187.7 (m)	1.5	1

(CF$_3$)$_3$COCF$_2$CF$_2$C(CF$_3$)$_2$OCF$_2$CF$_2$CF(CF$_3$)$_2$ 104 751(C$_{14}$F$_{29}$O$_2$), P–F
b d f   a   e   g h c

a = 69.8 (m)	24	6
b = 72.3 (t) (J = 9.5)	35	9
c = 74.8 (octet) (J = 5.8)	25	6
d = 79.3 (m)	7	2
e = 80.8 (m)	7.5	2
f = 115.7 (m)	7.5	2
g = 118.8 (septet) (J = 11.5)	8	2
h = 137.8 (m)	4	1

A series of rather unusual new perfluoro polyethers (42) has been prepared by the reaction of fluorine with the polyether:

$$\left[CH_2CH_2-O-\underset{\underset{CF_3}{|}}{\overset{\overset{CF_3}{|}}{C}}- \right]_n \xrightarrow[\substack{1.\ \text{ambient} \\ 2.\ \Delta}]{F_2} \left[-CF_2-CF_2O-\underset{\underset{CF_3}{|}}{\overset{\overset{CF_3}{|}}{C}} \right]_n$$

The precursor polymer was prepared by Tabata et al. (43) using the radiation-induced low-temperature copolymerization of ethylene and hexafluoroacetone:

$$CH_2{=}CH_2 + CF_3-\overset{\overset{O}{\|}}{C}-CF_3 \xrightarrow[T < -9^\circ C]{^{60}Co\ \gamma\text{-rays}} [CH_2CH_2OC(CF_3)_2]_n$$

The second fluorination step is extremely temperature sensitive in the 65–70°C range, with even 2°C making substantial differences in the range of polyethers produced. A number of interesting new perfluoro polyethers were obtained as may be seen in Table VI.

The high-molecular-weight perfluoropolyethers obtained in the first step have excellent thermal stability (TGA initial decomposition >370°C) and chemical resistance and are true fluorocarbon elastomers that are, in contrast to polytetrafluoroethylene, flexible. Other methods for the synthesis of perfluoropolyethers and functionalized perfluoropolyethers are under development in our laboratories.

In the next few years it will become clear that traditional problems with the physical properties, such as strength and durability of polymers produced by direct fluorination, are being solved largely by specific design of precursor hydrocarbon polymer morphology for the fluorination process. Polymers produced by direct fluorination have previously been satisfactory or outstanding with regard to their thermal and chemical properties. Stimulated by new efforts this field will expand rapidly.

3. Functionalization of Hydrocarbon Polymers by "Oxyfluorination" Techniques

Thermally and chemically resistant fluorocarbon polymers that support isolated reactive acid fluoride groups have been recently produced and may find utility as a catalyst support, as a template for biochemical or other synthetic applications, and as membranes and separators for chloro alkali cells and batteries by a process called "oxyfluorination" (44). In this reaction, existing hydrocarbon polymers are simultaneously fluorinated and functionalized using various mixtures of fluorine and oxygen to produce a "polytetrafluoroethylene-like" backbone with varying degrees of functionalization, ranging from to acid fluoride to monomer groups.

It can be seen from the F_2/O_2 ratios and the COF content (Fig. 8) that the degree of functionalization is directly related to the F_2/O_2 ratio.

The oxyfluorinated polymers would be expected to react with a variety of organic, inorganic, and organometallic compounds. Derivatives of each functionalized material have been prepared with CH_3OH, C_2H_5OH, n-BuOH, and aniline. The efficiency of the derivativization process ranged from 44 to 98%.

It is worthwhile to note that oxyfluorination of linear polyethylene fails to produce significant functionalization. This implies that acid fluoride groups are produced primarily by oxidation of pendant methyl or other alkyl groups in low-density polyethylene or in polypropylene rather than from cleavage of carbon–carbon bonds. Up to 60% of the pendant methyl groups in polypropylene are converted to acid fluoride sites. Although the degree of functionalization is about the same in the low-density polyethylene, the percentage of available pendant alkyl groups converted to acid fluorides is larger. It is probable that the amorphous nature of low-density polyethylene permits a higher fluorine and oxygen diffusion rate than the more crystalline polypropylene. Steric factors may also favor more complete functionalization of polyethylene.

The fact that functionalization of polymers and small molecules is observed to occur predominately on terminal (methyl) carbon atoms does not imply that the oxyfluorination reaction is truly selective. Although the reaction mechanism has not been studied in detail, it is undoubtedly a free-radical process. Molecular oxygen reacts spontaneously with the fluorocarbon–hydrogen radicals generated by fluorine during the fluorination process. Acid fluorides are retained on terminal carbon atoms because they are stable in 1 atm of elemental fluorine. Hypofluorites, which may be short-lived intermediates of oxygen reactions with methylene radical sites along the carbon chain, are not observed in the functionalized polymers. It is probable that, if they are intermediates, they are cleaved and removed by the excess elemental fluorine.

It should be possible, using this technique to prepare a number of different functionalized fluorocarbon polymers with varying physical, thermal, and chemical properties that should complement or compete in applications established for the sulfonic acid functionalized fluorocarbon polymer "Nafion", which has been reported previously (45).

There is considerable promise for the synthesis of useful functionalized fluorocarbon materials using other direct fluorination techniques (46), such as the fluorination of pendant polyesters and polymers containing pendant acyl fluoride units.

One of the more important and promising groups is comprised of the pendant polyesters, which can be fluorinated as in example a and preserve up to 80% of the ester linkages. Of ester linkages that do decompose, most end up as acid fluorides directly. The perfluoroesters are extremely hydrolytically unstable and are easily hydrolyzed to give the perfluoro acids. Specific examples of

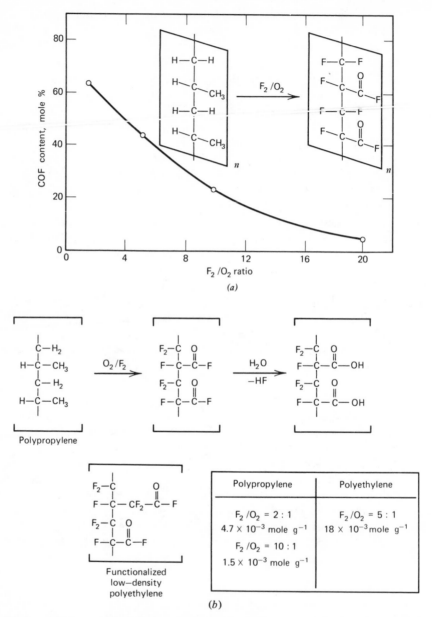

Fig. 8. (a) The relationship between oxygen introduced and the COF content (polypropylene). (b) Functionalization of polymer surfaces.

interest include methyl methacrylate:

$$
\begin{bmatrix} R_x-CH-R_y \\ \ \ \ | \\ C=O \\ | \\ O \\ | \\ C_2H_5 \end{bmatrix}_n \xrightarrow{F_2} \begin{bmatrix} FR_x-CF-FR_y \\ \ \ \ | \\ C=O \\ | \\ O \\ | \\ C_2F_5 \end{bmatrix}_n \xrightarrow{H_2O} \begin{bmatrix} FR_x-CF-FR_y \\ \ \ \ | \\ C=O \\ | \\ O \\ | \\ H \end{bmatrix}_n \quad \text{(a)}
$$

$$
(-CH_2-\overset{\overset{\displaystyle O}{\|}}{\underset{\underset{\displaystyle CH_3}{|}}{\overset{\displaystyle C-OCH_3}{\underset{\displaystyle |}{CH}}}})_n + F_2 \longrightarrow (CF_2-\overset{\overset{\displaystyle O}{\|}}{\underset{\underset{\displaystyle CF_3}{|}}{\overset{\displaystyle C-OCF_3}{\underset{\displaystyle |}{CF}}}})_n \xrightarrow{H_2O} (CF_2-\overset{\overset{\displaystyle O}{\|}}{\underset{\underset{\displaystyle CF_3}{|}}{\overset{\displaystyle C-OH}{\underset{\displaystyle |}{CF}}}})_n + CF_3\overset{\overset{\displaystyle O}{\|}}{C}OH \quad \text{(b)}
$$

It is also possible to fluorinate hydrocarbon polymers containing acid groups as a means toward functionalization. We have found that the most successful route is first to convert the acid to an acid fluoride and then to conduct the fluorination as shown in example (c):

$$
\begin{bmatrix} (CH_2)_n-CH \\ \ \ \ \ \ | \\ C=O \\ | \\ O \\ | \\ H \end{bmatrix}_m \xrightarrow{HF} \begin{bmatrix} (CH_2)_n-\overset{\overset{\displaystyle H}{|}}{\underset{\underset{\displaystyle F}{|}}{\overset{\displaystyle |}{C}-}} \\ C=O \end{bmatrix}_m \xrightarrow{F_2} \begin{bmatrix} -(CF_2)_n-\overset{\overset{\displaystyle F}{|}}{\underset{\underset{\displaystyle F}{|}}{\overset{\displaystyle |}{C}-}} \\ C=O \end{bmatrix}_m \quad \text{(c)}
$$

Several major chemical firms have active research programs designed to exploit the fluorination of polymeric materials. There are now at least five cases of special applications that have been successful in pilot plant operations and are being converted to production facilities.

B. The Synthesis of Inorganic Compounds by Direct Fluorination

The direct fluorination of inorganic compounds is an area still in its early stages of development. Enough work has been completed to establish that direct fluorination techniques will add a new dimension to inorganic fluorine chemistry.

1. Inorganic Hydrides as Precursors of New Fluorine Compounds

The direct fluorination of almost all categories of inorganic hydrides leads to fluorine-containing analogues of the hydride precursors. The fluorination of simple inorganic hydrides (47, 48) is an elementary example of such syntheses (see Fig. 9).

Reaction	Temp	Approx. yield, %
$NaBH_4 + 4F_2 \rightarrow NaBF_4 + 4HF$	20°C	~100
$KBH_4 + 4F_2 \rightarrow KBF_4 + 4HF$	20°C	~100
$LiBH_4 + 4F_2 \rightarrow LiBF_4 + 4HF$	20°C	~100
$LiAlH_4 + 4F_2 \rightarrow LiAlF_4 + 4HF$	20°C	~100
$Na_3AlH_6 + 6F_2 \rightarrow Na_3AlF_6 + 6HF$	20°C	~100
$Li_3AlH_6 + 6F_2 \rightarrow Li_3AlF_6 + 6HF$	20°C	~100

Fig. 9.   Direct fluorination of metal hydrides. Reactions are carried out at 20°C and yield is ~100% in all cases.

The only new compound produced in these studies was $LiAlF_4$; however, this work is significant because control was achieved even with the highly exothermic reactions of lithium aluminum hydride, alkali metal borohydrides, and alkali metal hexahydroaluminated with elemental fluorine. The reactions of alkali metal amides with fluorine have also been successfully controlled. These reactions provide extremely clean routes to the fluoride analogues.

Direct fluorination has also been successful in the preparation of perfluorocarboranes from the decarborane systems (49) and lower molecular weight carboranes (50).

The synthesis of $B_{10}C_2F_{12}$ from 1,2-decarborane:

$$1,2\text{-}B_{10}H_{10}C_2H_2 \xrightarrow[\substack{\text{room temp.} \\ \text{4-8 hr}}]{F_2} 1,2\text{-}B_{10}F_{10}C_2H_2$$

$$1,2\text{-}B_{10}F_{10}C_2H_2 \xrightarrow[\substack{\text{pressure + 100°C} \\ \text{1-5 days}}]{} 1,2\text{-}B_{10}F_{10}C_2F_2$$

was the first observed case of the relatively inert nature of certain "acidic" protons on highly fluorine substituted cages. The replacement of the hydrogen on boron proceeded smoothly at room temperature, but the replacement of protons on carbon is extremely difficult. This appears to be an electronic effect and is also a factor in the replacement of the last proton of hydrocarbon cages such as adamantane, as is seen later in the section on organic synthesis. The cages become highly substituted with the very electronegative fluorine and the last protons develop a partial positive charge. It is known that positively charged species are attacked slowly or not at all by elemental fluorine. It is, for example, extremely difficult to promote any reaction between the quaternary ammonium compounds $R_4N^+X^-$, such as $(CH_3)_4 N^+Cl$, and elemental fluorine.

The reaction of fluorine with lower molecular weight carboranes is more difficult to control but results in a number of new fluorine compounds, such as $5,6\text{-}B_3H_5C_2F_2$ and $1,3\text{-}B_3H_5C_2F_2$, with all structural features intact. Among the interesting by-products found was $H_2C(BF_2)_2$ (50). Recent studies producing fluorinated borazines, phosphorous–nitrogen ring systems, and new boron fluorine species from boron hydrides are yet to be reported.

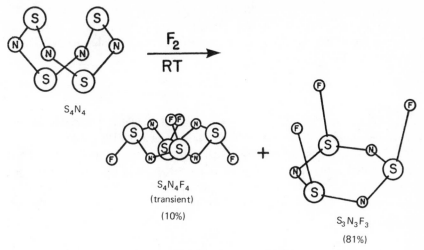

S_4N_4

$S_4N_4F_4$
(transient)
(10%)

$S_3N_3F_3$
(81%)

Fig. 10.   Direct fluorination of S_4N_4.

Studies on sulfur nitrogen systems such as S_4N_4 (51) have produced sulfur–nitrogen–fluorine ring systems in high yield. The shock-sensitive material, S_4N_4, gives an 81% yield of $S_3N_3F_3$ and up to 10% of $S_4N_4F_4$ with carefully controlled fluorination (see Fig. 10). More conventional approaches have resulted in extensive fragmentation of S_4N_4.

The reactions of elemental fluorine with inorganic compounds are exothermic and often have little or no reaction associated activation energies. Most often the major synthetic problem is kinetic and thermodynamic control of these vigorous reactions. It is therefore a very unusual synthetic situation when reactions must be activated by methods such as high temperatures, plasmas, or photochemical means. Examples of such cases are the synthesis of $NO^+BF_4^-$ by the photochemically activated reaction of fluorine and oxygen with boron nitride (52) and the plasma-activated synthesis of $(CF_{1.12})_n$ from graphite (53).

Although the literature often indicates catastrophic reactions of fluorine with oxides, nitrides, carbides, silicides, borides and the like, many such species may be reacted under controlled conditions to produce oxyfluorides, nitrofluorides such as TiNF, ZrNF, HFNF, ThNF, and UNF, and other interesting ternary systems.

2. Synthesis of Perfluoroalkyls by the Reaction of Elemental Fluorine with Metal and Metaloid Alkyls

One of the most interesting and, in fact, most surprising new applications of direct fluorination has been its recent use to convert metal alkyls $M(CH_3)_n$ to perfluoroalkyls $M(CF_3)_n$ without fracture of the carbon–metal or carbon–

metaloid bonds (54). The most interesting reactions reported to date involve the reactions of tetramethylgermanium with elemental fluorine at $100°C$ in the cryogenic zoned reactor (55, 56). Under the appropriate conditions $Ge(CF_3)_4$ is obtained in 63% yield on a 3 g scale.

$$Ge(CH_3)_4 + F_2/He \xrightarrow[-100°C]{} Ge(CF_3)_4 + \text{polyfluorotetramethylgermanes} + \text{fluorocarbons}$$
$$(63\%)$$

Under milder fluorination conditions a range of partially fluorinated methyl-germanium compounds is produced (see Tables VIIa and VIIb).

The reaction of tetramethylsilane with fluorine led to the isolation of several partially fluorine substituted tetramethylsilanes (see Table VIII) and preservation of over 80% of the silicon—carbon bonds in the initial tetramethyl-silane reactant. The stability of many of the partially fluorinated germanes and

TABLE VIIa

Fluorine NMR Spectra of Polyfluorotetramethylgermanes[a]

Compound	CF_3[b]	J_{FF}	CHF_2[c]	J_{HF}	J_{FF}	CH_2F[d]	J_{HF}	J_{FF}
$Ge(CF_3)_4$	−27.0							
$Ge(CF_3)_3(C_2F_5)$	−27.7	2.7[c]						
$Ge(CF_3)_3(CF_2H)$	−27.2	3.0[d]	49.0	45.5	3.1[f]			
$Ge(CF_3)_3(CFH_2)$	−25.8							
$Ge(CF_3)_2(CF_2H)_2$	−27.6	3.2[g]	49.4	46.0	3.1[h]			
$Ge(CF_3)_2(CF_2H)(CFH_2)$	−26.3	3.3[i]	50.6	45.5	3.0[j]	193.2	46.5	3.3[k]
$Ge(CF_3)(CF_2H)_3$	−27.9	3.2[h]	49.7	46.0	3.1[i]			
$Ge(CF_3)_2(CFH_2)_2$	−24.8	3.4[b]				193.0	47.0	3.5[l]
$Ge(CF_3)(CF_2H)_2(CFH_2)$	−26.8	3.2[m]	50.5	45.6	3.0[g]	193.0	46.0	2.9[j]
$Ge(CF_3)(CF_2H)_2(CH_3)$	−23.5	3.4[g]	53.0	46.5	3.2[i]			
$Ge(CF_3)(CF_2H)(CFH_2)_2$	−25.2	3.2[g]	51.8	45.5	3.0[m]	192.0	46.0	2.7[m]
$Ge(CF_2H)_3(CFH_2)$			50.5	46.2	2.2[c]	192.9	46.0	2.4[l]
$Ge(CF_2H)_2(CFH_2)_2$			51.4	46.0	2.5[g]	192.5	46.6	2.5[g]
$Ge(CF_3)(CF_2H)(CFH_2)(CH_3)$	−22.3	3.4[i]	53.7	46.0	3.0[g]	192.5	46.0	3.0
$Ge(CF_2H)(CFH_2)_3$			52.0	46.0	2.1[i]	191.9	46.6	1.8[d]
$Ge(CF_2H)(CFH_2)_2(CH_3)$			53.9	46.2	2.1[d]	192.5	46.5	2.0[d]
$Ge(CFH_2)_4$						191.4	46.7	
$Ge(CF_3)_3(OH)$	−21.6							

[a] Shifts in ppm from external TFA.
  + upfield from TFA.
  Coupling constants in hertz.
[b] Singlet.
[c] Doublet.
[d] Triplet.
[e] Basic quartet. C_2F_5 group:
  CF_3, 6.76 multiplet CF_2, 38.4, $J_{FF} = 3.3$, septet.
[f] Triplet.
  4 of 10 lines.

[g] Pentet.
[h] Septet.
[i] Quartet.
[j] 6 of 8 lines.
[k] 7 of 9 lines.
[l] 5 of 7 lines.
[m] Sextet.

TABLE VIIb
Proton NMR Spectra of Polyfluorotetramethylgermanes

Compound	CH_3, ppm^{a,b}	CH_2F, ppm^{b,c}	J_{HF}, Hz	CHF_2, ppm^{b,d}	J_{HF}, Hz
$Ge(CF_3)_3(CF_2H)$				6.10	45.0
$Ge(CF_3)_3(CFH_2)$					
$Ge(CF_3)_2(CF_2H)_2$				6.23	45.5
$Ge(CF_3)_2(CF_2H)(CFH_2)$		4.98	46.5	6.24	45.7
$Ge(CF_3)(CF_2H)_3$				6.25	45.5
$Ge(CF_3)_2(CFH_2)_2$					
$Ge(CF_3)(CF_2H)_2(CFH_2)$		4.89	46.0	6.15	45.5
$Ge(CF_3)(CF_2H)_2(CH_3)$	0.51			6.10	45.6
$Ge(CF_3)(CF_2H)(CFH_2)_2$		4.90	46.0	6.25	45.6
$Ge(CF_2H)_3(CFH_2)$		5.02	45.7	6.28	45.6
$Ge(CF_2H)_2(CFH_2)_2$		4.97	46.0	6.26	45.2
$Ge(CF_3)(CF_2H)(CFH_2)(CH_3)$	0.47	4.79	46.0	6.08	46.0
$Ge(CF_2H)(CFH_2)_3$		4.97	46.5	6.29	46.0
$Ge(CF_2H)(CFH_2)_2(CH_3)$	0.34	4.78	46.0	6.10	46.0
$Ge(CFH_2)_4$		4.87	47.0		
$Ge(CF_3)_3(OH)^e$					

^{a} Singlet.
^{b} Downfield from TMS (external).
^{c} Doublet.
^{d} Triplet.
^{e} OH (2.43), singlet.

silanes (some are stable to over 100°C) is very surprising, for the possibility of hydrogen fluoride elimination is obvious. Indeed, before the first reported synthesis of $Ge(CF_3)_4$ and $Sn(CF_3)_4$ (57) in 1975, many authors had given detailed reasons for the instability of these species as a rationale for the failure of conventional syntheses to produce them.

TABLE VIII
Melting Points of Polyfluorotetramethylsilanes

Compound	Melting point, °C
$Si(CH_3)_3(CH_2F)$	-86.5 to -85.0
$Si(CH_3)(CH_2F)_3$	-89.5 to -84.5
$Si(CH_3)(CH_2F)_2(CHF_2)$	-63.5 to -62.0
$Si(CH_2F)_4$	-18.0 to -16.6
$Si(CH_3)(CH_2F)(CHF_2)_2$	-58.7 to -56.8
$Si(CH_3)(CH_2F)_2(CF_3)$	-144 to -142.7
$Si(CH_2F)_3(CHF_2)$	-54.0 to -53.2
$Si(CH_3)(CH_2F)(CHF_2)(CF_3)$	-144 to -138
$Si(CH_2F)_2(CHF_2)_2$	-68.2 to -66.8
$Si(CH_2F)(CHF_2)_3$	-72.6 to -71.0
$Si(CH_3)_4$	-91.1

The reaction of fluorine with dimethylmercury:

$$Hg(CH_3)_2 \xrightarrow[-110°C]{He/F_2} Hg(CF_3)_2$$
$$(6.8\%)$$

to produce bistrifluoromethylmercury has been reported (54) along with a study of the reactions of fluorine with tetramethyltin (58).

That the carbon—metal or carbon—metalloid bonds are preserved at all in these reactions is quite surprising. With tetramethylgermanes, for example, this free radical reaction must be a 24 step process. The success in preserving carbon—germanium bonds must arise from very rapid molecular vibrational, rotational, and translational relaxation processes occurring on the cryogenically cooled surfaces such that the energy from the extremely exothermic reaction is smoothly dissipated.

3. Selective Oxidations.

Another development that promises applications to several areas of the Periodic Table is forecast by a study of the reactions of fluorine with carbon disulfide (59):

$$CS_2 \xrightarrow[-120°C]{F_2/He} SF_3CF_2SF_3 \qquad SF_3CF_2SF_3 \xrightarrow[-80°C]{F_2/He} SF_5CF_2SF_5$$
$$(60\%) \qquad\qquad\qquad (25\%)$$

By careful control of the conditions of the reaction one can obtain preferentially oxidation of sulfur to its four coordinate oxidation state and by using a second set of conditions one can obtain the oxidation to the six-coordinate sulfur species. The dynamic NMR study of $SF_3CF_2SF_3$ is currently in progress in collaboration with A. H. Cowley (60). This differentiation of oxidation states is extremely promising, and work in progress shows that this is not at all an isolated situation. Mercaptans and other organosulfur compounds definitely exhibit this capacity in fluorine reactions.

A somewhat related breakthrough in elemental fluorine chemistry by Bastian and Ruppert, which is extremely exciting, involves the selective oxidation of phosphorous (61), arsenic (62), antimony (62), and bismuth (62) without fluorine substitutions occurring on the phosphorane, arsane, and so on.

These reactions, which are conducted in CHF_3 at low temperatures, appear to be very general in nature. Examples of the capabilities are provided by the following reactions:

$$Ph_2P-(CH_2)_n-PPh_2 + 2F_2 \xrightarrow{CFCl_3} Ph_2\overset{\overset{\displaystyle F}{|}}{\underset{\underset{\displaystyle F}{|}}{P}}-(CH_2-)_n-\overset{\overset{\displaystyle F}{|}}{\underset{\underset{\displaystyle F}{|}}{P}}Ph_2$$

$n = 1-4$

$$Ph_2E-R + F_2 \xrightarrow{CFCl_3} Ph_2\overset{\overset{\displaystyle F}{|}}{\underset{\underset{\displaystyle F}{|}}{E}}-R$$

E = As; R = Me, Ph
E = Sb; R = Me, Ph
E = Bi; R = Ph

C. Applications of Direct Fluorination in Organic Chemistry

1. Direct Fluorination of Hydrocarbons

The fluorination, even of hydrocarbons, offers a new source of fluorine compounds. Some of the initial work on this fluorination process involved polynuclear hydrocarbons. White crystalline perfluoro alicyclic compounds (see Fig. 11) were obtained in high yield from the fluorination of anthracene, tetracene, pentacene, coronene, decacyclene, and ovalene (63). Yields for these syntheses ranged from 60 to over 90%. These compounds have structures very similar to that of poly(carbon monofluoride) (64) (see Fig. 12), an extremely stable thermogravimetric analysis shows an initial decomposition point of 670°C.), white polymeric product of the high-temperature reaction of fluorine with graphite. Polycarbonmonofluoride is now commercially manufactured for use as a high-temperature solid lubricant and as a cathode material in high-energy lithium batteries. In fact, our initial work on direct fluorination was begun to obtain lower molecular weight analogues of this useful polymeric material of approximate stoichiometry $CF_{1.12}$.

Fluorination of normal hydrocarbons is not difficult with the La-Mar fluorination process; however, fluorination studies also have been successful with structurally unusual hydrocarbon compounds (65–67a) (see Fig. 13). These studies were undertaken to establish that direct fluorination was useful even with some of the most sensitive hydrocarbon structures. While the initial studies of the successful direct fluorination of these species often resulted in yields as low as 10%, these same experiments, after additional technical developments in our laboratory, routinely give yields of 70–95%. Such syntheses have often been repeated in our laboratories to satisfy scientific needs for such compounds in other laboratories. The sterically crowded fluorocarbon compounds prepared in

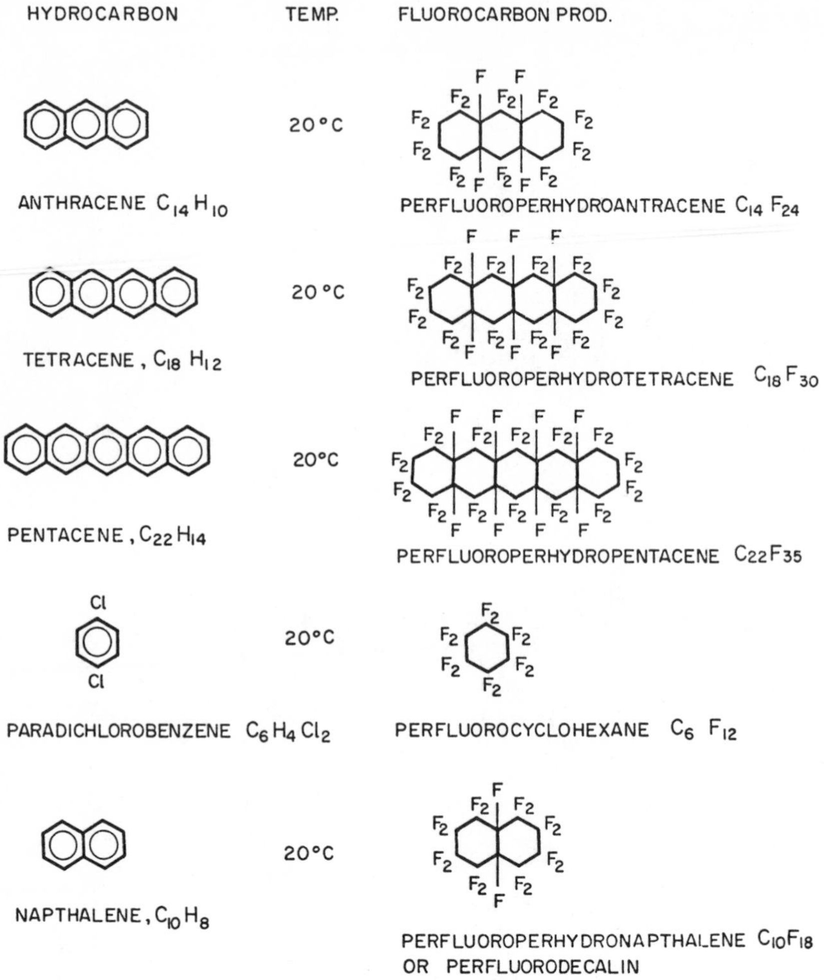

Fig. 11.   Fluorocarbon products.

this study have unusually high volatility for their molecular weight, as well as other unusual properties. Perfluoroneopentane (65), $C(CF_3)_4$, is of considerable interest because it is a nearly spherical molecule with its surface crowded with nonbonding fluorine electrons allowing very little intermolecular interaction. It melts at $73°C$ but is so volatile, as is perfluorohexamethylethene, $C_2(CF_3)_6$ (mp $108°C$), that it must be sealed in a glass container at room temperature to prevent immediate sublimation. Even pentadecafluoroadamantane and perfluoroadamantane have been prepared (66), and the eight-membered ring in cyclooctane was preserved in at least 30% yield.

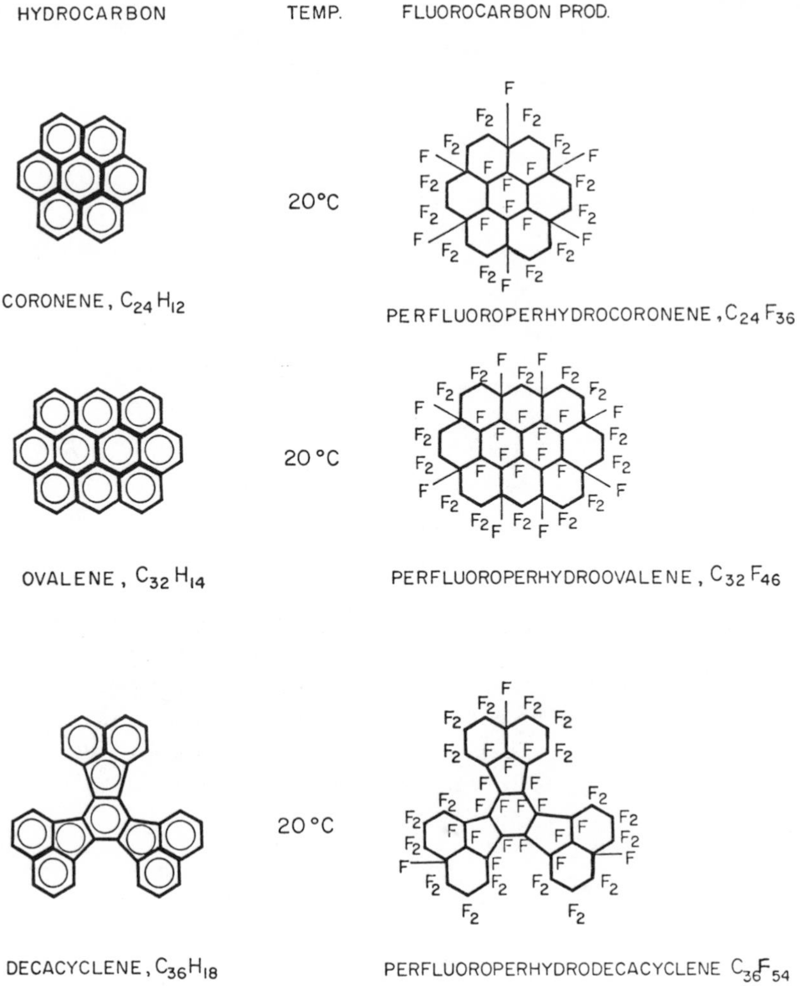

HYDROCARBON   TEMP.   FLUOROCARBON PROD.

CORONENE, $C_{24}H_{12}$ 20°C   PERFLUOROPERHYDROCORONENE, $C_{24}F_{36}$

OVALENE, $C_{32}H_{14}$ 20 °C   PERFLUOROPERHYDROOVALENE, $C_{32}F_{46}$

DECACYCLENE, $C_{36}H_{18}$ 20 °C   PERFLUOROPERHYDRODECACYCLENE $C_{36}F_{54}$

Fig. 11.   *continued*

Other recent work in the perfluoroadamantane area (67b) has been successful in producing 1,3,5,7-perfluorotetramethyladamantane, 1,3-perfluordimethyladamantane, and perfluoro-1,-difluoroamino adamantane (see Fig. 14).

The adamantane work was undertaken as a contribution to the artificial fluorocarbon blood substitute area. The use of fluorocarbons as artificial blood and as oxygen carriers in other applications is an area that is being developed by Leland Clark, Jr., of Cincinnati Children's Hospital and Paul Geyer of Peter Bent Brigham Hospital in Boston. The reader who is unfamiliar with this area of research will find reviews on the subject to be extremely interesting (68). The

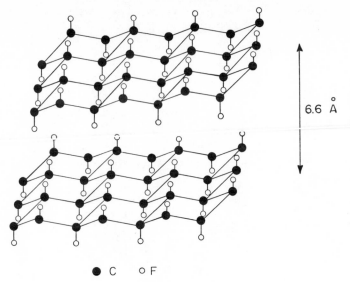

● C   ○ F

Fig. 12.   Rüdorff $(CF_x)_n$ structure.

blood of most mammals, including monkeys, has been completely replaced with oxygen-carrying fluorocarbon emulsions that satisfactorily support life over several-day periods. Other striking experiments include the ability of a mouse or rat to obtain oxygen from the fluorocarbon and live completely submerged in the fluorocarbon liquid for several days. These emulsions have also been used for artificial organ perfusion. The suitability of perfluoroadamantanes for such applications is under investigation by Clark.

Another series of compounds, the fluorine-substituted 2,2,4,4-tetra-methylpentanes, have proven to be a good artificial blood candidate (69). The compounds are prepared by the direct fluorination of the hydrocarbon precursor (70a) (see Fig. 15). The overall yield of the reaction can be seen from the figure to be in excess of 99%. The central protons are retained preferentially during the fluorination, a result that is attributed to steric rather than electronic factors (70a).

More recent work has been directed toward the synthesis of the next higher homologue, perfluoro-2,2,5,5-tetramethylhexane (see Fig. 16).

Under appropriate conditions, the perfluorocarbon $(CF_3)_3CCF_2CF_2$-$C(CF_3)_3$ is obtained in 89% yield (70b). As may be seen from the reaction scheme, a number of partially fluorinated products are also isolable and it is clear that the protons are readily accessible to fluorine. The physical properties of highly branched fluorocarbons (see Table IX) are most unusual.

Three recent papers directed toward the reaction of elemental fluorine with hexafluorobenzene (71), decafluorohexane (72), and pentafluoro-

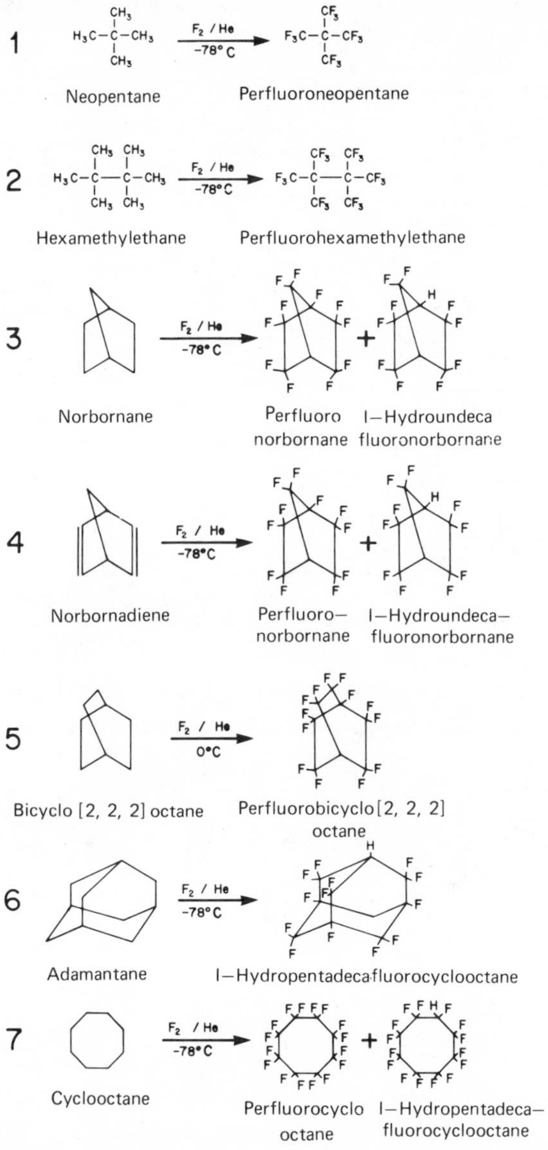

Fig. 13.   Synthesis of fluorocarbons of unusual structures.

pyridine (73) illustrate that even contemporary direct fluorination experiments are not successful in every laboratory, long after detailed experimental procedures have appeared in the literature. A clue to this problem lies in the reference sections of the manuscripts.

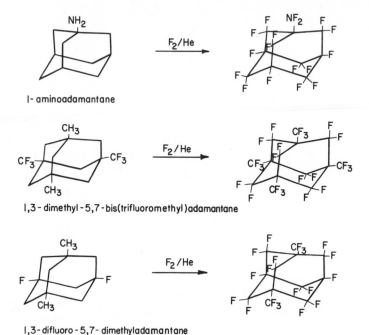

I- aminoadamantane

1,3- dimethyl-5,7-bis(trifluoromethyl)adamantane

1,3- difluoro-5,7- dimethyladamantane

Fig. 14.   Direct fluorination of substituted adamantane compounds.

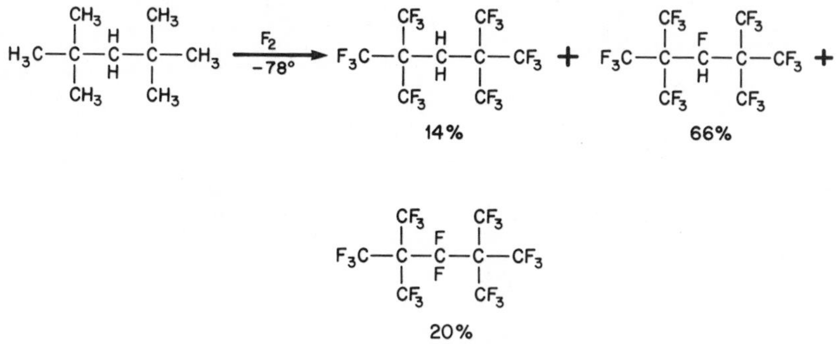

$$H_3C-\overset{\overset{\displaystyle CH_3}{|}}{\underset{\underset{\displaystyle CH_3}{|}}{C}}-\overset{\overset{\displaystyle H}{|}}{\underset{\underset{\displaystyle H}{|}}{C}}-\overset{\overset{\displaystyle CH_3}{|}}{\underset{\underset{\displaystyle CH_3}{|}}{C}}-CH_3 \xrightarrow[-78°]{F_2} F_3C-\overset{\overset{\displaystyle CF_3}{|}}{\underset{\underset{\displaystyle CF_3}{|}}{C}}-\overset{\overset{\displaystyle H}{|}}{\underset{\underset{\displaystyle H}{|}}{C}}-\overset{\overset{\displaystyle CF_3}{|}}{\underset{\underset{\displaystyle CF_3}{|}}{C}}-CF_3 \ + \ F_3C-\overset{\overset{\displaystyle CF_3}{|}}{\underset{\underset{\displaystyle CF_3}{|}}{C}}-\overset{\overset{\displaystyle F}{|}}{\underset{\underset{\displaystyle H}{|}}{C}}-\overset{\overset{\displaystyle CF_3}{|}}{\underset{\underset{\displaystyle CF_3}{|}}{C}}-CF_3 \ +$$

14%   66%

$$F_3C-\overset{\overset{\displaystyle CF_3}{|}}{\underset{\underset{\displaystyle CF_3}{|}}{C}}-\overset{\overset{\displaystyle F}{|}}{\underset{\underset{\displaystyle F}{|}}{C}}-\overset{\overset{\displaystyle CF_3}{|}}{\underset{\underset{\displaystyle CF_3}{|}}{C}}-CF_3$$

20%

Fig. 15.   Reaction scheme.

200

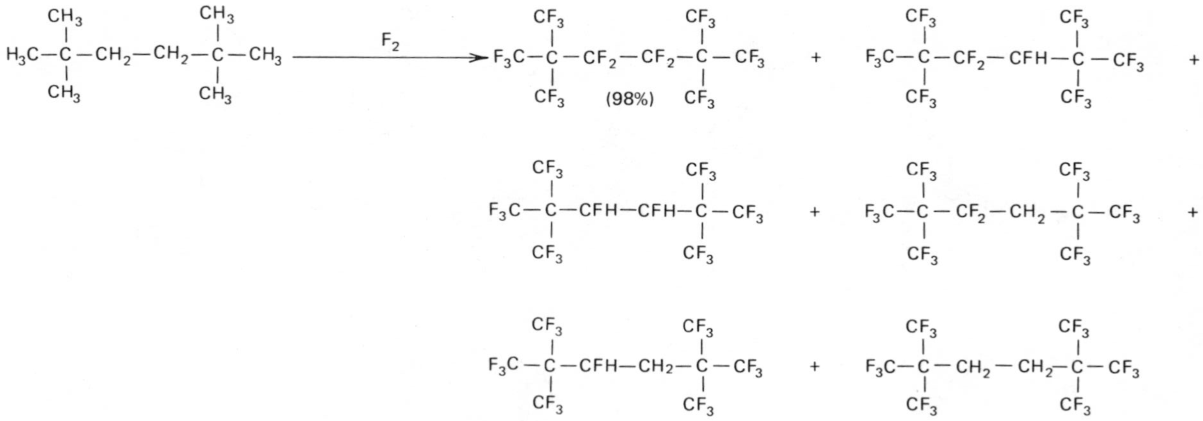

Fluorination of 2,2,5,5-tetramethylhexane reaction scheme (2,2,5,5-tetramethylhexane + F_2 →):

Starting material:
$$H_3C-C(CH_3)_2-CH_2-CH_2-C(CH_3)_2-CH_3$$

Products:

$$F_3C-C(CF_3)_2-CF_2-CF_2-C(CF_3)_2-CF_3 \quad (98\%)$$

$$F_3C-C(CF_3)_2-CFH-CFH-C(CF_3)_2-CF_3 \;+$$

$$F_3C-C(CF_3)_2-CFH-CH_2-C(CF_3)_2-CF_3 \;+$$

$$F_3C-C(CF_3)_2-CF_2-CFH-C(CF_3)_2-CF_3 \;+$$

$$F_3C-C(CF_3)_2-CF_2-CH_2-C(CF_3)_2-CF_3 \;+$$

$$F_3C-C(CF_3)_2-CH_2-CH_2-C(CF_3)_2-CF_3$$

Fig. 16. Fluorination of 2,2,5,5-tetramethylhexane.

TABLE IX

Physical properties

Compound	Melting point, °C	Boiling point, °C	Compound	Melting point °C[a]
$C(CH_3)_4$	-20	9.5	$C(CF_3)_4$	72.5 to 73
$(CH_3)_3CC(CH_3)_3$	100.7	106.5	$(CF_3)_3CC(CF_3)_3$	108[b]
$(CH_3)_3CCH_2C(CH_3)_3$	-67.0	122.3	$(CF_3)_3CCF_2(CF_3)_3$	-24 to -2
$(CH_3)_3CCH_2CH_2C(CH_3)_3$	-12.6	137	$(CF_3)_3CCF_2CF_2C(CF_3)_3$	59 to 60

[a] Sealed capillary.

[b] Sublimes, melting point not reported.

2. The Synthesis of Functionalized Fluorocarbons by Direct Fluorination

It was recently demonstrated that direct fluorination can make important contributions in the synthesis of functional oxygen-containing fluorocarbons (74–76). Early studies in our laboratory showed that gentle fluorination of sodium acetate produces trifluoroacetyl fluoride (77). It has been possible to prepare higher molecular weight perfluoroethers such as perfluoro-1,2-dimethoxyethane ("perfluoroglyme") and perfluorobis(2-methoxyethyl ether)("perfluorodiglyme") in yields as high as 60% from hydrocarbon starting materials (74). Perfluoro-1,4-dioxane (75) has been prepared in 40% yield from dioxane. The fluorination of acid fluorides, as shown in Fig. 17 has also been very successful. Even with crowded structures such as perfluoropivolyl fluoride,

$$(CF_3)_3\overset{\overset{\displaystyle O}{\|}}{C}F,$$ for which cobalt trifluoride and electrochemical-fluorination methods have not been successful, a yield of over 50% has been obtained. The direct fluorination of ethyl acetate (75), shown in Fig. 17, was the first example of conversion of a hydrocarbon ester to a perfluorocarbon ester by any fluorination technique.

The direct fluorination of dimethyl sulfone and dimethyl sulfate was also achieved recently (78). The sulfonyl groups and sulfate groups survive the reaction in yields that average 20–41%.

$$CH_3\overset{\overset{\displaystyle O}{\|}}{\underset{\underset{\displaystyle O}{\|}}{S}}CH_3 \xrightarrow[-100°C]{F_2/He} CF_3\overset{\overset{\displaystyle O}{\|}}{\underset{\underset{\displaystyle O}{\|}}{S}}CF_3 + CF_3\overset{\overset{\displaystyle O}{\|}}{\underset{\underset{\displaystyle O}{\|}}{S}}F$$

$$CH_3-O-\overset{\overset{\displaystyle O}{\|}}{\underset{\underset{\displaystyle O}{\|}}{S}}-O-CH_3 \xrightarrow[-100°C]{F_2/He} CF_3O-\overset{\overset{\displaystyle O}{\|}}{\underset{\underset{\displaystyle O}{\|}}{S}}-O-CF_3 + CF_3O\overset{\overset{\displaystyle O}{\|}}{\underset{\underset{\displaystyle O}{\|}}{S}}-F$$

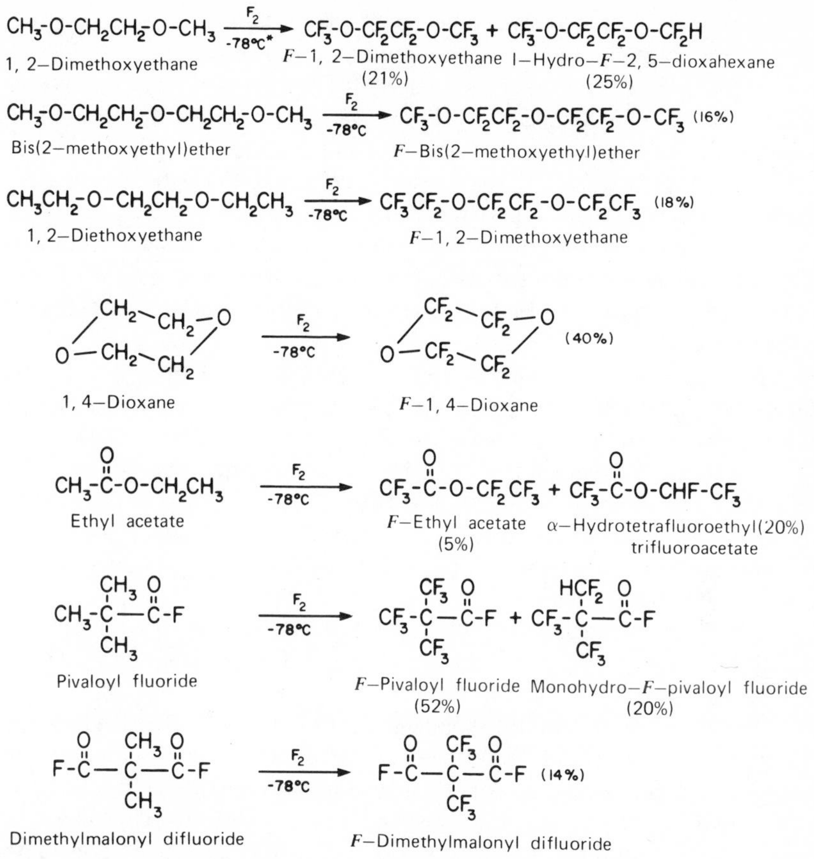

CH_3-O-CH_2CH_2-O-CH_3 $\xrightarrow[-78°C^*]{F_2}$ CF_3-O-CF_2CF_2-O-CF_3 + CF_3-O-CF_2CF_2-O-CF_2H

1, 2-Dimethoxyethane F-1, 2-Dimethoxyethane   1-Hydro-F-2, 5-dioxahexane
                                    (21%)   (25%)

CH_3-O-CH_2CH_2-O-CH_2CH_2-O-CH_3 $\xrightarrow[-78°C]{F_2}$ CF_3-O-CF_2CF_2-O-CF_2CF_2-O-CF_3 (16%)

Bis(2-methoxyethyl)ether F-Bis(2-methoxyethyl)ether

CH_3CH_2-O-CH_2CH_2-O-CH_2CH_3 $\xrightarrow[-78°C]{F_2}$ CF_3CF_2-O-CF_2CF_2-O-CF_2CF_3 (18%)

1, 2-Diethoxyethane F-1, 2-Dimethoxyethane

1, 4-Dioxane $\xrightarrow[-78°C]{F_2}$ F-1, 4-Dioxane (40%)

CH_3-$\overset{O}{\overset{\|}{C}}$-O-$CH_2CH_3$ $\xrightarrow[-78°C]{F_2}$ CF_3-$\overset{O}{\overset{\|}{C}}$-O-$CF_2CF_3$ + CF_3-$\overset{O}{\overset{\|}{C}}$-O-CHF-$CF_3$

Ethyl acetate F-Ethyl acetate α-Hydrotetrafluoroethyl(20%)
                            (5%)   trifluoroacetate

CH_3-$\overset{CH_3}{\underset{CH_3}{C}}$—$\overset{O}{\overset{\|}{C}}$-F $\xrightarrow[-78°C]{F_2}$ CF_3-$\overset{CF_3}{\underset{CF_3}{C}}$—$\overset{O}{\overset{\|}{C}}$-F + CF_3-$\overset{HCF_2}{\underset{CF_3}{C}}$—$\overset{O}{\overset{\|}{C}}$-F

Pivaloyl fluoride F-Pivaloyl fluoride Monohydro-F-pivaloyl fluoride
                            (52%)   (20%)

F-$\overset{O}{\overset{\|}{C}}$—$\overset{CH_3}{\underset{CH_3}{C}}$—$\overset{O}{\overset{\|}{C}}$-F $\xrightarrow[-78°C]{F_2}$ F-$\overset{O}{\overset{\|}{C}}$—$\overset{CF_3}{\underset{CF_3}{C}}$—$\overset{O}{\overset{\|}{C}}$-F (14%)

Dimethylmalonyl difluoride F-Dimethylmalonyl difluoride

Fig. 17.   Direct fluorination of oxygen-containing hydrocarbons. The temperatures given are the lowest gradient temperatures.

3. Prospects for Selective Fluorination of Organic and Inorganic Species with Elemental Fluorine

In view of the extremely exothermic nature of the reaction of elemental fluorine with hydrocarbons, inorganic hydrides, and so on, it might seem obvious that there should be little or no selectivity in the reactions of molecular or atomic fluorine.

The threshold energy or activation energy for the reaction of atomic fluorine with hydrocarbons (79, 80) or hydrogen (81) is near zero, and these reactions, in fact, produce a similar set of HF molecules in excited vibrational states. In a variety of studies by various techniques, the reactions of F and F_2 with such species are generally reported to have activation energies of

$0-0.1$ kcal mole^{-1}. The threshold energy of the reaction of F_2 is near zero and almost every collision of F_2 with hydrogen produces a reaction.

Indeed, there is apparently no selectivity in the reactions of fluorine with hydrocarbons at room temperature. However, at room temperature the reaction of fluorine with a molecule such as benzene initially may show a slight "steric preference" for reactions with the π system due only to the higher probability of a collision occurring there in preference to a proton site.

At significantly lower temperatures the prospects are completely different. Although the reasons are not well understood, there is a picture of rather amazing selectivity of the reactions of elemental fluorine below $-78°$C emerging in the literature.

Perhaps the earliest hint of such a possibility came from the study by Bockemüller of the low-temperature liquid-phase fluorination of n-butyric acid, where it was shown that β and γ isomers were selectively produced (82).

Subsequently the work of R. F. Merritt clearly established the selectivity of additions of fluorine to double bonds in $CFCl_3$ at $-78°$C in preference to reactions with protons. Merritt and co-workers studied such unsaturated systems as indenes and acenapthenes (83), Δ^4 cholestenone (84), 1,1-diphenyl-ethylene (85), Schiff's bases (86), and acetylenes (87).

The selective addition of fluorine to π systems had been studied previously by Miller (88) (propenylbenzene) and in Russian laboratories (89, 90) (vinyl acetate and fumaric acid).

Early reports of selective replacement of hydrogen by fluorine in solvents below room temperature include the discovery by Kamlet and Adolph (91) that dinitromethyl groups can be converted to fluoronitromethyl groups selectively with retention of the nitro groups and similar work by Grakauskas and Baum (92). Grakauskas also observed a selectivity in the fluorination of $CCl_3CO_2CH_3$ to $CCl_3CO_2CH_2F$ in $CF_3ClCFCl_2$ at $-10°$C (93).

An important special case of selective direct fluorination is the reaction of elemental fluorine with uracil to produce 5-fluorouracil in $55-80\%$ yield, which was first reported by Schuman, Tarrant et al. (94) and also studied by Hesse et al. (95).

PCR Incorporated (associated with the Schuman-Tarrant group) manufactures 5-fluorouracil by this process for use in cancer chemotherapy and has had annual sales of over a million dollars for the last several years. This fluorination process, which is selective even above room temperature and may be conducted in H_2O, CF_3CO_2H, or CF_3CH_2OH, is a curiosity. The selectivity observed at higher temperatures must involve a strong solvent effect rather than being primarily a feature of the fluorination process itself. A report by Cech and Holz describes the fluorination of nitrogenous bases and associated nucleosides to produce 2'-deoxy-5-fluorouridine derivatives (96). This study requires that a saturated room-temperature solution of molecular fluorine in acetic (or trifluoroacetic?) acid be used as a fluorination medium! The credibility of this report increases with the possibility of Czechoslovakian translation errors.

A recent report of the synthesis of 2-deoxy-2-fluoro-D-glucose by direct fluorination at $-78°$ C in $CFCl_3$ is quite interesting (97).

Recently Barton et al. (98) reported extremely selective high-yield reactions at $-78°$ C in $CFCl_3$ between fluorine and adamantane and substituted adamantanes. 1-Fluoroadamantane is produced in 84% yield from adamantane, and 3-fluoro-N-trifluoroacetyladamantane is produced from N-trifluoroacetyladamantane in 83% yield (98). In addition, Barton, Hesse and coworkers reported regioselective reactions of a number of steroids with yields ranging from 25 to 75% (99). Some of the selectivity in the steroids is rather remarkable and is thought to result from electrophyllic reactions that are strongly influenced by polar substituents on the molecules and by solvents (98, 99) p. 73. Although the yields are very impressive in these reports, no information is provided on the conversion or the scale.

A somewhat related report in which equally impressive selectivity and yield (60–90%) were obtained Cacou and Wolf have studied the low temperature direct fluorination of aromatic compounds in $CCl_3 F$ (100). They have established that under appropriate conditions molecular fluorination proceeds on substituted benzenes and tolulenes with similar selectivity and orientation as has been commonly observed for other halogens. For example reaction with nitrobenzene gives 80% meta substitution and reaction with toluene gives 60% ortho substitution. The conversions were reported in this study and found to average 0.01%.

Aside from the elegance of such reactions with steroids, perhaps the most striking selective fluorination reactions are those reported by Ruppert and Bastian (61, 62) which are discussed in the inorganic section VB4 of this chapter. Their work established 70–90% yields on a 2–3 gram scale for the selective oxidation of phosphorous, arsenic, antimony, and bismuth compounds at $-90°$ C in $CFCl_3$. At this low temperature, these workers are able to oxidize the central atom to the difluoride without significant fluorination occurring on the methyl and phenyl substituents attached to the oxidized metal or metalloid. The extremely low temperature is probably the key to this remarkable success.

The general utility of $CFCl_3$ as a solvent in this and most other selective fluorination reactions may indicate that solvent effects are not always a prerequisite for selectivity. $CFCl_3$ is a rather inert solvent that is not extremely polar. Of course, another factor in these reactions is that the chlorine in the solvent is not effectively replaced by fluorine at these low temperatures.

Considerable insight into the future of selective reactions of elemental fluorine may be gleaned from consideration of the thermodynamics and kinetic factors of the low-temperature reactions of elemental fluorine. As is stated earlier, there is, under normal circumstances, no selectivity of the reactions of fluorine with carbon–hydrogen bonds and most other bonds at room tempera-

TABLE X

Thermodynamic Data for Steps in Fluorination of CH_4 [a]

Step	Reaction	$\Delta H_{298}°K$ kcal mole^{-1}	$\Delta H_{598}°K$ kcal mole^{-1}	$\Delta G_{298}°K$ kcal mole^{-1}	$\Delta G_{598}°K$ kcal mole^{-1}
Initiation	1a $F_2 \rightarrow 2F\cdot$	+ 37.7	+ 38.5	+ 29.55	+ 20.9
	1b $F_2 + RH \rightarrow R\cdot + HF + F\cdot$	+ 3.9	+ 5.1	− 5.84	−18.904
Propagation	2a $RH + F\cdot \rightarrow R\cdot + HF$	−33.8	−33.4	−36.215	−37.51
	2b $R\cdot + F_2 \rightarrow RF + F\cdot$	−69.1	−69.5	−68.1	−64.15
Termination	3a $R\cdot + F\cdot \rightarrow RF$	−106.8	−108.0	−97.5	−85.091
	3b $R\cdot + R\cdot \rightarrow R − R$	−83.8	−83.06	−70.3	−57.5
Overall reaction	$R−H + F_2 \rightarrow RF + HF$	−102.9	−102.9	−103.4	−103.9

[a] Based on JANAF Table data for CH_4.

ture. The initiation step for the reaction of fluorine with hydrocarbons (1b), which was first proposed by Miller et al. (9) (see Table X), involves the reactions of F_2 rather than F· at low temperatures. This initial step is only exothermic by about 5–8 kcal (ΔH is endothermic by +3.9 kcal also) at room temperature and, as may be observed from Table X, it is exothermic by ~18 kcal at 598°K. Obviously this step becomes less exothermic as the temperature decreases.

If this trend is extrapolated to lower temperatures it is clear that at temperatures varying between −78 and −150°C for carbon–hydrogen bonds on various types of hydrocarbons, ΔF becomes zero and then endothermic. Under these conditions ($\Delta G \leqslant 0$) the reaction with elemental fluorine will not and should not proceed at all. We have observed this in many cases experimentally. At temperatures as low as −150°C the reaction of fluorine with all studied organic compounds either does not occur or is extremely slow. We have occasionally reported reactions in the cryogenic reactor as low as −120°C. It should be explained that the −120°C reported is the temperature of a single trap in a multizone reactor rather than that for the entire reactor. It is possible that fluorination occurs during the volatilization process at such low temperatures.

A truly selective fluorination does not just depend on finding conditions where an extremely small activation energy difference occurs among, for example, three types of hydrogen on an organic molecule, but requires the existence of at least 1–2 kcal differences among the three hydrogen sites. Otherwise, a preferential reaction rather than a selective reaction would occur. While, as stated above, no such activation energy differences exist at room temperature, the appropriate conditions to seek maximum selectivity are just above the temperature where $\Delta G = 0$ for the reaction. We recently have found a number of cases where such activation energy differences exist (101). Such selective reactions appear in many cases to be very sensitive to small differences in temperature. Of course, another limitation is that the temperature must be high enough so that the reaction proceeds at a significant rate. Such activation energy differences are probably responsible for the discovery of Bastian and Ruppert (61, 62) at $-90°C$.

Most of the previously reported selective fluorinations observed have been studied at the convenient solid CO_2 temperature of $-78°C$. It would be indeed fortuitous if this were for addition to double bonds (83–87) and for selective replacement of hydrogen (98, 99) the optimum thermal condition.

The ultimate goal in the new research direction in our laboratory is the possibility of developing an alternate fluorination process to our current method: a process based solely on temperature control. This process would not depend on low initial fluorine concentrations. Fluorinations might be conducted with as much as $\frac{1}{2}-1$ atm of pure fluorine in the initial stages. Such a possible procedure would proceed slowly toward complete fluorination as the temperature is gradually raised at a rate of less than $\frac{1}{2}°K\ hr^{-1}$. If suitable selectivity were found at the various stages of the reaction, it would then be possible to proceed through several steps of direct fluorination toward the perfluoro compound. Each step would consist of reactions of fluorine with unique proton sites of increasingly high activation energy. The process could be stopped at any given stage or it could simply be developed as an alternative method for perfluorination.

It may be true that presently the most important contribution of direct fluorination is in the preparation of fluorine compounds for which conventional fluorination methods have failed, but as direct fluorination techniques mature, they may be expected to compete in many syntheses where other techniques are also successful. Certainly the new procedures for direct fluorination discussed in this review will have a significant impact on fluorine chemistry.

Acknowledgments

Work in fluorine chemistry at the University of Texas at Austin, and previously as the Massachusetts Institute of Technology, has had as its principal sponsor the Air Force Office of Scientific Research, after initial support from the U.S. Air Force Materials Laboratory. We are grateful for other support from the National Science Foundation and the Office of Naval Research. Work in fluorine chemistry at Rice University has been supported over more than a decade by the U.S. Army Research Office, Durham, by the U.S. Atomic Energy Commission, by the National Science Foundation, by the U.S. Air Force Materials Laboratory, and the Robert A. Welsh Foundation.

References

1. A. M. Lovelace, D. A. Ranch, and W. Postelnek, *Aliphatic Fluorine Compounds*, Reinhold., New York, 1958, p. 20.

2. G. H. Cady, A. V. Grosse, E. J. Barker, I.. L. Burger, and Z. D. Sheldon, *Ind. Eng. Chem.*, *39*, 290 (1947).

3. L. A. Bigelow, *Fluorine Chemistry*, Vol. 1, J. H. Simmons, Ed., Academic Press, New York, 1970, p. 373; E. A. Tyczkowski and L. A. Bigelow, *J. Am. Chem. Soc.*, *77*, 3007 (1955).

4. C. M. Sharts, *J. Chem. Educ.*, *45*, 3 (1968).

5. R. E. Banks, *Fluorocarbons and Their Derivatives*, Oldbourne Press, London, 1964, p. 7.

6. R. J. Lagow and J. L. Margrave, *Proc. Natl. Acad. Sci. U.S.*, *67*, 4, 8A (1970).

7. R. J. Lagow and N. J. Maraschin, *J. Am. Chem. Soc.*, *94*, 8601 (1972); R. J. Lagow and J. L. Margrave, *Polym. Lett.*, *12*, 177 (1974).

8. R. E. Florin and L. A. Wall, *J. Chem. Phys.*, *57*, 4, 1191 (1972).

9. W. T. Miller, Jr., S. D. Koch, Jr., and F. W. McLafferty, *J. Am. Chem. Soc.*, *78*, 4992 (1956).

10. J. L. Franklin, J. Wang, R. H. Hauge, and J. L. Margrave, paper presented at 1st Winter Fluorine Conference, St. Petersberg, Florida, Jan. 1972, to be published.

11. R. N. Haszeldine and F. Smith, *J. Chem. Soc.*, *1950* 2689; *ibid.*, *1950*, 2787.

12. N. J. Maraschin and R. J. Lagow, *J. Am. Chem. Soc.*, *94*, 8601 (1972).

13. N. J. Maraschin and R. J. Lagow, *Inorg. Chem.*, *12*, 1459 (1973).

14. J. L. Adcock and R. J. Lagow, *J. Org. Chem.*, *38*, 3717 (1973).

15. *Chem. Eng. News 40* (January 12 1970).

16. R. J. Lagow and J. L. Margrave, *12*, 177 (1974).

17. A. J. Otsuka and R. J. Lagow, *J. Fluorine Chemistry*, *4*, 371 (1974).

18. D. T. Clark, W. T. Feast, W. K. R. Musgrave, and I. Ritchie, *J. Polym. Sci. Polym. Chem. Ed.*, *13*, 857 (1975); D. T. Clark, W. J. Feast, W. K. R. Musgrave, and I. Ritchie, *Advances in Friction and Wear*, Vol. 5A, L. H. Lee, Ed., Plennum Press, 1975, p. 373.

19. M. Okada and K. Makuuchi, *I.E.C. Prod. Res. Devel.*, *3*, 334 (1969).

20. H. Shinohara, M. Iwasaki, S. Tsujimuara, K. Watanabe, and S. Okazaki, *J. Polym. Sci.*, *A1*, 10, 2129 (1972).

21. Shimada and M. Hoshino, *J. Appl. Polym. Sci.*, *19*, 1439 (1975).

22. A. J. Rudge, British. Patent 710, 523 (1954).

23. H. Schonhorn, P. K. Gallager, J. P. Luongo, and F. J. Padden, Jr., *Macromolecules*, *3*, 800 (1970).

24. S. F. Joffre, U.S. Patent 2,811,468 (1957).

25. W. T. Miller, U.S. Patent 2,700,661 (1955).

26. J. Pinsky, A. Adakomis, and A. Nielsen, *Mod. Packag.*, *6*, 130 (1968).

27. W. R. Siegart, W. D. Blackley, H. Chaftez, R. C. Suber, U.S. Patent 3,380,983 (1968).

28. H. Schonhorn, R. H. Hansen, *J. Appl. Polym. Sci.*, *12*, 1231 (1968).

29. W. T. Miller and M. Prober, *J. Am. Chem. Soc.*, *70*, 2620 (1948).

30. L. J. Hayes and D. D. Dixon, *J. Fluorine Chem. 10*, 1 (1977).

31. L. J. Hayes and D. D. Dixon, *J. Fluorine Chem.*, *10*, 17 (1977).

32. L. J. Hayes and D. D. Dixon, *J. Appl. Polym. Sci.*, *22*, 1007 (1978).

33. L. J. Hayes, *J. Fluorine Chem.*, *8*, 69 (1976).

34. G. E. Gerhardt and R. J. Lagow, *Chem. Commun.*, *8*, 259 (1977).

35. G. E. Gerhardt and R. J. Lagow, *J. Org. Chem.*, *43*, 4505 (1978).

36. (a) A. L. Henne, S. B. Richter, *J. Am. Chem. Soc.*, *74*, 5420 (1952). (b) A. L. Henne, M. A. Smook, *J. Am. Chem. Soc.*, *72*, 4378 (1950). (c) R. E. Banks, *Fluorocarbons and Their Derivatives*, MacDonald and Company, Ltd., London 1970 pp. 162.

37. (a) G. V. D. Tiers, *J. Am. Chem. Soc.*, *77*, 4837 (1955). (b) G. V. D. Tiers, *J. Am. Chem. Soc.*, *77*, 6703 (1955). (c) G. V. D. Tiers, *J. Am. Chem. Soc.*, *77*, 6704 (1955).

38. (a) K. J. L. Paciorek, J. Kaufman, J. H. Nakahara, T. I. Ito, R. H. Kratzer, R. W. Rosser, and J. A. Parker, *J. Fluorine Chem.*, *10*, 277 (1977). (b) F. C. McGrew, *Chem. Eng. News*, *45*, 18 (Aug. 7, 1967). (c) H. S. Eleuterio, *J. Macromo. Sci. - Chem.*, *A6*, 1027 (1972).
39. (a) J. T. Hill, *J. Macromol. Sci. - Chem.*, *A8*, 499 (1974). (b) H. S. Eleuterio, *J. Macromol. Sci. - Chem.*, *A6*, 1027 (1972).
40. R. J. Lagow and Shoji Inoue, U.S. Patent No. 4,113,772 (1978).
41. G. E. Gerhardt and R. J. Lagow, *J. Chem. Soc.*, submitted for publication.
42. G. E. Gerhardt and R. J. Lagow, *J. Polym. Sci.*, *Polym. Chem. Ed.*, in press.
43. Y. Tabata, W. Ito, H. Matsbakashi, Y. Chatani and H. Tadokoro, *Polym. J. (Jap.)*, *9*, 2, 145 (1977).
44. J. L. Adcock, S. Inoue and R. J. Lagow, *J. Am. Chem. Soc.*, *100*, 6, 1950 (1978).
45. M. Lopez, B. Kipling, and H. L. Yeager, *Anal. Chem.*, *48*, 1120 (1976); M. F. Homer and G. B. Butler, *J. Polym. Sci.*, *C*, *45*, 1 (1974).
46. R. J. Lagow, U.S. Patent 4,076,916. (1978).
47. R. J. Lagow and J. L. Margrave, *Inorg. Acta*, *10*, 9 (1974).
48. S. D. Arthur, R. A. Jacob, and R. J. Lagow, *Inorg. Nucl. Chem.*, *35*, 3435 (1973).
49. R. J. Lagow and J. L. Margrave, *Inorg. Nucl. Chem.*, *35*, 2084 (1973).
50. N. J. Maraschin and R. J. Lagow, *Inorg. Chem. 14*, 1855 (1975).
51. N. J. Maraschin and R. J. Lagow, *J. Am. Chem. Soc.*, *94*, 8601 (1972).
52. J. L. Adcock and R. J. Lagow, *J. Fluorine Chem.*, *2*, 434 (1973).
53. R. J. Lagow, L. A. Shimp, D. K. Lam, and R. F. Baddour, *Inorg. Chem.*, *11*, 2568 (1972).
54. E. K. S. Liu and R. J. Lagow, *J. Am. Chem. Soc.*, *98*, 8270 (1976).
55. E. K. S. Liu and R. J. Lagow, *J. Organometal. Chem.*, *145*, 161 (1978).
56. E. K. S. Liu and R. J. Lagow, *Chem. Commun.*, *450* (1977).
57. J. A. Morrison, R. A. Jacob, L. L. Gerchman, and R. J. Lagow, *J. Am. Chem. Soc.*, *97*, 518 (1975).
58. E. K. S. Liu and R. J. Lagow, *Inorg. Chem. 17*, 618 (1978).
59. L. A. Shimp and R. J. Lagow, *Inorg. Chem, 16*, 2974 (1977).
60. A. H. Cowley, R. J. Lagow, R. Braun, and L. A. Shimp, to be submitted for publication.
61. I. Ruppert and U. Bastian, *Angew. Chem. Int. Ed. Engl.*, *16*, 718 (1978).
62. I. Ruppert and U. Bastian, *Angew. Chem. Int. Ed. Engl. 17*, 214 (1978).
63. R. J. Lagow and J. L. Margrave, "The Reaction of Polynuclear Hydrocarbons with Elemental Fluorine", to be submitted for publication.
64. R. J. Lagow, R. B. Baddachhape, J. L. Wood, and J. L. Margrave, *J. Chem. Soc.*, 1974, 1268; R. J. Lagow, R. B. Baddachhape, J. L. Wood, and J. L. Margrave, *J. Am. Chem. Soc.*, *96*, 2381 (1974).
65. N. J. Maraschin and R. J. Lagow, *Inorg. Chem.*, *12*, 1459 (1973).
66. N. J. Maraschin, B. D. Catsikis, L. H. Davis, G. Jarvinen, and R. J. Lagow, *J. Am. Chem. Soc.*, *97*, 513 (1975).
67. (a) R. B. Badachhape, A. P. Conroy, and J. L. Margrave, paper presented at American Chemical Society Regional Meeting, Dallas, Texas, 1972. (b) G. Robertson and E. K. S. Liu, *J. Org. Chem.*, *43*, 4981 (1978).
68. See, for example, L. C. Clark, *Sci. News*, *106*, 202 (1974) and *Science*, *9*, 669, Feb. (1973). J. G. Riess and M. LeBlanc, *Angew. Chem. Int.*, *17*, 621 (1978).
69. R. J. Lagow, L. C. Clark, and L. A. Shimp, U.S. Patent No. 4,110,474 (1978).
70 (a) L. A. Shimp and R. J. Lagow, *J. Org. Chem.*, *42*, 3437 (1976). (b) E. K. S. Liu and R. J. Lagow *J. Fluorine Chem.*, in press.
71. I. J. Hotchkins, R. Stephens, and J. C. Tatlow, *J. Fluorine Chem.*, *6*, 135, (1975).

72. I. J. Hotchkiss, R. Stephens, and J. C. Tatlow, *J. Fluorine Chem.*, *8*, 379 (1975).
73. I. J. Hotchkiss, R. Stephens, and J. C. Tatlow, *J. Fluorine Chem*, *10*, 541 (1977).
74. J. L. Adcock and R. J. Lagow, *J. Org. Chem.*, *38*, 2617 (1973).
75. J. L. Adcock and R. J. Lagow, *J. Am. Chem. Soc.*, *96*, 24, 7588, (1974).
76. J. L. Adcock, R. A. Beh, and R. J. Lagow, *J. Org. Chem.*, *40*, 3271 (1975).
77. (a) R. J. Lagow and J. L. Margrave, unpublished work, 1969. (b) R. B. Badachhape, A. P. Conroy, and J. L. Margrave, USAF AFML-TR.71-271, Final Report (1972).
78. L. A. Harmon and R. J. Lagow, *J. Chem. Soc.*, in press.
79. J. G. Moelman, J. J. Gleeves, J. W. Hudgens, and J. D. McDonald, *J. Chem. Phys.*, *62*, 4790 (1973), J. G. Moelman and J. D. McDonald, *J. Chem. Phys.*, *62*, 3061 (1975); *62*, 3052 (1975).
80. H. W. Chang and D. W. Setzer, *J. Chem. Phys.* *58*, 2298 (1973); *51*, 2310 (1973). K. C. Kim and D. W. Setzer, *J. Phys. Chem.*, *77*, 2493 (1973); *77*, 2499 (1973). K. C. Kim, D. W. Setzer, and C. M. Bogan, *J. Chem. Phys.*, *60*, 1837 (1974).
81. J. T. Muckerman, *J. Chem. Phys.*, *54*, 1155 (1971).
82. W. Bockemüller, *Annalen*, *506*, 20 (1933).
83. R. F. Merritt and F. A. Johnson, *J. Org. Chem.*, *31*, 1859 (1966).
84. R. F. Merritt and T. E. Stevens, *J. Am. Chem. Soc.*, *88*, 1822 (1966).
85. R. F. Merritt, *J. Org. Chem.*, *31*, 3871 (1966).
86. R. F. Merritt and F. A. Johnson, *J. Org. Chem.*, *32*, 416 (1967).
87. R. F. Merritt, *J. Org. Chem.*, *32*, 4124 (1967). R. F. Merritt, *J. Am. Chem. Soc.*, *89*, 609 (1967).
88. W. T. Miller, J. O. Stoffer, G. Fuller, and A. C. Currie, *J. Am. Chem. Soc.*, *86*, 51 (1964).
89. A. Y. Yakubovich, S. M. Rozenshtein, and U. A. Ginsberg, U.S.S.R. Patent 162,825 (1965).
90. A. Y. Yakubovich, U. A. Ginsberg, S. M. Rozenshtein, and S. M. Smirnov, U.S.S.R. Patent 165,162 (1965).
91. M. J. Kamlet and H. G. Adolph, *J. Org. Chem.*, *33*, 3073 (1968).
92. V. Grakauskas and K. Baum, *J. Org. Chem.*, *33*, 3080 (1968).
93. V. Grakauskas, *J. Org. Chem.*, *34*, 965 (1969).
94. P. D. Schuman, P. Tarrant, D. A. Warner and G. Westmoreland, *Chem. Eng. News*, *49*, 40 (1971). U.S. Patent Appl. Ser. 658,645 (Aug. 7, 1967). U.S. Patent 3,954,758 (1976).
95. D. H. R. Barton, R. H. Hesse, H. T. Toh, and M. M. Pechet, *J. Org. Chem.*, *37*, 329 (1972).
96. D. Cech and A. Holz, *Collect. Czech., Chem. Commun.*, *41*, 3335 (1976).
97. T. Ido, C. N. Wan, J. S. Fowler, and A. P. Wolf, *J. Org. Chem.*, *42*, 13 (1977).
98. D. H. R. Barton, R. H. Hesse, R. E. Markwell, M. M. Pechet, and H. T. Toh, *J. Am. Chem. Soc.*, *98*, 3034 (1976).
99. D. H. R. Barton, R. H. Hesse, R. E. Markwell, M. M. Pechet, and Shloma Rozen, *J. Am. Chem. Soc.*, *98*, 3037 (1976).
100. F. Cacou and A. P. Wolf, *J. Amer. Chem. Soc.*, *100*, 3639 (1978).
101. T. Blackburn, and R. J. Lagow, to be published.

Metal–Metal Bonds of Order Four

JOSEPH L. TEMPLETON

Department of Chemistry, University of North Carolina
Chapel Hill, North Carolina

CONTENTS

I. INTRODUCTION

A. General

The chemistry of compounds containing metal–metal bonds constitutes one of the major topics in inorganic chemistry currently attracting vigorous and diverse research efforts. The field developed at a relatively steady, but deliberate, pace during the years from 1960 to 1970 before the advent of the current decade brought a dramatic and ongoing increase in research activity and productivity. The rich and varied chemistry exhibited by compounds with metal–metal bonds portends the emergence of an independent discipline based on these polynuclear metal moieties.

The purpose of this review is to present a current description of the chemistry of compounds with quadruple metal–metal bonds and closely related species characterized through early 1978. Species containing metal–metal bonds of order four are considered to the exclusion of units containing single, double, and triple-bonds except where structural comparisons or chemical reactivity patterns dictate the presentation of data for the lower bond orders.

A number of excellent reviews have been written that address various aspects of metal–metal bonding. In particular, Cotton has contributed several eminent early reviews commensurate with his pioneering efforts in this area (54, 59, 60). More recently he reviewed quadruple bonds (56), molybdenum–molybdenum bonds (57), and multiple bond studies in his laboratory (55) in separate articles. Earlier reports discuss multiple metal–metal bonds in the context of general metal–metal bonding schemes and clusters at a time when the area was more limited in scope and therefore amenable to an encompassing survey (13, 128, 129, 164, 188, 189, 252). The redox behavior of metal–metal bonds has been reviewed by Meyer (193), and Walton has recently described ligand-induced redox reactions of strong multiple metal–metal bonds (253).

Chisholm and Cotton have summarized their successful efforts characterizing metal–metal triple bonds in molybdenum(III) and tungsten(III) compounds containing no bridging ligands (46). This realm of strong homonuclear metal–metal bonds was unknown prior to the 1970s, and the rapid development and exploitation of the chemistry of these dimers provide an excellent example of fundamental and fruitful research involving new compounds with strong multiple metal–metal bonds.

The impetus for understanding multiple metal–metal bonds in particular and metal clusters in general derives from several sources. Muetterties has espoused and lucidly delineated the concepts that link homogeneous studies of discrete clusters with heterogeneous surface processes (195, 196). Sinfelt's research in the area of surface clusters (231) and Ozin's matrix isolation studies of metal atom aggregates (208) contribute to a developing picture that may correlate the behavior of discrete molecules containing metal–metal bonds with metal surface behavior. Multiple electron transfer can occur by means of redox reactions involving metal–metal localized orbitals and electrons where the metal–metal interaction serves as an electron source or sink (145). The realm of organometallic reactions is vast, and processes such as oxidative addition are promoted by a different set of molecular properties than in the mononuclear case (46, 86). The theoretical impact of bond orders greater than three is evident relative to the classical multiple bonds of the second row elements, which cannot exceed a bond multiplicity of three. For such reasons the field of multiple metal–metal bonds will no doubt continue to flourish as our understanding of these units improves and suggests new insights and applications.

B. Scope of the Review

The simple molecular orbital definition of bond order as the number of electron pairs occupying bonding molecular orbitals minus the number of electron pairs in antibonding molecular orbitals is sufficient to define the range of metal–metal bonds considered in this review. A bond order of four thus requires an excess of eight bonding electrons between the two centers. Although no arbitrary criteria with respect to the contribution of any given component to the total bond strength is imposed, it is clear that a quadruple bond must exhibit properties consistent with a bond strength in excess of that characterizing lesser bond orders in similar instances. Cases where strong multiple bonds are present with nonintegral bond orders are included regardless of the detailed molecular orbital description where the possible gradation of molecular orbitals from nonbonding to antibonding or bonding presents a continuum of schemes that defy rigorous classification. Within the realm of metal–metal bonds characterized to date, only dimers provide sufficient internuclear electron density and appropriate orbital combinations for inclusion under this title. In other words, this definition is naturally restrictive in that only two-center interactions are known to be of multiplicity four.

Hence the many extant metal clusters containing three or more metal atoms do not fall within the scope of this review. In addition to the metal halide derivatives with three or more metal atoms (13, 168, 211), the chemistry of metal carbonyl clusters and related derivatives with three or more metal atoms is thus excluded (3). This logical limitation provides an opportunity to understand the simplest case where metal–metal bonding can occur, namely, in the prototype two-metal-center cases.

II. STRONG METAL–METAL MULTIPLE BOND PROPERTIES

A. Characteristic Properties

The general features of compounds containing multiple metal–metal bonds resemble those of other metal clusters that have been expounded previously (13, 54, 227). A brief survey of characteristic properties of compounds containing metal–metal bonds emphasizing those points most relevant to multiple metal–metal bonds is nonetheless appropriate.

Firstly, the metal is invariably found in a relatively low oxidation state, typically in the range of 2+ to 3+ for species without π-acceptor ligands. When ligands such as carbon monoxide are present, the removal of electron density from the metal that occurs alters this criterion such that lower oxidation states such as 0+ and 1+ are common.

A second feature of metal halide cluster chemistry is that the early transition metals are more prone to form metal–metal bonds than are the later noble metals and coinage metals. Again the polynuclear metal carbonyls differ in this facet of metal–metal bond behavior, and, in fact, metal carbonyl clusters become more common on going from the left to the right of the Periodic Table.

Thirdly, it has generally been conceded that the heavier transition elements exhibit a greater propensity toward metal–metal bond formation than the corresponding first-row congener. The stability of the hexanuclear metal clusters of Nb, Ta $[(M_6 X_{12}^{n+})$ $n = 2, 3, 4]$, Mo, and W $(M_6 X_8^{4+})$ contributed heavily to this concept, since no analogous V or Cr units have been reported. However, in the case of multiple metal–metal bonds between only two centers, the extensive chemistry of the homonuclear chromium–chromium quadruply bound fragment that has developed contradicts this generalization. Noteworthy in this regard is the paucity of tungsten–tungsten quadruple bonds established to date. It is thus premature to conclude that first row transition metals are recalcitrant reagents in the formation of metal–metal multiple bonds.

Two additional points that are self-evident are discussed below for the sake of completeness. The coordination number of a metal ion counting all ligands other than the adjacent metal involved in multiple metal bonding is less than the maximum coordination number possible for that metal center. Furthermore, the number of d-electrons on the metal is nearly equal to the number of metal valence orbitals not involved in metal–ligand bonding to optimize metal–metal bond formation by filled bonding MOs.

The chemical rationale invoked to account for the above characteristics is based on several premises. One aspect of importance is considered to be the radial distribution of the metal d orbitals that combine to form the bonding and antibonding molecular orbitals responsible for metal–metal bonds (54, 129, 230). The consensus that larger orbitals are preferable for bond formation between metals is consistent with the observed low-oxidation-state metals common to metal–metal bound species, since higher oxidation states cause orbital contraction to occur as the effective nuclear charge at the metal is increased. The expanded d orbitals present in the lower oxidation states relative to the higher oxidation state ions are capable of overlapping significantly at distances where the underlying internuclear repulsion, which is always present, is not dominant, and hence bond formation is promoted. The trend for early transition metals to form metal halide clusters is also consistent with the importance of orbital size. The increased effective nuclear charge of metals to the right of the Periodic Table that results as protons are added to the nucleus and electrons are placed in d orbitals that do not provide effective screening from the additional positive charge causes the orbital contraction that is observed on progressing to the right of the Periodic Table. In this case the orbital extension of the early transition metals up to and including Group VIIB is more favorable for metal–metal attractive interactions than are the smaller noble metal radial distributions.

An additional feature favoring early transition metal involvement in metal clusters is the accessibility of high coordination numbers for these metals. Seven- and eight-coordinate geometries are common for Nb, Ta, Mo, and W. For metal clusters with several metal centers and bridging ligands the total coordination number can be quite high. For example, in $M_6X_8^{4+}$ clusters each metal atom is bound to five ligands (including one at the terminal site) and four metal atoms for a total of nine coordinated atoms. However, for dimeric species with multiple metal–metal bonds the number of coordinated atoms tends to decrease inversely with the bond multiplicity, so the ability to coordinate a large number of atoms is not a prerequisite for metal–metal bonding in these compounds.

The optimal electronic configuration for triple bonds is clearly d^3, consistent with the observed behavior of Mo^{3+} and W^{3+}. On the other hand, the isoelectronic ions Nb^{2+} and Ta^{2+} have not been identified in metal–metal triple bonds to date, and this sort of analogy would seem to provide an opportunity for expansion of elements forming strong multiple metal–metal bonds. Similar considerations suggest that the d^4 configuration should be optimal for quadruple-bond formation and this is reflected in the extensive chemistries of Cr_2^{4+}, Mo_2^{4+}, and Re_2^{6+}. Metal ions with more than four valence electrons can still form strong multiple bonds if the MO scheme allows for filling nonbonding orbitals or antibonding orbitals while maintaining an excess of $2n$ bonding electrons for a bond of order n.

Application of thermodynamic principles to low-oxidation-state metal halides indicates that the enthalpy of metal–metal bond formation is required in lieu of additional metal halide bonds if an exothermic reaction is to occur relative to the elements when the heat of sublimation of the metal is high (54, 227). Although this argument is valid for metal halides such as $MoCl_2$, where bridging halides and hexanuclear metal cluster formation furnish stabilizing bond energies, ions such as $Mo_2Cl_8^{4-}$ are not as easily analyzed by means of thermodynamic cycles. The formation of additional terminal M–Cl bonds in anionic derivatives would no doubt be favorable relative to neutral species with fewer M–Cl interactions.

B. Characterization of Multiple Metal–Metal Bonds

The widespread use of x-ray techniques in characterizing polynuclear metal complexes has provided definitive information regarding multiple metal–metal bonds. The metal–metal distance found in single-crystal x-ray studies is certainly one of the key structural parameters to be considered, but the absolute value of the internuclear distance must be analyzed with caution. Comparison of metal atom separations with that of the elemental metal structure provides a basis for qualitative conclusions regarding overall attraction or repulsion. More detailed analyses are now possible owing to the wealth of structural data available for compounds containing metal–metal bonds. In addition to the metal–metal distance the distribution of ligands is a key feature that offers insight into the total molecular orbital scheme and often reflects the impact of metal–metal bonding. Finally, it should be noted that distortional parameters have been employed quite effectively in the presence of bridging ligands, where idealized geometries are considered as a standard prior to metal–metal attraction or repulsion (115).

Other techniques have been employed to infer the presence of metal–metal bonds. As an example, the magnetic properties of dimeric molecules containing two d^3 metal ions can be informative. The presence of six unpaired spins in the ground state is definitive evidence for the absence of metal–metal bonding. It is possible to interpret diamagnetism in this same molecule as the result of metal–metal triple-bond formation that pairs all the electrons. The influence of geometric distortions or spin–orbit coupling can also lead to diamagnetic behavior, however, and therefore considerable caution must be employed when magnetic data are cited as evidence for metal–metal bonding. Paramagnetic behavior indicative of electron delocalization over more than one metal ion is also a source of information to support the existence of metal–metal bonds in a given complex.

High electron density in the internuclear region of multiple metal bonds leads to a large polarizability and thus the vibrational features are amenable to Raman studies. Resonance Raman effects are common owing to the presence of electronic absorption bands in the visible region for many metal–metal bound fragments, and the Raman spectrum can be extremely informative, particularly when compared with related systems.

A variety of tools address the stoichiometry and molecular weight of compounds. The necessary condition that at least two metals be present for multiple metal bond formation is a simple sorting method for initial studies: the molecular unit as determined by any type of molecular weight study must correspond to that of at least two metals per molecule. Conductivity measurements supply similar data for ions, and mass spectral data can indicate the presence of at least two metals per molecule. Analytical data with nonintegral ligand-to-metal ratios require that some multiple number of metal centers be present in order to formulate a stoichiometric compound. This array of techniques only eliminates the possibility of metal–metal bonds for mononuclear metal complexes, and further studies are always necessary to confirm the presence of an attractive metal–metal interaction.

III. METAL–METAL QUADRUPLE BONDS

A. Recognition of a Quadruple Bond

In 1964 Cotton et al. presented the first report elaborating a bond of order greater than three (62). The historical aspects of this event and the repercussions of postulating a δ component to the bond in addition to the classical σ- and two π-orbital contributions in triple bonds have been reviewed (56).

The correct crystal structure of the rhenium chloride complex of stoichiometry $KReCl_4 \cdot H_2O$ revealed a dimeric structure (Fig. 1) with three outstanding features (82). The rhenium–rhenium separation was extremely short (2.24 ± 0.01 Å), no bridging chloride ligands were present, and the two sets of four chlorides were eclipsed relative to one another when viewed down the Re–Re axis.

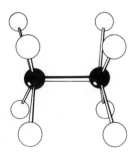

Fig. 1.   The structure of the $Re_2Cl_8^{2-}$ ion as determined by Cotton and Harris (82).

The metal atom separation of 2.24 Å in the absence of bridging ligands is definitive evidence that a strong metal–metal bond has formed. Comparison with the elemental metal separation of 2.74 Å (209) qualitatively confirms that an unusually strong attractive force is present.

An eclipsed conformation for the chlorides cannot be rationalized in terms of steric factors, which would obviously favor the staggered structure resulting from a rotation of 45° for one $ReCl_4^-$ unit relative to the other. An electronic explanation must consider the bonding between the rhenium atoms, since this is the only link connecting the two fragments. The qualitative molecular orbital ground state description originally put forth by Cotton (58) for the metal–metal interaction with each rhenium(III) contributing four electrons to the resultant molecular orbitals has remained unaltered in its essential features in spite of the proliferation of relevant data that has occurred since that time. A simple scheme that represents only the fundamental bonding features is as follows. Establish a molecular orbital scheme appropriate for the D_{4h} point of the molecule with the Re–Cl σ bonds formed using the s, p_x, p_y, and $d_{x^2-y^2}$ valence orbitals of the metal as shown in the coordinate system of Fig. 2. The remaining metal orbitals (p_z, d_{z^2}, d_{xz}, d_{yz}, and d_{xy}) are nonbonding with respect to the metal–ligand σ framework, and appropriate symmetry analysis identifies the p_z and d_{z^2} orbitals as σ, the d_{xz} and d_{yz} orbitals as π, and the d_{xy} orbitals as δ relative to the homonuclear rhenium axis. A hybrid orbital consisting of p_z and d_{z^2} components can be considered to be optimal for σ-bond formation; this bonding contribution is independent of rotation about the internuclear axis as is typical of σ bonds. Furthermore, filling both π-bonding orbitals of the form $[d_{xz}(1) + d_{xz}(2)]$ and $[d_{yz}(1) + d_{yz}(2)]$ creates an overall cylindrical electron density distribution along the rhenium–rhenium axis similar to that in the acetylene case where a cylindrical bond results from filling the two perpendicular p-π orbitals. In terms of the angular dependence inherent in the spherical harmonics, the imaginary forms for atomic d orbitals with $l = 2$ and $m = \pm1$ are equally valid here because of the equivalence of the (d_{xz}, d_{yz}) and $[d(m = +1),$ $d(m = -1)]$ pairs, and the cylindrical symmetry of the degenerate orbitals with $m = \pm1$ is retained in the total π bond. Thus neither σ nor π effects influence the angular orientation of the two $ReCl_4^-$ units.

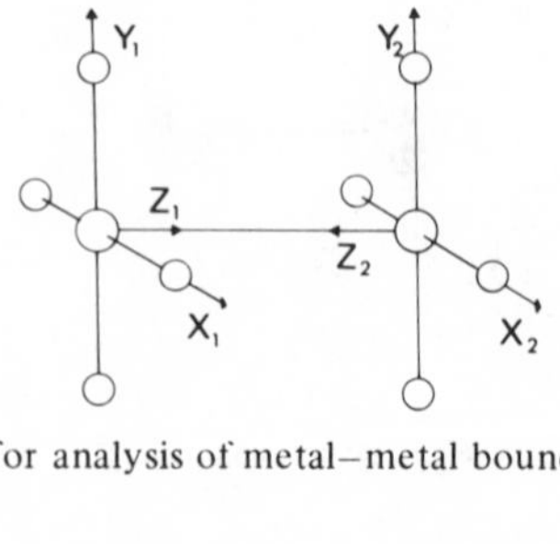

Fig. 2.   Coordinate system for analysis of metal–metal bound dimers with D_{4h} symmetry.

The beauty of the $Re_2Cl_8^{2-}$ anion structure as the prototype for understanding quadruple bonds has been elaborated by Cotton (54). The presence of a δ bond is the logical explanation for the observed eclipsed orientation, since the δ bond formed by overlap of $d_{xy}(1)$ with $d_{xy}(2)$ is optimal in the eclipsed structure and nonexistent as a result of zero overlap in the staggered form. Since this is the only component of the metal–metal bond that varies with rotation of the two $ReCl_4^-$ fragments, the energy of this δ bond must be sufficient to overcome any steric repulsion that exists among the chloride ligands in the solid state. The absence of bridging ligands or bulky inequivalent ligands in this case strongly supports the δ-bond concept, since the eclipsed ligating atoms found later in other dimers containing bridging bidentate ligands or both bulky ligands and halides in the coordination sphere could favor an eclipsed structure for steric reasons or as a result of constraints imposed by the ligand geometries.

The generation of one σ-, two π-, and one δ-bonding molecular orbital is accompanied by formation of the respective antibonding molecular orbital counterparts, and eight electrons nicely fill the bonding orbitals while no antibonding molecular orbitals are occupied. The resultant bond order is four. One aspect of the above scheme that has been neglected in this qualitative treatment is the location of the two molecular orbitals of σ symmetry present in addition to the σ and σ^* levels already considered. Since four atomic orbitals are of σ symmetry relative to the internuclear axis, a total of four molecular orbitals must result. The location of these two remaining orbitals has been the subject of further analysis and discussion, but no reassessment of the basic quadruple bond postulate has been required.

Major contributions to the synthetic and structural chemistry of low-oxidation-state rhenium compounds resulted from preparative studies reported by Kotel'nikova and Tronev (171, 172) and x-ray data published by Kuznetzov and Koz'min (186) during the late fifties and early sixties. Characterization of a compound formulated as $(pyH)HReCl_4$ based on an oxidation-state determination indicative of Re^{2+} required the presence of "free H" to account for the rhenium(II) oxidation state (11). The eclipsed structure determined for this ion (186) and closely related rhenium dimers involving chloride ligands (175, 176), bromide ligands (174), acetic acid derivatives (173, 177, 178), and formic acid derivatives (170, 179) all fit nicely into the scheme proposed by Cotton for quadruple-bond formation between two rhenium(III) d^4 centers (56). Nonetheless, the initial proposal that free hydrogen and rhenium(II) are present in the pyridinium salt has been extended to the various rhenium derivatives reported in succeeding publications as noted above. An accurate assessment of the chemical information relevant to this review is somewhat difficult to incorporate in light of the continued description of these strong multiply metal–metal bound complexes as Re^{2+} derivatives even though the necessity to postulate the presence of free hydrogen no longer seems to be required. For purposes of consistency in presentation, the possible presence of an extra hydrogen per metal atom in rhenium dimers reported by these workers is discounted in the following discussions. This is in accord with selecting the rhenium(III) oxidation state option, which is consistent with the bulk of germane experimental evidence to date. (Note that the presence of hydrogen in $[Mo_2 Cl_8 H]^{3-}$ remained undetected for 7 years following the crystal structure determination (86). This example serves as a reminder that definitive identification of hydrogen in such species cannot be considered a routine experimental procedure.)

Studies of the magnetic behavior of $Cu_2(O_2 CCH_3)_4 \cdot 2H_2 O$ and $Cr_2(O_2 CCH_3)_4 \cdot 2H_2 O$ during the fifties led Figgis and Martin (130) to postulate that the diamagnetic behavior of the chromium dimer was attributable to overlap of metal orbitals of δ symmetry in addition to σ and π overlap. Although the metal–metal bond was considered to be relatively weak, a conclusion largely based on the inaccurate chromium–chromium distance of 2.64 Å in the literature at that time (251) [redetermined to be 2.39 Å (66)], the presence of a bond of order four was implied by the pairing of eight electrons in the σ, two π, and δ orbitals. Differing interpretations of the nature and extent of metal–metal bonding in diamagnetic dimers of chromium(II) have appeared as recently as 1977 (112, 133). It now seems that a consensus has been reached among researchers involved in assessing the ground-state electronic structure of carboxylate-bridged chromium(II) dimers (16, 112, 139), and the accepted description tends to support the qualitative bonding scheme published by Figgis and Martin in 1956 (130).

B. Preparation of a Quadruple Bond Between Group VI Elements

The recognition of a quadruple metal–metal bond in rhenium chemistry was soon followed by the report that a molybdenum compound displayed similar properties suggestive of a bond of order four. The isolation of $[Mo(O_2 CCH_3)_2]_2$ by Bannister and Wilkinson (14) from the reaction of molybdenum hexacarbonyl with refluxing acetic acid produced the prototype molybdenum dimer for the extensive chemistry that followed. The structure of the dimolybdenum tetraacetate reported by Lawton and Mason in 1965 (187) was classified as that of the copper acetate type (Fig. 3). The molybdenum–molybdenum atom separation of 2.11 Å found was approximately 0.6 Å less than the elemental metal separation of 2.73 Å (209). Even though the bridging acetates necessarily hold the metal centers together, the separation of 2.64 Å found previously in $Cu_2(O_2 CCH_3)_4(H_2 O)_2$ (250) indicates that much greater distances between the two metals are possible within the constraints of the tetraacetate structure, and thus strong attractive forces between the molybdenum centers are evident. The analogy between the $Mo^{2+}(d^4)$ system and the

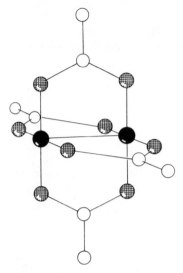

Fig. 3. The structure of $Mo_2(O_2CCH_3)_4$ as determined by Lawton and Mason (187).

$Re^{3+}(d^4)$ dimer is clear, but the point made earlier concerning the eclipsed $Re_2Cl_8^{2-}$ geometry in light of a δ bond is apparent when arguments of bond order are applied to $Mo_2(O_2CCH_3)_4$, where an eclipsed structure is favored by the acetate bridges. The earlier synthesis of $Mo_2(O_2CC_6H_5)_4$ (2) and later preparations of higher carboxylate analogues were assumed to produce compounds with similar structures (235).

The reaction of $Mo(CO)_6$ and acetic acid is still the most convenient preparative route known for condensing two monomeric units into a compound containing a quadruple bond. In fact, facile conversion of $Mo_2(O_2CCH_3)_4$ to chloro species occurs in concentrated hydrochloric acid to form the $Mo_2Cl_8^{4-}$ ion (36), which is isostructural with the $Re_2Cl_8^{2-}$ ion. The eclipsed configuration of $Mo_2Cl_8^{4-}$ and the molybdenum atom separation of 2.14 Å are entirely consistent with a quadruple bond in the anion as expected (33).

C. Syntheses That Mold Monomers Into Quadruply Bound Dimers

The reactions discussed in this section include those in which monomeric reagents are converted into dimeric species containing quadruple bonds. Later sections of this review consider reactions in which substitution or redox processes occur with retention, alteration, or cleavage of the metal–metal multiple bond. In very general terms, it is the initial formation of the quadruple bond that is common to the reactions below, and the number and variety of such condensations is not overwhelming. Nonetheless, the limited number of known reactions of this type have been adequate to expose a vast chemistry associated with compounds having strong multiple metal–metal bonds, since extensive derivatization has been possible once the dimeric unit has been welded. The reactions considered below are classified according to the metal involved in the synthesis of homonuclear dimers.

1. Rhenium

The origin of the majority of rhenium–rhenium quadruple bonds can be traced directly to the parent $Re_2Cl_8^{2-}$ anion. Several routes to the octachlorodirhenate(III) ionic complex have been compared (42). The method of choice is the reaction of trirhenium nonachloride with molten diethylammonium chloride (12). Yields of up to 65% are attainable when the reaction mixture is extracted with hydrochloric acid and the anion is precipitated as the salt of a large cation (i.e., tetrabutylammonium cation).

Other routes to the octachlorodirhenate(III) ion include reduction of perrhenate ion in acid solution using either hydrogen gas under pressure or hypophosphorous acid as the reducing agent (63). Yields of 30–40% are obtained, so the previous method offers the advantages of improved yield and convenience.

Quadruple bonds are found in compounds of the type $Re_2(O_2CR)_4Cl_2$, which are conveniently prepared from the $Re_2Cl_8^{2-}$ ion by substitution (42). However, several other preparations of the tetracarboxylatodichlorodirhenium(III) from different rhenium compounds are known. The reaction of Re_3Cl_9 with RCO_2H led to isolation of $Re_2(O_2CR)_4Cl_2$ (239), as did the reaction of trans-$ReOCl_3[P(C_6H_5)_3]_2$ with RCO_2H (219). Walton et al. have presented a slight modification of the Rouschias and Wilkinson (219) preparation of $Re_2(O_2CCH_3)_4Cl_2$ from $ReOCl_3(PPh_3)_2$ which they consider to be the method of choice for synthesis of this dimer (136).

More recently the methyl analogue of $Re_2Cl_8^{2-}$ was prepared from monomeric rhenium pentachloride, as well as from the rhenium benzoate dimer, $Re_2(O_2CC_6H_5)_4Cl_2$, which has the quadruple bond already established (80). Addition of excess methyllithium to $ReCl_5$ in diethyl ether at $-78°C$ followed by low-temperature isolation of the resultant red crystals produced $Li_2Re_2(CH_3)_8 \cdot 2(C_2H_5)_2O$ in a 16% yield.

The above data establish the prominence of $Re_2Cl_8^{2-}$ as the reagent of choice for expanding the chemistry of quadruply bound rhenium dimers by ligand substitution reactions. Alternate routes are available for synthesis of dimeric carboxylates of rhenium(III), but high conversions of $Re_2Cl_8^{2-}$ or $Re_2Br_8^{2-}$ to $Re_2(O_2CR)_4X_2$ are equally convenient. Only for $Re_2(CH_3)_8^{2-}$ is a reaction employing $ReCl_5$ as a reagent clearly preferable to substitution of ligands on a quadruply bound rhenium dimer previously formed from the octachlorodirhenate anion.

2. Molybdenum

The tetracarboxylatodimolybdenum dimers that result from the reaction of molybdenum hexacarbonyl with a wide variety of carboxylic acids are the initial source of quadruply bound molybdenum units for many further chemical investigations. The initial isolation of $Mo_2(O_2CC_6H_5)_4$ (2), $Mo_2(O_2CCH_3)_4$ (14), and the higher alkyl carboxylate analogues (235) by Wilkinson et al. established a reactivity pattern that has been successfully exploited to prepare many related molybdenum dimers. In addition to those carboxylates enumerated above, a series of substituted aromatic carboxylate derivatives has been prepared by refluxing the appropriate acid with molybdenum hexacarbonyl in diglyme (155).

Air stability of solid molybdenum carboxylate dimers varies greatly. Although the electronic structure of the coordinated carboxylate no doubt plays a role in determining the air sensitivity of these compounds, it is clear that steric factors that impede approach along the axial molybdenum coordination sites trans to the quadruple metal–metal bond are effective in protecting the dimeric species from rapid oxidation (155). An illustrative example is the comparison of *ortho-* and *para-*toluenecarboxylic acid derivatives of Mo^{2+}, where oxidation of the molybdenum dimer, $Mo_2(p\text{-}MeC_6H_4CO_2)_4$, which has readily accessible terminal sites, occurs in a matter of hours, while the ortho methyl groups, which necessarily inhibit axial attack at molybdenum, result in air stability for $Mo_2(o\text{-}MeC_6H_4CO_2)_4$ over a period of months.

An isoelectronic nitrogen analogue of the carboxylate ligand $[R'C(NR)_2^-$, cf. $R'CO_2^-]$ has been successfully employed to prepare $Mo_2(amidinato)_4$ from both N,N'-diphenylbenzamidine and N,N'-ditolylbenzamidine by reaction with molybdenum hexacarbonyl (84).

The above reactions, which produce high yields of quadruply bound molybdenum dimers from the readily available molybdenum hexacarbonyl, remain unchallenged as the principal source of compounds containing a molybdenum quadruple bond. Conversion to the octachlorodimolybdate ion (36) with concentrated hydrochloric acid, as mentioned earlier, furnishes the $Mo_2Cl_8^{4-}$ anion, which is isoelectronic with the $Re_2Cl_8^{2-}$ ion and undergoes substitution and further ligand exchange under various reaction conditions.

Organometallic derivatives have been synthesized from monomeric molybdenum precursors, as well as from tetraacetatodimolybdenum (114). The reaction of $MoCl_3\cdot3THF$ with methyllithium in ether produced $Li_4Mo_2(CH_3)_8\cdot2Et_2O$ (152), which was structurally similar to the $Mo_2Cl_8^{4-}$ ion. The ionic $Li_3Mo(C_6H_5)_6$ reacted with $MoCl_3\cdot3THF$ to produce a compound of empirical formula $Li_2Mo(C_6H_5)_2H_2 \cdot 2THF$, which may contain a quadruple bond with a formulation involving the dimeric $Mo_2(C_6H_5)_4H_4^{4-}$ ion (153).

3. Technetium

The isolation of strong multiply bound dimers of technetium from technetium monomers has been limited in scope. In spite of expectations that species analogous to those of the heavier third row rhenium dimers could be synthesized, the prototype technetium dimer that was initially prepared was the paramagnetic $Tc_2Cl_8^{3-}$ ion, and this was obtained by formation of $TcCl_6^{2-}$ from pertechnetate followed by zinc reduction in hydrochloric acid (121). The extra electron relative to $Re_2Cl_8^{2-}$ did not alter the essential structural features of the $Tc_2Cl_8^{3-}$ anion, which conformed to the D_{4h} symmetry found previously for $Re_2Cl_8^{2-}$ with an eclipsed arrangement of chloride ligands (29). In more recent work, preparation and purification of the $Tc_2Cl_8^{2-}$ ion (228) has been reported. Reduction of pertechnetate with hypophosphorous acid produces both $TcCl_6^{2-}$ and $Tc_2Cl_8^{2-}$. Precipitation as the tetrabutylammonium salt followed by a careful purification and recrystallization sequence produced pure $[(n\text{-}C_4H_9)_4N]_2Tc_2Cl_8$, which was isomorphous with the rhenium analogue as judged by x-ray determination of the cell constants.

4. Chromium

As is mentioned in the introductory material, chromium exhibits a propensity to form quadruple bonds that is unmatched by any other first-row transition metal. Although $Cr_2(O_2CCH_3)_4(H_2O)_2$ was first reported over 100 years ago (210), it was not promoted as a quadruple-bond-containing unit until Cotton and co-workers carried out a redetermination of the x-ray structure and reported a change in the chromium atom separation from the accepted value of 2.64 Å initially reported in 1953 (251) to 2.3855(5) Å (66, 67). Reduction of chromium trichloride to chromium(II) in aqueous solution with zinc followed by addition of sodium acetate is a convenient route to $Cr_2(O_2CCH_3)_4(H_2O)_2$ (204), which can be dehydrated by heating under vacuum to form $Cr_2(O_2CCH_3)_4$.

Organometallic chromium dimers have been prepared from $CrCl_2$, $CrCl_3(THF)_3$, and $CrBr_2(THF)_2$ in different reactions. Allyl Grignard reagents form $Cr_2(allyl)_4$ (254), while methyllithium in ether produces $Li_4[Cr_2(CH_3)_8](Et_2O)_4$ (184). Other alkylation reactions have resulted in dimers such as $Li_4Cr_2(C_4H_8)_4 \cdot 4THF$ (181, 183), $Cr_2[(CH_2)_2P(CH_3)_2]_4$ (185), and $Cr_2(o\text{-}C_6H_4OCH_3)_4$ (141).

5. Tungsten

Unlike the lighter Group VI transition metals tungsten is reluctant to form homonuclear quadruple bonds. No system analogous to the extensive series of $Mo_2(O_2CR)_4$ carboxylate dimers has been characterized for tungsten, although compounds of similar stoichiometry have been isolated (85, 237). The first confirmation of a tungsten–tungsten quadruple bond resulted from isolation of $Li_4[W_2(CH_3)_8](Et_2O)_4$ by reduction of tungsten tetrachloride with excess methyllithium in ether (89). A crystal structure of a sample containing the statistically disordered $W_2(CH_3)_{8-x}Cl_x^{4-}$ anion confirmed the existence of a quadruple bond in this tungsten dimer (51).

More recently the $W_2(C_8H_8)_3$ species was prepared from tungsten tetrachloride and potassium cyclooctatetraenide in THF (88). The crystal structure of this air-stable dimer has been interpreted in terms of a quadruple bond, albeit the tungsten–tungsten separation is a relatively lengthy 2.38 Å as compared to the 2.26 Å distance found in the $W_2Me_{8-x}Cl_x^{4-}$ ion noted above. Earlier reports of $W_2(C_8H_8)_3$ did not include structural or preparative details (32).

6. Ruthenium

The odd-electron dimer $Ru_2(O_2CC_3H_7)_4Cl$ exhibits a surprisingly short ruthenium–ruthenium separation of 2.28 Å (19), and in view of the structural similarities that exist with quadruply bound carboxylate dimers of Groups VI and VII, it is appropriate to include the ruthenium species in this section. The first report of the series $Ru_2(O_2CR)_4Cl$ was that of Stephenson and Wilkinson (238) where refluxing ruthenium trichloride with the carboxylic acid led to the observed dimeric product. The initial report of three unpaired electrons per dimeric unit (238) has been confirmed by a thorough temperature-dependent study of the magnetic behavior (102). These ruthenium dimers are unique in that 11 electrons are available to populate the metal orbitals that are not involved in metal–ligand bonding, and further understanding of the role played by the excess electrons relative to the optimal number of 8 found for the other dimeric carboxylates discussed above will provide a clearer description of the bonding in other cases of importance.

7. Rhodium

No report involving incorporation of rhodium into a quadruple bond has been published. For the sake of completeness the dimeric rhodium carboxylates are noted here. The $Rh_2(O_2CCH_3)_4$ dimer can be conveniently synthesized from rhodium(III) chloride and sodium acetate (215), although routes from

$Rh(OH)_3 H_2 O$ (161, 236) and hexachlororhodate(III) have also been reported (19). The carboxylates are initially isolated as the solvent adduct, but heating under vacuum is adequate to prepare the anhydrous product. A variety of donors can occupy the terminal positions of the copper acetate type structure found in these rhodium dimers. No simple explanation of the bond order in these compounds seems adequate to describe the metal–metal attractive force in view of the number of electrons available and the prominent role of axial ligands in the total bonding scheme (200).

8. Heteronuclear Quadruple Bonds

Several dimers containing two different metal atoms joined by a bond of order four have been isolated. Addition of $Mo(CO)_6$ to a refluxing solution of $Cr_2(O_2 CCH_3)_4(H_2 O)_2$ yields the mixed-metal dimer $CrMo(O_2 CCH_3)_4$, which was characterized by spectroscopic and analytical data including a set of parent ion peaks in the mass spectrum (134).

McCarley et al. first succeeded in introducing tungsten into a carboxylate-bridged metal–metal quadruple bond by synthesizing $MoW[O_2 CC(CH_3)_3]_4$ as the major component of a mixture containing both the heteronuclear dimer and the tetrapivalatodimolybdenum (191). A mole ratio of 6:1 $W(CO)_6/Mo(CO)_6$ was employed with pivalic acid in refluxing chlorobenzene to attain nearly 80 mole % mixed dimer in the isolated solid. Later chemical oxidation and reduction reactions effected complete separation of the mixture and produced pure tetrapivalatomolybdenumtungsten.

D. Ligand Substitution Reactions

The proliferation of quadruple bonds that has occurred since the recognition of such a structural unit in 1964 (62) has primarily resulted from substitution reactions in which the strong multiple bond retains its integrity. As is emphasized in the previous section, there are relatively few synthetic routes commonly employed to generate quadruple bonds from monomeric reagents. However, retention of multiple metal–metal bonding throughout a series of ligand substitutions under a variety of conditions has produced a vast array of quadruply bound metal compounds from the parent dimers. A compilation of quadruply bound dimers and substituted derivatives is presented in Table I.

1. Rhenium

The octachlorodirhenate(2–) anion undergoes halogen exchange in methanol with aqueous hydrogen bromide at $60°C$ to form $[(n\text{-}C_4 H_9)_4 N]_2 Re_2 Br_8$ in 98% yield (63). Kinetic data have been obtained for the reverse reaction of HCl

TABLE I
Representative Quadruply Bound Metal Dimers and Substituted Derivatives

Compound	Remarks	Ref.
A. RHENIUM		
$(Bu_4N)_2 Re_2 Cl_8$	$Re_3 Cl_9 + Et_2 NH_2 Cl;$	11, 12, 42, 63,
	$ReO_4^- + H_2 (g) + HCl (aq)$	64, 172
	$ReO_4^- + H_3 PO_2 + HCl (aq)$	
$(Bu_4N)_2 Re_2 Br_8$	$Re_2 Cl_8^{2-} + HBr;$	42, 63, 64
	$ReO_4^- + H_3 PO_2 + HBr (aq)$	
$(Bu_4N)_2 Re_2 (NCS)_8$	$Re_2 Cl_8^{2-} + SCN^-$	108
$(Bu_4N)_2 Re_2 (NCSe)_8$	$Re_2 Cl_8^{2-} + SeCN^-$	142, 143
$Na_2 Re_2 (SO_4)_4$	$Re_2 Cl_8^{2-} + SO_4^{2-} + H_2 SO_4$ in glyme	76
$Li_2 Re_2 (CH_3)_8 \cdot 2ET_2 O$	$Re_2 (O_2 CPh)_4 Cl_2 + MeLi; ReCl_5 + MeLi$	80, 226
$(NH_4)_2 [Re_2 Cl_6 (O_2 CH)_2]$	$Re_2 Cl_8^{2-} + NH_4 Cl + HCO_2 H$	170
$Re_2 (O_2 CR)_4 X_2$	$ReOCl_3 (PPh_3)_2 + CH_3 CO_2 COCH_3$	136, 173, 219
$\quad$ X = Cl; R = Me, Et	$Re_2 Cl_8^{2-} + RCO_2 H$	63
$\quad$ X = Cl; R = Me, Et,		
$\quad\quad$ $n-Pr, i-Pr, C_7 H_{15}$	$Re_3 Cl_9 + RCO_2 H$	239
$\quad$ X = Cl, Br;		
$\quad\quad$ R = Me, Et, $n-Pr$, Ph	$Re_2 X_8^{2-} + RCO_2 H$	98
$\quad$ X = Cl; R = $CH_2 Cl, CH_2 Ph,$		
$\quad\quad$ CMe_3, o, m and		
$\quad\quad$ $p-C_6 H_4 Me$	$Re_2 Cl_8^{2-} + RCO_2 H$	98
$\quad$ X = Cl, Br; R = Ph,		
$\quad\quad$ $p-C_6 H_4 Cl, p-C_6 H_4 Br$	$Re_2 (O_2 CMe)_4 X_2 + ArCO_2 H$	42
$\quad$ X = I; R = $n-Pr$	$Re_2 (O_2 CPr)_4 (H_2 O)_2^{2+} + HI$	28
$\quad$ X = SCN; R = $n-Pr$	$Re_2 (O_2 CPr)_4 Cl_2 + AgSCN$	239
$Re_2 (O_2 CMe)_2 (O_2 CCCl_3)_2 Cl_2$	$Re_2 (O_2 CMe)_4 Cl_2 + CCl_3 CO_2 H$	99
$[Re_2 (O_2 CPr)_4 (H_2 O)_2] [SO_4]$	$Re_2 (O_2 CPr)_4 Cl_2 + Ag_2 SO_4$	239
$Re_2 (O_2 CR)_2 X_4 \cdot L_2$		
$\quad$ X = Cl, Br; R = Me; L = $H_2 O$	$Re_2 X_8^{2-} + RCO_2 H$	99, 178
$\quad$ X = Cl; R = Me; L = py	$Re_2 (O_2 CR)_2 Cl_4 \cdot 2H_2 O + py$	99
$Re_2 (O_2 CCl_3)_2 Cl_4$	$Re_2 Cl_8^{2-} + CCl_3 CO_2 H$	99
$Re_2 (O_2 CPh)_2 I_4$	$Re_2 (O_2 CPr)_4 I_2 + PhCO_2 H$	28
$Re_2 (PhNC(Ph)NPh)_2 Cl_4$	$Re_2 Cl_8^{2-} + HN(Ph)C(Ph)NPh$	109
$Re_2 X_6 L_2$	$Re_2 X_8^{2-} + L$	
$\quad$ X = Cl, Br; L = PPh_3		64, 124
$\quad$ X = Cl; L = PPr_3, PBu_3		221
$\quad$ X = Cl, Br; L = PEt_3, $PEt_2 Ph$		20, 71
$\quad$ X = Cl; L = $PMePh_2$,		
$\quad\quad$ $PEtPh_2$, $PEt_2 Ph$		124
$\quad$ X = Cl, Br; L = tmtu		99
$[ReX_3 (dth)]_2$	$Re_2 X_8^{2-} + dth$	99, 158
$\quad$ X = Cl, Br		
$\{Re_2 [O_2 C(i-Pr)]_3 Cl_2\}[ReO_4]$	$Re_3 Cl_9 + HO_2 C(i-Pr) + O_2 (g)$	43
$[ReX_3 (dppe)]_2$	Magnetically dilute Re^{3+} centers	64, 158, 159, 160
$ReCl_3 (bipy)$	$Re_2 Cl_8^{2-} + bipy$	99, 158

TABLE I *contd*

Compound	Remarks	Ref.
$(Bu_4N)_2Re_2(NCS)_8L_2$ $L = PEt_2Ph, PPr_3, PPh_3,$ dppm, dppe	Magnetically dilute Re^{3+} centers	108, 198

B. MOLYBDENUM

Compound	Remarks	Ref.
$Mo_2(O_2CR)$	$Mo(CO)_6 + RCO_2H$	
R = Me		14, 42, 120
R = Ph, C_6H_4Me, C_6H_4OMe		2, 42

R = Et, n-Pr, i-Pr, C_6H_{11},
    $C_7H_{15}, C_3F_7, C_6H_4Me$,
    C_6H_4F, C_6H_4OH
		235

R = CF_3		96
R = H		1, 97, 155
R = CMe_3		68, 191
$Mo_2(DMP)_4$	$Mo_2(O_2CMe)_4 + LiDMP$	90
$[Mo_2(O_2CCH_2NH_3)_4](SO_4)_2$	$Mo_2Cl_8^{4-} + HO_2CCH_2NH_2$	116
$Mo_2(PhNC(Ph)NPh)_4$	$Mo(CO)_6 + PhC(NPh)(NHPh)$	84
$Mo_2(S_2COEt)_4$	$Mo_2(O_2CMe)_4 + K[S_2COEt]$	218, 234
$Mo_2(S_2CNR_2)_4$		
R = Et, i-Pr	$Mo_2(O_2CMe)_4 + Na[S_2CNR_2]$	216, 234
$[Mo(S_2PPh_2)_2]_n$	$Mo_2(O_2CMe)_4 + NH_4[S_2PPh_2]$	234
$Mo_2(O_2CPh)_4(PBu_3)_2$	$Mo_2Br_4(PBu_3)_4 + PhCO_2H$	223, 225
$Mo_2(O_2CPh)_4(diglyme)_2$	$Mo_2(O_2CPh)_4 + diglyme$	53
$Mo_2(O_2CR)_4$	R aromatic, $Mo(CO)_6 + RCO_2H$	155

O_2CR = 1-naphthoate,
    2-naphthoate,
    p-chlorobenzoate,
    p-fluorobenzoate,
    o-chlorobenzoate,
    anthracene-9-carboxylate,
    thiophene-2-carboxylate,
    furan-2-carboxylate		

$Mo_2[S(O)CPh]_4$	$Mo_2(O_2CMe)_4 + PhC(O)S^-$	234
$Mo_2Cl_8^{4-}$	$Mo_2(O_2CMe)_4 + HCl$ (aq)	33, 34, 35, 36
$Mo_2Br_8^{4-}$	$Mo_2(SO_4)_4^{4-} + HBr$ (aq)	40
$K_4Mo_2(SO_4)_4$	$Mo_2Cl_8^{4-} + SO_4^{2-}$	26, 27, 75, 78
$[Mo_2(en)_4]Cl_4$	$Mo_2Cl_8^{4-} + en$	27
$Mo_2(O_3SCH_3)_4$	$Mo_2(O_2CMe)_4 + HSO_3CH_3$	154
$(NH_4)_2Mo_2(NCS)_8 \cdot 4DME$	$Mo_2(O_3SCH_3)_4 + NH_4SCN$	154
$Mo_2(O_3SCH_3)_2X_4^{2-}$		
X = Cl, Br, I	$Mo_2(O_3SCH_3)_4 + X^-$	154

TABLE I *contd*

Compound	Remarks	Ref.
$Mo_2(O_3SCF_3)_4$	$Mo_2(O_2CMe)_4 + HSO_3CF_3$	1
$[Mo_2(CH_3CO_2Et)_4][O_3SCF_3]_4$	$Mo_2(O_2CMe)_4 + HSO_3CF_3$ in ethyl acetate	1
$Mo_2(CH_3)_8^{4-}$	$Mo_2(O_2CMe)_4 + MeLi;$ $MoCl_3(THF)_3 + MeLi$	114, 152
$Mo_2(allyl)_4$	$Mo_2(O_2CMe)_4 + Li\ allyl$	104, 114, 254
$(Li_2MoH_2Ph_2 \cdot 2THF)_n$	$Li_3MoPh_6 + MoCl_3 \cdot 3THF$	153
$Mo_2(O_2CMe)_2(LL)_2$		
$LL = Bpz_2Et_2^-, HBpz_3^-$	$Mo_2(O_2CMe)_4 + LL$	52
$Mo_2(Bpz_2Et_2)_4$	$Mo_2(O_2CMe)_4 + Bpz_2Et_2^-$	52
$Mo_2[(CH_2)_2PMe_2]_4$	$Mo_2Me_8^{4-} + Me_4PCl$	182
$Mo_2X_4L_4$		
$X = Cl; L = PEt_3, P(n-Pr)_3,$ $P(n-Bu)_3, PMe_2Ph,$ $P(OMe)_3$	$Mo_2Cl_8^{4-} + L$	136, 221
$X = Cl; L = SMe_2, SEt_2,$ $S_2C_4H_8$	$Mo_2Cl_8^{4-} + L$	224
$X = Cl; L = CH_3CN, PhCN$	$Mo_2Cl_4(SMe_2)_4 + RCN$	224
$X = Cl; L = DMF, py$	$Mo_2Cl_4(SR_2)_4 + L$	37, 224, 225
$X = Br; L = py, P(n-Bu)_3$	$Mo_2Br_8H^{3-} + L$	224, 225
$X = Br; L = SMe_2, DMF$	$Mo_2Br_4L_4 + L$	224
$X = Cl; L = PEtPh_2,$ $PMePh_2, PEt_2Ph,$	$Mo_2Cl_8^{4-} + L$	136
$X = Br; L = PEt_3, P(n-Pr)_3$	$\beta - MoBr_2 + L$	136
$X = I; L = P(n-Pr)_3$	$\beta - MoI_2 + L$	137
$X = I; L = py, NH_3, \gamma - pic$	$Mo_2I_6(H_2O)_2^{2-} + L$	39
$X = Br; L = py, NH_3$	From $Cs_3Mo_2Br_7 \cdot 2H_2O$	38
$Mo_2X_4(LL)_2$		
$X = Cl; LL = dth, dtd, dtdd,$ bipy, tmedp	$Mo_2Cl_8^{4-} + LL$	224
$X = Br; LL = dtd, bipy$	$Mo_2Br_4L_4 + LL$	224
$X = Cl; LL = dppm, dppe,$ arphos, dpae, diars	From $Mo_2Cl_8^{4-}$ and $Mo_2Cl_4L_4$	23
$X = Br; LL = dppm, dppe$	$Mo_2Br_4(PEt_3)_4 + LL$	23
$X = Cl; LL = dipy, phen$	$Mo_2Cl_4py_4 + LL$	37
$X = I; LL = bipy$	$Mo_2I_4py_4 + bipy$	39
$Mo_2X_2(O_2CPh)_2(PBu_3)_2$	$Mo_2X_4(PBu_3)_4 + PhCO_2H$	213, 223, 225
$Mo_2Br_2(2,4,6-Me_3C_6H_2CO_2)_2(PBu_3)_2$	$Mo_2Br_4(PBu_3) + 2,4,6-Me_3C_6H_2CO_2H$	223, 225
$Mo_2Cl_2(7-azaindolyl)_2(PEt_3)_2$	$Mo_2Cl_4(PEt_3)_4 + 7-azaindolyl$	91
$(picH)_2[Mo_2X_6(H_2O)_2]$		
$X = Br, I$	$Mo_2Br_6(H_2O)^{2-} + HI$	39, 41
$Mo_2(C_8H_8)_3$	$Na_2C_8H_8$ preparation	32

C. CHROMIUM

$Cr_2(O_2CMe)_4 \cdot 2H_2O$	$CrCl_2(aq) + NaO_2CMe$	169, 204, 210

TABLE I *continued*

Compound	Remarks	Ref.
$Cr_2(O_2CMe)_4$	Heat $Cr_2(O_2CMe)_4 \cdot 2H_2O$ at 100 °C under vacuum	105, 204
$Cr_2(O_2CR)_4$ $R = Me, Et, Pr, Bu, C_5H_{11},$ $C_6H_{13}, C_7H_{15}, C_{11}H_{23},$ $C_{18}H_{37}$	$Cr^{2+} + RCO_2^-$	148
$R = CMe_3$		69
$Cr_2(O_2CR)_4L_2$		
$R = Ph; L = PhCO_2H$	$Cp_2Cr + PhCO_2H$	69
$R = CF_3 ; L = Et_2O$	$Cr_2(CO_3)_4^{4-} + CF_3CO_2H$	69
$R = H; L = py$	$Cr_3(O_2CH)_6(H_2O)_2 + py$	69, 149
$R = $ 9-anthracenecarboxylate;		
$L_2 = DME$	$Cp_2Cr + RCO_2H$	69
$[Cr(O_2CH)_2 \cdot XH_2O]_n$	$Cr^{2+} + HCO_2^-$	69, 122, 150, 151, 241, 242
$Cr_2(DMP)_4$	$Cr_2(O_2CMe)_4 + LiDMP$ $\left(Li-\text{(2,6-dimethoxyphenyl, OMe above and below)} \right)$	90
$Cr_2(\sigma-C_6H_4OMe)_4$	$CrCl_3 \cdot 3THF + Li-\text{(2-methoxyphenyl, OMe)}$	141
$Cr_2(CH_3)_8^{4-}$	$CrCl_2 + MeLi$	184
$Cr_2(allyl)_4$	$CrCl_2 + allyl\ MgX$	4, 232, 254
$Cr_2(C_4H_8)_4^{4-}$	$CrCl_3 + Li(CH_2)_4Li$	181, 183
$Cr_2(C_5H_{10})_4^{4-}$	$CrCl_3 + Li(CH_2)_5Li$	183
$Cr_2[(CH_2)_2PMe_2]_4$	$Cr_2Me_8^{4-} + Me_4PCl$	185
$Cr_2(C_8H_8)_3$	$Na_2C_8H_8$ preparation	31, 32
$Mg_2Cr_2(CO_3)_4 \cdot 6H_2O$		207
$Cr_2(O_2CCH_2CH_2CO_2)_2 \cdot 4H_2O$	$Cr^{2+} + (O_2CCH_2)_2^{2+}$	229
$Cr_2R_2(PMe_3)(\mu-R)_2$	$R = CH_2SiMe_3$	7

D. TUNGSTEN

Compound	Remarks	Ref.
$W_2Me_8^{4-}$	$WCl_4 + MeLi$	89
$W_2(C_8H_8)_3$	$WCl_4 + K_2C_8H_8$	32, 88
$W_2Me_{8-x}Cl_x^{4-}$		51

E. TECHNETIUM

Compound	Remarks	Ref.
$Tc_2Cl_8^{2-}$	$TcO_4^- + H_3PO_2 + HCl\ (aq)$	228
$Tc_2(O_2CCMe_3)_4Cl_2$	$Tc_2Cl_8^{3-} + HO_2CCMe_3$	79

with $Re_2 Br_8^{2-}$ in methanol (157). The rate expression has two terms with a zero and first-order dependence, respectively, on the concentration of hydrochloric acid times the concentration of the rhenium bromide dimer, suggesting a solvent-assisted pathway in addition to chloride attack.

The reaction of $Re_2 Cl_8^{2-}$ with excess acetic acid produces $Re_2(O_2 CCH_3)_4 Cl_2$ in nearly quantitative yields (63). Other carboxylic acids also form $Re_2(O_2 CR)_4 X_2$ compounds under exchange conditions where the acid is present in excess (98). The formation of $Re_2(O_2 CR)_4 X_2$ from $Re_2 X_8^{2-}$ (X = Cl, Br) was demonstrated for R = CH_3, $C_2 H_5$, $C_3 H_7$, and $C(CH_3)_3$, and aromatic analogues were best prepared by exchange of the acetate ligands in $Re_2(O_2 CCH_3)_4 X_2$. These reactions are reversible as shown by the conversion of $Re_2(O_2 CC_2 H_5)_4 Cl_2$ to $[(C_6 H_5)_4 As]_2 [Re_2 Cl_8]$ upon treatment with hydrochloric acid followed by addition of $[(C_6 H_5)_4 As] Cl$ (63). Thus the ability of the rhenium quadruple bond to withstand ligand exchange reactions was established in the early reports of $Re_2 X_8^{2-}$ reactivity patterns.

The conversion of $Re_2 X_8^{2-}$ to $Re_2(O_2 CR)_4 X_2$ represents the limiting product of ligand exchange reactions, and species with the intermediate composition $Re_2(O_2 CR)_2 X_4$ have been isolated. The $Re_2(O_2 CCH_3)_2 Cl_4 \cdot 2H_2 O$ dimer exhibits a cis orientation with respect to the four ligand atoms bound to each rhenium (177, 178), and $Re_2(O_2 CC_6 H_5)_2 I_4$ adopts the trans configuration about both metal centers (28). The eclipsed structure associated with quadruple bonds is found in both cases (Figs. 4 and 5).

Replacement of two halides in $Re_2 X_8^{2-}$ by two neutral donor ligands under mild conditions produces neutral dimers of the type $Re_2 X_6 L_2$, as has been established for L = PPh_3 (64), PEt_3 (20, 71), $P(C_3 H_7)_3$, $P(C_4 H_9)_3$ (221), and L = tmtu (tetramethylthiourea) (99). It is interesting to note that monomeric $ReX_3(tu)_3$ (tu = thiourea) results when thiourea reacts with $Re_2 X_8^{2-}$ in

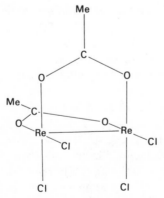

Fig. 4. The cis-type geometry of the $Re_2(O_2 CCH_3)_2 Cl_4$ unit in $Re_2(O_2 CCH_3)_2 Cl_4 \cdot 2H_2 O$ (178).

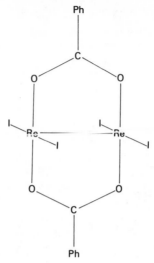

Fig. 5.   The trans-type geometry of $Re_2(O_2CC_6H_5)_2I_4$ (28).

methanol, so cleavage of the quadruple bond is facile in this case as opposed to the replacement of two halides that occurs with the tetramethylthiourea ligand, where the quadruple bond is retained. Phosphine substitution of $Re_2X_8^{2-}$ follows a second-order rate expression for $X = Cl$ while the bromo analogue has a first-order term as well as a second-order term in the rate expression (157). Ligand-induced reduction of rhenium(III) is common under more vigorous reaction conditions as discussed in Section III.I.2.

Complete replacement of the chloride ligands in $Re_2Cl_8^{2-}$ is effected with NaSCN in methanol acidified with acetic acid (108). The resulting $Re_2(NCS)_8^{2-}$ ion exhibited infrared spectral behavior associated with N-bound thiocyanate ligands. The $Re_2(NCSe)_8^{2-}$ analogue has also been prepared from $Re_2Cl_8^{2-}$ (142, 143). Isolation of the sulfato derivative $Re_2(SO_4)_4^{2-}$ as $Na_2Re_2(SO_4)_4 \cdot 8H_2O$ in 90% yield from $Re_2Cl_8^{2-}$ and Na_2SO_4/H_2SO_4 has been reported (76). Fusion of $(Bu_4N)_2Re_2Cl_8$ in N,N'-diphenylbenzamidine (mp 144°C) leads to $Re_2(amidine)_2Cl_4$, which can be recrystallized from chloroform to produce two crystalline modifications, one form containing a weakly coordinated THF molecule and chloroform of solvation in the unit cell (109). The structure is similar to that of $Re_2(O_2CC_6H_5)_2I_4$ in that the distribution of ligand atoms on each rhenium atom is trans (Fig. 6). Methyl substitution of the ligands in $Re_2(O_2CC_6H_5)_4Cl_2$ with methyllithium in diethyl ether at low temperatures produces $Li_2Re_2(CH_3)_8 \cdot 2Et_2O$ (80, 226). The reactions described above are selected examples in which the quadruple bond is retained and do not exhaust the possible reaction mechanisms available to multiply bound metal dimers. Bond cleavage to form monomers (146) or a change in metal—metal bond

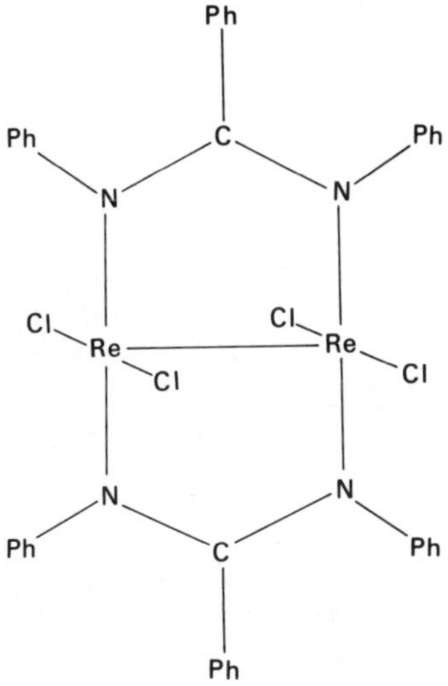

Fig. 6. The trans-type geometry determined for $Re_2[PhNC(Ph)NPh]_2Cl_4$ (109).

multiplicity can also occur as has been demonstrated by Nimry and Walton using various phosphine ligands (198). Reagents that induce redox processes are considered in a later section.

2. Molybdenum

Interconversions among molybdenum compounds containing quadruple bonds are numerous, and $Mo_2(O_2CCH_3)_4$ is the most common reagent. Carboxylate exchange reactions occur, as in the case of $Mo_2(O_2CCH_3)_4$ being converted to $Mo_2(O_2CCF_3)_4$ (96). The isolation of $Re_2Cl_8^{2-}$ and $Mo_2(O_2CCH_3)_4$ suggested that $Mo_2Cl_8^{4-}$ might exist and, indeed, Brencic and Cotton (33) determined the appropriate experimental conditions for conversion of $Mo_2(O_2CCH_3)_4$ to $K_4(Mo_2Cl_8 \cdot 2H_2O)$ and succeeded in isolating this material as a crystalline solid. The reaction of tetraacetatodimolybdenum(II) with hydrochloric acid is sensitive to the exact conditions employed and complicated reactions involving oxidation or hydrolysis can also occur, as was indicated by the original synthetic efforts of Sheldon et al. (5, 6, 8, 249). Preparative details for $K_4Mo_2Cl_8 \cdot 2H_2O$, $K_4Mo_2Cl_8$, $K_3Mo_2Cl_7 \cdot 2H_2O$, and $Rb_3Mo_2Cl_7 \cdot 2H_2O$ have been reported by Brencic and Cotton (36). Isolation of $[enH_2]_2[Mo_2Cl_8] \cdot 2H_2O$ from a hydrochloric acid solution of $Mo_2(O_2CCH_3)_4$ and ethylenediamine dihydrochloride (34) also demonstrated retention of the quadruple bond in the resulting $Mo_2Cl_8^{4-}$ ion of eclipsed geometry. A compound of stoichiometry $(NH_4)_5Mo_2Cl_9 \cdot H_2O$ has been characterized following precipitation of the chloride-substituted anion with ammonium as the counterion (35).

The basic molybdenum unit in this compound is again $Mo_2 Cl_8^{4-}$, with dimensions similar to those found in previous studies. The unusual stoichiometry merely reflects the presence of extraneous NH_4^+, Cl^-, and $H_2 O$ in the unit cell, which do not influence the basic structure of the dimeric molybdenum anion. The plethora of trimeric and tetrameric halide complexes reported during the late 1960s (5, 6, 8, 249) can all be rationalized as species with dimeric anions with metal halide stoichiometries resulting from $[Mo_2 X_8]^{4-}$, $[Mo_2 X_8(H)]^{3-}$, and $[NH_4(Mo_2 Cl_8)Cl]^{4-}$. Oxidation-state determinations on solutions of $Mo_2(O_2 CCH_3)_4$ in hydrohalic acids by Sheldon et al. in 1969 (5) indicated that oxidation to $Mo(3+)$ occurred; in retrospect this is seen to be consistent with the now well-established formation of $Mo_2 X_8 H^{3-}$ (86) as is discussed in Section III.I.3.

Isolation of the $Mo_2 Br_8^{4-}$ ion has proven more elusive than that of the chloride derivative. Brencic et al. have recently reported the characterization and crystal structure of $(NH_4)_4 Mo_2 Br_8$ formed by substitution of the sulfato ligands of $Mo_2(SO_4)_4^{4-}$ in hydrobromic acid solution (40). Vibrational data for $Cs_4 Mo_2 Br_8$ prepared from $Mo_2(O_2 CCH_3)_4$ and aqueous HBr was published in 1973 by Ketteringham and Oldham (165).

Bowen and Taube first reported substitution of $Mo_2 Cl_8^{4-}$ with $K_2 SO_4$ to form the pink $K_4 Mo_2(SO_4)_4$ in 1971 (26) and later refined the procedure using potassium sulfate in $0.2M$ $HSO_3 CF_3$ (27). The $Mo_2(en)_4^{4+}$ cation has been isolated as the chloride salt (27) upon heating $K_4 Mo_2 Cl_8$ in neat ethylenediamine followed by recrystallization of the solid product from aqueous acid. The electronic spectrum of the aquodimolybdenum(II) cation was recorded by Bowen and Taube (27) after precipitation of $BaSO_4$ from a solution of $K_4 Mo_2(SO_4)_4$ in $0.01M HSO_3 CF_3$. The initial report of Mo_2^{4+} as a stable species in aqueous solution (26) promotes the general concept of developing a coordination chemistry of these multiply bound metal fragments along the lines of classical monomeric complexes.

The unique acceptor behavior of the Mo_2^{4+} moiety has been explored by Abbott et al. (1). Isolation of the Mo_2^{4+} cation with a noncoordinating anion has not yet been achieved. This observation is consistent with the covalency that is evident in $Mo_2(O_3 SCF_3)_4$ (sublimes at $200°C$) and contrasts dramatically with the poor ligating behavior commonly associated with the trifluoromethanesulfonate ion. The availability of two cationic centers at a separation that correlates nicely with the donor atom locations on the trifluoromethanesulfonate ion promotes the bidentate binding interaction and forms a relatively strain-free five-membered ring of surprising stability. Dissolution of $Mo_2(O_3 SCF_3)_4$ in ethanol followed by addition of ligand produces derivatives of Mo_2^{4+} such as $Mo_2(O_2 CH)_4$, which formed instantaneously when formic acid was added to the ethanol solution.

Reactions of $Mo_2(O_2 CCH_3)$ with mono- and dithioacid anions produce a variety of products. Complexes with empirical formulas $Mo(S_2COEt)_2$, $Mo(S-\underset{\underset{O}{\|}}{C}-C_6H_5)_2$, and $Mo(S_2CNR_2)_2$ (R = Et, i-Pr) have been synthesized

(234). However, in light of a structural determination by Weiss et al. (216, 217) on the compound of stoichiometry $Mo(S_2CNR_2)_2$ (R = n-Pr), it is clear that one cannot assume retention of the quadruple bond with bridging ligands as in the acetate precursor. The stoichiometry of $Mo(S_2CNPr_2)_2$ was indeed confirmed, but ligand integrity was not maintained during the reaction, and the molecular structure includes two bridging sulfide ions with a thiocarboxamido attached to each molybdenum, as well as a normal chelating dithiocarbamate. This result can be viewed as an oxidative addition of one of the C–S bonds across the molybdenum d^4 center to form Mo(IV). The thiocarboxamido that results from the C–S bond cleavage is bound to molybdenum through both the C and the remaining S atoms of the original dithiocarbamate ligand. The dimeric structure of this sulfido-bridged compound is shown in Fig. 7. Generalizations regarding oxidation by similar ligands are not appropriate, since it is possible for simple substitution of the acetate ligands to occur with dithioacid anions, as revealed by the structure of $Mo_2(S_2COEt)_4$ · $2C_4H_8O$, where the dimolybdenum tetraacetate type structure is retained (218).

Phosphine and phosphite complexes of dinuclear molybdenum(II) units of the general form $Mo_2Cl_4L_4$ were first prepared by San Filippo (221). The $Mo_2Cl_8^{4-}$ anion present in $(NH_4)_5Mo_2Cl_9 \cdot H_2O$ (35) provided the quadruply bound substrate, which reacted with monodentate phosphous donor ligands in methanol to form the neutral $Mo_2Cl_4L_4$ derivatives with L = PEt_3, $P(C_3H_7)_3$, $P(C_4H_9)_3$, $P(C_6H_5)(CH_3)_2$, and $P(OCH_3)_3$. The ^1H and ^{31}P NMR data were interpreted in terms of a trans distribution of ligands on each $MoCl_2L_2$ component of the dimer. The relative orientation of the two molybdenum centers could then generate either a D_{2h} or a D_{2d} structure depending on

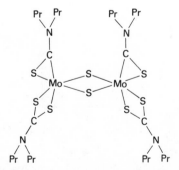

Fig. 7. The structure of $[MoS(SCNR_2)(S_2CNR_2)]_2$ (216).

whether identical ligands eclipse one another or whether phosphine and chloride eclipse one another.

An extensive investigation of tetrahalodimolybdenum(II) compounds with sulfur, nitrogen, and oxygen donor ligands has been completed (224). Electronic spectra, infrared spectra, Raman spectra, and elemental analyses were employed to characterize the isolated products. Pentaammonium nonachlorodimolybdenum(II) monohydrate was allowed to react with mono- and bidentate alkyl sulfides in methanol solution to form $Mo_2 Cl_4 L_4$, where L = $Et_2 S$ and $Me_2 S$, and $Mo_2 Cl_4 (LL)_2$ with LL = 2,5-dth, 4,7-dtd, and 5,8-dtdd. Replacement of the alkyl sulfide ligands in $Mo_2 Cl_4 L_4$ with py or bipy took place in refluxing methanol and produced $Mo_2 Cl_4 (py)_4$ and $Mo_2 Cl_4 (bipy)_2$. Although $(NH_4)_5 Mo_2 Cl_9 \cdot H_2 O$ was also a suitable reactant for preparation of the bipy derivative, it led to ill-defined products with pyridine and DMF. The DMF derivative $Mo_2 Cl_4 (DMF)_4$ was prepared by reaction of $Mo_2 Cl_4 (LL)_2$ bidentate sulfur derivatives with DMF. Acetonitrile was also introduced as a ligand by dissolving $Mo_2 Cl_4 (SEt_2)_4$ in acetonitrile solution and concentrating to dryness. Tetrachlorotetrakis(benzonitrile)dimolybdenum(II) was prepared by a similar reaction of the neat ligand with $Mo_2 Cl_4 (SMe_2)_4$ followed by precipitation with ether. Infrared similarities for the monodentate derivatives $Mo_2 Cl_4 L_4$ in the Mo—Cl stretching region consisted of two bands at 337 ± 15 and 285 ± 10 cm^{-1}. The NMR data accumulated in an earlier study (221), the structure reported for $Re_2 Cl_4 [P(C_2 H_5)_3]_4$ (72), and infrared data were cited in support of a D_{2d} structure of the type shown in Fig. 8. With bidentate ligands the Mo—Cl stretching region of the infrared spectrum differed significantly from the complexes containing only monodentate ligands. A cis-disposition of halide

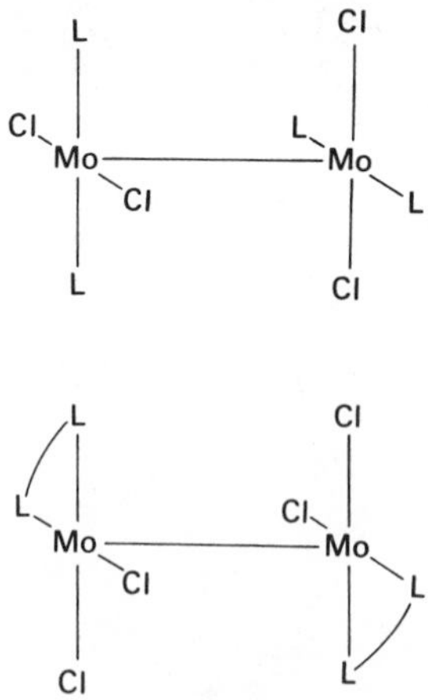

Fig. 8. Probable geometries for neutral Mo^{2+} derivatives of the type $Mo_2 Cl_4 L_4$ and $Mo_2 Cl_4 (LL)_2$ (224).

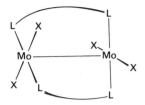

Fig. 9. Staggered structure proposed for $Mo_2X_4(LL)_2$ with LL = arphos, dpae and dppe (β form) (23).

ligands on each of the molybdenum centers was suggested as the origin of observed infrared differences, and a structure of the form shown in Fig. 8 was considered the most likely geometry, although other isomers could not be definitively eliminated.

Recently Walton et al. (23) explored bidentate ligand derivatives of the type $Mo_2Cl_4(LL)_2$, where LL is a bidentate phosphine or arsine. Precursors included $K_4Mo_2Cl_8$ (36), $Mo_2Cl_4(PEt_3)_4$ (221), $Mo_2Cl_4[P(n\text{-Bu})_3]_4$ (221), and $Mo_2Br_4(PEt_3)_4$ (136) and ligands incorporated into $Mo_2Cl_4(LL)_2$ units included dppm, dppe, arphos, dpae, and diars. Bromide analogues were isolated with dppm and dppe. Three types of behavior were observed. (1) The spectral characteristics of Mo_2X_4 (dppm)$_2$ were suggestive of bridging dppm ligands joining the two molybdenum atoms rather than chelating dppm. (2) The products formed with dppe, arphos, and dpae were spectroscopically similar to rhenium analogues such as $Re_2Cl_4(dppe)_2$ and $Re_2Cl_4(arphos)_2$ (123). An unusual staggered structure was proposed with the bidentate ligands bridging the two molybdenum atoms (Fig. 9) (3). For $Mo_2Cl_4(dppe)_2$ two isomers were isolated as a function of solvent and reactants, and the possibility of forming a chelate-containing dimer initially with the bridged ligand dimer thermo-dynamically favored as the second isomer was put forth.

Polypyrazolylborate complexes of quadruply bound dimolybdenum units have been prepared as a clever probe of the tendency of the axial positions trans to the multiple metal–metal bond to resist coordination even when the geometry of the ligand favors such binding (52). With dipyrazolyldiethylborate two products were obtained depending on the mole ratio of ligand to $Mo_2(O_2CCH_3)_4$ employed. A mixed-ligand complex, $Mo_2(O_2CCH_3)_2$-$[(pz)_2Et_2B]_2$, was isolated at a low ligand-to-metal ratio, while $Mo_2[(pz)_2Et_2B]_4$ resulted from excess ligand. A similar synthesis of $Mo_2[(pz)_3BH]_2(O_2CCH_3)_2$ from $KHB(pz)_3$ and $Mo_2(O_2CCH_3)_4$ in glyme was reported.

The octamethyldimolybdate(4–) anion can be prepared from the reaction of $Mo_2(O_2CCH_3)_4$ and $LiCH_3$ in ether (114). The tetraallyldimolybdenum compound prepared by Wilke et al. (254) is also readily synthesized from tetraacetatodimolybdenum and allyllithium or allylmagnesium bromide (114). A neutral organometallic derivative with phosphine ligands of the type $Mo_2(CH_3)_4[P(CH_3)_3]_4$ has been isolated from the reaction of $Mo_2(O_2CCH_3)_4$ with methyllithium in the presence of trimethylphosphine (7).

The recent report of the reaction of $Mo_2(O_2CCH_3)_4$ with 2,6-dimethoxy-phenyllithium to form $Mo_2(DMP)_4$ is highly significant in view of the resulting value determined for the molybdenum–molybdenum separation in this dimer (90).

Reaction of gaseous hydrogen halides with molybdenum(II) acetate at $300°C$ affords neutral β-MoX_2 phases for $X = Cl$ (136, 156), Br (136), and I (137). Thermolysis of $(NH_4)_4 Mo_2 Cl_8$ and $Mo(CO)_4 Cl_2$ also produces isomeric molybdenum dichloride (156), which differs from the well-known α-$MoCl_2$, where hexanuclear molybdenum units of the type $[Mo_6 Cl_8]^{4+}$ are present. Walton et al. have found that the β-MoX_2 phases react with trialkyl phosphines to form $Mo_2 X_4 L_4$ dimers and thus support the presence of strong multiple metal–metal bonds between pairs of molybdenum atoms in the solid (136, 137). Formulation as $[Mo_2 X_4]_n$ would seem to be appropriate for these β-MoX_2 phases as the parent halides of numerous discrete quadruply bound molybdenum(II) dimers.

3. Chromium

Although many quadruply bound dichromium units have now been characterized, the syntheses are generally from monomers or from $Cr_2(O_2CCH_3)_4$. No extensive series of substituted compounds with halide or phosphine ligands has been synthesized. The 1964 preparation of phenyl derivatives of the type $Cr_2(o$-$C_6 H_4 OCH_3)_4$ (141) has recently become an extremely noteworthy synthesis. The efforts of Cotton et al., who have prepared $Cr_2(2,6$-dimethoxy-phenyl)_4$ from tetraacetatodichromium and 2,6-dimethoxyphenyllithium (90), have led to identification of the chromium–chromium quadruple bond in this particular dimethoxyphenyl derivative as an example par excellence of strong homonuclear bonding among all of the elements.

It is well-known that an extensive chemistry of mononuclear chromium(II) exists in addition to the dimeric chemistry relevant to this review. Furthermore, oxidation of chromous acetate with a variety of oxidants has been shown to produce monomeric chromium(III) species. The rate law for disappearance of the chromous dimer has been interpreted as indicating that dissociation of the acetate dimer precedes the oxidation step (45).

4. Other Metals

The extensive derivative chemistry accessible to rhenium and molybdenum through ligand substitution has not been exploited in a similar manner for any other metal. The quadruply bound units for tungsten and technetium have yet to display the diversity of reactions characterizing the chemistry of their columnar transition metal analogues, molybdenum and rhenium, respectively.

E. X-ray Crystallographic Structural Data

A number of structures have been determined for compounds containing metal–metal quadruple bonds. Although various parameters are of interest in these structures, variation of the metal–metal bond length is of utmost interest and is therefore the only data given in Table II for all compounds. The structures invariably involve coordination of four equatorial ligating atoms relative to the metal–metal axis as defining the axial direction. Coordination of additional ligands at the terminal axial positions varies greatly as is discussed in the text.

A comparison of metal–metal bond-length variations for a given metal in differing environments among structures of homonuclear rhenium, molybdenum, and chromium dimers reveals that the flexibility of strong multiple chromium–chromium bonds is distinctly different from the behavior of the two heavier metals. Both rhenium and molybdenum display relatively constant quadruple bond lengths in most structures examined to date, while chromium exhibits a wide range of metal–metal distances.

Rhenium(III) dimers with eight electrons available for metal–metal bond formation generally exhibit metal–metal distances in the relatively narrow range of 2.22 ± 0.04 Å (see Table II). A systematic correlation of the nature of the equatorial ligand with the observed quadruple bond length is not evident, and, indeed, no such relationship necessarily exists. Where comparative data are available for similar dimeric units with differing coordination at the terminal sites trans to the multiple metal–metal bond, an increase in the quadruple bond length is observed with more tightly bound terminal ligands. Examples of such dimeric pairs include $[Re_2Cl_8]^{2-}$, $[Re_2Cl_8(H_2O)_2]^{2-}$ (81), and Re_2Cl_4 $[PhNC(Ph)NPh]_2$, $Re_2Cl_4[PhNC(Ph)NPh]_2 \cdot THF$ (109). Binding two axial H_2O ligands to the $Re_2Cl_8^{2-}$ ion (Re–O 2.66 Å) increases the metal–metal separation by 0.015(4) Å, while an increase of 0.032(6) Å is observed upon coordination of THF to one rhenium (Re–O 2.54 Å) in the Re_2 $[PhNC(Ph)NPh]_2Cl_4$ dimer. Taken by themselves these small variations are not sufficient evidence to present conclusions with regard to the interplay between metal–metal bond strengths and axial coordination, but the trends observed in these rhenium dimers are in accord with more abundant data for other quadruply bound metals.

A particularly significant structure is that of $Re_2Cl_4(PEt_3)_4$, which contains two Re(II) ions with a total of 10 electrons to fill the metal-based orbitals in this dimer (72). The structure of this dimer does not differ in any significant geometrical feature from the pattern established by numerous Re_2^{6+}

TABLE II
Structural Data for Quadruply Bound Metal–Metal Dimers and Related Derivatives

Compound	M–M Distance, Å	Comments	Ref.
A. RHENIUM DIMERS			
1. Re_2^{6+}		All dimers are eclipsed unless noted otherwise	
$K_2 Re_2 Cl_8 2H_2O$	2.241 (7)		82
$[(n-Bu)_4 N]_2 Re_2 Cl_8$	2.222 (2)		77
$Cs_2 Re_2 Cl_8 \cdot H_2O$	2.237 (2)	$Re_2 Cl_8^{2-}$	81, 176
	2.252 (2)	$Re_2 Cl_8 \cdot 2H_2O^{2-}$	
$(pyH)_2 Re_2 Cl_8$	2.244 (15)		28, 186
$(pyH)_2 Re_2 Br_8$	2.207 (3)		174
$Cs_2 Re_2 Br_8$	2.228 (4)		65
$Li_2 Re_2 (CH_3)_8 \cdot 2Et_2O$	2.178 (1)		80
$Re_2 Cl_6 (PEt_3)_2$	2.222 (3)		20, 71
$Re_2 (O_2 CPh)_4 Cl_2 \cdot 2CHCl_3$	2.235 (2)	17 $Cl-Re-Re-Cl$, $Re-Cl_{ax}$ 2.49Å	
$[Re_2 (O_2 CC_3 H_7)_4][ReO_4]_2$	2.251 (2)	ReO_4^- serves as the counterion	44
$Na_2 Re_2 (SO_4)_4 \cdot 8H_2O$	2.214 (1)		76
$Re_2 (O_2 CPh)_2 I_4$	2.198 (1)	trans I^- ligands	28
$[Re_2 (O_2 CC_3 H_7)_3 Cl_2][ReO_4]$	2.259 (3)	$Re-Re$, $Re-O_{ax}$ 2.28Å	43
$Re_2 Cl_4 (O_2 CCH_3)_2 \cdot 2H_2O$	2.224 (5)	178 $H_2O-Re-Re-OH_2$ cis Cl^- ligands, $Re-O_{ax}$ 2.50	

TABLE II *contd*

Compound	M–M Distance, Å	Comments	Ref.
$Re_2Cl_4(PhNC(Ph)NPh)_2$	2.177 (2)	Trans Cl^- ligands	109
$Re_2Cl_4(PhNC(Ph)NPh)_2 \cdot THF$	2.209 (1)	THF in one axial position, $Re-O_{ax}$, 2.54 Å	109
$(NH_4)_2Re_2Cl_6(O_2CH)_2$	2.260 (5)	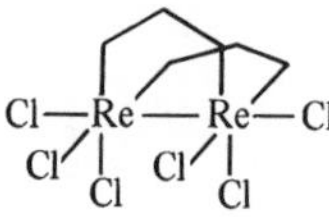 $Cl—Re——Re—Cl, Re—Cl_{ax}$ 2.71Å	170, 179

2. Re_2^{5+}

Compound	M–M Distance, Å	Comments	Ref.
$Re_2Cl_5(dth)_2$	2.293 (2)	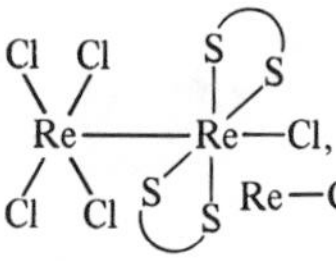 $Re——Re—Cl$, staggered, $Re—Cl_{ax}$ 2.51 Å	21, 22
$Re_2Cl_5(dppm)_2$	2.263 (1)	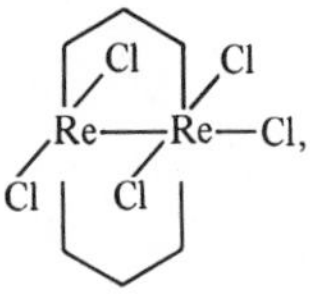 $Re—Re—Cl, Re—Cl_{ax}$ 2.58 Å	111

3. Re_2^{4+}

Compound	M–M Distance, Å	Comments	Ref.
$Re_2Cl_4(PEt_3)_4$	2.232 (5)	Trans Cl^- ligands 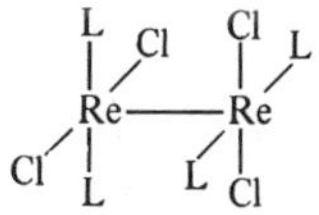	72

B. MOLYBDENUM DIMERS

1. Mo_2^{4+}

Compound	M–M Distance, Å	Comments	Ref.
$Mo_2(O_2CCH_3)_4$	2.093 (1)		94
$Mo_2(O_2CCF_3)_4$	2.090 (4)		96
$Mo_2(O_2CCF_3)_4 \cdot 2py$	2.129 (2)		95
$Mo_2(PhNC(Ph)NPh)_4$	2.090 (1)		84
$Mo_2(O_2CPh)_4$	2.096 (1)		68
$Mo_2(O_2CPh)_4 \cdot (diglyme)_2$	2.100 (1)		53
$Mo_2(O_2CCMe_3)_4$	2.088 (1)		68
$Mo_2(S_2COEt)_4 \cdot 2THF$	2.125 (1)		218
$Mo_2(C_3H_5)_4$	2.183 (2)	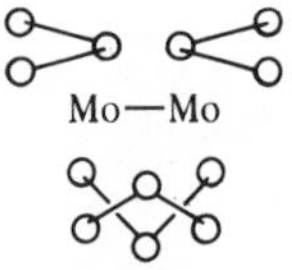 $Mo—Mo$	104
$K_4Mo_2Cl_8 \cdot 2H_2O$	2.139 (4)		33
$K_4Mo_2(SO_4)_4 \cdot 2H_2O$	2.111 (3)		75

241

TABLE II   *continued*

Compound	M–M Distance, Å	Comments	Ref.
$Li_4Mo_2(CH_3)_8 \cdot 4THF$	2.147 (3)		114
$(enH_2)_2Mo_2Cl_8 \cdot 2H_2O$	2.134 (1)		34
$(NH_4)_5Mo_2Cl_9 \cdot H_2O$	2.150 (5)	The $Mo_2Cl_8^{4-}$ unit is present	35
$(NH_4)_4Mo_2Br_8$	2.135 (2)		40
$[Mo_2(O_2CCH_2NH_3)_4](SO_4)_2(H_2O)_4$	2.115 (1)		116
$Mo_2(O_2CH)_4$	2.091 (2)		97
$Mo_2[(pz)_2BEt_2]_2(O_2CCH_3)_2$	2.129 (1)		52
$Mo_2[(pz)_3BH]_2(O_2CCH_3)_2$	2.147 (3)	One $Mo-N_{ax}$ 2.45 Å	52
$(picH)_2[Mo_2Br_6(H_2O)_2]$	2.122 (2)	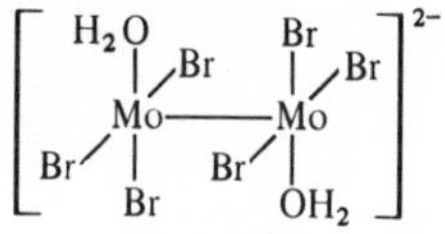	41
$Mo_2(O_2CPh)_2[(n-Bu)_3P]_2Br_2$	2.091 (3)	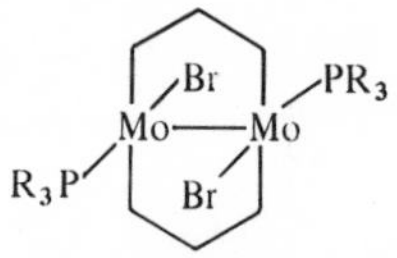	213
$Mo_2(C_7H_5N_2)_2(PEt_3)_2Cl_2$	2.125 (1)	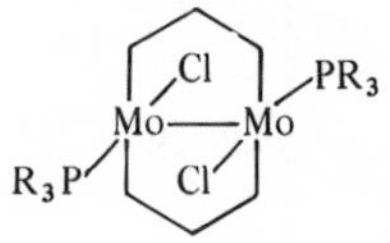	91
$Mo_2[2,6-C_6H_3(OMe)_2]_4$	2.064 (1)		90
2. Mo_2^{5+}			
$K_3Mo_2(SO_4)_4 \cdot 3.5H_2O$	2.164 (2)		75, 78
3. MoW^{5+}			
$MoW(O_2CCMe_3)_4I \cdot CH_3CN$	2.194 (2)		163
C. CHROMIUM DIMERS (Cr_2^{4+})			
$Cr_2(O_2CCH_3)_4 \cdot 2H_2O$	2.3855 (5)		66, 67
$Mg_2Cr_2(CO_3)_4 \cdot 6H_2O$	2.22		207
$Li_4Cr_2(CH_3)_8 \cdot 4THF$	1.980 (5)		180
$Li_4Cr_2(C_4H_8)_4 \cdot 4THF$	1.975 (5)	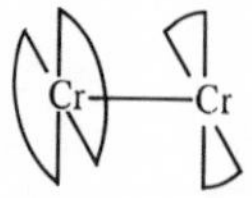	181

242

TABLE II   *continued*

Compound	M–M Distance, Å	Comments	Ref.

		Cr——Cr	
$Cr_2(C_3H_5)_4$	1.97	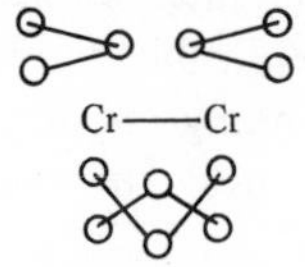	4, 10

$Cr_2(O_2CCH_3)_4$	2.288 (2)		105
$Cr_2(O_2CCF_3)_4(Et_2O)_2$	2.541 (1)	$Cr-O_{ax}$ 2.244 Å	69
$Cr_3(O_2CH)_6(H_2O)_2$	2.451 (1)	$Cr_2(O_2CH)_4$ unit is present with terminal ligands, $Cr-O_{ax}$ 2.224 Å	69
$Cr_2(O_2CH)_4(C_5H_5N)_2$	2.408 (1)	$Cr-N_{ax}$ 2.308 Å	69
$Cr_2(O_2CCMe_3)_4$	2.388 (4)		69
$Cr_2(O_2CPh)_4(PhCO_2H)_2$	2.352 (3)	$Cr-O_{ax}$ 2.295 Å	69
$Cr_2(O_2CC_{14}H_9)_4(CH_3OCH_2CH_2OCH_3)$	2.283 (2)	$Cr-O_{ax}$ 2.283 Å ($C_{14}H_9$ = 9-anthracenecarboxylate)	69

| $Cr_2(CH_2SiMe_3)_2(Me_3P)_2(\mu\text{-}CH_2SiMe_3)_2$ | 2.100 (1) | 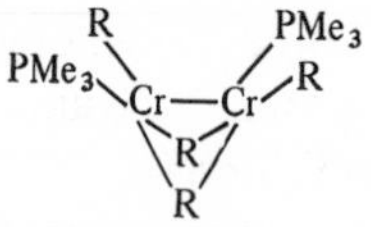 | 7 |
| $Cr_2[2,6\text{-}C_6H_3(OMe)_2]_4$ | 1.847 (1) | | 90 |

| $Cr_2(C_8H_8)_3$ | 2.214 (1) | 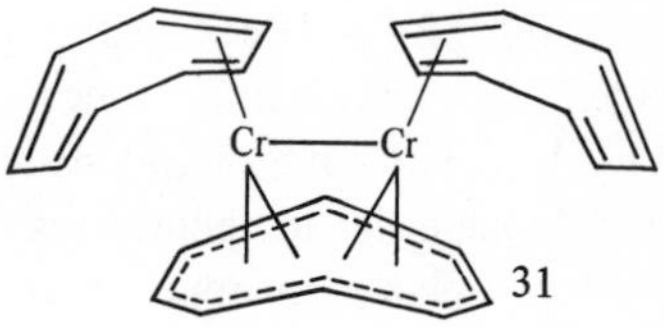 | 31 |

D. TUNGSTEN DIMERS (W_2^{4+})

| $W_2(C_8H_8)_3$ | 2.375 (1) | | 88 |
| $Li_4W_2(CH_3)_{8-x}Cl_x\cdot4THF$ | 2.261 | | 51 |

E. TECHNETIUM DIMERS

1. Tc_2^{5+}

| $(NH_4)_3Tc_2Cl_8\cdot2H_2O$ | 2.13 (1) | | 29 |
| $K_3Tc_2Cl_8\cdot nH_2O$ | 2.117 (2) | | 110 |

2. Tc_2^{6+}

| $Tc_2(O_2CCMe_3)_4Cl_2$ | 2.192 (2) | | 79 |

quadruply bound units. The eclipsed conformation of $Re_2 Cl_4 (PEt_3)_4$ may well reflect the steric impact of the bulky phosphine ligands being located so as to minimize repulsive interactions, and thus this geometry is not a particularly surprising or inexplicable result. The more interesting aspect of the structure with regard to bonding implications is the observed metal–metal bond length of 2.23 Å, which is near the midpoint of the range established for quadruple rhenium–rhenium bonds. It then follows that the excess two electrons present in the Re_2^{4+} case must enter molecular orbitals that are neither strongly bonding nor antibonding with respect to the metal–metal interaction. This result suggests that the δ component contributes very little to the quadruple bond strength as reflected in bond length if, indeed, the δ^* level is occupied by the two available electrons (see Section III.F for a more complete discussion of molecular orbital descriptions in quadruply bound dimers). Alternatively the excess electrons could enter nonbonding energy levels above the filled δ bonding orbital, but no calculations for the Re^{2+} dimers have been reported to suggest the presence or absence of such orbitals. The currently accepted MO scheme for rhenium(III) dimers with δ^* as the LUMO has been supported by extensive electronic spectral investigations (see Section III.G.1), but extrapolation to the rhenium(II) dimer discussed above produces an obvious discrepancy that requires resolution. (There is no a priori reason to assume that a similar ordering of energy levels should obtain in the two cases.)

Two intermediate oxidation-state rhenium dimers containing the Re_2^{5+} moiety have been structurally characterized with metal–metal separations of 2.26 and 2.29 Å [$Re_2 Cl_5 (dppm)_2$ (111) and $Re_2 Cl_5 (dth)_2$ (22), respectively], which are significantly longer than the corresponding distance in $Re_2 Cl_4$ $(PEt_3)_4$. The staggered structure of $Re_2 Cl_5 (dth)_2$ with chelating 2,5-dithiahexane ligands eliminates any possible δ contribution to the metal–metal bond and places an upper limit of three on the bond order (22). The relatively long Re–Re distance of 2.29 Å in this dimer fits nicely into the anticipated pattern for triple bonds as a result of the 0.07 Å elongation relative to typical quadruple bond lengths. The $Re_2 Cl_5 (dppm)_2$ dimer has an eclipsed conformation with bridging bidentate phosphine ligands (111). If the bond length of 2.26 Å is a reflection of electronic factors due to the added electron weakening the metal–metal multiple bond slightly and hence shifting the metal atom separation toward the high side of observed quadruple bond values, a compounding of this effect would be anticipated in the more reduced $Re_2 Cl_4 (PEt_3)_4$ case. Such is not observed. The separation actually decreases to 2.23 Å in the Re_2^{4+} derivative as discussed above.

Molybdenum(II) dimers display a slightly larger span of metal–metal distances than the quadruply bound rhenium(III) units. Certain ligand types can be associated with fairly specific metal–metal bond lengths in the molybdenum species. Thus molybdenum quadruple bonds differ in this respect from the distribution found for rhenium distances, which fail to correlate with any simple ligand classification. Simple $Mo_2(O_2CR)_4$ structures invariably exhibit a separation near 2.09 Å. Anionic $Mo_2X_8^{4-}$ ions (X = Cl, Br, Me) have quadruple bond lengths of 2.14 ± 0.01 Å [d(Mo–Mo) in various salts containing the $Mo_2Cl_8^{4-}$ ion varies from 2.134 to 2.139 to 2.150 Å]. Numerous dimeric molybdenum(II) compounds containing sulfate, xanthate, pyrazolylborate, phosphine, and other ligands in the equatorial positions exhibit quadruple bond lengths in the intermediate region, as well as near the extremes of the 2.09–2.15 Å range (see Table II for a compilation of Mo–Mo distances). The bonding distinction between carboxylate-bridged molybdenum dimers and the anionic $Mo_2X_8^{4-}$ species evident in the differing bond distances is confirmed by spectroscopic studies (see Section III.G and III.H).

Two exceptions to the generalizations presented for molybdenum quadruple bond distances of 2.09–2.15 Å should be noted. A length of 2.183 Å characterizes the metal–metal bond in tetraallyldimolybdenum (104). This is significantly longer (0.03 Å) than the next closest value of 2.150 Å given in Table II for any Mo_2^{4+} unit [$Mo_2Cl_8^{4-}$ in $(NH_4)_5Mo_2Cl_9 \cdot H_2O$ (35)]. Since this compound is the only molybdenum dimer containing unsaturated organic ligands bound in a π fashion, it is tempting to speculate on the existence of a cause and effect relationship. The extremely short metal–metal distance found in $Cr_2(\eta^3\text{-}C_3H_5)_4$ (1.97 Å) (4, 10) precludes extension of such a hypothesis to relevant derivatives. No satisfactory explanation of the observed bond distance in $Mo_2(C_3H_5)_4$ is evident.

The second exception is notable in that it exhibits a distance (2.067 Å) significantly below the next closest value given in Table II [2.088 Å reported for $Mo_2(O_2CMe_3)_4$ (68)]. The structure of the 2,6-dimethoxyphenyl (DMP) derivative reported by Cotton et al. (90) is the only molybdenum(II) dimer yet investigated that has a quadruple bond substantially shorter than the numerous carboxylate-bridged compounds. Although the unique ligand properties of DMP responsible for enhancing the strength of the metal–metal bond have not been completely elucidated, the orientation of two unbound ortho methoxy groups over the vacant terminal positions of the dimer is certainly an important factor (Fig. 10). Further consideration of the dimethoxyphenyl ligand is reserved for inclusion in the discussion of chromium dimers that follows later in this section.

Molybdenum is reluctant to coordinate axial ligands along the internuclear metal–metal axis of quadruply bound dimers. Collins et al. (52) employed the $[(pz)_3BH]^-$ ligand in an effort to promote strong axial coordination of one of the three pyrazolyl groups based on the preference exhibited by $[(pz)_3BH]^-$ to bind to three mutually cis positions in octahedrally coordinated metal complexes. The structure of $Mo_2[(pz)_3BH]_2(O_2CCH_3)_2$ fulfilled the geometrical prerequisites for possible axial coordination by binding two of the nitrogen donor atoms of the potentially tridentate pyrazolylborate in a chelating

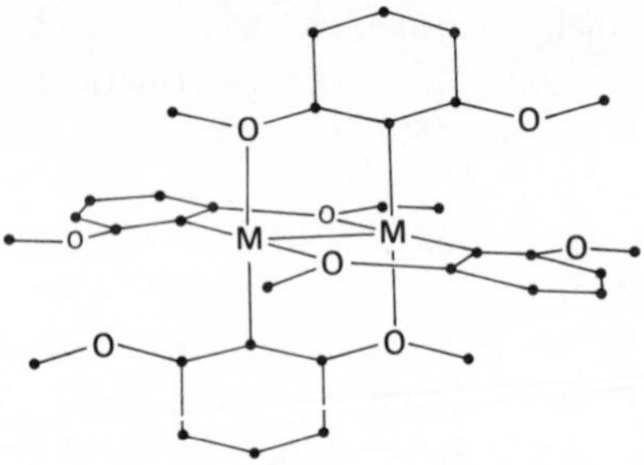

JOSEPH L. TEMPLETON

Fig. 10.   The molecular structure of $M_2(DMP)_4$, which exhibits the shortest known quadruple bond length for M = Cr and Mo (90).

fashion that should leave the third nitrogen free to bond to the terminal position. In fact, one of the axial positions is occupied by such a nitrogen (Mo—N, 2.45 Å), but the second axial site is vacant and the pyrazolyl ring that could bond is located well away from the metal centers. The conformation of the bidentate ligand excludes similar behavior by the second $[(pz)_3 BH]^-$ ligand as a result of steric congestion that prohibits identical orientations of the two, that is, at least one of the two must place a ring in position to bond axially to a molybdenum. The fact that neither pyrazolylborate ligand is prevented from tridentate coordination by steric considerations and yet only one does so (and this can be rationalized as a steric effect) is strong evidence for a very high trans influence associated with the molybdenum—molybdenum quadruple bond.

Data available for dimers of molybdenum(II) that differ only in the coordination of coaxial ligands suggests that the length of the quadruple bond is relatively unresponsive to changes in the weakly bound trans ligand. The largest shift induced by axial ligands is an increase of about 0.04 Å upon introduction of pyridine to the axial sites of $Mo_2(O_2 CCF_3)_4$ (95, 96). The benzoate derivative $Mo_2(O_2 CPh)_4$ has a quadruple bond length of 2.096(1) Å and a weak axial Mo—O interaction (2.88 Å), while the diglyme adduct exhibits a nearly identical metal—metal distance of 2.100(1) Å in spite of tighter Mo—O terminal bonding (2.66 Å) (68). The direction of the shift in quadruple bond length with more tightly bound trans ligands, the general reluctance of molybdenum to coordinate ligands in these positions, and additional germane data from rhenium and chromium dimers leads to the conclusion than an inverse relationship between axial ligand bond strength and metal—metal bond strength exists. Competition for the d_{z^2} orbital may be the major factor involved as electron donation from the incoming axial ligand places electron density in the metal—metal σ^* molecular orbital and hence sacrifices some metal—metal bond stabilization energy to establish a metal—ligand linkage.

One-electron oxidation of the Mo_2^{4+} unit in $Mo_2(SO_4)_4^{4-}$ leads to an increase of about 0.05 Å in the molybdenum—molybdenum bond length (78).

This lengthening is in accord with the effect anticipated for removal of an electron from a metal–metal bonding orbital, in this case producing a bond of order 3.5 based on the $\sigma^2 \pi^4 \delta^1$ configuration. A formal bond order of 3.5 is also appropriate for the metal–metal bond in the heteronuclear $MoW(O_2 CCMe_3)_4 I \cdot CH_3 CN$ dimer where a Mo–W distance of 2.19 Å obtains (163).

Chromium exhibits by far the greatest variation in metal–metal distances among quadruply bound homonuclear dimers of the elements Cr, Mo, and Re. The influence of axial ligands is reflected in the following structurally related pairs: (1) $Cr_2(O_2 CCH_3)_4$ (2.29 Å) (105) and $Cr_2(O_2 CCH_3)_4 \cdot 2H_2 O$ (2.39 Å) (66); (2) $Cr_2(O_2 CH)_4 \cdot 2py$ (2.41 Å) (69) and $Cr_2(O_2 CH)_4 \cdot Cr(O_2 CH)_2(H_2 O)_2$ (2.45 Å) (69). A trend toward increased chromium–chromium distances with more tightly bound terminal ligands is evident in the data of Table II for compounds of the form $Cr_2(O_2 CR)_4$ and $Cr_2(O_2 CR)_4 L_2$, which span a wide range of distances from 2.22 to 2.54 Å [this includes $Cr_2(CO_3)_4^{4-}$ at 2.22 Å]. No quantitative relationship has been proposed and in view of the several factors contributing to determination of the metal–metal separation in these dimers, no such numerical correlation is expected to apply.

Significantly shorter quadruple bond lengths characterize organometallic derivatives of the Cr_2^{4+} unit. Cr–Cr distances of 1.98 Å are found in the σ-bound alkyl anionic dimers $Cr_2(CH_3)_8^{4-}$ (180) and $Cr_2(C_4 H_8)_4^{4-}$ (181), while the π-allyl ligands in $Cr_2(allyl)_4$ also promote a short Cr–Cr separation of 1.97 Å (4, 10). The cyclooctatetraene derivative $Cr_2(COT)_3$ has a substantially longer metal–metal bond (2.21 Å) (31). Brauer and Krüger noted that the chromium atoms in $Cr_2(CH_3)_8^{4-}$, $Cr_2(C_4 H_8)_4^{4-}$, and $Cr_2(C_3 H_5)_4$ attain a 16-electron configuration, while an 18-electron configuration obtains in $Cr_2(COT)_3$. These same authors suggested that $Mo_2(COT)_3$ and $W_2(COT)_3$ should have structures similar to that of the chromium analogue, that is, a formal quadruple-bond description should apply (31). The isolation of single crystals of $W_2(COT)_3$ and succeeding x-ray structural determination carried out by Cotton and Koch confirmed the presence of a multiple metal–metal bond of order four in this tungsten dimer (88).

The unique ligating behavior of the bridging 2,6-dimethoxyphenyl ligand with respect to promoting a substantial decrease in the metal atom separation for molybdenum(II) dimers is even more prominent in the case of chromium. The chromium–chromium distance of 1.847(1) Å in $Cr_2(DMP)_4$ (90) is more than 0.1 Å less than the corresponding value in any other chromous dimer yet reported. To compare homonuclear multiple bonds among elements with inherently different atomic radii, Cotton, Koch, and Millar proposed a normalized value for internuclear distances based on Pauling's atomic radius of the element in question (209). A simple definition of "formal shortness" as $d(M–M)/2r(M)$ then follows as a measure of the relative compactness of the attractive interaction (90). The formal shortness ratio of 0.778 for the quadruple bond in

$Cr_2(DMP)_4$ not only falls below comparable ratios for other multiply bound metal dimers but also is slightly less than the formal shortness ratio calculated for the strongest homonuclear bonds known, for example, $N_2(0.786)$, C_2 (0.783), and $P_2(0.860)$. The utility of the formal shortness concept as a guide to the assessment of homonuclear attractive forces makes it a valuable addition to comparative chemistry of elements forming such bonds.

F. Molecular Orbital Schemes for Quadruple Bonds

The first elaboration of a quadruple bond between two atoms was put forth by Cotton et al. in 1964 (62). A qualitative molecular orbital scheme to account for the observed structure of the $Re_2Cl_8^{2-}$ anion followed in 1965 (58). The essential features of the electron distribution among the metal d orbitals of the original qualitative scheme have not been substantially altered by molecular orbital calculations that have since followed. To establish a basis for further discussion the qualitative MO scheme for $Re_2Cl_8^{2-}$ is presented here. It is most convenient to orient the z axis of the dimer along the Re—Re axis, consistent with the observed D_{4h} symmetry of the ion. Consider this line to be the axis of quantization for each rhenium as in Fig. 2. Utilization of metal s, p_x, p_y, and $d_{x^2-y^2}$ orbitals for Re—Cl σ bonds leaves four d atomic orbitals on each rhenium to interact in forming metal—metal molecular orbitals. Symmetry considerations lead to bonding and antibonding σ, π, and δ molecular orbitals where the p_z orbital of each rhenium has been included and two σ-nonbonding levels result as is discussed earlier. These orbitals are not strictly d in character, and ligand orbitals of appropriate symmetry can mix according to overlap and energy-matching considerations.

The earliest quantitative molecular orbital calculation was of the extended Hückel type and was presented by Cotton and Harris (83). The results of this treatment were in agreement with the original molecular orbital scheme as proposed by Cotton (58), but also discussed some semiquantitative features. The $^1A_{1g}$ ground state derived from an $(a_{1g})^2(e_u)^4(b_{2g})^2$ distribution of the eight metal electrons from the two Re^{3+} ions accounts for the observed ground-state structural and magnetic behavior. The quantitative accuracy of such Hückel methods was considered unsatisfactory by the authors, and electronic spectral assignments were tentatively made on the basis of the limited experimental data available at the time. This calculation provided further support for the existence of a δ component in addition to the strong π-bonding and σ-bonding overlap present between the two rhenium atoms. The eclipsed configuration and short metal—metal separation of 2.24 Å found in the $Re_2Cl_8^{2-}$ ion were considered consistent with this result.

Dubicki and Martin have examined the electronic spectra of several quadruply bound molybdenum carboxylates in solution and also in the solid state by means of diffuse reflectance measurements (120). A self-consistent charge and configuration semiempirical molecular orbital calculation was presented to aid in spectral assignments. Although the molybdenum acetate dimer is similar to the octachlorodirhenate anion in that both are d^4 metal ions involved in strong metal–metal bonding, the differences in metals, ligands, and oxidation states nullify the validity of any *a priori* prediction of identical molecular orbital diagrams. In the one-electron energy diagram reported for $Mo_2(O_2CCH_3)_4$, the highest filled molecular orbital was $b_{2g}(\delta)$, as had been proposed for the rhenium dimer. The vacant a_{2u} and a_{2g} nonbonding σ orbitals were reversed, the $b_{1g}(x^2-y^2)$ metal–ligand antibonding orbital was located between $b_{2g}(\delta)$ and $b_{1u}(\delta^*)$, and nonbonding ligand π-orbitals were abundant in the region of the σ and π metal–metal bonding orbitals, with regard to major differences in the MO scheme put forward for $Mo_2(O_2CCH_3)_4$ (120) compared to that of $Re_2Cl_8^{2-}$.

Application of the SCF–X_α–SW method to complex systems containing heavy transition metals bound to one another has been pursued in recent years as a first-principles calculation that seems to provide a basis for discussion of the electronic structures in such cases.

Norman and Kolari published an SCF–X_α–SW calculation for the $Mo_2Cl_8^{4-}$ anion and correlated the observed electronic spectrum with the predicted one-electron transitions (199, 201). The energy levels found corresponded well with the classical expectation for a transition metal complex in that the energy of the valence levels increased steadily going through Cl $3s$, Mo–Cl bonding, nonbonding Cl, Mo–Mo bonding, Mo–Mo antibonding, and Mo–Cl antibonding levels. Once more the highest filled orbital was identified as $2b_{2g}$, the δ-bonding orbital, with 89% Mo character according to this calculation. Indeed the σ, π, and δ orbitals of the quadruple bond were readily identified here as molecular orbitals of the appropriate symmetry with greater than 75% Mo character. The HOMO (highest occupied molecular orbital) and LUMO (lowest unoccupied molecular orbital) were assigned as δ and δ^*, respectively. One interesting feature of the bonding scheme for $Mo_2Cl_8^{4-}$ is that the Mo contribution to all the orbitals listed is largely from the $4d$ atomic orbitals with $a_{1g}(\sigma)$ and $a_{2u}(\sigma^*)$ the only orbitals with significant $5s$ or $5p$ character, and even here the contribution is less than 20%. Thus metal–ligand covalent bonding is derived from $d_{x^2-y^2}$ and d_{z^2} for the σ bonding and from d_{xy}, d_{xz}, and d_{yz} for the π bonding. The absence of $5s$ and $5p$ orbital contributions in $Mo_2Cl_8^{4-}$ eliminates the two nonbonding σ levels proposed in $Re_2Cl_8^{2-}$ (83) and $Mo_2(O_2CCH_3)_4$ (120), which are discussed earlier. However, between the δ^* and π^* levels are two metal–ligand σ antibonding orbitals (b_{1g} and b_{2u}) consisting largely of Mo $d_{x^2-y^2}$ character, which would be nonbonding relative to the metal–metal interaction. These orbitals are thus similar to the earlier proposed σ nonbonding orbitals of a_{1g} and a_{2u} symmetry

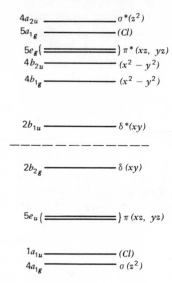

$4a_{2u}$ ———————— $\sigma^*(z^2)$

$5a_{1g}$ ———————— (Cl)

$5e_g\{$ ══════════ $\}\pi^*(xz,\ yz)$
$4b_{2u}$ ———————— $(x^2 - y^2)$
$4b_{1g}$ ———————— $(x^2 - y^2)$

$2b_{1u}$ ———————— $\delta^*(xy)$

———————————————

$2b_{2g}$ ———————— $\delta\ (xy)$

$5e_u\{$ ══════════ $\}\pi\ (xz,\ yz)$

$1a_{1u}$ ———————— (Cl)
$4a_{1g}$ ———————— $\sigma\ (z^2)$

Fig. 11. Molecular orbital diagram for the $Mo_2Cl_8^{4-}$ ion based on SCF–X_α–SW calculations; the highest filled level is $2b_{2g}(\delta)$ (199).

in that filling them should not effect the metal–metal bond substantially, since only the metal–ligand bonds should be weakened by the metal–ligand antibonding character of these orbitals (see Fig. 11).

Similar SCF–X_α–SW calculations have been performed on $Mo_2(O_2CH)_4$ (202, 203). The basic scheme found for molybdenum–molybdenum bonding orbitals in $Mo_2Cl_8^{4-}$ (199) is retained in the formate-bridged dimer, but interesting and significant comparisons emerge when $Mo_2Cl_8^{4-}$, $Mo_2(O_2CH)_4$, Mo_2^{4+}, and Mo_2 are considered collectively. The removal of excess positive charge from the hypothetical Mo_2^{4+} ion by ligand association promotes orbital expansion and greater metal–metal bonding by increased overlap. The increased covalency found for Mo–O bonds in the formate derivative relative to the Mo–Cl bonds in $Mo_2Cl_8^{4-}$ (deduced from greater Mo character in the metal–ligand bonding orbitals) effectively removes positive charge from the formal Mo_2^{4+} unit, and this is reflected in the decreased metal separation for the formate compound.

Another important piece of information that emerges from the $Mo_2(O_2CH)_4$ calculation is the extent of ligand contribution to stabilization of the D_{4h} structure. Mixing of ligand character into orbitals that are bonding between the two metal atoms promotes bonding character between ligands that are adjacent along the metal–metal axis. Furthermore, some metal–metal bonding character is mixed into molecular orbitals that are largely ligand in character, and filling δ^* may not cancel the total δ-bonding contribution exactly owing to the δ-bonding contributions from orbitals which supplement the contribution from the most prominent δ level of largely metal d_{xy} character. In summary then, the total picture is one in which complete separation of the metal orbitals from the ligand orbitals does not apply, and stabilization of the eclipsed geometry may be the result of other factors in addition to the δ-bonding component of the metal–metal interaction.

An SFC–X_α analysis for the $Re_2Cl_8^{2-}$ ion (194) produced an energy level pattern similar to that found in $Mo_2Cl_8^{4-}$. The HOMO and LUMO were the $b_{2g}(\delta)$ and $b_{1u}(\delta^*)$, respectively. The distribution of metal d character into the molecular orbitals is more dispersed in the Re(3+) dimer than in Mo(2+), and in particular there is extensive mixing of metal–ligand orbitals in the molecular orbitals of a_{1g} symmetry. An *ab initio* calculation of the bonding in the $Re_2Cl_8^{2-}$ ion was performed utilizing an effective core potential for all but the valence electrons in this dimeric anion (140). Representation of the d-like orbitals involved in metal–metal bonding as four pairs of two-configuration functions based on bonding- and antibonding-type combinations indicated that the overlap decreases from 0.65 to 0.51 to 0.11 for σ, π, and δ metal–metal bonding, respectively. The orbitals containing the eight electrons of the quadruple bond were largely $5d$ in character.

The $Mo_2Cl_8^{4-}$ and $Re_2Cl_8^{2-}$ complexes thus offer a firm basis for interpretation of the d^4 quadruply bound molecular orbital scheme applicable to Group VI and Group VII metals. The chemical behavior of these elements does not extrapolate to the remaining heavy metal congener in each group as neatly as one might anticipate, however. The location of the odd electron in the paramagnetic $Tc_2Cl_8^{3-}$ ion is of utmost importance in considering the nature of the LUMO for other quadruply bound systems. A CNDO calculation by Biagini Cingi and Tondello (24) indicated that the unpaired electron resides in the δ^* orbital rather than in a nonbonding level between the δ and δ^* orbitals.

Cotton and Kalbacher have calculated the ground-state electronic structure of both the paramagnetic $Tc_2Cl_8^{3-}$ ion, which exists with one electron in excess of the number required for quadruple bond formation, and of the $W_2Cl_8^{4-}$ ion, which has not been isolated to date (87). Once more the general energy level pattern resembled previous results for $Mo_2Cl_8^{4-}$ and $Re_2Cl_8^{2-}$. The extent to which metal–ligand orbital mixing occurs increases systematically from the M^{2+} to the M^{3+} with the $M^{2.5+}$ of intermediate covalent character. This is in accord with the expected increase in covalency as the oxidation state of the metal increases. One result of this orbital mixing is that $Mo_2Cl_8^{4-}$ has four orbitals that are predominantly derived from four metal d orbitals and correlate nicely with the qualitative quadruple bond scheme of σ, π, and δ metal–metal bonds. In $Tc_2Cl_8^{3-}$ the amount of metal character in each of these orbitals ($4a_{1g}$, $5e_u$, and $2b_{2g}$) is considerably less, and correspondingly greater metal contributions are found in lower lying orbitals, which are largely ligand based in the molybdenum case. This trend is continued in the $Re_2 Cl_8^{2-}$ ion, where each metal is formally $Re(3+)$ and the metal–metal bonding orbitals equivalent to those of $Mo_2 Cl_8^{4-}$ have even less metal character with increased metal contribution to other orbitals of lower energy.

The variation observed for absolute energies suggests that $W_2 Cl_8^{4-}$ would be more susceptible to oxidation than the known $M_2 Cl_8^{n-}$ quadruply bound dimers. The MO energies decrease from $Mo_2 Cl_8^{4-}$ to $Tc_2 Cl_8^{3-}$ with $Re_2 Cl_8^{2-}$ lying lowest. The $W_2 Cl_8^{4-}$ calculation does not identify any particularly unusual feature that would preclude isolation of this ion as a stable entity. The structural characterization of the $W_2 (CH_3)_x Cl_{8-x}^{4-}$ ($x = 4$ to 6) ion of course confirms the ability of tungsten to form homonuclear quadruple bonds (89).

While a consistent picture of the bonding in multiply bound dimers of Mo, W, Tc, and Re with D_{4h} symmetry has evolved smoothly from the early qualitative schemes through the more recent and extensive molecular orbital calculations, the analogous carboxylato-bridged chromium dimers have been a subject of greater controversy with regard to an accurate description of the energy levels and bonding scheme. One of the first publications on this topic was an *ab initio* MO calculation by Garner et al. (133) for the $Cr_2 (O_2 CH)_4 \cdot 2H_2 O$ dimer at a Cr–Cr separation of 2.362(1) Å that led to the initial conclusion that no net Cr–Cr bonding exists in this molecule. The eight electrons localized in the metal orbitals of the two Cr^{2+} ions were predicted to occupy completely two bonding and two antibonding levels of σ, δ, δ^*, and σ^* character with no resultant bond formation. Interpretation of the photoelectron spectrum was based on this assignment to support the calculation.

An extension of this work to other chromium dimers, including $Cr_2 (CH_3)_8^{4-}$, was reported in a later article (139). The SCF calculation for $Cr_2 (CH_3)_8^{4-}$ indicated no net bond for this nonbridged dimer, which has a Cr–Cr separation of less than 2.0 Å. Guest, Hillier, and Garner therefore concluded that correlation effects were important in these systems (139), and, indeed, the correlated wave function containing a large contribution from the quadruple bond configuration was found to lie below the SCF "no bond" ground state in every case analyzed by these same authors. Other *ab initio* calculations have been reported that agree with the conclusions reached by Garner et al. in their second publication (139) regarding the Cr–Cr interaction.

SCF calculations by Benard and Veillard (16) were consistent with an absence of Cr–Cr bonding in $Cr_2 (O_2 CH)_4$ before inclusion of configuration interaction (CI) completely altered the predicted ground-state configuration. The CI expansion led to a ground-state bonding configuration with the bonding metal–metal π orbitals containing four electrons in the $(\sigma)^2 (\pi)^4 (\delta)^2$ arrangement, in contrast to the $(\sigma)^2 (\delta)^2 (\delta^*)^2 (\sigma^*)^2$ configuration of the SCF ground-state prediction. The authors concluded that the SCF approximation alone is inadequate to correctly describe the $Cr_2 (O_2 CH)_4$ ground state. The leading term in a limited CI analysis produces a lower energy configuration corresponding to allocation of eight electrons to Cr–Cr bonding orbitals in accord with the descriptions generally accepted for the heavier metals in similar dimeric environments.

Cotton and Stanley (112) reported an SCF–X_α–SW calculation on $Cr_2(O_2CH)_4$ in support of the quadruple bond description of this system, in which the presence of metal–metal attraction was emphasized in terms of comparison with related systems in which far greater metal atom separations are realized in the absence of bonding forces. The bond multiplicity of four in this case is not inconsistent with the relatively long metal–metal distance, since the allocation of an excess of eight electrons to metal–metal bonding orbitals will fulfill the formal requirement for a bond order of four regardless of the actual strength of the bond as reflected in the metal–metal distance. Substantive differences between the SCF–X_α–SW results for $Cr_2(O_2CH)_4$ and $Mo_2(O_2CH)_4$ are evident in the metal character of the $4a_{1g}$ and $5a_{1g}$ orbitals: 18% Cr versus 75% Mo in $4a_{1g}$ and 99% Cr versus 48% Mo in $5a_{1g}$. The σ contribution to bonding between the metals is thus stronger in the molybdenum case and yet the observed "weak" Cr–Cr bond retains the electronic features required to formulate a formal quadruple bond.

It is appropriate to note here that Norman and Kolari have presented SCF–X_α–SW calculations for $Rh_2(O_2CH)_4(H_2O)_2$ and $Rh_2(O_2CH)_4$ (200). The absence of nonbonding levels in the energy region of metal–metal bonding and antibonding orbitals produces a bond order of one as the two d^7 metal ions populate three antibonding levels with six electrons beyond the eight bonding electrons utilized in quadruply bound dimers. The influence of the Rh–Rh interaction with trans H_2O ligands in the terminal positions of the bridged dimer is analyzed in detail. Competition between the metal–metal bond and the trans metal–ligand bond results in weaker metal–metal bonding as the strength of the metal–ligand bond increases. It is perhaps surprising that for these Rh^{2+} dimers the simplest formulation as a $\sigma^2\pi^4\delta^2\pi^{*4}\delta^{*2}$ configuration is basically confirmed by this calculation.

G. Electronic Spectra of Quadruply Bound Dimers

Investigations of the electronic structure of quadruply bound dimers have relied heavily on electronic spectra to reveal the nature of the excited-state configurations. The availability of quantitative molecular orbital calculations coupled with single-crystal polarized electronic absorption spectral studies of quadruply bound dimers at low temperature has firmly established certain features of the excited electronic states of these compounds. A discussion of electronic spectra follows and a tabulation of energies associated with $\delta \rightarrow \delta^*$ transitions in quadruply bound dimers is given in Table III.

TABLE III
Electronic Transition Energies Associated with the $\delta \to \delta^*$ Transition
in Metal–Metal Bound Dimers

Complex	$\delta \to \delta^*$ assignment (10^3 cm^{-1}) ($^1A_{1g} \to {}^1A_{2u}$)	Ref.
A. RHENIUM		
$(Bu_4N)_2 Re_2 Cl_8$	14.7	118, 243
$(Bu_4N)_2 Re_2 Br_8$	14.1	118, 119
$Re_2 Cl_6 (PEt_3)_2$	14.4	118, 246
$Li_2 [Re_2 (CH_3)_8] \cdot 2Et_2O$	18.6	226
B. TECHNETIUM		
$(Bu_4N)_2 Tc_2 Cl_8$	14.3	228
C. MOLYBDENUM		
$K_4 Mo_2 Cl_8 \cdot 2H_2O$	18.8	127, 199
$(NH_4)_4 Mo_2 Br_8$	18.0	50
$Mo_2 Cl_4 (PEt_3)_4$	17.0	119
$Li_4 [Mo_2 (CH_3)_8] \cdot 4Et_2O$	19.5	226
$K_4 [Mo_2 (SO_4)_4] \cdot 2H_2O$	19.4	92
$Mo_2 (O_3 SCH_3)_2 Cl_4^{2-}$	19.0	154
$Mo_2 (O_3 SCH_3)_2 Br_4^{2-}$	18.5	154
$Mo_2 (O_3 SCH_3)_2 I_4^{2-}$	17.8	154
$Mo_2 (aq)^{2+}$	19.8	27
$[Mo_2 (en)_4] Cl_4$	20.9	27
$Mo_2 Cl_2 (O_2 CPh)_2 (PBu_3)_2$	19.1	225
$Mo_2 Br_2 (O_2 CPh)_2 (PBu_3)_2$	19.0	225
$Mo_2 Br_2 (2,4,6-Me_3 C_6 H_2 CO_2)_2 (PBu_3)_2$	19.3	225
$Mo_2 (O_2 CH)_4$	30.8	203
D. CHROMIUM		
$Li_4 [Cr_2 (CH_3)_8] \cdot 4Et_2O_3$	22.0	226
E. TUNGSTEN		
$[W_2 (CH_3)_8]$	16.8	89
	($^2B_{2g} \to {}^2B_{1u}$)	
F. MOLYBDENUM		
$K_3 [Mo_2 (SO_4)_4] \cdot 3.5H_2O$	7.1	126
	($^2B_{1u} \to {}^2B_{2g}$)	
TECHNETIUM		
$K_3 [Tc_2 Cl_8] \cdot 2H_2O$	5.9	70

1. $Re_2 X_8^{2-}$ and Related Rhenium Dimers

Early characterization of $Re_2 Cl_8^{2-}$ included the electronic spectrum of the tetra-n-butylammonium salt in acetone, acetonitrile, and acidified methanol (63). The location of the observed band maxima (cm^{-1}), molar extinction coefficients, and oscillator strength ($f = (4.6 \times 10^{-9}) \epsilon_{max} \Delta \nu$, where ϵ is the decadic molar extinction coefficient and $\Delta \nu$ is the width of the absorption band in cm^{-1}) listed in that order were 14,500, 1530, 0.023; 32,800, 5650, 0.31; and 39,200, 8840, 0.65, respectively in acidified methanol. A substantial body of spectroscopic evidence has been accumulated that leads to the assignment of the low-energy absorption near 14,500 cm^{-1} as the $\delta \rightarrow \delta^*$ transition in this dimer (243). Cowman and Gray measured the polarized electronic absorption spectrum of the $Re_2 Cl_8^{2-}$ ion and found the lowest energy band at 14,700 cm^{-1} to be polarized parallel to the Re–Re axis (118). The intensity of this absorption was independent of temperature, implying that the transition is electric dipole allowed rather than vibronic in nature. In addition, the very slight red shift observed upon substitution of bromide for chloride suggests that the transition is metal localized. The absence of an A term in the MCD spectrum near 14,700 cm^{-1} is consistent with a nondegenerate excited state (119). All these observations support the $\delta \rightarrow \delta^*$ assignment for this transition, which should be z-polarized [$^1 A_{1g} \rightarrow {}^1 A_{2u}$ for $\delta(b_{2g}) \rightarrow \delta^*(b_{1u})$], electric dipole allowed, metal based, and thus approximately independent of the ligand and nondegenerate in the excited state.

A very complete analysis of the electronic spectral properties of $Re_2 Cl_8^{2-}$ and $Re_2 Br_8^{2-}$ has been published by Trogler et al. (243). Previous results are incorporated with low-temperature polarization measurements of thick crystals to assign the major bands, as well as weak absorptions present between the intense bands at 14,700 and 32,600 cm^{-1}. Observation of a moderately intense band in the region near 14,000 cm^{-1} for $(n\text{-}Bu_4 N)_2 [Re_2 Cl_8]$, $(n\text{-}Bu_4 N)_2 [Re_2 Br_8]$, and $Re_2 Cl_6 (PEt_3)_2$ that is similar for all three cases, with intensity independent of temperature and exhibiting a vibrational progression in a_{1g} (Re–Re stretch), provides a firm basis for accepting the $\delta \rightarrow \delta^*$ assignment of Cowman and Gray (118). Definitive assignment of this transition is of great interest in this and related quadruply bound dimers since these systems provide the only experimental opportunity yet available to characterize the δ component of a quadruple bond. Accepting the $\delta \rightarrow \delta^*$ assignment involves rationalizing the intensity of this band, which is of the $N \rightarrow V$ diatomic-type transition described by Mulliken (197). Very intense absorptions are characteristic of this general class of transitions, but here the oscillator strength of the $\delta \rightarrow \delta^*$ absorption depends on the square of the overlap integral S, which is only calculated to be about 0.1 for such bonds (248). This accounts for the relatively low value of the oscillator strength (0.023) reported for $Re_2 Cl_8^{2-}$ (63).

Quantitative application of Mulliken's equation $(f = (1.096 \times 10^{11})\bar{\nu}S^2 r^2$, where $\bar{\nu}$ is the transition energy in reciprocal centimeters, r is the internuclear distance in centimeters, and S is the orbital overlap) leads to predictions of $f \approx 10^{-2}$ for $\delta \to \delta^*$ transitions.

The energy region between 15,000 and 30,000 cm^{-1} displays four weak absorptions in acetonitrile at room temperature for the $Re_2 Cl_8^{2-}$ ion. A single crystal study at 15°K has identified the vibrational fine structure of these bands. Comparison of bands with respect to location and intensity for $Re_2 Cl_8^{2-}$ and $Re_2 Br_8^{2-}$ is of value in identifying metal-localized and ligand-based orbitals that are involved in electronic transitions in this study (243).

A band at 17,675 cm^{-1} in $Re_2 Cl_8^{2-}$ exhibits three vibrational progressions with an average frequency of 225 cm^{-1}, which is assigned to $a_{1g}(Re-Re)$. This compares to the ground-state vibrational frequency for $a_{1g}(Re-Re)$ of 274 cm^{-1} (30) and the $\delta \to \delta^*$ value of 248 cm^{-1} (194) and requires greater weakening of the metal–metal bond in this excited state than in the $\delta \to \delta^*$ transition excited state. The presence of a similar band at 17,475 cm^{-1} in the spectrum of $Re_2 Br_8^{2-}$ suggests that this transition is metal localized. The multiple vibronic origins, mixed polarization, and small oscillator strength are all consistent with a $\delta \to \pi^*$ ($^1A_{1g} \to {^1E_g}$) assignment. The possibility of a $\pi \to \delta^*$ transition is considered by the authors, but in view of the greater ligand character calculated for the π orbitals (194), a substantial halide dependence would be predicted for the $\pi \to \delta^*$ transition in contrast to the observed insensitivity that is characteristic of this band.

The second of the four weak bands in $Re_2 Cl_8^{2-}$ is centered at 20,940 cm^{-1} with multiple vibronic origins and a common progressional frequency of about 400 cm^{-1}. All the components are perpendicularly polarized and an intensity increase with temperature accompanies a slight red shift to indicate the vibronic character of the allowedness. The 400 cm^{-1} frequency was associated with a Re–Cl stretching mode of a_{1g} symmetry [ground state $a_{1g}(Re-Cl)$ 359 cm^{-1} (30)]. The analogous bromide band at 19,150 cm^{-1} in $Re_2 Br_8^{2-}$ supported the $\delta \to d_{x^2-y^2}$ assignment, since the red shift of about 1800 cm^{-1} was near the observed shift found for $d_{xy} \to d_{x^2-y^2}$ upon halogen exchange in the (n-Bu$_4$N)[OsNX$_4$] monomer (X = Cl, Br; red shift about 1400 cm^{-1}). In the D_{4h} dimer the two metal $d_{x^2-y^2}$ orbitals form linear combinations of b_{1g} and b_{2u} symmetry. Only the $^1A_{1g} \to {^1A_{1u}} [\delta(b_{2g}) \to d_{x^2-y^2}(b_{2u})]$ transition is limited to perpendicular polarization by vibronic coupling, hence the assignment. A very weak band at 23,645 cm^{-1} with a progressional frequency of 185 cm^{-1} is present in $Re_2 Cl_8^{2-}$. This has been tentatively assigned as $\pi \to \delta^*$, with the fine structure resulting from a bending mode, $a_{1g}(Re-Re-Cl)$. As is the case for all these weak bands, the absorption is electric-dipole forbidden.

The fourth and final band system between the more intense absorptions in $Re_2 Cl_8^{2-}$ exhibits two well-defined maxima at 27,000 and 28,100 cm^{-1} with red shifts of about 6000 cm^{-1} occurring upon bromide substitution. The halide sensitivity indicates that LMCT characterizes these transitions and the authors suggest transitions to the spin–orbit components of the LMCT state, $^1A_{1g} \rightarrow {}^3A_{2u}$ (a^3E_u and b^3E_u) account for these bands.

The more intense absorptions at 30,870 cm^{-1} for $Re_2Cl_8^{2-}$ and 23,630 cm^{-1} for $Re_2Br_8^{2-}$ are LMCT-based on the halide dependence. The substitution of bromide for chloride causes a red shift of about 9000 cm^{-1}, which is consistent with a ligand-based orbital contribution. Assignment of these bands for $Re_2Cl_8^{2-}$ and $Re_2Br_8^{2-}$ relies partially on the A terms present in solution MCD spectra at 32,300 and 23,800 cm^{-1}, respectively, which implies a degenerate excited state (119). The $^1A_{1g} \rightarrow {}^1E_u$ transition originating from ligand $e_g(\pi)$ orbitals meets this criterion. Furthermore, the observed polarization perpendicular to the metal–metal axis is also consistent with this electric-dipole allowed $X(\pi) \rightarrow \delta^*$ assignment (243).

Also relevant to the excited electronic states of $Re_2X_8^{2-}$ ions is the location of the triplet state derived from the $\delta \rightarrow \delta^*$ transition. After the singlet $^1A_{2u}$ state resulting from the $\delta \rightarrow \delta^*$ transition was determined to be the lowest energy electric-dipole-allowed transition by absorption spectroscopy, the emission spectra of these dimeric anions were investigated to probe the luminescent states resulting from excitation of the $\delta \rightarrow \delta^*$ transition (247). A broad emission at 13,020 cm$_{,}^{-1}$ results at 1.3°K when $Re_2Cl_8^{2-}$ is irradiated at 650 nm. Two key observations relevant to this emission are (1) the absorption and emission are not mirror images and (2) no overlap occurs between the emission and the 0–0 transition of the $^1A_{1g} \rightarrow {}^1A_{2u}$ ($\delta \rightarrow \delta^*$) excitation. These features suggest that emission occurs from the $^3A_{2u}$ state, since fluorescence from $^1A_{2u}$ would not be anticipated to exhibit either feature described above in the absence of a large geometrical distortion. The singlet–triplet separation of the $^1A_{2u}$ and $^3A_{2u}$ excited states would then fall into the range of 1000–3000 cm^{-1} based on such an analysis assigning the emission to one of the spin–orbit components, $A_{1u}(^3A_{2u})$ or $E_u(^3A_{2u}) \rightarrow {}^1A_{1g}$. This relatively small separation is similar to that found in related systems by analysis of the emission spectra of $Re_2Br_8^{2-}$ and $Mo_2(O_2CCF_3)_4$ (248).

Photochemical reactivity of the octachlorodirhenate(2−) anion has been investigated in acetonitrile (131, 135). Monomeric $ReCl_3(CH_3CN)_3$ ($\mu = 1.60$ BM) forms with a quantum yield of 0.017 ± 0.005 when $Re_2Cl_8^{2-}$ is irradiated at 313 nm. Cleavage of the quadruple bond to form $ReCl_4(CH_3CN)_2^-$ is followed by rapid conversion to the neutral monomer by means of acetonitrile replacement of chloride. No reaction occurs in the absence of irradiation at reflux, and the $\delta\delta^*$ excited state is not responsible for the observed photochemistry, since 632.8 nm radiation does not lead to cleavage of the rhenium dimer (135). Flash-photolysis studies suggest that the dominant photochemical process involves internal conversion to a transient $\delta\delta^*$ excited state, with a minor solvent-assisted pathway accounting for the formation of monomeric rhenium(III) products (131). The absence of photochemically induced redox reactions or quadruple bond cleavage in these rhenium(III) investigations is noteworthy.

2. $Mo_2X_8^{4-}$ $(X = Cl, Br)$ and $Mo_2(SO_4)_4^{4-}$

The extensive series of known quadruply bound molybdenum dimers has also been the subject of detailed experimental investigations to unravel the electronic properties characterizing excited states in these Mo_2^{4+} derivatives. Consideration of the $Mo_2X_8^{4-}$ spectra is presented first since the $Mo_2(O_2CR)_4$ compounds exhibit significantly different features in their electronic spectra.

Norman and Kolari correlated the solid-state spectrum of $K_4Mo_2Cl_8 \cdot 2H_2O$ with the predicted electronic transitions based on their SCF–X_α–SW calculation for $Mo_2Cl_8^{4-}$ (199). Two strong bands with maxima at 18,800 and 31,400 cm^{-1} were found between 4800 and 40,000 cm^{-1}, with an intense absorption rising above 34,000 cm^{-1} without maximizing. A weak band near 24,000 cm^{-1} was the only other absorption feature in the region studied. The three strong transitions are assigned as $\delta \rightarrow \delta^*$, $\pi \rightarrow d_{x^2-y^2}$, and $\pi \rightarrow \pi^*$ or $Cl \rightarrow \delta^*$ for $Mo_2Cl_8^{4-}$ (18,800, 31,400, >34,000 cm^{-1}), all of which are dipole allowed. The weaker absorption near 24,000 cm^{-1} is assigned to one or both of the dipole-forbidden $\pi \rightarrow \delta^*$ or $\delta \rightarrow d_{x^2-y^2}$ transitions. Quantitative agreement between the calculated and experimental transition energy for the $\delta \rightarrow \delta^*$ band has not been achieved based on one-electron energy diagrams to date (13,700 calculated versus 18,800 cm^{-1} observed), but a logical pattern that matched calculated and observed transitions was evident and led to a consistent interpretation of the molecular orbital scheme and electronic spectrum.

In 1973 Cowman and Gray reported that $K_4Mo_2Cl_8$ exhibited a band at 17,897 cm^{-1} with a progressional frequency of 351 cm^{-1} (118), but polarization data were not published until 1977 when Fanwick et al. measured single-crystal spectra of $K_4Mo_2Cl_8 \cdot 2H_2O$ at 300 and 3.7°K (127). The statistical disorder present in the crystal structure of $K_4Mo_2Cl_8 \cdot 2H_2O$ (33) indicated that 7% of the ions have the Mo–Mo axis perpendicular to the orthorhombic c axis and 93% have the internuclear axis aligned along c. These data were extrapolated to the single-crystal spectral analysis to account for the polarized absorption behavior of the sample. Polarization of the band near 19,000 cm^{-1} along the metal–metal bond was established in this study and a progression exhibiting somewhat variable peak separations was observed at 3.7°K. Assigning 18,083 cm^{-1} as the 0–0 transition for the $^1A_{1g} \rightarrow {}^1A_{2u}$ transition in the single crystal led to a value of 381 cm^{-1} for the first peak separation, while the final five separations averaged only 327 cm^{-1}. The three A_{1g} vibrations involving Mo–Mo stretch, Mo–Cl stretch, and Mo–Mo–Cl bending should form the basis for progressions observed in the $^1A_{2u}$ excited state, with the normal mode of Mo–Mo stretch expected to have the greatest intensity and a frequency below that of the ground-state value [345 cm^{-1} (9)] owing to the $\delta^1\delta^{*1}$ configuration. The irregularity of the observed separations was attributed to contributions from more than one vibrational mode (127).

The higher energy absorption near 31,400 cm^{-1} was polarized perpendicular to the metal–metal axis in agreement with the $^1A_{1g} \rightarrow {}^1E_u$ assignment of Norman and Kolari (199) for a $\pi \rightarrow d_{x^2-y^2}$ transition. A weak absorption at 28,800 cm^{-1} increased in height at low temperature and was interpreted as being indicative of a static dipole-allowed transition rather than a vibronically allowed band. The spin-forbidden $^1A_{1g} \rightarrow {}^3E_u$ transitions led to prediction of a weak absorption for $A'_1 \rightarrow A'_2$ in the D'_4 double group, which is appropriate because of spin–orbit coupling and thus supports the earlier assignment (199). The detailed analysis presented by Fanwick et al. (127) is consistent with the molecular orbital diagram generated by Norman and Kolari (199).

Excitation of $Mo_2Cl_8^{4-}$ at 1.3°K produces a broad emission centered at 14,950 cm^{-1} and no overlap of this emission band with the 0–0 transition near 18,000 is present (247). Arguments rationalizing the $Re_2X_8^{2-}$ emission properties are also applicable to $Mo_2Cl_8^{4-}$ and the increased energy separation between absorption and emission for $Mo_2Cl_8^{4-}$ provides further support of the assignment of the emitting state as one of the $^3A_{2u}$ spin–orbit components.

A structured absorption band near 6000 cm^{-1} reported in the spectra of some solid salts of $Mo_2Cl_8^{4-}$ was considered as a possible candidate for assignment as the spin-forbidden $\delta \rightarrow \delta^*$ transition, $^1A_{1g} \rightarrow {}^3A_{2u}$ (48), but a more recent publication, also from Clark's laboratory (50), suggests attributing this transition to $\delta \rightarrow \delta^*$ absorption by an oxidized surface impurity based on corresponding near infrared absorptions characteristic of $Mo_2(SO_4)_4^{3-}$ (126) and $Tc_2Cl_8^{3-}$ (70) (see Section III.I for properties of these odd-electron dimers).

Comparison of spectral data for $Mo_2Cl_8^{4-}$ and $Mo_2Br_8^{4-}$ was made possible by the isolation of $(NH_4)_4Mo_2Br_8$ in 1976 by Brencic et al. (40). Clark and D'Urso (50) assigned a strong band at 18,000 cm^{-1} with a vibrational progression averaging 320 cm^{-1} to the $\delta(b_{2g}) \rightarrow \delta^*(b_{1u})$ transition based on analogy with previous assignments for quadruply bound metal halide complexes (246). The small decrease of 16 cm^{-1} in the frequency of the Mo–Mo stretching mode upon excitation to the $^1A_{2u}$ state suggests that the contribution of the δ component to the total metal–metal bond energy is small in the $Mo_2Br_8^{4-}$ ion.

The $Mo_2(SO_4)_4^{4-}$ ion exhibits a 19,400 cm^{-1} absorption that is polarized along z in accord with the $\delta \rightarrow \delta^*$ transition (92). Thus $Mo_2(SO_4)_4^{4-}$ fits the pattern of $Re_2X_8^{2-}$ and $Mo_2X_8^{4-}$ in that the lowest energy absorption band characteristic of the quadruply bound dimers is assigned to the dipole allowed $\delta \rightarrow \delta^*$ transition.

3. $Mo_2(O_2CR)_4$ Compounds

The $Mo_2(O_2CR)_4$ dimers do not exhibit the electronic absorption pattern established by the quadruply bound dimers lacking carboxylato bridges. Dubicki and Martin reported electronic spectra for $Mo_2(O_2CCH_3)_4$ and three related carboxylato-bridged molybdenum dimers (120). A weak absorption centered near 23,000 cm^{-1} in solution showed extensive fine structure at 77°K with a progressional frequency of about 350 cm^{-1} for $Mo_2(O_2CCH_3)$. Although several possibilities for this band were considered and the origin of the progression was not firmly established, the authors concluded that the band was not due to the $\delta \rightarrow \delta^*$ transition. This early analysis, which became more anomalous as later studies of related dimers without carboxylate ligands consistently attributed the lowest energy band to the $\delta \rightarrow \delta^*$ transition, has remained valid and has, in fact, been supported by more recent investigations.

The second major absorption feature in $Mo_2(O_2CCH_3)_4$ was a strong band near 34,000 cm^{-1} with an oscillator strength of about 0.2 that was considered to be consistent with an $N \rightarrow V$ type transition, such as the one-electron excitation $\delta(2b_{2g}) \rightarrow \delta^*(2b_{1u})$, which would be localized on the metals.

The isolation and crystal structure of $Mo_2(O_2CCH_2NH_3)_4(SO_4)_2 \cdot 4H_2O$ (116) provided and identified an ideal tetracarboxylatodimolybdenum dimeric cation for polarization studies. Detailed analysis of the single-crystal absorption between 20,000 and 25,000 cm^{-1} indicated that, indeed, a $\delta \rightarrow \delta^*$ assignment would be incompatible with the experimental data collected (93). Two overlapping progressions were observed at 15°K with both (z) and (xz) polarization. The spectrum fit very nicely with assignment of a z-allowed transition in the S_4 symmetry of the crystal field with the 0—0 transition at 20,570 cm^{-1}. The weak intensity of this absorption reflects the virtual D_{4h} symmetry that applies to the molecule until the NH_3 orientations and the site symmetry are considered. The four progressions are all based on frequencies of 340—345 cm^{-1}, which are associated with the normal vibrational mode largely consisting of Mo—Mo stretching character. Vibronically allowed transitions at higher frequencies populate asymmetric vibrations in the excited state with frequencies of 940, 1220, and 1360 cm^{-1} to account for the total of four observed progressions.

The $\delta \rightarrow \delta^*$ assignment is of particular importance in describing the electronic structure of these dimers. The fundamental difference in the origin of the lowest energy absorption band established for $(Mo_2(O_2CCH_2NH_3)_4 - (SO_4)_2 \cdot 4H_2O$ (93) as compared to $Mo_2Cl_8^{4-}$ and $Mo_2(SO_4)_4^{4-}$ has been extended to $Mo_2(O_2CH)_4$ (92) and $Mo_2(O_2CCH_3)_4$ (248) by single-crystal studies.

The weak band near 23,000 cm^{-1} in $Mo_2(O_2CR)_4$ compounds exhibits extensive fine structure at low temperatures. In addition to the sophisticated experimental and theoretical analyses of $Mo_2(O_2CCH_2NH_3)_4^{4+}$ (93) that have been performed, the spectrum of $Mo_2(O_2CCH_3)_4$ has also been analyzed in detail (248). The oscillator strength of 0.0011 found for the 22,700 cm^{-1} band in $Mo_2(O_2CCF_3)_4$ is an order of magnitude below the 0.015 value of f reported for the $K_4Mo_2Cl_8$ band at 19,300 cm^{-1}, which is well established as the $\delta \rightarrow \delta^*$ transition. Emission from $Mo_2(O_2CCF_3)_4$ shows vibronic structure similar to that characterizing the absorption and lies 1800 cm^{-1} below the 0–0 absorption, suggesting relaxation to a relatively long-lived triplet state is responsible for the observed lifetime and energy of emission. The observed temperature dependence and two types of perpendicular polarization absorption data were consistent with an electric-dipole-forbidden transition to a degenerate electronic state split by low-site symmetry. Specific assignment of the 23,000 cm^{-1} band features for $Mo_2(O_2CCH_3)_4$ was based on a progressional spacing of about 370 cm^{-1} built on each of three vibronic origins. The 370 cm^{-1} spacing was attributed to the excited state a_{1g}(Mo–Mo) stretch, which is reduced by roughly 35 cm^{-1} relative to the ground state value of 406 cm^{-1} as established by Raman data (30). The splitting of the x and y components of an excited ^1E_g state coupled with an a_{2u} normal vibrational mode was suggested as the dominant vibronically allowed excitation process. The relative insensitivity of the transition energy to the nature of the carboxylate ligand argues against charge-transfer assignments. A $\delta \rightarrow \pi^*$ or $\pi \rightarrow \delta^*$ transition fits both the metal-localized requirement and the degeneracy criterion. Calculations of the SCF–X_α–SW type (203) place the $\delta \rightarrow \pi^*$ energy below the $\pi \rightarrow \delta^*$ energy, but even stronger support for the $\delta \rightarrow \pi^*$ assignment derives from the metal character of δ (89%) and π^* (96%) when compared to π (65%) and δ^* (86%), since this band is metal localized. The authors conclude that a $^1A_{1g} \rightarrow {}^1E_g$ assignment is appropriate for the 23,000 cm^{-1} band in $Mo_2(O_2CCH_3)_4$.

4. $M_2(CH_3)_8^{n-}$ Anions (M = Cr, Mo, W and n = 4; M = Re and n = 2)

The mixing of ligand and metal orbitals in the δ and δ^* molecular orbitals of b_{2g} and b_{1u} symmetry complicates the spectral interpretation somewhat for carboxylate and halide ligands. Sattelberger and Fackler have presented a correlation between the quadruple bond length and the position of the $\delta \rightarrow \delta^*$ transition for a series of compounds where metal–ligand π interactions should not be a significant factor, notably $Cr_2(CH_3)_8^{4-}$, $Mo_2(CH_3)_8^{4-}$, and $Re_2(CH_3)_8^{2-}$ (226). The absence of ligand lone-pair electrons of π character on the methyl ligands minimizes the ligand contribution to the δ and δ^* orbitals. The $Mo_2(CH_3)_8^{4-}$ spectrum in diethyl ether is similar to that of the $Mo_2Cl_8^{4-}$ anion, and the $\delta \rightarrow \delta^*$ band can accordingly be easily identified at 19,500 cm^{-1} (blue shifted approximately 700 cm^{-1} from the chloro analogue) with an upper limit on the extinction coefficient of $1500\ M^{-1}\ cm^{-1}$. A weak absorption at $25,000\ cm^{-1}$ and a high-energy absorption at $30,800\ cm^{-1}$ were also reported for the methylated molybdenum dimer.

The $Re_2(CH_3)_8^{2-}$ species has a low-energy band assigned to $\delta \rightarrow \delta*$ at $18,600\ cm^{-1}$, a blue shift of more than $4000\ cm^{-1}$ relative to the chloro analogue. The increased energy gap between the δ and $\delta*$ levels found for the methyl derivative is consistent with greater metal character in these molecular orbitals. The trend from Mo^{2+} to Re^{3+} is as expected, since greater halogen character is present in the higher oxidation state rhenium case originally, and thus greater change occurs upon elimination of the ligand π contribution. However, Sattelberger and Fackler (226) point out that the magnitude of the observed shift in the two cases differs greatly and hence requires explanation. The crystal structures of $[Li(THF)]_4Mo_2(CH_3)_8$ (114) and $(NH_4)_4Mo_2Cl_8 \cdot NH_4Cl \cdot H_2O$ (35) reveal metal–metal bond lengths that are essentially equal [2.148(2) and 2.150(5) Å, respectively], while $[Li(Et_2O)]_2$-$Re_2(CH_3)_8$ (80) and $[(C_4H_9)_4N]_2Re_2Cl_8$ (77) differ by almost 0.05 Å [2.178(1) and 2.224(1) Å, respectively]. The much larger shift observed for the $\delta \rightarrow \delta*$ transition in the rhenium dimers is certainly a reflection of this increased metal–metal overlap.

Only a single band maximum was observed for $Cr_2(CH_3)_8^{4-}$ in solution from 230 to 2000 nm. This $22,000\ cm^{-1}$ band was assigned to the $\delta \rightarrow \delta*$ transition, with the relatively large gap evident here resulting from metal–metal overlap at a metal–metal separation of 1.980(5) Å in this first-row transition metal dimer (180).

A plot of metal–metal bond distance versus the energy of the $^1A_{1g} \rightarrow$ $^1A_{2u}$ transition for these three dimers, one from each transition metal series, suggested a reasonable correlation between decreasing metal–metal separation and increasing metal–metal overlap as evidenced in greater separation of the δ and $\delta*$ orbitals (226).

Although no theoretical basis exists to quantify the relationship observed in this study, the predictive value of this correlation was strikingly confirmed by the preparation of $W_2(CH_3)_8^{4-}$, which exhibited a band at $16,800\ cm^{-1}$, quite reasonably assigned as $\delta \rightarrow \delta*$ (89). Assuming the W–W bond length in $W_2(CH_3)_8^{4-}$ to be nearly equal to that determined for $W_2(CH_3)_{8-x}Cl_x^{4-}$ ($x \simeq 3.2$, 2.26 Å) (51) allows the placement a fourth point on the plot proposed by Sattleberger and Fackler (226). The location of this additional point strongly supports the correlation as originally presented (Fig. 12).

H. Raman Vibrational Data

The past decade has produced a dramatic increase in the utilization of low-frequency vibrational data to characterize strong metal–metal bonds. The

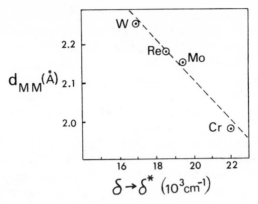

Fig. 12. A plot of the $\delta \to \delta^*$ transition energy versus the metal–metal distance in $M_2(CH_3)_8^{n-}$ dimers ($n = 4$ for Cr, Mo, and W; $n = 2$ for Re) (89).

proliferation of informative spectroscopic studies concerned with quadruply bound metal dimers is due, in large part, to the availability of improved experimental techniques for obtaining Raman and resonance Raman spectra of these compounds.

The importance of vibrational analyses and force-constant calculations as a means of exploring the strength of the metal–metal interaction in metal dimers has been recognized for some time (233). A comprehensive early study of the infrared and Raman spectra of several dirhenium and dimolybdenum dimers was reported by Bratton et al. in 1971 (30). Vibrational spectra of $Re_2X_8^{2-}$, $Re_2(O_2CCH_3)_4X_2$ (X = Cl, Br), $Mo_2Cl_8^{4-}$, and $Mo_2(O_2CCH_3)_4$ were recorded, although the amount of Raman data was limited by experimental difficulties.

Low-frequency vibrational spectra for 16 rhenium dimers of the form $Re_2(O_2CR)_2X_4$, $Re_2(O_2CR)_4X_2$, or $Re_2X_8^{2-}$ (X = Cl, Br) led to empirical assignments for stretching modes involving Re–Re, Re–X, and Re–O bonds in a preliminary publication by Oldham et al. in 1971 (205). The Raman data collected in these investigations suggested that the potential for recognizing and evaluating the strength of multiple metal–metal bonds by use of Raman spectroscopy could be exploited successfully. The expectation that Raman spectroscopy could play a unique role in the characterization of highly symmetrical dimers containing quadruple bonds has been fulfilled by recent studies discussed in more detail below.

1. Rhenium

The presence of a strong Raman band associated with the totally symmetric normal coordinate consisting largely of rhenium–rhenium stretching was recognized in early reports as a characteristic feature of the Raman spectra of compounds containing quadruply bound rhenium atoms (30, 205). Oldham et al. (205) summarized four general spectral features observed for an extensive series of rhenium dimers: (1) an intense Raman absorption in the range of 285 ± 11 cm^{-1} for each of the 16 compounds investigated, (2) an infrared absorption at 230 ± 10 cm^{-1} for all $Re_2(O_2CR)_4Cl_2$ dimers that was not present in the case of the bromo analogues, (3) infrared absorptions at 332 ± 1 cm^{-1} for $Re_2(O_2CCH_3)_2Cl_4 \cdot 2H_2O$ and $Re_2Cl_8^{2-}$ with mated absorptions at 230 ± 2 cm^{-1} in the corresponding bromo derivatives, (4) two strong infrared absorptions in the range from 350 to 540 cm^{-1} for carboxylate derivatives. The interpretation of this data was based on assignment of the strong band consistently located near 285 cm^{-1} to the metal–metal stretching mode. No polarization data were included and the assignment was partly based on the intensity of the band, which was high, as is expected for a multiple metal–metal bond stretching mode. Another consideration cited involved an extrapolation from the frequency of 128 cm^{-1} associated with the rhenium–rhenium single-bond stretch in $Re_2(CO)_{10}$ (132) (reported at 120 cm^{-1} in Refs. 117 and 214) to a quadruple bond utilizing the harmonic diatomic approximation to predict a value near 260 cm^{-1} (i.e., roughly twice the single-bond frequency). The observed frequency of 285 cm^{-1} was near this value, although the approximation that the force constant will increase by a factor of 4 while the reduced mass remains constant in extrapolating from $Re_2(CO)_{10}$ to $Re_2X_8^{2-}$ is obviously extremely crude.

The metal halide vibrational modes were cleanly divided into two sets: those due to ligands trans to the metal–metal bond and those due to halides cis to the metal–metal bond. In other words, for $Re_2Cl_8^{2-}$ all the chlorides are cis to the metal–metal bond (referred to as equatorial by the authors) and $\nu(Re-Cl)$ was 332 or 333 cm^{-1} depending on the cation. For $Re_2(O_2CCH_3)_2Cl_4 \cdot 2H_2O$ the Re–Cl stretch was also assigned at 332 cm^{-1} in agreement with the location of these chlorine atoms in equatorial positions. For $Re_2(O_2CCR)_4Cl_2$ compounds the two chlorines are trans to the multiple metal–metal bond (these are axial sites that are occupied by the chloride ligands) and the Re–Cl stretch was assigned at 230 ± 10 cm^{-1}, approximately 100 cm^{-1} below the frequency found for equatorial chloride ligands. The corresponding bromo derivatives displayed infrared absorptions at 230 ± 2 cm^{-1} for the dimers containing equatorial bromine ligands, while the axial Re–Br stretching modes were predicted to lie below the range of the instrument.

Force-constant (F) calculations for the dimeric rhenium anion $Re_2Br_8^{2-}$ were performed using both a valence force field and a Urey-Bradley force field (30). Neither analysis was considered entirely satisfactory by the authors, but both treatments led to very similar values for F_{ReBr} of 1.63 ± 0.02 mdyne Å^{-1}. Of more interest with regard to the strength of the quadruple bond is the fact that the results of both calculations suggested that F_{ReRe} lies somewhere in the range of 3.1–3.4 mdyne $Å^{-1}$. Very limited data available for the $Re_2Cl_8^{2-}$ anion were utilized in a valence force field calculation that led to F_{ReRe} = 3.77 mdyne $Å^{-1}$; similar analysis of data pertaining to $Re_2(O_2CCH_3)_4Cl_2$ indicated F_{ReRe} = 4.47 mdyne $Å^{-1}$. The concluding discussion emphasized the implications of the range of values found for the rhenium–rhenium force constant rather than the actual numerical value appropriate for any one dimer. A force constant of 3.0–4.5 mdyne $Å^{-1}$ is, in fact, strong support for a high-order bond multiplicity in these metal–metal dimers. The exceptional intensity of the Raman band due to the metal–metal stretching normal mode contribution was noted as a basis for further vibrational investigations of related compounds.

In 1973 San Filippo and Sniadoch reported Raman data for a variety of homonuclear transition metal dimers containing metal–metal bonds (222). Rhenium dimers were included in this study, and data obtained for $Re_2X_8^{2-}$, $Re_2(O_2CCH_3)_4X_2$, and $Re_2(O_2CCH_3)_2X_4 \cdot 2H_2O$ were in basic agreement with earlier values reported for the band assigned to Re–Re stretching (30, 205). Raman data for $Re_2Cl_6[P(C_3H_7)_3]_2$ and $Re_2Cl_6(tmtu)_2$ were also reported in this article. The compilation of data in Table IV reflects the consistency of the

TABLE IV
Raman Data for Multiply Bound Metal Dimers

Compound	ν (M–M), cm^{-1}	Ref.
A. RHENIUM		
$(n\text{-}Bu_4N)_2[Re_2Cl_8]$	273	30
$(pyH)_2[Re_2Cl_8]$	275	30
$(Ph_4As)_2[Re_2Cl_8]$	275	205
$(n\text{-}Bu_4N)_2[Re_2Br_8]$	277	30
$Cs_2[Re_2Br_8]$	275	30
$(Ph_4As)_2[Re_2Br_8]$	278	205
$Re_2(O_2CMe)_4Cl_2$	288	205
$Re_2(O_2CMe)_4Br_2$	281	205
$Re_2(O_2CEt)_4Cl_2$	288	205
$Re_2(O_2CEt)_4Br_2$	277	205
$Re_2(O_2CPr)_4Cl_2$	289	205
$Re_2(O_2CPr)_4Br_2$	280	205
$Re_2(O_2CC_6H_{11})_4Cl_2$	293	205
$Re_2(O_2CC_6H_{11})_4Br_2$	278	205
$Re_2(O_2CPh)_4Cl_2$	295	205
$Re_2(O_2CPh)_4Br_2$	280	205
$Re_2(O_2CMe)_2Cl_4 \cdot 2H_2O$	274	205
$Re_2(O_2CMe)_2Br_4 \cdot 2H_2O$	278	205
$Re_2Cl_6(PPr_3)_2$	278	222
$Re_2Cl_6(tmtu)_2$	276	222
$Re_2Cl_6(PPh_3)_2$	278	206
$Re_2Br_6(PPh_3)_2$	285	206

TABLE IV *continued*

Compound	$\nu(M-M)$, cm^{-1}	Ref.
B. MOLYBDENUM		
$Mo_2(O_2CMe)_4$	406	30
$Mo_2(O_2CCF_3)_4$	397	96
$Mo_2(O_2CCF_3)_4 \cdot 2py$	367	95
$Mo_2(O_2CCF_3) \cdot 2MeOH$	386	165
$Mo_2(O_2CCF_3) \cdot 2PPh_3$	377	165
$Mo_2(O_2CEt)_4$	400	165
$Mo_2(O_2CPr)_4$	402	165
$Mo_2(2,4,6\text{-}Me_3C_6H_2CO_2)_4$	404	223
$Mo_2(O_2CPh)_4$	404	165
$Mo_2(O_2CC_6H_4CN)_4$	397	223
$Mo_2(O_2CC_6H_{11})_4$	397	165
$Mo_2(O_2CC_6H_4OMe)_4$	402	223
$K_4Mo_2Cl_8$	350	165
$K_4Mo_2Cl_8 \cdot 2H_2O$	349	165
$(enH_2)_2Mo_2Cl_8$	349	165
$(NH_4)_5Mo_2Cl_9 \cdot H_2O$	339	222
$Cs_4Mo_2Br_8$	335	165
$(NH_4)_4Mo_2Br_8$	336	50
$K_4Mo_2(SO_4)_4$	370	9
$Mo_2Cl_4[P(n\text{-}Bu)_3]_4$	350	222
$Mo_2Br_4[P(n\text{-}Bu)_3]_4$	342	224
$Mo_2Cl_4[P(OMe)_3]_4$	347	222
$(Me_4N)_2Mo_2Cl_4(O_3SCH_3)_2$	372	154
$(Bu_4N)_2Mo_2Br_4(O_3SCH_3)_2$	364	154
$(Bu_4N)_2Mo_2I_4(O_3SCH_3)_2$	362	154
$Mo_2Cl_4(SMe_2)_4$	358	224
$Mo_2Br_4(SMe_2)_4$	350	224
$Mo_2Cl_4(SEt_2)_4$	348	224
$Mo_2Cl_4(dth)_2$	359	224
$Mo_2Cl_4(dtd)_2$	353	224
$Mo_2Br_4(dtd)_2$	345	224
$Mo_2Cl_4(dtdd)_2$	351	224
$Mo_2Cl_4(S_2C_4H_8)_2$	357	224
$Mo_2Cl_4(py)_4$	348	224
$Mo_2Br_4(py)_4$	335	224
$Mo_2Cl_4(bipy)_2$	338	224
$Mo_2Br_4(bipy)_2$	330	224
$Mo_2Cl_4(CH_3CN)_4$	347	224
$Mo_2Cl_4(PhCN)_4$	352	224
$Mo_2Br_4(DMF)_4$	348	224
$Mo_2Cl_4(DMF)_4$	352	224
$Mo_2Cl_4(tmedp)_2$	349	224
$Mo_2Cl_2(O_2CPh)_2(PBu_3)_2$	392	223
$Mo_2Br_2(O_2CPh)_2(PBu_3)_2$	383	223
$Mo_2[PhNC(Ph)NPh]_4$	410	84

intense band located at 285 ± 10 cm^{-1} for all the quadruply bound rhenium dimers listed. A feature common to the Raman spectra of dimers with axial chloride ligands was a band of moderate intensity near 365 cm^{-1}, while bromo analogues displayed a comparable, but more intense, band near 210 cm^{-1}. The frequency and intensity of these lines were considered appropriate for association with the totally symmetric Re–X stretching modes.

Qualitative observations regarding variations in both the relative and absolute intensities of Raman bands reflected the role of the resonance Raman effect in these highly colored species (222). The absorption of light in the visible region that characterizes the Re$_2^{6+}$ unit is sufficiently close to the exciting line used in some cases to obtain Raman spectra that the resonance denominator becomes a dominant factor. The normal intensity pattern for Raman spectra obtained when the exciting frequency differs greatly from any molecular absorption frequency is replaced by an intensity pattern that is highly dependent on the excitation frequency. The enhancement of intensities is usually greatest for totally symmetric modes, and strong overtones exhibit increasing half widths as characteristics of resonance Raman effects.

The utility of identifying strong metal–metal multiple bonds by the experimental detection of an intense spike in the Raman spectrum of the material in question has been unquestionably established. Sample decomposition due to absorption of intense laser light remains a problem in many cases, but spinning sample techniques (167) and the availability of a range of exciting frequencies (49) have enabled researchers to obtain data on a variety of highly absorbing compounds that were previously intractable for Raman studies.

Infrared spectroscopy is not as inherently informative with regard to metal interactions in highly symmetrical metal–metal bound dimers as is Raman spectroscopy, since the totally symmetric metal–metal stretch is a forbidden absorption in the infrared experiment. Oldham and Ketteringham have prepared mixed-halide dimers of the type Re$_2$Cl$_x$Br$_{8-x}^{2-}$ to lower the symmetry and hence introduce some infrared allowedness into the Re–Re stretching mode (206). Indeed, the appearance of a medium-intensity band at 274 cm^{-1} in the infrared spectrum of the mixed-halo species was considered to be the result of absorption by the metal–metal stretching vibration, which was also observed in the Raman spectrum at 274 cm^{-1}.

The definitive resonance Raman spectra obtained by Clark and Franks for $[(C_4H_9)_4N]_2Re_2Cl_8$ and $[(C_4H_9)_4N]_2Re_2Br_8$ were particularly informative (49). The observation of resonance enhancement in the Raman spectrum recorded with excitation frequencies in the range of the lowest electronic absorption near 14,000 cm^{-1} not only provided useful vibrational data, but also confirmed that this electronic absorption was electric dipole allowed in accord with the $\delta \rightarrow \delta^*$ assignment. Normal Raman spectra were obtained with excitation energies differing substantially from the absorption maximum near

680 nm (i.e., 457.9, 488.0, and 514.5 nm produced normal Raman spectra). Strong fluorescence limited the choice of excitation frequencies somewhat, but for both $Re_2X_8^{2-}$ (X = Cl, Br) ions it was possible to observe overtones of ν_1 (Re–Re) a_{1g} up to and including $4\nu_1$ that displayed the intensity decrease and half-bandwidth increase indicative of resonance Raman effects. A weaker progression corresponding to $\nu_2 + n\nu_1$ was also present for both dimers with $n = 0$, 1, and 2 where ν_2 is the Re–X stretching mode of a_{1g} symmetry that exhibited a polarization ratio of 0.01 in acetone solution for both $Re_2Cl_8^{2-}$ (356.5 cm^{-1}) and $Re_2Br_8^{2-}$ (210 cm^{-1}). Extraction of harmonic frequencies and anharmonicity constants was possible by utilization of the overtone frequencies observed experimentally and the anharmonic oscillator description (147) as reflected in Eq. 1:

$$\nu(n) = n w_1 - \chi_{11}(n^2 + n) \tag{1}$$

(where ν is the overtone frequency and n is the vibrational quantum number), which can be rearranged to plot $\nu(n)/n$ versus n to obtain $\omega_1 - \chi_{11}$ as the intercept and $-\chi_{11}$ as the slope. The results of such a plot for ν_1(Re–Re) a_{1g} with overtones measured up to and including $n = 4$ for $[(C_4H_9)_4N]_2Re_2Cl_8$ and $[(C_4H_9)_4N]_2Re_2Br_8$ were $\omega_1 = 272.6 \pm 0.4$ and 276.2 ± 0.5 cm^{-1} and $\chi_{11} = 0.35 \pm 0.05$ and 0.39 ± 0.06 cm^{-1}, respectively. The small value of χ_{11} is indicative of nearly harmonic behavior for this fundamental ν_1(Re–Re) a_{1g} mode. Intensity measurements as a function of the exciting wavelength were also reported. From a series of intensity measurements utilizing wavelengths between 457.9 and 630.0 nm with the 980 cm^{-1} band of the sulfate ion as an internal standard, the intensity of ν_1(Re–Re) in $Re_2Br_8^{2-}$ increased roughly thirtyfold, while bands at 210, 241, and 192 cm^{-1} increased only slightly (roughly factors of 1–5). A similar preferential enhancement of ν_1(Re–Re) was evident for the $Re_2Cl_8^{2-}$ ion. Also characteristic of resonance Raman effects is an increase in the relative intensity of the first overtone relative to the fundamental one as the exciting frequency approaches the electronic absorption maximum. For $Re_2Cl_8^{2-}$ ($\lambda_{max} \simeq 690$ nm) the intensity ratio of $2\nu_1$ to ν_1 increased monotonically from 0.09 at 580 nm to 0.22 at 615 nm, while $Re_2Br_8^{2-}$ ($\lambda_{max} \simeq 710$ nm) exhibited a ratio change from 0.12 at 585 nm to 0.31 at 647 nm. Thus both ions display properties reflecting the resonance Raman effect as selectively evident in the fundamental mode most closely involved in the $\delta \rightarrow \delta^*$ electronic absorption near 14,000 cm^{-1}, notably ν_1(Re–Re) of a_{1g} symmetry.

An interesting estimate of the Re–Re bond dissociation energy for the $Re_2X_8^{2-}$ anions is based on a Birge-Sponer extrapolation of the type employed for anharmonic diatomic molecules (243). This approximate method for determining diatomic bond dissociation energies from Eq. 2 (147) often overestimates the actual bond energy. Application to more complex molecules is necessarily suspect owing to mixing of modes of appropriate symmetries.

$$\text{bond dissociation energy} = \frac{\omega_1^2}{4x_{11}} - \frac{\omega_1}{2} \qquad (2)$$

However, the bond dissociation energies calculated, $152 \pm 20\,\text{kcal mole}^{-1}$ for $Re_2Cl_8^{2-}$ and $139 \pm 25\,\text{kcal mole}^{-1}$ for $Re_2Br_8^{2-}$, are indicative of strong multiple metal–metal bonds even if a 15–20% overestimate is assumed to be inherent in the Birge-Sponer approximation. Thus Trogler et al. (243) conclude that a range of $115–130\,\text{kcal mole}^{-1}$ is reasonable for the rhenium–rhenium quadruple bond energy in these anionic dimers.

2. Molybdenum

Raman spectra of quadruply bound molybdenum dimers show a much greater frequency variation for the metal–metal stretching mode than is found in related rhenium compounds. The $Mo_2X_8^{4-}$ anions and other halo derivatives are considered first, then a review of tetracarboxylatodimolybdenum vibrational data is given, and finally comparisons for various molybdenum dimers are presented.

Spectra of compounds containing the $Mo_2Cl_8^{4-}$ ion have been obtained using the spinning sample technique to avoid decomposition of the highly colored solids (9). Lines of nearly equal intensity near 300 and 350 cm^{-1} were present with an excitation wavelength of 647.1 nm, and general assignments to Mo–Cl stretch and Mo–Mo stretch were proposed. Excitation at 488.0 nm resulted in resonance enhancement of the band attributed to Mo–Mo stretch near 350 cm^{-1} and overtones were observed out to $3\nu_1$ in some compounds. The frequency attributed to the ν_1(Mo–Mo) vibrational mode was reported for $K_4Mo_2Cl_8 \cdot 2H_2O$, $K_4Mo_2Cl_8$, $(NH_4)_5Mo_2Cl_9 \cdot H_2O$, and $(enH_2)_2Mo_2Cl_8 \cdot 2H_2O$, although the authors emphasized that extensive mixing of Mo–Cl and Mo–Mo modes prohibited a simple interpretation of the spectral data (9).

The infrared and Raman spectra of $Mo_2X_8^{4-}$ (X = Cl, Br) displayed two infrared bands associated with Mo–X stretching modes and a prominent Raman band for the Mo–Mo stretch (165). Two infrared-allowed M–X stretching modes are predicted for $M_2X_8^{n-}$ species, and absorptions at 301 ± 2 and 273 ± 2 cm^{-1} for three compounds containing the $Mo_2Cl_8^{4-}$ unit were in the appropriate region for these assignments. The most intense Raman band (349 ± 1 cm^{-1}) was attributed to ν(Mo–Mo) in $Mo_2Cl_8^{4-}$. The authors stated that preliminary calculations were indicative of extensive mixing of ν(Mo–Mo) and ν(Mo–Cl) with the observed 350 cm^{-1} value for ν(Mo–Mo) actually higher than the uncoupled value would be. Infrared frequencies of 251 and 223 cm^{-1} for Mo–Br stretching modes and a Raman value of 335 cm^{-1} for ν(Mo–Mo) were reported with the Mo–Mo stretching frequency virtually free of significant coupling to Mo–Br modes. Comparison of the metal–metal stretching frequency found for $Mo_2X_8^{2-}$ ions with the corresponding frequency of 193 cm^{-1} for the $[(\eta^5\text{-}C_5H_5)Mo(CO)_3]_2$ dimer, which contains a metal–metal single bond [Mo–Mo separation of 3.22 Å (255)], was considered support for the presence of a strong molybdenum–molybdenum bond in the halide derivatives.

Substituted molybdenum dimers of the type $Mo_2Cl_4L_4$ [$L = P(C_4H_9)_3$ and $P(OCH_3)_3$] display prominent bands at 350 and 347 cm^{-1}, respectively, in Raman spectra obtained with 514.5 nm excitation (222). This result was assumed to be due to the Mo–Mo stretching mode with the first and second overtones observable in both cases. The value of 339 cm^{-1} reported for ν_1(Mo–Mo) of $(NH_4)_5Mo_2Cl_9 \cdot H_2O$ (222) differed by 11 cm^{-1} from the 350 cm^{-1} figure reported by Angell et al. (9). A weak band at 282 ± 10 cm^{-1} was attributed to the totally symmetric Mo–Cl stretching vibration in Mo_2Cl_4-$[P(C_4H_9)_3]_4$, $Mo_2Cl_4[P(OCH_3)_3]_4$, and $(NH_4)_5Mo_2Cl_9 \cdot H_2O$.

More recent publications have extended resonance Raman data for $Mo_2Cl_8^{4-}$ ions to include overtones up to $11\nu_1$ for $Cs_4Mo_2Cl_8$ and from $6\nu_1$ to $9\nu_1$ for salts with other cations (47, 48). The absorption maximum at 517 nm for solid $K_4Mo_2Cl_8$ and similar absorptions for other $Mo_2Cl_8^{4-}$ salts leads to resonance Raman spectra when the 514.5 nm line of an argon laser is used for excitation. The detailed study by Clark and Franks (47) of five $Mo_2Cl_8^{4-}$ salts nicely illustrates the experimental techniques and the data interpretation applicable to these systems. The harmonic frequency ω_1 and anharmonicity constant χ_{11} were extracted from least-squares analyses according to a plot of the ν_1 fundamental and overtone frequencies $[\nu(n)/n$ versus $n]$ as described for the rhenium dimers. A range of values from 338.8 to 348.7 cm^{-1} was found for ω_1 as a function of the cation ($Rb^+ < NH_4^+ < Cs^+ < K^+ < enH_2^{2+}$), with χ_{11} in the range of 0.59 ± 0.17 cm^{-1} for all five salts. A second progression in ν_1(Mo–Mo) a_{1g} was observed built on the fundamental labeled ν_4 in this chapter, which is attributed to metal–chlorine stretching and occurs in the range of 270–280 cm^{-1} in all of the salts studied. This second progression was much weaker, with $\nu_4 + 4\nu_1$ being the highest order overtone observed. Values for ω_1 and χ_{11} were determined from a plot of $[(\nu_4 + n\nu_1) - \nu_4]/n$ versus n for the more limited data available for this second progression. The results were in agreement with values obtained from the main progression, although the accuracy of the determination was lower.

The normal Raman spectrum obtained with 647.1 nm excitation serves as a comparison for the Raman spectra obtained with excitation frequencies of 488.0 and 514.5 nm, which lie within the $\delta \rightarrow \delta^*$ absorption band. The tremendous enhancement of the ν_1(Mo–Mo) a_{1g} mode, the high overtone progression in ν_1, the increase in overtone bandwidth with increasing vibrational quantum number, and the increased intensity of the overtones relative to the fundamental as the excitation frequency approaches the electronic absorption maximum are all attributable to the resonance Raman effect. Polarization measurements to assist in assignments were prohibited by instability of the $Mo_2Cl_8^{4-}$ ion towards hydrolysis in solution. Concentrated hydrochloric acid solutions displayed Raman bands at 355.7, 708.5, 1056.8, and 1404.7 cm^{-1} that were considered appropriate for ν_1 and overtones, but the frequency differences from those of $Mo_2Cl_8^{4-}$ were sufficiently large to indicate that the Mo_2^{4+} unit was not present as $Mo_2Cl_8^{4-}$. Electronic spectra recorded over a period of months indicated that a new band appears at 23,600 cm^{-1} and the 19,500 cm^{-1} absorption decays with time.

Comparison of the fundamental frequency of 346 cm^{-1} for the ground-state $\nu(Mo-Mo)$ a_{1g} frequency with the spacing of 351 cm^{-1} observed in the vibrational progression of the 19,000 cm^{-1} absorption in the low-temperature electronic spectrum of $K_4Mo_2Cl_8$ (118) suggests that no significant difference exists between the ground-state and excited-state frequencies. Clark and Franks mention a possible variation in coupling of the Mo–Mo and Mo–Cl modes in the two electronic states as a rationale for the similar frequencies. The observation of resonance Raman spectra for excitation frequencies near the 19,000 cm^{-1} absorption band in $Mo_2Cl_8^{4-}$ salts is strong support for the $\delta \rightarrow \delta^*$ assignment proposed on the basis of polarized electronic absorption data (118), since such effects are expected only for electric-dipole-allowed transitions. It is generally accepted that the lowest electric-dipole-allowed absorption in $Mo_2X_8^{4-}$ ions is indeed the $\delta \rightarrow \delta^*$ transition, thus confirming the assignment. The metal-localized nature of the transition is reflected in the enhancement of $\nu(Mo-Mo)$, while the other two totally symmetric vibrations in $Mo_2Cl_8^{4-}$ involving Mo–Cl stretching and bending are not resonance enhanced in this manner.

Application of the anharmonic diatomic approximation as discussed for the $Re_2X_8^{2-}$ ions was also employed to estimate a bond dissociation energy for $Mo_2Cl_8^{4-}$ (243). Employing the ω_1 and χ_{11} values obtained by Clark and Franks (47) as data for the Birge-Sponer extrapolation led to a calculated energy of 172 ± 20, 190 ± 25, and 127 ± 15 kcal mole^{-1} for the molybdenum–molybdenum bond in $Mo_2Cl_8^{4-}$ with the K$^+$, Rb$^+$, and Cs$^+$ cations, respectively. Trogler et al. allowed for an error of 15–20% over the actual bond dissociation energy to arrive at a range of 110–160 kcal mole^{-1} for the molybdenum–molybdenum quadruple bond in $Mo_2Cl_8^{4-}$. Although the qualitative nature of such a treatment is evident, there is a definite need to establish bond energies for strong multiple metal–metal bonds, and further efforts along these lines are required and promoted by these results. The variation in the calculated bond dissociation energy from 127 to 190 kcal mole^{-1} as the cation is changed from Cs$^+$ to Rb$^+$ is somewhat disconcerting in terms of the reliability of the extrapolation in these complex ions.

Resonance Raman effects are evident in the spectrum of $(NH_4)_4Mo_2Br_8$ upon excitation near the $\delta \rightarrow \delta^*$ transition energy of 18,000 cm^{-1} (50). Overtones in $\nu_1(Mo-Mo)$ are observed to $6\nu_1$ at room temperature and to $11\nu_1$ at about 80°K. A second progression in the a_{1g} molybdenum–molybdenum stretching mode is built on the totally symmetric ν_2 MoBr stretching mode at 168.8 cm^{-1}. Calculations indicate that the ν_1 mode is very nearly a harmonic oscillator with $\omega_1 = 336.9$ cm^{-1} and $X_{11} = -0.48 \pm 0.09$ cm^{-1}. The Birge-Sponer extrapolation predicts a bond dissociation energy of nearly 170 kcal mole^{-1} for $Mo_2 Br_8^{4-}$. Allowance for a 20% overestimate and inclusion of data from previous extrapolations of this type (243) led these authors to conclude that a substantial bond energy, in the region of 120 kcal mole^{-1}, is associated with rhenium and molybdenum quadruple bonds in the $M_2 X_8^{n-}$ ions (50).

The ν_1(Mo–Mo) a_{1g} fundamental, which was first associated with the intense band at 406 cm^{-1} in the Raman spectrum of $Mo_2(O_2 CCH_3)_4$ (30), is well above the vibrational frequency found for the Mo–Mo stretching mode in the $Mo_2 X_8^{4-}$ ions. Normal coordinate analyses of the $Mo_2(O_2 CCH_3)_4$ infrared and Raman vibrational data produced a calculated Mo–Mo force constant of 3.6 to 3.9 mdyne Å^{-1} for several possible assignments of the Mo–O stretching and bending modes (30).

Considerable variation in the Mo–Mo stretching frequency of $Mo_2(O_2 CCF_3)$ occurs as a function of the environment of the sample (96): 397 (solid), 383 (ether solution), 343 (pyridine solution), and 367 cm^{-1} for $Mo_2(O_2 CCF_3)_4 \cdot 2py$ as the solid. The strength of the interaction of the molybdenum centers with ligands in the terminal positions obviously influences the Raman spectra in these dimers significantly, with stronger metal–terminal ligand interactions causing a decrease in the frequency of the fundamental Mo–Mo stretching mode.

San Filippo and Sniadoch included $Mo_2(O_2 CCH_3)$, $Mo_2(O_2 CCF_3)_4$, and $Mo_2(O_2 CCF_3) \cdot 2py$ in their Raman study of metal–metal bound molecular units (222). The location of ν(Mo–Mo) at 406 cm^{-1} in $Mo_2(O_2 CCH_3)_4$ was in agreement with the earlier work of Bratton et al. (30); the other two carboxylate derivatives also had Raman bands attributable to the metal–metal stretching mode at energies within 2 cm^{-1} of previous literature values. Other aspects of the Raman spectra for these trifluoroacetate-bridged dimers (222) were not in agreement with all facets of earlier reports (30). Several comparative features for rhenium and molybdenum dimers were presented: (1) ν(Mo–Mo) is well above ν(Re–Re), as would be expected from the mass ratio of roughly 0.5 for Mo to Re, (2) ν(Mo–Mo) varies over a range of nearly 70 cm^{-1} in quadruply bound dimers, while ν(Re–Re) remains surprisingly constant in various quadruple-bond environments.

A series of $Mo_2(O_2 CR)_4$ dimers, with R = CH_3, CF_3, Et, n-Pr, Ph, and $C_6 H_{11}$, were characterized by infrared and Raman spectroscopy by Ketteringham and Oldham (165). All these derivatives displayed ν(Mo–Mo) in the narrow range of 400 ± 4 cm^{-1}, approximately 50 cm^{-1} above the location of the molybdenum–molybdenum stretching vibrational frequency in the $Mo_2 X_8^{4-}$

ions. The $Mo_2(O_2CCF_3)_4$ dimer followed the above pattern for carboxylates $[\nu(Mo–Mo) = 398\ cm^{-1}]$, with lower frequencies resulting upon ligand coordination at the axial sites as in $Mo_2(O_2CCF_3)_4 \cdot 2L$ ($L = MeOH$, $386\ cm^{-1}$; $L = PPh_3$, $377\ cm^{-1}$). The authors concluded that the two distinct vibrational energy ranges for $Mo_2X_8^{4-}$ and $Mo_2(O_2CR)_4$ were indicative of weaker molybdenum–molybdenum bonding in the halide derivatives. This is consistent with the slightly greater metal atom separation typically found in crystal structures of the halo compounds, where 2.14 Å is representative of $Mo_2Cl_8^{4-}$ and 2.10 Å is the approximate distance found for $Mo_2(O_2CR)_4$ compounds (35, 68). Recently force constant calculations were presented that indicated two distinct force fields are required to describe the two general classes of molybdenum dimers (166), that is, the force constant for molybdenum carboxylates is about 4.5 mdyne $Å^{-1}$, while for $Mo_2X_8^{4-}$ F_{MoMo} is near 3.5 mdyne $Å^{-1}$.

The conclusion that variations in the coupling of vibrational modes in these dimers is not responsible for the entire frequency shift is an important point in that it justifies the search for a chemical rationale of the bonding differences in $Mo_2X_8^{4-}$ and $Mo_2(O_2CR)_4$. The authors suggest that δ bonds may make an important bonding contribution in rhenium dimers and molybdenum carboxylates to account for F_{MM} near 4.5 mdyne $Å^{-1}$. Furthermore, it is proposed that such a δ component would probably not contribute significantly in the molybdenum halides where F_{MM} is near 3.5 mdyne $Å^{-1}$, possibly reflecting the difference in ligand constraints of the molecular geometry according to these authors (166). Since both dimers exhibit eclipsed structures, the basis for more favorable δ-bond formation in the bridged species is not clear.

Vibrational spectra of sulfato-bridged Mo_2^{4+} and Mo_2^{5+} units have been reported and assigned in a general manner (190). Observation of $\nu(Mo–Mo)\ a_{1g}$ at $370\ cm^{-1}$ had been established by Angell et al. (9) for the $Mo_2(SO_4)_4^{4-}$ ion, and thus falls between the 350 and 400 cm^{-1} values typically found for $Mo_2X_8^{4-}$ and $Mo_2(O_2CR)_4$. Loewenschuss et al. (190) cited the monotonic increase in $\nu_1(Mo–Mo)$ from $Mo_2Cl_8^{4-}$ through $Mo_2(SO_4)_4^{4-}$, $Mo_2(O_2CCF_3)_4$ and $Mo_2(O_2CCH_3)_4$ as a reflection of increased ring stress opposing the metal–metal stretch without invoking a change in metal–metal bond strength. This rationale contrasts with the force-constant variation presented by other workers (166). The oxidized Mo_2^{5+} unit with four bridging sulfate ligands displays two bands for $\nu_1(Mo–Mo)$ at 373 and 386 cm^{-1} in accord with the presence of two crystallographically distinct sites for the anion in $K_3Mo_2(SO_4)_4 \cdot 3.5H_2O$ (75). The decreased bond order in $Mo_2(SO_4)_4^{3-}$ relative to $Mo_2(SO_4)_4^{4-}$ (3.5 versus 4.0) supports the importance of the ligand geometry in determining the Raman frequency of the metal–metal stretch, since the increase in this frequency observed for the Mo_2^{5+} unit must be due to greater bridging-ligand distortions opposing and overcoming the tendency for the decreased metal–metal bond order to lower $\nu_1(Mo–Mo)$. No resonance Raman character was evident in the

4880 Å excitation spectrum of $K_3Mo_2(SO_4)_4 \cdot 3.5H_2O$ in accord with the absence of an allowed electronic absorption in this region.

I. Redox Reactions of Quadruple Bonds

The HOMO and LUMO of quadruply bound metal dimers invariably have their provenance in metal-localized orbitals. As a consequence the addition to or removal of electrons from these dimers necessarily alters the metal electron configuration, and investigation of the reduced or oxidized species should provide information about the bonding, nonbonding, or antibonding character of the orbital involved.

The application of modern electrochemical techniques to strong multiply bound metal–metal dimers has recently proved that redox processes are indeed a common feature of the chemistry of these compounds. Chemical oxidation and reduction have now been observed with retention of a strong metal–metal bond in many instances, and isolation of closely related dimers with differing electron configurations has provided new insight into the bonding role of various orbitals. Electron spin resonance spectroscopy of paramagnetic dimers provides a powerful tool for probing the nature of the singly occupied molecular orbital.

1. Technetium

The original isolation of $(NH_4)_3Tc_2Cl_8 \cdot 2H_2O$ (121) preceded the postulate of a quadruple bond in the closely related rhenium dimers. In view of the eight electrons available for metal–metal bond formation in the stable dimers of Mo^{2+} and Re^{3+} that were characterized in the following years, the anomalous nature of the paramagnetic $Tc_2Cl_8^{3-}$ ion with an "extra" electron and an average technetium oxidation state of 2.5+ was confirmed. That the stoichiometry first reported by Eakins et al. (121) was indeed correct was evidenced by the x-ray crystal structure determination of $(NH_4)_3Tc_2Cl_8 \cdot 2H_2O$ (29, 61). The structural details for the $Tc_2Cl_8^{3-}$ ion with virtual D_{4h} symmetry were very similar to those previously established for $Re_2Cl_8^{2-}$ (82) and $Mo_2Cl_8^{4-}$ (34). The eclipsed geometry, absence of bridging ligands, and short technetium–technetium separation of 2.13(1) Å were all features common to the quadruply bound $Re_2Cl_8^{2-}$ ion. A more recent x-ray study of the isostructural potassium salt of the octachloroditechnetate ion improved the accuracy of the metal–metal distance in a $Tc_2Cl_8^{3-}$ salt [Tc–Tc, 2.117(2) Å] (110).

The magnetic susceptibilities of $Tc_2Cl_8^{3-}$ salts are consistent with one unpaired electron per dimeric unit (101). An early report of 2.0 ± 0.2 BM for the magnetic moment of $(NH_4)_3Tc_2Cl_8 \cdot 2H_2O$ (29) was probably slightly high as a result of traces of $(NH_4)_2TcCl_6$ in the sample. More recently temperature-dependent susceptibility measurements over the range from 80 to 300°K with samples of $Y[Tc_2Cl_8] \cdot 9H_2O$ and $(NH_4)_3Tc_2Cl_8 \cdot 2H_2O$ that were free of hexa-chlorotechnetate(IV) determined magnetic moments of 1.78 ± 0.03 BM for both complexes based on the observed Curie behavior (101). A g factor of 2.06 ± 0.03 results from the logical assumption that the ground state is a spin doublet with $\mu_{eff} = g\sqrt{s(s+1)}$. The ESR data discussed below are consistent with this average value of the g factor.

Frozen-solution ESR spectra of $Tc_2Cl_8^{3-}$ in mixed aqueous hydrochloric acid and ethanol provided data consistent with equal coupling of the unpaired electron to both technetium nuclei (101). Isotopically pure ^{99}Tc ($I = 9/2$) in $^{99}Tc_2Cl_8^{3-}$ leads to a large number of lines in the X-band spectrum owing to second-order effects, in addition to the hyperfine lines presence for this dimeric axially symmetric system. The Q-band spectrum obtained at $77°K$ with a microwave frequency of 35.56 GHz exhibited fewer lines, and computer-simulated spectra were generated to correspond to the experimental spectrum with $g_{\parallel} = 1.912$, $g_{\perp} = 2.096$, $|A_{\parallel}| = 166 \times 10^{-4}$ cm^{-1}, $|A_{\perp}| = 67.2 \times 10^{-4}$ cm^{-1}, and $g_{av} = 2.035$.

Cotton and Pedersen utilized second-order perturbation theory to rationalize the observed g values in terms of the ground-state configuration appropriate for $Tc_2Cl_8^{3-}$ (100). Expressions for $g_{\parallel}$ and $g_{\perp}$ can be derived within this theory (192) and can be qualitatively correlated with molecular orbital descriptions and experimental g values.

$$g_{\parallel} = 2.0023 - 4 \sum_n \frac{\langle \psi_n | H_{ls} | \psi_0 \rangle \langle \psi_0 | L_z | \psi_n \rangle}{E_n - E_0} \tag{3}$$

$$g_{\perp} = 2.0023 - 4 \sum_n \frac{\langle \psi_n | H_{ls} | \psi_0 \rangle \langle \psi_0 | L_y | \psi_n \rangle}{E_n - E_0} \tag{4}$$

Symmetry considerations for identifying nonzero matrix elements in the correction term were based on the A_{2g} and E_g transformation properties of the angular momentum operator L in the D_{4h} point group appropriate for the $Tc_2Cl_8^{3-}$ anion analysis. The key point that was addressed was comparison of the predicted g values for a $^2A_{2u}$ ground state, as would be consistent with the location of the excess electron in a σ-nonbonding level of a_{2u} symmetry (83), versus the g values anticipated for location of the excess electron in the δ^* orbital of b_{1u} symmetry, as suggested by SCF–X_α calculations (201). The resultant correlation of g values supported the $\sigma^2 \pi^4 \delta^2 \delta^{*1}$ configuration as the ground state of the technetium dimer and thus added another substantive factor to the list of experimental and theoretical results favoring the δ^* orbital as the lowest lying molecular orbital above the δ level. A $^2A_{2u}$ ground state would involve excited states of A_{1u} symmetry in determining the value of $g_{\parallel}$, and since no low-lying excited A_{1u} state is accessible, $g_{\parallel}$ should be near 2. For $g_{\perp}$ excited states of E_u symmetry, such as the state resulting from $e_u(\pi) \rightarrow b_{1u}(\delta)$ excitation, should increase $g_\perp$ to a value greater than 2. The experimental $g_{\parallel}$ of 1.912 is therefore not consistent with this molecular-orbital description.

The assumption of a $^2B_{1u}$ ground state leads to $g_{\parallel}$ depending on B_{2u} excited states and to $g_\perp$ depending on E_u. As before the excited E_u state $a_{1g}(\sigma)^2 e_u(\pi)^3 b_{2g}(\delta)^2 b_{1u}(\delta^*)^2$ predicts $g_\perp > 2$ in accord with experiment. An energetically accessible empty b_{2u} metal-localized molecular orbital leads to an excited state that will decrease $g_{\parallel}$ below 2, also in accord with experiment.

Electronic spectra of $Tc_2Cl_8^{3-}$ in hydrochloric acid solution and in the solid state are qualitatively similar, indicating retention of the basic dimeric unit upon dissolution. The band maximum at 615 nm ($\epsilon = 185\ M^{-1}\ cm^{-1}$), originally reported as the major visible absorption (121), has not been reproduced in more recent publications, which locate this band at 638 nm (29, 70). Other solution maxima were present at 507 and 320 nm (29).

A more detailed spectroscopic study of the $Tc_2Cl_8^{3-}$ ion has been reported by Cotton et al. (70). Of particular interest is the band in the near infrared, between 6000 and 8000 cm^{-1}, which the authors assign as the $\delta \rightarrow \delta^*$ transition. Values for absorption energies were calculated from SCF–X_α analysis, and numerical agreement with assigned bands was superior to the agreement attained in related even-electron calculations for $Re_2Cl_8^{2-}$ (194) and $Mo_2Cl_8^{4-}$ (199).

The low-energy band near 6000 cm^{-1} shows vibrational fine structure both at room temperature and at 5°K. The progressional spacing of about 320 cm^{-1} is not inconsistent with an excited-state Tc–Tc stretching frequency assignment based on comparison with data for related molybdenum [335 cm^{-1} (127)] and rhenium [248 cm^{-1} (118)] dimers, but the absence of ground-state vibrational data for $Tc_2Cl_8^{3-}$ and the complex progressional pattern indicative of overlapping absorptions limit quantitative interpretation in this regard. The integrated intensity of the near infrared absorption is independent of temperature ($\pm10\%$) between 300 and 5°K as expected for an electric-dipole-allowed transition. The calculated energy of 6.0×10^3 cm^{-1} for the $\delta \rightarrow \delta^*$ transition is in excellent agreement with the observed energy of the first component of the vibrational progression at 5.9×10^3 cm^{-1}. In conjunction with intensity and vibrational structure considerations, the calculated $\delta \rightarrow \delta^*$ transition energy leads to a convincing set of data in support of a $^2B_{1u} \rightarrow {}^2B_{2g}$ assignment for the near infrared band.

One point worthy of comment is the excellent agreement between calculated and observed energies for the $\delta \rightarrow \delta^*$ transition in $Tc_2Cl_8^{3-}$, where no change in spin multiplicity is possible as opposed to poor numerical agreement for the $\delta \rightarrow \delta^*$ energy in diamagnetic systems (see Table V). The difficulties that arise in calculating transition energies for singlet excitations using spin-restricted configurations that lead to a weighted average for excitation to the singlet and triplet states are evident in these results. The best agreement occurs in the $Tc_2Cl_8^{3-}$ case, where spin multiplicity is not a factor, and the poorest agreement

TABLE V

Observed $\delta \rightarrow \delta^*$ Assignments and SCF–Xα–SW Calculation Comparisons

Complex	$\delta \rightarrow \delta^*$ observed, cm^{-1}	Ref.	$\delta \rightarrow \delta^*$ calculated, cm^{-1}	Ref.
$Re_2Cl_8^{2-}$	14.2×10^3	118	4.5×10^3	194
$Mo_2Cl_8^{4-}$	18.1×10^3	127	13.7×10^3	199
$Tc_2Cl_8^{3-}$	5.9×10^3	70	6.0×10^3	70

results in the $Re_2Cl_8^{2-}$ case, where a spin-restricted calculation was used. For $Mo_2Cl_8^{4-}$ the agreement is intermediate, as might be expected for adjustment of the spin-restricted weighted-average transition energy prediction by including a spin-unrestricted calculation to determine the triplet energy and obtaining the singlet transition energy by difference.

The main visible absorption between 600 and 700 nm was deconvoluted into two bands (13,600 cm^{-1}, $\epsilon = 35$; 15,700 cm^{-1}, $\epsilon = 172$), with the more intense band assigned as $\delta^* \rightarrow \pi^*$, since this is the only other allowed transition calculated to lie below 20,000 cm^{-1}. It should be noted that no such bands are present in closed-shell quadruple-bond configurations where no δ^* electron is present in the ground state. The weaker band is assigned to the forbidden $\pi \rightarrow \delta^*$ transition, although the authors are not certain that the solution spectrum requires the existence of this band (70).

A weak band at 20,000 cm^{-1} is assumed to be a forbidden transition with little charge-transfer character, and several possible transitions could contribute partially or totally to this absorption. Ligand-to-metal charge transfer [$Cl(\pi) \rightarrow \delta^*$], $\pi \rightarrow \pi^*$ and $\pi \rightarrow d_{x^2-y^2}$ transitions are compatible with the intense band maximizing at 31,400 cm^{-1} prior to the onset of increasingly intense absorptions at higher energy.

The SCF–X$_\alpha$–SW description of $Tc_2Cl_8^{3-}$ that served as a basis for the calculated electronic spectrum (70) was reported in detail by Cotton and Kahlbacher (87). The general pattern of energy levels closely resembled the molecular orbital distribution calculated for $Mo_2Cl_8^{4-}$ (199) and $Re_2Cl_8^{2-}$ (194) by similar methods, with the excess electron in $Tc_2Cl_8^{3-}$ located in the δ^* orbital of b_{1u} symmetry. Comparison of these systems is presented in Section III.F. The same ground-state configuration had been reported in 1974 based on a CNDO calculation by BiaginiCingi and Tondello (24). The formal bond order in the $Tc_2Cl_8^{3-}$ ion is thus 3.5 for the metal–metal linkage.

The presence of an antibonding δ^* electron in $Tc_2Cl_8^{3-}$ coupled with the stability of the Re_2^{6+} unit suggests that oxidation to form $Tc_2Cl_8^{2-}$ should be practical. Bratton and Cotton followed the course of oxidation of $Tc_2Cl_8^{3-}$ in hydrochloric acid spectrophotometrically (29). The initial turquoise solution passed through a green color to form a yellow solution of $TcCl_6^{2-}$ as the final product. Only one isosbestic point was observed and no evidence for intermediate species was found.

The electrochemical behavior of $Tc_2 Cl_8^{3-}$ has been investigated by Cotton and Pedersen (101). Rotating-disk platinum electrode polarograms displayed a one-wave oxidation at $E_{1/2} = 0.140$ V versus SCE with a limiting slope of E versus $\log[(i_L - i)/i]$ equal to 60 ± 1 mV, consistent with a relatively slow one-electron oxidation. Cyclic voltammetry indicated that a quasi-reversible oxidation was forming a product with a lifetime greater than 300 sec, since the ratio of the anodic and cathodic peak current remained equal to 1 for a 300 sec interval between the two measurements. The potential separation between peaks varied from 70 to 210 mV as a function of sweep rate. The data were interpreted in terms of Scheme I with the decomposition products A and B not identified. Attempts to isolate a salt of the oxidized product, $Tc_2 Cl_8^{2-}$, were unsuccessful. The reduction potential for the transformation of $Tc_2 Cl_8^{2-}$ to the trinegative anion is consistent with the exclusive isolation of the more highly reduced dimer from the hydrochloric acid and zinc reaction mixtures employed in the preparation of these anions. Nonetheless, based on the data collected it would seem that chemical oxidation of $Tc_2 Cl_8^{3-}$ to $Tc_2 Cl_8^{2-}$ should be a feasible route to the quadruply bound technetium analogue of $Re_2 Cl_8^{2-}$.

$$[Tc_2 Cl_8]^{2-} + e^- \rightleftharpoons [Tc_2 Cl_8]^{3-} \qquad E_{1/2} = 0.140 \text{ V}$$

$$[Tc_2 Cl_8]^{2-} \longrightarrow A \qquad k \ll 3 \times 10^{-3} \text{ sec}^{-1}$$

$$[Tc_2 Cl_8]^{3-} \longrightarrow B \qquad \text{very slow}$$

The recent isolation of $[(n\text{-}C_4 H_9)_4 N]_2 [Tc_2 Cl_8]$ (228) and $Tc_2 [O_2\text{-}CC(CH_3)_3]_4 Cl_2$ (79) attests to the stability of the Tc_2^{6+} moiety in spite of the more than 10 year hiatus between isolation and characterization of the "anomalous" $Tc_2 Cl_8^{3-}$ ion (61, 121) and the preparation of the more elusive $Tc_2 Cl_8^{2-}$ dimer, which has the classical d^4 metal-ion configuration appropriate for quadruple-bond formation. Reports of an average oxidation state of 2.67+ for technetium in salts containing $[Tc_2 Cl_8]_3^{8-}$ are compatible with cocrystallization of $Tc_2Cl_8^{3-}$ and $Tc_2 Cl_8^{2-}$ units in a stoichiometric ratio of 2:1 (138).

2. Rhenium

Electrochemical studies exploring the reduction of $Re_2 X_8^{2-}$ ions were promoted by the observed stability of the $Tc_2 Cl_8^{-3}$ ion (121) and by an early molecular orbital description that suggested a σ-nonbonding level as the LUMO of $Re_2 X_8^{2-}$ (83). Although the presence of the δ^* orbital now seems to be firmly established as the LUMO in $Re_2 Cl_8^{2-}$ (243), the redox chemistry of multiply bound Re_2^{n+} fragments has nonetheless proven to be an extremely fertile area of research.

Initial reports of the polarographic behavior of $Re_2Cl_8^{2-}$ and $Re_2(NCS)_8^{2-}$ at a dropping mercury electrode suggested that two reduction steps of one electron each occurred to form $Re_2X_8^{3-}$ and $Re_2X_8^{4-}$ (107). The measured $E_{1/2}$ values of -0.82 and -1.44 V versus SCE for the two reduction steps in the polarogram of $Re_2Cl_8^{2-}$ as published in 1967 (107) were reproduced by Hendriksma and van Leeuwen within experimental error (-0.83 and -1.40 V) in a publication in 1973 (144). In addition to polarographic data, cyclic voltammetry and coulometric measurements were also performed using acetonitrile as a solvent (144). The reduced rhenium species that are generated have a short lifetime and spectral properties are difficult to measure (125). It is important to note that more recently Cotton and Pedersen found only one reduction wave for $Re_2Cl_8^{2-}$ in the range -2 to $+2$ V using platinum, gold, and carbon electrodes (101). An $E_{1/2}$ of -0.840 V was measured. This potential was the same as that observed for a quasi-reversible reduction in the cyclic voltammogram, where the ratio of the peak anodic current to the peak cathodic current varied from 0 to 1 as a function of the sweep rate. The reduced product $Re_2Cl_8^{3-}$ was assumed to undergo rapid decomposition to unidentified products; a first-order rate constant of 0.5 sec^{-1} was estimated from the voltammetric data. The failure of various attempts to isolate salts of $Re_2Cl_8^{3-}$ is explained by the reported rapid decomposition of this ion. The authors point out in concluding remarks that the electrochemistry and respective stability of $Re_2Cl_8^{2-}$ and $Tc_2Cl_8^{3-}$ are consistent with the increased stability of higher oxidation states generally associated with the heavier transition metals in any particular vertical column, where here the comparison involves Tc^{2-5+} and Re^{3+}.

Electrochemical generation of dimers more stable than $Re_2Cl_8^{3-}$ but, nonetheless, containing the Re_2^{5+} unit has been observed upon reduction of $Re_2(O_2CC_6H_5)_4Cl_2$ and upon oxidation of $Re_2Cl_4[P(C_2H_5)_3]_4$ (100). Rotating-disk polarograms and cyclic voltammetry in dichloromethane and acetonitrile indicated that $Re_2(O_2CC_6H_5)_4^{1+}$ forms upon reduction of $Re_2(O_2CC_6H_5)_4^{2+}$ (the nature of the actual species present upon dissolution of $Re_2(O_2CC_6H_5)_4$ Cl_2 is not known) and has a lifetime on the order of hours. Conversely the $Re_2Cl_4[P(C_2H_5)_3]_4$ dimer undergoes a one-electron oxidation to form the $Re_2Cl_4[P(C_2H_5)_3]_4^{1+}$ cation, which also has an average rhenium oxidation state of 2.5+ and is stable for hours.

For $Re_2(O_2CC_6H_5)_4Cl_2$ the first reduction wave near -0.3 V is quasi-reversible. The product of this one-electron oxidation was not isolated as a solid, but ESR spectra were obtained on frozen dichloromethane solutions at 77°K at both X- and Q-band frequencies. Both naturally occurring rhenium isotopes have a nuclear spin of 5/2 with slightly different nuclear magnetic moments. For dimeric rhenium species there are three distinct isotopic isomers and each generates a unique ESR spectrum as a result of differing hyperfine interactions. The observed ESR spectrum is a superposition of these lines with the intensity ratio of resolved signals indicative of delocalization of the unpaired electron resulting in equivalent coupling to both rhenium nuclei in the dimer. Preliminary estimates of the ESR parameters for $Re_2(O_2CC_6H_5)_4^+$ were obtained from the X-band spectrum, and those were employed to generate more refined values in accord with the observed Q-band spectrum. The resultant spin Hamiltonian parameters $(g_\| = 1.71(3)$, $g_\perp = 2.13(6)$, $|A_\|| = 573 \times 10^{-4}$ cm^{-1}, and $|A_\perp| = 272 \times 10^{-4}$ cm^{-1}) were in qualitative agreement with a $^2B_{1u}$ ground state with the unpaired electron located in the δ^* orbital. The theory employed to analyze the value of g in terms of likely configurations for $Tc_2Cl_8^{3-}$ should be equally applicable to the $Re_2(O_2CC_6H_5)_4^+$ ion (see Section III.I.1). The deviation of both $g_\|$ and $g_\perp$ from 2 is larger for rhenium than for technetium in accord with the larger spin–orbit coupling constant for the heavier metal.

The second reduction wave for $Re_2(O_2CC_5H_5)_4Cl_2$ was irreversible, occurring near -1.1 V.

Two quasi-reversible one-electron oxidation steps were observed for $Re_2Cl_4[P(C_2H_5)_3]_4$ in dichloromethane at -0.36 and 0.87 V versus SCE, respectively. Decomposition of the doubly oxidized species, postulated to be $Re_2Cl_4[P(C_2H_5)_3]_4^{2+}$, was more rapid in acetonitrile $(k > 1$ sec$^{-1})$ than in dichloromethane $(k \ll 3 \times 10^{-3}$ sec$^{-1})$. Generation of the monocation by electrolysis at 0.0 V in dichloromethane produced greenish yellow solutions used to obtain ESR data for $Re_2Cl_4[P(C_2H_5)_3]_4^+$. The presence of ^{31}P hyperfine coupling in the already complex Re_2^{5+} ESR pattern was cited as a likely cause of resolution and interpretation difficulties with regard to extracting accurate spin Hamiltonian parameters in this case. Values suggested were $g_\| < 2$, $g_\perp \approx 2.2$, $|A_\perp^{Re}| \approx 155$ cm^{-1}, and $|A_\perp^P| \approx 85$ cm^{-1}. The high value of $|A_\perp^P|$ was considered to be consistent with the $^2B_{1u}$ ground state having greater electron interaction with the ligands than would a $^2A_{2u}$ ground state, where the electron is localized in a σ-nonbonding metal orbital directed away from the phosphine ligands.

The identification of three stable oxidation states for $Re_2Cl_4[P(C_2H_5)_3]_4^{n+}$ $(n = 0, 1, 2)$ provides an example of a multiple metal–metal bond serving as a redox center while maintaining a strong bonding interaction in each oxidation state (100).

Electrochemical investigations of a series of $Re_2X_4(PR_3)_4$ dimers with $X = Cl$, Br, or I and $R = C_2H_5$, n-C_3H_7, or n-C_4H_9 revealed two one-electron reversible oxidations (220). Coulometry followed by cyclic voltammetry was used to unravel the interesting interplay between electrochemical and chemical reaction processes that characterize this system. Cyclic voltammetry established $E_{1/2}$ values of -0.42 and 0.80 V versus SCE for formation of the mono- and dication, respectively, from $Re_2Cl_4[P(C_3H_7)_3]_4$. Electrochemical oxidation at 0.0 V followed by cyclic voltammetry indicated that the major product was $Re_2Cl_4[P(C_3H_7)_3]_4^+$ as expected, but additional waves were evident at 0.31 and -0.88 V. These two new waves were indicative of the formation of

$Re_2 Cl_5 [P(C_2 H_5)_3]_3$ based on identical $E_{1/2}$ values measured for an analytically pure sample. Electrolysis at +1.0 V added one more wave to the cyclic voltammogram in addition to the two present for the $Re_2 Cl_4 [P(C_2 H_5)_3]_4^{n+}$ ($n = 0, 1, 2$) system and the two associated with $Re_2 Cl_5 [P(C_2 H_5)_3]_3$. The additional feature at -0.11 V was identified as being due to $Re_2 Cl_6 [P(C_2 H_5)_3]_2$ based on cyclic voltammetry of a pure sample. Formation of $Re_2 Cl_6 [P(C_2 H_5)_3]_2$ also occurs upon exhaustive electrolysis at 0.5 V.

The mechanism leading to production of the neutral Re(III) dimer depends on the applied oxidation potential as follows. At 1.0 V two electrons are electrochemically removed from the original dimer to form the dication, which then adds chloride ion and loses one phosphine ligand to form $Re_2 Cl_5 [P(C_2 H_5)_3]_3^+$. A similar second substitution reaction produces the final neutral product. The authors classify this scheme as EECC in accord with two electrode processes (each a one-electron oxidation in this case) followed by two chemical reactions (Cl^- substitution) to form the observed product (220). A potential of 0.5 V is not adequate to oxidize the $Re_2 Cl_4 [P(C_2 H_5)_3]_4^+$ formed to the dication, but a chemical reaction converts this monocation into the neutral $Re_2 Cl_5 [P(C_2 H_5)_3]_3$. Since this neutral dimer is oxidized at an applied potential of 0.5 V ($E_{1/2} = 0.31$ V), the monocation precursor, $Re_2 Cl_5 [P(C_2 H_5)_3]_3^+$, forms in an electrochemical step and then goes on to product. In summary this is then an ECEC scheme, where chemical reaction precedes the second electrochemical oxidation step, as opposed to the first pathway described, where a potential of 1.0 V causes both oxidation steps to precede the chemical steps.

Chemical oxidation and reduction of dimers containing the Re_2^{6+} quadruply bound moiety have been employed to synthesize multiply bound rhenium dimers with geometries and electronic configurations differing from those of the parent compounds. Halogen oxidation of $Re_2 X_8^{2-}$ ($X = Cl, Br$) produces dimeric Re(IV) species, $Re_2 X_9^-$, that can be isolated as the tetrabutylammonium salt (25). Dissolution in methanol under various conditions led to formation of the violet $Re_2 Cl_9^{2-}$ mixed oxidation-state dimer from the green $Re_2 Cl_9^-$ ion. Previously the $Re_2 Cl_9^{2-}$ ion had been produced by reacting $ReCl_4$ with chloride ion in an attempt to form $Re_2 Cl_9^-$ (106). Reduction of $Re_2 X_9^-$ to $Re_2 X_9^{2-}$ was easily accomplished with various reducing agents, and elemental halogen effected the reverse oxidation almost instantaneously in methylene chloride solution (25). Thus the only observed product from combining X_2 and $Re_2 X_8^{2-}$ is $Re_2 X_9^-$, with no $Re_2 X_9^{2-}$ evident. The violet $Re_2 Cl_9^{2-}$ formed upon heating $Re_2 Cl_9^-$ briefly in methanol is an intermediate that can be converted to $Re_2 Cl_8^{2-}$ by continued heating of the solution. Figure 13 is a diagram depicting the interconversions of these anionic rhenium dimers.

Physical characterizations of the $Re_2 X_9^{n-}$ ($n = 1, 2$) ions included room-temperature susceptibility ($\mu_{eff} = 1.5$ BM, $Re_2 Cl_9^{2-}$; $\mu_{eff} = 1.7$ BM, $Re_2 Br_9^{2-}$), far infrared absorption data (353, 250 cm^{-1}, $Re_2 Cl_9^-$; 321 cm^{-1}, $Re_2 Cl_9^{2-}$), and

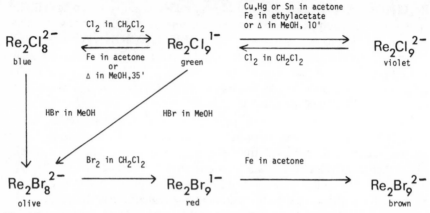

Fig. 13.   Chemical interconversions among $Re_2X_8^{2-}$ and $Re_2X_9^{n-}$ ($n = 1, 2$) anions (25).

electronic absorption parameters (25). X-ray photoelectron spectra have been obtained for these $Re_2Cl_9^{n-}$ dimers (240).

Ligand-induced reduction of the Re_2^{6+} unit present in $Re_2X_8^{2-}$ has been observed with sulfides and phosphines. The reaction of $[(C_4H_9)_4N]_2Re_2Cl_8$ with excess 2,5-dithiahexane in refluxing acetonitrile or acidified methanol over a period of several days leads to precipitation of black crystals of $Re_2Cl_5(dth)_2$ (99). The mean oxidation state of 2.5+ for the two rhenium atoms is assumed to result from reduction by the excess sulfide (21). The structure of this unusual dimer has been interpreted as reflecting a net rhenium–rhenium bond order of three with the δ bond absent. As seen in Fig. 14 there are two distinct rhenium environments that can be viewed as indicative of differing oxidation states for the two sites, that is, Re^{2+} and Re^{3+} (21, 22). The staggered configuration of the two square planar rhenium fragments ensures the absence of any δ-bonding contribution, since the δ overlap vanishes as a result of symmetry considerations. Allocation of the nine valence electrons remaining on the two metals after metal–ligand σ-bonds are formed no doubt involves placing six electrons in the metal–metal σ- and two π-bonding molecular orbitals. The remaining three electrons are then proposed to occupy the d_{xy} orbitals of δ symmetry on the two metal atoms, two becoming localized on the rhenium with the lower d_{xy} orbital energy to account for a formal Re(II) and the remaining electron going to the other metal atom, then formally Re(III). The bond length of 2.29 ± 0.01 Å is consistent with loss of the relatively weak δ bonding component from the quadruple bond present in $Re_2Cl_8^{2-}$, where a separation of 2.24 Å is observed (81). The magnetic moment of 1.71 BM is independent of temperature from 80 to 300°K (22).

Ebner and Walton have investigated the reduction of octahalodi-rhenate(III) dimers with a variety of phosphines (124). Simple substitution reactions are observed under mild conditions to form $Re_2X_6(PR_3)_2$ as is noted

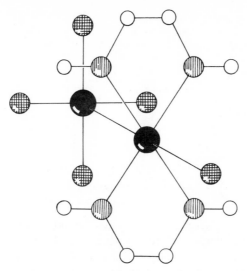

Fig. 14. The molecular geometry of $Re_2Cl_5(dth)_2$ showing the staggered conformation (21).

in Section III.D. However, it has now been shown that ligand-induced reduction can be promoted by an increase in the reaction temperature, prolonged reaction periods, and decreased phenyl substitution on the phosphine.

Formation of the rhenium(II) dimers $Re_2Cl_4(PR_3)_4$ results from the reduction of $Re_2Cl_8^{2-}$ with PMe_3, PEt_3, $P(n\text{-}Pr)_3$, and PEt_2Ph (124). Yields of greater than 50% are typical. It should be noted that similar products have also been obtained from reactions of these tertiary phosphines with the trimeric rhenium(III) species Re_3Cl_9.

Increased phenyl substitution on the phosphine led to more limited reduction as evidenced by the isolation of $Re_2Cl_5(PR_3)_3$ from the reaction of $PMePh_2$ or $PEtPh_2$ with $Re_2Cl_8^{2-}$ or Re_3Cl_9. Similar products were isolated from the reduction of $Re_2Br_8^{2-}$. The dimeric nature of these compounds was deduced from a molecular weight measurement indicative of $Re_2Cl_5(PEtPh_2)_3$, an intense polarized Raman line at 277 cm^{-1} that was attributed to the ν(Re–Re) mode, an intense electronic absorption near 1400 nm, and a magnetic moment appropriate for one unpaired electron per dimeric unit (μ_{eff} = 2.0 BM) (124).

A preliminary report of the x-ray structure of $Re_2Cl_4(PEt_3)_4$ (72) was of great interest, since this reduced dimer has a total of 10 electrons available for the valence orbitals of the two metals. The molecular $Re_2Cl_4P_4$ skeleton was found to be of the eclipsed dimeric form, resembling that found for quadruply bound rhenium(III) dimers. The D_{2d} symmetry of the dimer is seen in Fig. 15, where two trans-$ReCl_2(PEt_3)_2$ fragments are joined in such a fashion as to minimize steric repulsion among the bulky phosphine ligands.

Accurate molecular parameters were obtained for $Re_2Cl_4(PEt_3)_4$ in spite

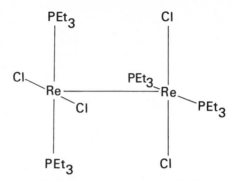

Fig. 15.   The eclipsed geometry of $Re_2Cl_4(PEt_3)_4$ (73).

of an unusual crystallographic disorder (73). The rhenium–rhenium distance of 2.232(6) Å is surprisingly close to the value of 2.22 Å found for $Re_2Cl_8^{2-}$ (77) and $Re_2Cl_6(PEt_3)_2$ (71). Although the eclipsed structure can be rationalized on the basis of steric factors, since it seems quite reasonable that the phosphine ligands will dominate the molecular geometry, the short rhenium–rhenium distance seems inconsistent with occupation of the δ^* orbital by the two excess electrons that would lead to a formal bond order of three (73). Cotton and coworkers concluded that the "additional" electrons do not occupy the δ^* orbital, since comparison with relevant data suggests that a lengthening of the rhenium–rhenium bond by 0.06–0.10 Å would be expected for loss of the δ bond. The distances cited to support this prediction include $Re_2Cl_5(dth)_2$ data with a triple bond and a Re–Re distance of 2.293 Å (21) (increased by ca. 0.07 Å from typical quadruple-bond lengths), and the $Mo_2(SO_4)_4^{n-}$ ($n = 3, 4$) couple (75), where the metal–metal distance increases by 0.05 Å when one of the δ-bonding electrons is removed from $Mo_2(SO_4)_4^{4-}$.

A summary of the effect of the two electrons beyond those required for formation of a quadruple bond between the metal atoms includes (1) no qualitative change in the dimeric structure, (2) no significant difference in the metal–metal bond length, and (3) perhaps a slight weakening of the metal ligand bonds. With regard to the metal–ligand bond lengths found for $Re_2Cl_4(PEt_3)_4$, $Re_2Cl_6(PEt_3)_2$, and $Re_2Cl_8^{2-}$, it should be noted that the increase in Re–Cl bond length in the Re(II) dimer is not statistically significant. Furthermore, it is in the direction expected for a more reduced metal center and thus provides no firm basis for analyzing the location of added electron density. The absence of an adequate molecular orbital description of rhenium(II) dimers with 10 metal valence electrons prohibits conclusive deductions about the correct ground-state configuration for this molecule.

An exemplary series of phosphines that illustrates the ligand-reducing tendency toward $Re_2Cl_8^{2-}$ as a function of phenyl substitution is PEt_3, PEt_2Ph,

$PEtPh_2$, and PPh_3. Reaction with $Re_2Cl_8^{2-}$ forms $Re_2Cl_4(PEt_3)_4$, $Re_2Cl_4(PEt_2Ph)_4$, $Re_2Cl_5(PEtPh_2)_3$, and $ReCl_6(PPh_3)_2$, respectively, as the final product (124). Steric effects due to the large phenyl groups are considered to play an important role in determining the ultimate product. The increased basicity of alkyl phosphines relative to phenyl substituted analogues may also promote reduction of the metal.

The reactivity patterns of $Re_2X_4(PR_3)_4$ and $Re_2X_5(PR_3)_3$ are consistent with the presence of a strong multiple metal–metal bond. Oxidation of these dimers with chlorocarbon solvents or methanol acidified with hydrochloric acid led to Re(III) dimer formation in several cases. The product of methanolic HCl oxidation of $Re_2Cl_4(PEt_3)_4$ was $Re_2Cl_6(PEt_3)_2$, while carbon tetrachloride oxidation produced $[Et_3PCl]_2Re_2Cl_8$. Mixed-halide dimers resulted from carbon tetrachloride oxidation of $Re_2Br_4(PEt_3)_4$, which produced $Re_2Cl_4Br_2(PEt_3)_2$ and $[Et_3PCl]_2Re_2Cl_4Br_4$. The analytical data were consistent with the stoichiometric halide content indicated rather than suggestive of variable composition. Reduction of $Re_2Cl_5(PEtPh_2)_3$ with PEt_3 formed $Re_2Cl_4(PEt_3)_4$, and oxidation with CH_2Cl_2/CCl_4 formed $Re_2Cl_8^{2-}$. Interconversions among these rhenium phosphine dimers are summarized in the schematic diagram in Fig. 16.

Spectral characterization of the phosphine-substituted rhenium dimers included electronic spectra and infrared data (124). Detailed assignments of electronic absorption bands were not proposed by the authors, but characteristic structure–spectra features were identified. For the Re_2^{6+} unit the lowest energy absorption was in the range of 695–760 nm, similar to results in previous Re_2^{6+}

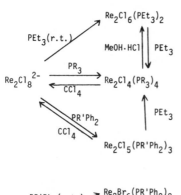

Fig. 16. Chemical interconversions among rhenium dimers (Re_2^{6+}, Re_2^{5+}, and Re_2^{4+}) (124).

investigations (243). The mixed-oxidation-state dimers $Re_2X_5(PR_3)_3$ exhibited an intense absorption in the near infrared region at approximately 1400 nm. It is interesting to note that the $Tc_2Cl_8^{3-}$ anion, also a mixed-valence-state dimer, exhibits a similar band in the near infrared that has been assigned to the $\delta \rightarrow \delta^*$ transition (70) as is discussed in Section III.I.1. Observation of a similar low-energy transition in some spectra of $Re_2Cl_4(PR_3)_4$ dimers was attributed to oxidation.

Infrared data were consistent with the presence of coordinated tertiary phosphines and the absence of rhenium—oxygen contaminants. The low-frequency bands assigned to Re—Cl stretching modes displayed the normal shift to lower frequency upon reduction of the metal in the series $Re_2Cl_6(PR_3)_2 > Re_2Cl_5(PR_3)_3 > Re_2Cl_4(PR_3)_4$ (124).

Reduction of rhenium also occurs with 2,2'-bipyridyl upon refluxing $Re_2Cl_8^{2-}$ with excess ligand in n-butanol for 2 days. A product with the stoichiometry $[ReCl_{2.5}(bipy)]_n$ was reported (158). This brown product is insoluble in most noncoordinating solvents and only slightly soluble in strongly coordinating solvents.

Observation of rhenium reduction from Re(III) to Re(II) with tertiary phosphines led Hertzer and Walton to attempt reduction of the Re(IV) dimer $Re_2Cl_9^-$ under similar conditions (145). Reduction to rhenium(III) was promoted by triphenylphosphine in MeOH—HCl to form $Re_2Cl_6(PPh_3)_2$. More highly reduced species resulted from the reaction of $PEtPh_2$ with $[(C_4H_9)_4N]-[Re_2Cl_9]$ in refluxing acetone to produce $Re_2Cl_5(PEt_2Ph)_3$. Reduction of the rhenium(IV) dimer with PEt_3 led to a four-electron reduction of the dimer and the rhenium(II) species $Re_2Cl_4(PEt_3)_4$ was isolated. This four-electron reduction with retention of a strong multiple metal—metal bond provides unprecedented support for the existence of a rich redox chemistry in these systems, where the integrity of the metal—metal bond is retained in a variety of oxidation states.

The reactions of $Re_2X_8^{2-}$ with bidentate phosphine ligands are somewhat more complex than those observed for monodentate phosphines (123). Reduction does not occur in the room-temperature reaction of dppe with $Re_2Cl_8^{2-}$, but rather the chlorine-bridged conlateral bioctahedral dimer $[ReCl_3dppe]_2$ is isolated (159). This dimer contains two dilute rhenium(III) centers separated by 3.809(1) Å (160), that is, the metal—metal bond has not been retained. Prolonged reflux produces a reduced species of stoichiometry $[ReCl_2(dppe)]_n$ in low yields (~6%), but an improved yield of 80% is possible from reaction of $Re_2Cl_4(PEt_3)_4$ and dppe. Similar products were obtained with the bidentate arphos ligand in reactions with $Re_2Cl_8^{2-}$ and $Re_2Cl_4(PEt_3)_4$. Chemical reactivity of the $[ReCl_2(LL)]_n$ (LL = dppe or arphos) species suggested that these compounds were indeed dimeric; prolonged reflux in chlorocarbon solvents led to formation of the $Re_2Cl_8^{2-}$ ion. The spectral and magnetic properties of

$Re_2X_4(LL)_4$ differ substantially from those of $Re_2Cl_4(PR_3)_4$ (123), and a molecular geometry involving bridging dppe ligands with staggered $ReCl_2P_2$ units in a dimeric molecule was suggested. The room-temperature magnetic moments of 1.7(1) and 1.3(1) BM measured for $Re_2Cl_4(dppe)_2$ and $Re_2Cl_4(arphos)_2$ were below values expected for two unpaired electrons and, since these are even-electron compounds, a possible singlet–triplet equilibrium may obtain. Variable-temperature susceptibility studies should clarify this point.

The dppm ligand leads to different reaction products than dppe under similar reaction conditions. The rhenium(II) dimer $Re_2Cl_4(dppm)_2$ is formed from $Re_2Cl_4[P(n\text{-}Pr)_3]_4$, but $Re_2Cl_4(PEt_3)_4$ undergoes only limited substitution to produce $Re_2Cl_4(PEt_3)_2(dppm)$. These dppm complexes resemble $Re_2Cl_4(PR_3)_4$ in that electronic absorption bands grow in the region from 1400 to 1500 nm upon exposure to oxygen. This behavior is not observed for the dppe or arphos derivatives. The ability to bridge two rhenium atoms joined in a strong metal–metal bond may be unique to the dppm ligand for these bidentate phosphines in contrast to the behavior of dppe (113). Thus ring size and ligand bite may account for the different reactivity and products observed for the reactions of dppe and dppm with rhenium dimers.

Confirmation of the bridging behavior of dppm in rhenium dimer chemistry has been obtained by x-ray analysis of the minor product of the reaction of dppm with $Re_2Cl_8^{2-}$ in acetone (111). The brown, air-stable solid recrystallized from toluene and having the stoichiometry $Re_2Cl_5(dppm)_2 \cdot 2C_6H_5CH_3$ has the eclipsed structure with two bridging dppm ligands and one terminal chloride as shown in Fig. 17. This dimer thus differs from the Re_2Cl_5

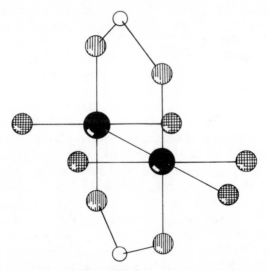

Fig. 17.   The eclipsed geometry of $Re_2Cl_5(dppm)_2$ with bridging dppm ligands (111).

(dth)$_2$ species where both bidentate dithiahexanes chelate the same rhenium atom. The distribution of charge in Re$_2$Cl$_5$(dppm)$_2$ is not obvious from the structural data. The rhenium with higher positive charge density would perhaps be expected to bind the terminal chloride ligand. Countering this argument are the observed metal–ligand distances, which would be predicted to decrease with increasing oxidation state of the metal, yet the rhenium with the terminal chloride exhibits bond lengths to phosphorus and equatorial chlorine atoms that are 0.04 and 0.05 Å longer than for the adjacent rhenium. Perhaps the most consistent picture results from steric considerations with coordination of a fifth ligand to one rhenium causing bond lengths to increase slightly at that metal center. A similar rationale was evident in the Re$_2$(Ph$_3$CN$_2$)$_2$Cl$_4$ dimers with and without THF coordinated to one of the terminal sites (109). The metal–metal separation of 2.263(1) Å in Re$_2$Cl$_5$(dppm)$_2$ is about 0.03 Å longer than that found in Re$_2$Cl$_6$(PEt$_3$)$_2$ (71) and Re$_2$Cl$_4$(PEt$_3$)$_4$ (72, 73). This increase in rhenium–rhenium distance is the same as that induced in Re$_2$(Ph$_3$CN$_2$)$_2$Cl$_4$ upon coordination of THF at an axial site, and thus occupation of one terminal position in Re$_2$Cl$_5$(dppm)$_2$ by chloride is logically correlated with the observed multiple metal–metal bond length.

3. Molybdenum

The electrochemical behavior of dimers containing the Mo$_2^{4+}$ unit has been reported for several derivatives. Oxidation of Mo$_2$(SO$_4$)$_4^{4-}$ in acid solution formed Mo$_2$(SO$_4$)$_4^{3-}$ ($E_{1/2}$ = 0.22 V versus SCE), but rapid decomposition of the oxidized product ensued with a rate constant estimated from cyclic voltammetry to be in the range of 10^{-3}–10^{-1} sec^{-1} (75). Reproducibility of the electrochemical behavior depended on acid concentration, electrode history, and sweep rate and was generally indicative of complex processes not amenable to rigorous interpretation.

Frozen-solution ESR data were obtained for the paramagnetic Mo$_2$(SO$_4$)$_4^{3-}$ ion at both X- and Q-band frequencies. The existence of two naturally abundant molybdenum nuclei with a nuclear spin of 5/2 (^{95}Mo, 15.7%, ^{97}Mo, 9.5%) produced a hyperfine splitting pattern consistent with delocalization of the unpaired electron over two equivalent molybdenum sites. Computer-simulated spectra were generated with values of $g_{\parallel}$ = 1.891, $g_{\perp}$ = 1.909, $|A_{\parallel}|$ = 45.2 x 10^{-4} cm^{-1}, and $|A_{\perp}|$ = 22.9 x 10^{-4} cm^{-1} providing a satisfactory fit (75). The theory applied to Tc$_2$Cl$_8^{3-}$ (see Section III.I.1) to correlate the electronic structure with the observed g values should also be applicable to the molybdenum dimer. Removal of one of the two δ bonding electrons from Mo$_2$(SO$_4$)$_4^{4-}$ would be expected to produce a $^2B_{2g}$ ground state. Excited states of B_{1g} and E_g symmetry would then influence the $g_{\parallel}$ and $g_{\perp}$

values, respectively. The location of empty orbitals of B_{1g} and E_g symmetry at approximately the same energy above the ground states according to the SCF–X_α–SW molecular orbital calculation for $Mo_2Cl_8^{4-}$ (199) would suggest that $g_\parallel$ and $g_\perp$ might be of comparable magnitude, with both reduced below 2.0, if the electronic structure of the sulfato-bridged dimers resembles that of $Mo_2Cl_8^{4-}$. The g values calculated from ESR data are in accord with this description of the $Mo_2(SO_4)_4^{3-}$ ion (100).

Electrochemical oxidation of the Mo_2^{4+} unit has also been observed for $Mo_2Cl_8^{4-}$ and $Mo_2(O_2CC_3H_7)_4$ (103). Cyclic voltammetry on $Mo_2Cl_8^{4-}$ in hydrochloric acid solution and methanol showed an oxidation at 0.5 V versus SCE, presumably indicative of formation of the short-lived $Mo_2Cl_8^{3-}$ ion. The corresponding reduction peak could only be observed with a rapid sweep rate of 500 mV sec^{-1} in $6M$ hydrochloric acid, and further characterization of the oxidized material was not undertaken (103). The electrochemical behavior of the tetrabutyratodimolybdenum compound was assumed to be representative of the many carboxylate-bridged molybdenum(II) dimers. A quasi-reversible one-electron oxidation was evident in cyclic voltammograms obtained in acetonitrile ($E_{1/2} = 0.39$ V), dichloromethane ($E_{1/2} = 0.45$ V), and ethanol ($E_{1/2} = 0.30$ V). Decomposition of the monocation was fairly rapid ($t_{1/2} \approx 1$ min), but frozen-solution ESR data were obtained on this odd-electron dimer. The spectrum resembled that of $Mo_2(SO_4)_4^{3-}$, with the various magnetic isomers again classified as having a total molybdenum nuclear spin of 0, 5/2, or 10/2 contributing in accordance with the respective natural abundances of the magnetically active and inactive nuclei. Interpretation of the ESR data followed the pattern established for $Mo_2(SO_4)_4^{3-}$ (75) to account for the axial symmetry, coupling to two equivalent molybdenum nuclei, and g values below 2.0 ($g_\parallel = 1.941$, $g_\perp = 1.941$, $|A_\parallel| = 35.6 \times 10^{-4}$ cm^{-1}, and $|A_\perp| = 17.8 \times 10^{-4}$ cm^{-1}) (100).

Isolation of crystalline salts of $Mo_2(SO_4)_4^{4-}$ and $Mo_2(SO_4)_4^{3-}$ suitable for x-ray diffraction analysis provided the first opportunity to compare the structures of a pair of isostructural dimers differing by one electron (75). The $\sigma^2\pi^4\delta^2$ configuration of $Mo_2(SO_4)_4^{4-}$ appropriate for a formal bond order of four is in accord with the observed Mo–Mo distance of 2.111(1) Å and the eclipsed structure. The $\sigma^2\pi^4\delta^1$ configuration describing $Mo_2(SO_4)_4^{3-}$ does not alter the gross geometrical structure of the dimer; it is still eclipsed with four bidentate bridging sulfato ligands. An increase of about 0.05 Å in the Mo–Mo distance to 2.164(2) Å in the oxidized dimer is consistent with a decrease in the formal metal–metal bond order to 3.5. The observed shortening of the average Mo–O bond lengths by approximately 0.08 Å in the more highly oxidized dimer is in the direction expected for comparison of $Mo^{2.5+}$ and Mo^{2+} metal–ligand distances. Stable aqueous solutions of Mo_2^{5+} have been generated from air oxidation of $Mo_2(SO_4)_4^{4-}$ in sulfuric acid (212). Pernick and Ardon associated the solution stability of the $Mo^{2.5+}$ species with coordinated sulfate ions presumably bridging the Mo_2^{5+} moiety, since other acids led to disproportionation reactions.

Isolation of stable $Mo_2(O_2CR)_4^+$ salts with the triiodide counterion has been reported for $R = C_2H_5$, C_6H_5, and $C(CH_3)_3$ (191). Characterization of these salts included comparative infrared data for $Mo_2(O_2CR)_4$ and $[Mo_2(O_2CR)_4]I_3$, with nearly identical spectra for $R = C(CH_3)_3$ above 700 cm^{-1} and a slight shift to higher frequency for vibrations in the oxidized species in the 400–700 cm^{-1} region. Bands at 420, 450, and 615 cm^{-1} in the neutral dimer increased by 25, 15, and 10 cm^{-1} in the oxidized dimer as would be anticipated for Mo–O modes as a function of the increased metal oxidation state. Regeneration of the neutral $Mo_2(O_2CR)_4$ dimers was observed upon thermolysis of the triiodide salts, also suggesting that the dinuclear configuration is retained in the ionic products. Temperature-dependent susceptibility studies on the oxidized pivalate derivative indicated Curie behavior with $\mu = 1.66$ BM per dimer in accord with the g value of 1.93(1) derived from ESR measurements on undiluted crystalline salts ($\mu_{calc} = 1.67$ based on $\mu_{calc} = g_{obsd}\sqrt{s(s + 1)}$ with $s = 1/2$).

A particularly interesting chemical oxidation of Mo^{2+} has been observed in the reaction of dimolybdenum tetraacetate with hydrohalic acids. Cotton and Kalbacher have recently presented data in support of an oxidative addition of HX to intermediate dinuclear Mo^{2+} species to form Mo^{3+} dimers, $Mo_2X_8H^{3-}$ (86). Since this report firmly establishes the presence of a hydride in the product, as shown by infrared studies on protio and deutero analogues, in addition to tritium labeling of $Rb_3Mo_2Cl_8H$, literature references to $Mo_2X_8^{3-}$ prior to 1976 are best interpreted in terms of $Mo_2X_8H^{3-}$. The history of the $Mo_2X_8H^{3-}$ dimers is informative in that these compounds were formulated as $Mo_2X_8^{3-}$ ions based on analytical data and an x-ray crystal structure of $Rb_3Mo_2Cl_8$. The $Mo_2Cl_8^{3-}$ anion conformed to the confacial bioctahedral structural class of M_2X_9 compounds, but a statistical occupancy factor of 0.67 in each of the three bridging positions resulted in the observed stoichiometry of $Mo_2Cl_8^{3-}$ (18). The structure of $Cs_3Mo_2Br_8$ was reported in 1973 and again a $W_2Cl_9^{3-}$ type structure with one of the bridging bromine atoms missing from each anion was found (74). Reformulation of the $Mo_2X_8^{3-}$ ions as $Mo_2X_8H^{3-}$ was prompted by the surprising results of experimental magnetic susceptibility measurements on the postulated $Cs_3Mo_2Br_8$ salt. The diamagnetism observed for this compound required that it contain an even number of electrons, and therefore the $Mo_2Br_8^{3-}$ formulation had to be reconsidered and revised (86). The presence of a hydrogen atom would not have been revealed by either elemental analysis or x-ray diffraction, and Cotton and Kalbacher proceeded to reinvestigate these compounds in light of the possible presence of a bridging hydride ligand in addition to the eight halides bound to the dinuclear molybdenum fragment (86).

The $Mo_2X_8H^{3-}$ formulation was an intuitively pleasing proposal that resolved many of the anomalous and somewhat inexplicable properties of the $Mo_2X_8^{3-}$ species. The identity of the oxidant remained a mystery as long as the product was assumed to be $Mo_2X_8^{3-}$, but the actual $Mo_2X_8H^{3-}$ product is quickly recognized to be derived from a formal oxidative addition of HX to the dimeric Mo(II) system, perhaps best visualized as $Mo_2X_7^{3-} + HX \rightarrow Mo_2X_8H^{3-}$. This example of oxidative addition to a multiply bound metal–metal dimer suggests an area of untapped potential for future exploration, since this fundamental mode of reaction will have entirely different criteria in the case of dimers than in the thoroughly investigated monomeric case. As an example, the coordination number and oxidation state of a metal in a monomeric complex will both increase by 2 upon oxidative addition of XY, but oxidative addition of the same XY molecule across a multiple metal–metal bond will only increase the coordination number and oxidation state of each metal by 1.

Reexamination of the magnetic susceptibility of $Rb_3Mo_2Cl_8$ indicated that it was diamagnetic and observation of an absorption due to bridging hydride stretch in the infrared spectrum confirmed the identity of $Mo_2Cl_8H^{3-}$.

Katović and McCarley have synthesized the heteronuclear $MoWCl_8H^{3-}$ anion from $MoW(O_2CMe_3)_4$ (162). These same authors reported a detailed infrared study in which the symmetric and asymmetric M–H–M stretching vibration frequencies for the molybdenum dimers were coupled with the known Mo–Mo distance to deduce the location of the hydrogen atom. Several straightforward assumptions were required (such as symmetric bonding of the hydride to both metals) to calculate the Mo–H–Mo bridge angle and hence the Mo–H distance, but all the postulates are quite reasonable and the resulting consistency for chloride, bromide, and deuteride derivatives serves to validate the calculation.

Ultraviolet irradiation of $Mo_2(SO_4)_4^{4-}$ in aqueous sulfuric acid yields the oxidized molybdenum dimer $Mo_2(SO_4)_4^{3-}$ and molecular hydrogen gas (126). The $Mo_2(SO_4)_4^{3-}$ ion has a characteristic absorption maximum at 1405 nm ($\epsilon = 143$) that displays a 350 cm^{-1} vibrational progression at room temperature. This band is attributed to the $\delta \rightarrow \delta^*$ ($^2B_{2g} \rightarrow {}^2B_{1u}$) transition (7.1 x 10^3 cm^{-1}) coupled with ν(Mo–Mo), which compares to the $\delta \rightarrow \delta^*$ ($^1A_{1g} \rightarrow {}^1A_{2u}$) transition energy of 19.4 x 10^3 cm^{-1} in the $Mo_2(SO_4)_4^{4-}$ ion. The quantum yield for disappearance of $Mo_2(SO_4)_4^{4-}$ is 0.17 at 254 nm while irradiation at longer wavelengths is ineffective with respect to hydrogen production.

Photooxidation of the Mo_2^{4+} moiety has also been demonstrated for $Mo_2(aq)^{4+}$, $Mo_2Cl_8^{4-}$, $Mo_2Br_8^{4-}$ (244), and $Mo_2Cl_4(PR_3)_4$ (245). Molybdenum(III) products are observed upon 254 nm irradiation of these complexes as opposed to the molybdenum (2.5+) species formed upon photolysis of the $Mo_2(SO_4)_4^{4-}$ dimer. Molecular hydrogen and a hydroxy bridged Mo^{3+} dimer are the products of UV photolysis of both $Mo_2(aq)^{4+}$ and $Mo_2X_8^{4-}$ (X = Cl, Br). As is discussed above, it is known that treatment of the $Mo_2X_8^{4-}$ ion with HX(aq) forms $Mo_2X_8H^{3-}$ in the absence of irradiation, but Gray et al. have shown that this oxidation can be photochemically promoted (244). Thermal decomposition of $Mo_2X_8H^{3-}$ then follows the photooxidation step to produce $H_2(g)$ and $Mo_2(\mu\text{-OH})_2^{4+}$. It is interesting that $Mo_2(O_2CCH_2NH_3)_4^{4+}$ shows no photochemical reactivity under similar circumstances. Trichlorobridged molybdenum(III) dimers result when $Mo_2Cl_4(PR_3)_4$ (PR_3 = PEt_3, $P(n\text{-Bu})_3$, and $PEtPh_2$) is photolyzed at 254 nm in chlorocarbon solvents (245). As was the case for quadruply bound rhenium dimers, no photoactivity is associated with the $\delta \rightarrow \delta^*$ transition.

Oxidation of the Mo_2^{4+} fragment is also induced by reacting $Mo_2(O_2CCH_3)_4$ with $[NH_4][S_2CN(C_3H_7)_2]$ as is discussed in Section III.D.2. Oxidative addition across one of the dithiocarbamate C–S linkages leads to an oxidation state of 4+ for the molybdenum atoms in the resultant sulfur-bridged dimer (216, 217).

Reduction of $Mo_2(O_2CCF_3)_4$ has been reported in a pulse radiolysis study (15). The formation of $Mo_2(O_2CCF_3)_4^-$ was postulated based on a broad ESR signal in a frozen glass with $g = 1.91$. An electronic absorption band at 780 ± 20 nm with $\epsilon = 2.6 \pm 0.3 \times 10^3$ 1 mole^{-1} cm^{-1} was observed for the product.

Selective oxidation of the heteronuclear molybdenum–tungsten dimer $MoW[O_2CC(CH_3)]_4$ in the presence of the dimolybdenum contaminant invariably present in the reaction mixture produces a clean chemical separation of these isomorphous compounds (191). One-electron oxidation of $MoW(O_2CR)_4$ was accomplished with a limited amount of I_2 in benzene to form a precipitate consisting of $[MoW(O_2CR)_4]I$, and unreacted $Mo_2(O_2CR)_4$ remained in solution. Generation of pure $MoW(O_2CR)_4$ was accomplished by zinc reduction of the monocation (163). The greater reactivity of the tungsten derivative towards oxidation is in accord with the general trend towards greater stability of the higher oxidation states seen on descending the Group VIB metal triad. The structure of $[MoW(O_2CR)_4]I \cdot CH_3CN$ (163) reveals a Mo–W distance of 2.194(2) Å, well within the range expected for a bond of order 3.5 based on comparison with the $Mo_2(SO_4)_4^{3-}$ ion, where a Mo–Mo separation of 2.164(2) Å was found (75). Weak terminal coordination of the iodide anion to the tungsten atom [3.054(2) Å] of each dimer orders the structure and couples with the average W–O bond distance of 2.064(13) Å, as compared to the M–O average distance of 2.081(13) Å, to suggest that greater positive character resides on the tungsten atom than on molybdenum.

Acknowledgements

The author is indebted to Professors M. H. Chisholm, F. A. Cotton, and H. B. Gray for providing him with preprints in advance of their publication.

Cheerful and efficient preparation of the manuscript by Ms. Sue Hester is also gratefully acknowledged.

Abbreviations

arphos	1-Diphenylphosphino-2-diphenylarsinoethane
bipy	2,2'-Bipyridyl
Bu	Butyl
diars	o-Phenylenebis(dimethylarsine)
diglyme	$CH_3 OCH_2 CH_2 OCH_2 CH_2 OCH_3$
DME	$CH_3 OCH_2 CH_2 OCH_3$
DMP	2,6-Dimethoxyphenyl
dpae	Bis(diphenylarsino)ethane
dppe	Bis(diphenylphosphino)methane
dppm	Bis(diphenylphosphino)methane
dtd	4,7-Dithiadecane
dtdd	5,8-Dithiadodecane
dth	2,5-Dithiahexane
en	Ethylenediamine
Et	Ethyl
HOMO	Highest occupied molecular orbital
LUMO	Lowest unoccupied molecular orbital
Me	Methyl
MO	Molecular orbital
Ph	Phenyl
phen	1,10-Phenanthroline
pic	γ-Picoline
Pr	Propyl
py	Pyridine
pz	Pyrazolyl
tmtu	Tetramethylthiourea

References

1. E. H. Abbott, F. Schoenewolf, Jr., and T. Backstrom, *J. Coord. Chem.*, *3*, 255 (1974).
2. E. W. Abel, A. Singh, and G. Wilkinson, *J. Chem. Soc., 1959*, 3097.
3. E. W. Abel and F. G. A. Stone, *Q. Rev., 23*, 325 (1969).
4. G. Albrecht and D. Stock, *Z. Chem., 7*, 321 (1967).
5. G. B. Allison, I. R. Anderson, W. Van Bronswyk, and J. C. Sheldon, *Aust. J. Chem., 22*, 1097 (1969).
6. G. B. Allison, I. R. Anderson, and J. C. Sheldon, *Aust. J. Chem., 20*, 869 (1967).
7. R. A. Andersen, R. A. Jones, G. Wilkinson, M. B. Hursthouse, and K. M. A. Malik, *Chem. Commun., 1977*, 283.
8. I. R. Anderson and J. C. Sheldon, *Aust. J. Chem., 18*, 271 (1965).
9. C. L. Angell, F. A. Cotton, B. A. Frenz, and T. R. Webb, *Chem. Commun., 1973*, 399.

10. T. Aoki, A. Furusaki, Y. Tomiie, K. Ono, and K. Tanaka, *Bull. Chem. Soc. Jap.*, *42*, 545 (1969).
11. G. K. Babeshkina and V. G. Tronev, *Zh. Neorg. Khim.*, *7*, 215 (1962); *Russ. J. Inorg. Chem.*, *7*, 108 (1962).
12. R. Bailey and J. McIntyre, *Inorg. Chem.*, *5*, 1940 (1966).
13. M. C. Baird, *Prog. Inorg. Chem.*, *9*, 1 (1968).
14. E. Bannister and G. Wilkinson, *Chem. Ind. (Lond)*, *1960*, 319.
15. J. H. Baxendale, C. D. Garner, R. G. Senior, and P. Sharpe, *J. Am. Chem. Soc.*, *98*, 637 (1976).
16. M. Benard and A. Veillard, *Nouv. J. Chim.*, *1*, 97 (1977).
17. M. J. Bennett, W. K. Bratton, F. A. Cotton, and W. R. Robinson, *Inorg. Chem.*, *7*, 1570 (1968).
18. M. J. Bennett, J. V. Brencic, and F. A. Cotton, *Inorg. Chem.*, *8*, 1060 (1969).
19. M. J. Bennett, K. G. Caulton, and F. A. Cotton, *Inorg. Chem.*, *8*, 1 (1969).
20. M. J. Bennett, F. A. Cotton, B. M. Foxman, and P. F. Stokely, *J. Am. Chem. Soc.*, *89*, 2759 (1967).
21. M. J. Bennett, F. A. Cotton, and R. A. Walton, *J. Am. Chem. Soc.*, *88*, 3866 (1966).
22. M. J. Bennett, F. A. Cotton, and R. A. Walton, *Proc. Roy. Soc. (Lond.)*, *A303*, 175 (1968).
23. S. A. Best, T. J. Smith, and R. A. Walton, *Inorg. Chem.*, *17*, 99 (1978).
24. B. Biagini Cingi and E. Tondello, *Inorg. Chim. Acta*, *11*, L3 (1974).
25. F. Bonati and F. A. Cotton, *Inorg. Chem.*, *6*, 1353 (1967).
26. A. R. Bowen and H. Taube, *J. Am. Chem. Soc.*, *93*, 3287 (1971).
27. A. R. Bowen and H. Taube, *Inorg. Chem.*, *13*, 2245 (1974).
28. W. K. Bratton and F. A. Cotton, *Inorg. Chem.*, *8*, 1299 (1969).
29. W. K. Bratton and F. A. Cotton, *Inorg. Chem.*, *9*, 789 (1970).
30. W. K. Bratton, F. A. Cotton, M. Debeau, and R. A. Walton, *J. Coord. Chem.*, 1, 121 (1971).
31. D. J. Brauer and C. Krüger, *Inorg. Chem.*, *15*, 2511 (1976).
32. H. Breiland and G. Wilke, *Angew. Chem.*, *78*, 942 (1966).
33. J. V. Brencic and F. A. Cotton, *Inorg. Chem.*, *8*, 7 (1969).
34. J. V. Brencic and F. A. Cotton, *Inorg. Chem.*, *8*, 2698 (1969).
35. J. V. Brencic and F. A. Cotton, *Inorg. Chem.*, *9*, 346 (1970).
36. J. V. Brencic and F. A. Cotton, *Inorg. Chem.*, *9*, 351 (1970).
37. J. V. Brencic, D. Dobcnik, and P. Segedin, *Monatsh. Chem.*, *105*, 142 (1974).
38. J. V. Brencic, D. Dobcnik, and P. Segedin, *Monatsh. Chem.*, *105*, 944 (1974).
39. J. V. Brencic, D. Dobcnik, and P. Segedin, *Monatsh. Chem.*, *107*, 395 (1976).
40. J. V. Brencic, I. Leban, and P. Segedin, *Z. Anorg. Allg. Chem.*, *427*, 85 (1976).
41. J. V. Brencic and P. Segedin, *Z. Anorg. Allg. Chem.*, *423*, 266 (1976).
42. A. B. Brignole and F. A. Cotton, *Inorg. Synth.*, *13*, 81 (1972).
43. C. Calvo, N. C. Jayadevan, and C. J. L. Lock, *Can. J. Chem.*, *47*, 4231 (1969).
44. C. Calvo, N. C. Jayadevan, C. J. L. Lock, and R. Restivo, *Can. J. Chem.*, *48*, 219 (1970).
45. R. D. Cannon and J. S. Lund, *Chem. Commun.*, *1973*, 904.
46. M. H. Chisholm and F. A. Cotton, *Acc. Chem. Res.*, *11*, 356 (1978).
47. R. J. H. Clark and M. L. Franks, *Chem. Commun.*, *1974*, *316*.
48. R. J. H. Clark and M. L. Franks, *J. Am. Chem. Soc.*, *97*, 2691 (1975).
49. R. J. H. Clark and M. L. Franks, *J. Am. Chem. Soc.*, *98*, 2763 (1976).
50. R. J. H. Clark and N. R. D'Urso, *J. Am. Chem. Soc.*, *100*, 3088 (1978).
51. D. M. Collins, F. A. Cotton, S. Koch, M. Millar, and C. A. Murillo, *J. Am. Chem. Soc.*, *99*, 1259 (1977).

52. D. M. Collins, F. A. Cotton, and C. A. Murillo, *Inorg. Chem.*, *15*, 1861 (1976).
53. D. M. Collins, F. A. Cotton, and C. A. Murillo, *Inorg. Chem.*, *15*, 2950 (1976).
54. F. A. Cotton, *Acc. Chem. Res.*, *2*, 240 (1969).
55. F. A. Cotton, *Acc. Chem. Res.*, *11*, 225 (1978).
56. F. A. Cotton, *Chem. Soc. Rev.*, *4*, 27 (1975).
57. F. A. Cotton, *J. Less-Common Met.*, *54*, 3 (1977).
58. F. A. Cotton, *Inorg. Chem.*, *4*, 334 (1965).
59. F. A. Cotton, *Q. Rev.*, *20*, 389 (1966).
60. F. A. Cotton, *Rev. Pure Appl. Chem.*, *17*, 25 (1967).
61. F. A. Cotton and W. K. Bratton, *J. Am. Chem. Soc.*, *87*, 921 (1965).
62. F. A. Cotton, N. F. Curtis, C. B. Harris, B. F. G. Johnson, S. J. Lippard, J. T. Mague, W. R. Robinson, and J. S. Wood, *Science*, *145*, 1305 (1964).
63. F. A. Cotton, N. F. Curtis, B. F. G. Johnson, and W. R. Robinson, *Inorg. Chem.*, *4*, 326 (1965).
64. F. A. Cotton, N. F. Curtis, and W. R. Robinson, *Inorg. Chem.*, *4*, 1696 (1965).
65. F. A. Cotton, B. G. DeBoer, and M. Jeremic, *Inorg. Chem.*, *9*, 2143 (1970).
66. F. A. Cotton, B. G. DeBoer, M. D. LaPrade, J. R. Pipal, and D. A. Ucko, *J. Am. Chem. Soc.*, *92*, 2926 (1970).
67. F. A. Cotton, B. G. DeBoer, M. D. LaPrade, J. R. Pipal, and D. A. Ucko, *Acta Crystallogr.*, *B27*, 1664 (1971).
68. F. A. Cotton, M. Extine, and L. D. Gage, *Inorg. Chem.*, *17*, 172 (1978).
69. F. A. Cotton, M. Extine, and G. W. Rice, *Inorg. Chem.*, *17*, 176 (1978).
70. F. A. Cotton, P. E. Fanwick, L. D. Gage, B. J. Kalbacher, and D. S. Martin, Jr., *J. Am. Chem. Soc.*, *99*, 5642 (1977).
71. F. A. Cotton and B. M. Foxman, *Inorg. Chem.*, *7*, 2135 (1968).
72. F. A. Cotton, B. A. Frenz, J. R. Ebner, and R. A. Walton, *Chem. Commun.*, *1974*, 4.
73. F. A. Cotton, B. A. Frenz, J. R. Ebner, and R. A. Walton, *Inorg. Chem.*, *15*, 1630 (1976).
74. F. A. Cotton, B. A. Frenz, and Z. C. Mester, *Acta Crystallogr.*, *B29*, 1515 (1973).
75. F. A. Cotton, B. A. Frenz, E. Pedersen, and T. R. Webb, *Inorg. Chem.*, *14*, 391 (1975).
76. F. A. Cotton, B. A. Frenz, and L. W. Shive, *Inorg. Chem.*, *14*, 649 (1975).
77. F. A. Cotton, B. A. Frenz, B. R. Stults, and T. R. Webb, *J. Am. Chem. Soc.*, *98*, 2768 (1976).
78. F. A. Cotton, B. A. Frenz, and T. R. Webb, *J. Am. Chem. Soc.*, *95*, 4431 (1973).
79. F. A. Cotton and L. D. Gage, *Nouv. J. Chim.*, *1*, 441 (1977).
80. F. A. Cotton, L. D. Gage, K. Mertis, L. W. Shive, and G. Wilkinson, *J. Am. Chem. Soc.*, *98*, 6922 (1976).
81. F. A. Cotton and W. T. Hall, *Inorg. Chem.*, *16*, 1867 (1977).
82. F. A. Cotton and C. B. Harris, *Inorg. Chem.*, *4*, 330 (1965).
83. F. A. Cotton and C. B. Harris, *Inorg. Chem.*, *6*, 924 (1967).
84. F. A. Cotton, T. Inglis, M. Kilner, and T. R. Webb, *Inorg. Chem.*, *14*, 2023 (1975).
85. F. A. Cotton and M. Jeremic, *Synth. Inorg. Metal-Org. Chem.*, *1*, 265 (1971).
86. F. A. Cotton and B. J. Kalbacher, *Inorg. Chem.*, *15*, 522 (1976).
87. F. A. Cotton and B. J. Kalbacher, *Inorg. Chem.*, *16*, 2386 (1977).
88. F. A. Cotton and S. A. Koch, *J. Am. Chem. Soc.*, *99*, 7371 (1977).
89. F. A. Cotton, S. Koch, K. Mertis, M. Millar, and G. Wilkinson, *J. Am. Chem. Soc.*, *99*, 4989 (1977).
90. F. A. Cotton, S. Koch, and M. Millar, *J. Am. Chem. Soc.*, *99*, 7372 (1977).
91. F. A. Cotton, D. G. Lay, and M. Millar, *Inorg. Chem.*, *17*, 186 (1978).

92. F. A. Cotton, D. S. Martin, Jr., P. E. Fanwick, T. J. Peters, and T. R. Webb, *J. Am. Chem. Soc., 98*, 4681 (1976).

93. F. A. Cotton, D. S. Martin, Jr., T. R. Webb, and T. J. Peters, *Inorg. Chem., 15*, 1199 (1976).

94. F. A. Cotton, Z. C. Mester, and T. R. Webb, *Acta Crystallogr., B30*, 2768 (1974).

95. F. A. Cotton and J. G. Norman, Jr., *J. Am. Chem. Soc., 94*, 5697 (1972).

96. F. A. Cotton and J. G. Norman, Jr., *J. Coord. Chem., 1*, 161 (1971).

97. F. A. Cotton, J. G. Norman, Jr., B. R. Stults, and T. R. Webb, *J. Coord. Chem., 5*, 217 (1976).

98. F. A. Cotton, C. Oldham and W. R. Robinson, *Inorg. Chem., 5*, 1798 (1966).

99. F. A. Cotton, C. Oldham, and R. A. Walton, *Inorg. Chem., 6*, 214 (1967).

100. F. A. Cotton and E. Pedersen, *J. Am. Chem. Soc., 97*, 303 (1975).

101. F. A. Cotton and E. Pedersen, *Inorg. Chem., 14*, 383 (1975).

102. F. A. Cotton and E. Pedersen, *Inorg. Chem., 14*, 388 (1975).

103. F. A. Cotton and E. Pedersen, *Inorg. Chem., 14*, 399 (1975).

104. F. A. Cotton and J. R. Pipal, *J. Am. Chem. Soc., 93*, 5441 (1971).

105. F. A. Cotton, C. E. Rice, and G. W. Rice, *J. Am. Chem. Soc., 99*, 4704 (1977).

106. F. A. Cotton, W. R. Robinson, and R. A. Walton, *Inorg. Chem., 6*, 223 (1967).

107. F. A. Cotton, W. R. Robinson, and R. A. Walton, *Inorg. Chem., 6*, 1257 (1967).

108. F. A. Cotton, W. R. Robinson, R. A. Walton, and R. Whyman, *Inorg. Chem., 6*, 929 (1967).

109. F. A. Cotton and L. W. Shive, *Inorg. Chem., 14*, 2027 (1975).

110. F. A. Cotton and L. W. Shive, *Inorg. Chem., 14*, 2032 (1975).

111. F. A. Cotton, L. W. Shive, and B. R. Stults, *Inorg. Chem., 15*, 2239 (1976).

112. F. A. Cotton and G. G. Stanley, *Inorg. Chem., 16*, 2668 (1977).

113. F. A. Cotton and J. M. Troup, *J. Am. Chem. Soc., 96*, 4422 (1974).

114. F. A. Cotton, J. M. Troup, T. R. Webb, D. H. Williamson, and G. Wilkinson, *J. Am. Chem. Soc., 96*, 3824 (1974).

115. F. A. Cotton and D. A. Ucko, *Inorg. Chim. Acta, 6*, 161 (1972).

116. F. A. Cotton and T. R. Webb, *Inorg. Chem., 15*, 68 (1976).

117. F. A. Cotton and R. M. Wing, *Inorg. Chem., 4*, 1328 (1965).

118. C. D. Cowman and H. B. Gray, *J. Am. Chem. Soc., 95*, 8177 (1973).

119. C. D. Cowman, W. C. Trogler, and H. B. Gray, *Isr. J. Chem., 15*, 308 (1976/77).

120. L. Dubicki and R. L. Martin, *Aust. J. Chem., 22*, 1571 (1969).

121. J. D. Eakins, D. G. Humphreys, and C. E. Mellish, *J. Chem. Soc., 1963*, 6012.

122. A. Earnshaw, L. F. Larkworthy, and K. S. Patel, *Proc. Chem. Soc., 1963*, 281.

123. J. R. Ebner, D. R. Tyler, and R. A. Walton, *Inorg. Chem., 15*, 833 (1976).

124. J. R. Ebner and R. A. Walton, *Inorg. Chim. Acta, 14*, 1987 (1975).

125. J. R. Ebner and R. A. Walton, *Inorg. Chim. Acta, 14*, L45 (1975).

126. D. K. Erwin, G. L. Geoffroy, H. B. Gray, G. S. Hammond, E. I. Solomon, W. C. Trogler, and A. A. Zagars, *J. Am. Chem. Soc., 99*, 3620 (1977).

127. P. E. Fanwick, D. S. Martin, F. A. Cotton, and T. R. Webb, *Inorg. Chem., 16*, 2103 (1977).

128. J. E. Fergusson, *Coord. Chem. Rev., 1*, 459 (1966).

129. J. E. Fergusson, *Prep. Inorg. React., 7*, 93 (1971).

130. B. N. Figgis and R. L. Martin, *J. Chem. Soc., 1956*, 3837.

131. R. H. Fleming, G. L. Geoffroy, H. B. Gray, A. Gupta, G. S. Hammond, D. S. Kliger, and V. M. Miskowski, *J. Am. Chem. Soc., 98*, 48 (1976).

132. H. M. Gager, J. Lewis, and M. J. Ware, *Chem. Commun., 1966*, 616.

133. C. D. Garner, I. H. Hillier, M. F. Guest, J. C. Green, and A. W. Coleman, *Chem. Phys. Lett.*, *41*, 91 (1976).

134. C. D. Garner and R. G. Senior, *Chem. Commun.*, *1974*, 580.

135. G. L. Geoffroy, H. B. Gray, and G. S. Hammond, *J. Am. Chem. Soc.*, *96*, 5565 (1974).

136. H. D. Glicksman, A. D. Hamer, T. J. Smith, and R. A. Walton, *Inorg. Chem.*, *15*, 2205 (1976).

137. H. D. Glicksman and R. A. Walton, *Inorg. Chem.*, *17*, 200 (1978).

138. M. I. Glinkina, A. F. Kuzina, and V. I. Spitsyn, *Zh. Neorg. Chem.*, *18*, 403 (1973); *Russ. J. Inorg. Chem.*, *18*, 210 (1973).

139. M. F. Guest, I. H. Hillier, and C. D. Garner, *Chem. Phys. Lett.*, *48*, 587 (1977).

140. P. Jeffrey Hay, *J. Am. Chem. Soc.*, *100*, 2897 (1978).

141. F. Hein and D. Tille, *Z. Anorg. Allg. Chem.*, *329*, 72 (1964).

142. R. R. Hendriksma, *Inorg. Nucl. Chem. Lett.*, *8*, 1035 (1974).

143. R. R. Hendriksma, *J. Inorg. Nucl. Chem.*, *34*, 1581 (1972).

144. R. R. Hendriksma and H. P. Van Leeuwen, *Electrochim. Acta.*, *18*, 39 (1973).

145. C. A. Hertzer and R. A. Walton, *Inorg. Chim. Acta*, *22*, L10 (1977).

146. C. A. Hertzer and R. A. Walton, *J. Organomet. Chem.*, *124*, C15 (1977).

147. G. Herzberg, '*Molecular Spectra and Molecular Structure*,' Vol. I, Van Nostrand, Princeton, N.J., 1950.

148. S. Herzog and W. Kalies, *Z. Anorg. Allg. Chem.*, *329*, 83 (1964).

149. S. Herzog and W. Kalies, *Z. Anorg. Allg. Chem.*, *351*, 237 (1967).

150. S. Herzog and W. Kalies, *Z. Chem.*, *4*, 183 (1964).

151. S. Herzog and W. Kalies, *Z. Chem.*, *5*, 273 (1965).

152. B. Heyn and C. Haroske, *Z. Chem.*, *12*, 338 (1972).

153. B. Heyn and H. Still, *Z. Chem.*, *13*, 191 (1973).

154. E. Hochberg and E. H. Abbott, *Inorg. Chem.*, *17*, 506 (1978).

155. E. Hochberg, P. Walks, and E. H. Abbott, *Inorg. Chem.*, *13*, 1824 (1974).

156. G. Holste and H. Schäfer, *J. Less Common-Met.*, *20*, 164 (1970).

157. M. J. Hynes, *J. Inorg. Nucl. Chem.*, *34*, 366 (1972).

158. J. A. Jaecker, D. P. Murtha, and R. A. Walton, *Inorg. Chim. Acta*, *13*, 21 (1975).

159. J. A. Jaecker, W. R. Robinson, and R. A. Walton, *Chem. Commun.*, *1974*, 306.

160. J. A. Jaecker, W. R. Robinson, and R. A. Walton, *J. Chem. Soc. Dalton Trans.*, *1975*, 698.

161. S. A. Johnson, H. R. Hunt, and H. M. Neuman, *Inorg. Chem.*, *2*, 960 (1963).

162. V. Katović and R. E. McCarley, *Inorg. Chem.*, *17*, 1268 (1978).

163. V. Katović, J. L. Templeton, R. J. Hoxmeier, and R. E. McCarley, *J. Am. Chem. Soc.*, *97*, 5300 (1975).

164. D. L. Kepert and K. Vrieze, in *Halogen Chemistry*, Vol. 3, V. Gutmann, Ed., Academic Press, New York, 1967, pp. 1–54.

165. A. Ketteringham and C. Oldham, *J. Chem. Soc. Dalton Trans.*, *1973*, 1067.

166. A. P. Ketteringham, C. Oldham, and C. J. Peacock, *J. Chem. Soc. Dalton Trans.*, *1976*, 1640.

167. W. Kiefer and H. J. Bernstein, *Appl. Spectrosc.*, *25*, 500, 609 (1971).

168. R. B. King, *Prog. Inorg. Chem.*, *15*, 287 (1972).

169. W. R. King and C. S. Garner, *J. Chem. Phys.*, *18*, 689 (1950).

170. A. S. Kotel'nikova, P. A. Koz'min, and M. D. Surazhskaya, *Zh. Strukt. Khim.*, *10*, 1128 (1969); *J. Struct. Chem.*, *10*, 1012 (1969).

171. A. S. Kotel'nikova and V. G. Tronev, *Zh. Neorg. Khim.*, *3*, 1008 (1958).

172. A. S. Kotel'nikova and V. G. Tronev, *Zh. Neorg. Khim.*, *3*, 1016 (1958).
173. A. S. Kotel'nikova and G. A. Vinogradova, *Zh. Neorg. Khim.*, *9*, 307 (1964); *Russ. J. Inorg. Chem.*, *9*, 168 (1964).
174. P. A. Koz'min, V. G. Kuznetsov, and Z. V. Popova, *Zh. Strukt. Khim.*, *6*, 651 (1965); *J. Struct. Chem.*, *6*, 624 (1965).
175. P. A. Koz'min, G. N. Novitskaya, V. G. Kuznetsov, and A. S. Kotel'nikova, *Zh. Strukt. Khim.*, *12*, 933 (1971); *J. Struct. Chem.*, *12*, 861 (1971).
176. P. A. Koz'min, G. N. Novitskaya, and V. G. Kuznetsov, *Zh. Strukt. Khim.*, *14*, 680 (1973); *J. Struct. Chem.*, *14*, 629 (1973).
177. P. A. Koz'min, M. D. Surazhdaya, and V. G. Kuznetsov, *Zh. Strukt. Khim.*, *8*, 1107 (1967); *J. Struct. Chem.*, *8*, 983 (1967).
178. P. A. Koz'min, M. D. Surazhskaya, and V. G. Kuznetsov, *Zh. Strukt. Khim.*, *11*, 313 (1970); *J. Struct. Chem.*, *11*, 291 (1970).
179. P. A. Koz'min, M. D. Surazhskaya, and T. B. Larina, *Zh. Strukt. Khim.*, *15*, 64 (1974); *J. Struct. Chem.*, *15*, 56 (1974).
180. J. Krausse, G. Marx, and G. Schödl, *J. Organomet. Chem.*, *21*, 159 (1970).
181. J. Krausse and G. Schödl, *J. Organomet. Chem.*, *27*, 59 (1971).
182. E. Kurras, H. Mennenga, G. Oehme, U. Rosenthal, and G. Engelhardt, *J. Organomet. Chem.*, *84*, C13 (1975).
183. E. Kurras and J. Otto, *J. Organomet. Chem.*, *3*, 479 (1965).
184. E. Kurras and J. Otto, *J. Organomet. Chem.*, *4*, 114 (1965).
185. E. Kurras, U. Rosenthal, H. Mennenga, G. Oehme, and G. Engelhardt, *Z. Chem.*, *14*, 160 (1974).
186. V. G. Kuznetzov and P. A. Koz'min, *Zh. Strukt. Khim.*, *4*, 55 (1965), *J. Struct. Chem.*, *4*, 49 (1963).
187. D. Lawton and R. Mason, *J. Am. Chem. Soc.*, *87*, 921 (1965).
188. J. Lewis, *Pure Appl. Chem.*, *10*, 11 (1965).
189. J. Lewis and R. S. Nyholm, *Sci. Prog.*, *52*, 557 (1964).
190. A. Loewenschuss, J. Shamir, and M. Ardon, *Inorg. Chem.*, *5*, 238 (1976).
191. R. E. McCarley, J. L. Templeton, T. J. Colburn, V. Katović, and R. J. Hoxmeier, *Adv. Chem. Ser.*, *150*, 318 (1976).
192. B. R. McGarvey, *Transition Metal Chem.*, *3*, 89 (1966).
193. T. J. Meyer, *Prog. Inorg. Chem.*, *19*, 1 (1975).
194. A. P. Mortola, J. W. Moskowitz, N. Rösch, C. D. Cowman, and H. B. Gray, *Chem. Phys. Lett.*, *32*, 283 (1975).
195. E. L. Muetterties, *Bull. Soc. Chim. Belg.*, *84*, 959 (1975).
196. E. L. Muetterties, *Science*, *196*, 839 (1977).
197. R. S. Mulliken, *J. Chem. Phys.*, *7*, 20 (1939).
198. T. Nimry and R. A. Walton, *Inorg. Chem.*, *16*, 2829 (1977).
199. J. G. Norman, Jr., and H. J. Kolari, *J. Am. Chem. Soc.*, *97*, 33 (1975).
200. J. G. Norman, Jr., and H. J. Kolari, *J. Am. Chem. Soc.*, *100*, 791 (1978).
201. J. G. Norman, Jr., and H. J. Kolari, *Chem. Commun.*, *1974*, 303.
202. J. G. Norman, Jr., and H. J. Kolari, *Chem. Commun.*, *1975*, 649.
203. J. G. Norman, Jr., H. J. Kolari, H. B. Gray, and W. C. Trogler, *Inorg. Chem.*, *16*, 987 (1977).
204. L. R. Ocone and B. P. Block, *Inorg. Syn.*, *8*, 125 (1966).
205. C. Oldham, J. E. D. Davies, and A. P. Ketteringham, *Chem. Commun.*, *1971*, 572.
206. C. Oldham and A. P. Ketteringham, *J. Chem. Soc. Dalton Trans.*, *1973*, 2304.
207. R. Ouahes, Y. Maouche, M. C. Perucaud, and P. Herpin, *C. R. Hebd. Seances Acad. Sci.*, *C276*, 281 (1973).

208. G. A. Ozin, *Acc. Chem. Res.*, *10*, 21 (1977).
209. L. Pauling, *The Nature of the Chemical Bond*, 3rd ed., Cornell Univ. Press, 1960, p 40.
210. E. Peligot, *Compt. Rend. Hebd. Seances Acad. Sci.*, *19*, 609 (1844).
211. B. R. Penfold, in *Perspectives in Structural Chemistry*, Vol. 2, J. D. Dunitz and J. A. Ibers, Eds., Wiley, New York, 1968, pp. 71–149.
212. A. Pernick and M. Ardon, *J. Am. Chem. Soc.*, *97*, 1254 (1975).
213. J. A. Potenza, R. J. Johnson, and J. San Filippo, Jr., *Inorg. Chem.*, *15*, 2215 (1976).
214. C. O. Quicksall and T. G. Spiro, *Inorg. Chem.*, *9*, 1045 (1970).
215. G. A. Rempel, P. Legzdins, H. Smith, and G. Wilkinson, *Inorg. Syn.*, *13*, 90 (1971).
216. L. Ricard, J. Estienne, and R. Weiss, *Inorg. Chem.*, *12*, 2182 (1973).
217. L. Ricard, J. Estienne, and R. Weiss, *Chem. Commun.*, *1972*, 906.
218. L. Ricard, P. Karagiannidis, and R. Weiss, *Inorg. Chem.*, *12*, 2179 (1973).
219. G. Rouschias and G. Wilkinson, *J. Chem. Soc. A.*, *1965*, 465.
220. D. J. Salmon and R. A. Walton, *J. Am. Chem. Soc.*, *100*, 991 (1978).
221. J. San Filippo, Jr., *Inorg. Chem.*, *11*, 3140 (1972).
222. J. San Filippo, Jr. and H. J. Sniadoch, *Inorg. Chem.*, *12*, 2326 (1973).
223. J. San Filippo, Jr. and H. J. Sniadoch, *Inorg. Chem.*, *15*, 2209 (1976).
224. J. San Filippo, Jr., H. J. Sniadoch, and R. L. Grayson, *Inorg. Chem.*, *13*, 2121 (1974).
225. J. San Filippo, Jr., H. J. Sniadoch, and R. L. Grayson, *J. Less-Common Met.*, *54*, 13 (1977).
226. A. P. Sattelberger and J. P. Fackler, *J. Am. Chem. Soc.*, *99*, 1258 (1977).
227. H. Schäfer and H. G. Schnering, *Angew. Chem.*, *76*, 833 (1964).
228. K. Schwochau, K. Hedwig, H. J. Schenk, and O. Greis, *Inorg. Nucl. Chem. Lett.*, *13*, 77 (1977).
229. P. Sharrock, T. Thiopanides, and F. Brisse, *Can. J. Chem.*, *51*, 2963 (1973).
230. J. C. Sheldon, *Aust. J. Chem.*, *17*, 1191 (1964).
231. J. H. Sinfelt, *Acc. Chem. Res.*, *10*, 15 (1977).
232. R. P. A. Sneeden and H. H. Zeiss, *J. Organomet. Chem.*, *28*, 259 (1971).
233. T. G. Spiro, *Prog. Inorg. Chem.*, *11*, 1 (1970).
234. D. F. Steele and T. A. Stephenson, *Inorg. Nucl. Chem. Lett.*, *9*, 777 (1973).
235. T. A. Stephenson, E. Bannister, and G. Wilkinson, *J. Chem. Soc.*, *1964*, 2538.
236. T. A. Stephenson, S. M. Morehouse, A. R. Powell, J. P. Heffer, and G. Wilkinson, *J. Chem. Soc.*, *1965*, 3632.
237. T. A. Stephenson and D. Whittaker, *Inorg. Nucl. Chem. Lett.*, *5*, 569 (1969).
238. T. A. Stephenson and G. Wilkinson, *J. Inorg. Nucl. Chem.*, *28*, 2285 (1966).
239. F. Taha and G. Wilkinson, *J. Chem. Soc.*, *1963*, 5406.
240. D. G. Tisley and R. A. Walton, *J. Chem. Soc. Dalton Trans.*, *1973*, 1039.
241. W. Traube, E. Burmeister, and R. Stahn, *Z. Anorg. Allg. Chem.*, *147*, 50 (1925).
242. W. Traube and A. Goodson, *Chem. Ber.*, *49*, 1679 (1916).
243. W. C. Trogler, C. D. Cowman, H. B. Gray, and F. A. Cotton, *J. Am. Chem. Soc.*, *99*, 2993 (1977).
244. W. C. Trogler, D. K. Erwin, G. L. Geoffroy, and H. B. Gray, *J. Am. Chem. Soc.*, *100*, 1160 (1978).
245. W. C. Trogler and H. B. Gray, *Nouv. J. Chim.*, *1*, 475 (1977).
246. W. C. Trogler and H. B. Gray, *Acc. Chem. Res.*, *11*, 232 (1978).
247. W. C. Trogler, E. I. Solomon, and H. B. Gray, *Inorg. Chem.*, *16*, 3031 (1977).
248. W. C. Trogler, E. I. Solomon, I. Trajberg, C. J. Ballhausen, and H. B. Gray, *Inorg. Chem.*, *16*, 828 (1977).
249. W. Van Bronswyk and J. C. Sheldon, *Aust. J. Chem.*, *20*, 2323 (1967).
250. J. N. Van Niekerk and F. R. L. Shoening, *Acta Crystallogr.*, *6*, 227 (1953).

251. J. N. Van Niekerk, F. R. L. Shoening, and J. F. deWet, *Acta Crystallogr.*, *6*, 501 (1953).
252. R. A. Walton, *Prog. Inorg. Chem.*, *16*, 1 (1972).
253. R. A. Walton, *Prog. Inorg. Chem.*, *21*, 105 (1976).
254. G. Wilke, B. Bogdanović, P. Hardt, P. Heimbach, W. Keim, M. Kröner, W. Oberkirch, K. Tanaka, E. Steinrücke, D. Walter, and H. Zimmermann, *Angew. Chem. Int. Ed.*, *5*, 151 (1966).
255. F. C. Wilson and D. P. Shoemaker, *J. Chem. Phys.*, *27*, 809 (1957).

The Chemistry of the Dithioacid
and 1,1-Dithiolate Complexes, 1968–1977

DIMITRI COUCOUVANIS

Department of Chemistry
University of Iowa, Iowa City, Iowa

CONTENTS

I. INTRODUCTION

A major class of sulfur-containing ligands is obtained by the general reaction of carbon disulfide with various nucleophiles. These ligands can be classified into two categories: the uninegative 1,1-dithioacids and the dinegative 1,1-dithiolates (Fig. 1).

The 1,1-dithioacids are obtained readily by the addition of uninegative nucleophiles to CS_2 under a variety of experimental conditions. Major types of ligands that have been obtained, which show exciting coordination properties,

Fig. 1.   The 1,1-dithio ligands.

are the dithiocarbamates $(X = NR_2$, Fig. 1), the xanthates $(X = OR)$, thio-xanthates $(X = SR)$, and the dithioaliphatic or dithioaromatic acid anions $(X = CR_3$ or aryl).

The addition of dinegative nucleophiles to CS_2, or at times the deprotona-tion of the S_2CNHR or S_2CCHR_2 dithioacid ligands, gives rise to the dinegative 1,1-dithiolates.

In a previous review (153) I reported on the developments in the rich coordination chemistry of the 1,1-dithio ligands until 1969. Since then a remarkable increase of interest and research activity in the chemistry of these ligands and complexes has become apparent.

The field has grown from a primarily synthetic exploratory stage to a more sophisticated stage where extensive studies of reactivities, electron structures, and molecular structures are vigorously pursued.

With the advances in x-ray crystallographic instrumentation and computa-tional facilities, the structural studies on the 1,1-dithio complexes have pro-gressed at an amazing rate. Since the first reports of "unusual" oxidation states of the dithiocarbamate complexes, the electrochemical properties of these molecules have been investigated extensively. New ligands have been designed to test various hypotheses regarding structural and electronic properties of the complexes.

This chapter avoids the extensive tabulations of physical and spectroscopic properties that characterized its predecessor. Instead, emphasis is placed on the structural and electronic aspects of the coordination chemistry of the 1,1-dithio ligands.

The "unusual" (and "normal") oxidation states of the dithiocarbamate complexes were reviewed recently (627). Part of one review also deals with the dynamic behavior of the dithiocarbamate complexes in solution (509). Standard treatises on the ligands and their complexes already have been cited (153). The author regrets neglect of any pertinent references that may have been overlooked.

In the various tables of crystallographic data presented throughout this chapter, the *mean* values of the obviously *chemically equivalent* structural parameters are reported. The larger of the standard deviations from the mean $([\Sigma_{i=1}^{N}(x_i - \bar{x})^2 / (N - 1)]^{1/2})$ or the reported standard deviations (as derived from the inverse matrix) are given.

Often the standard deviations from the mean for a set of parameters exceed the reported standard deviations. In such cases the retention of three significant figures in the reported mean values implies reported standard deviations of ≤ 0.009 Å.

A list of abbreviations used in the text can be found at the end of the chapter.

II. DITHIOACIDS

A. 1,1-Dithio Ligands and Alkali Metal Salts

The reaction of carbon disulfide with various nucleophiles such as NR_2^-, CR_3^-, OR^-, and SR^- gives rise to the general class of 1,1-dithio ligands of the general formula $^-S_2C-X$. These ligands have been discussed up to 1969 in a previous review (153). Since then various new reactions of CS_2 and nucleophiles and new 1,1-dithio ligands have appeared.

The reaction of CS_2 with diamines proceeds with CS_2 addition at the less basic site, as the more basic site accepts the proton in the formation of the zwitterion. In the case of β-aminoethylpiperazine, the primary amino group condenses with CS_2 and the secondary amine undergoes zwitterion formation (254).

In the presence of CS_2 and base, piperazine forms a bisdithiocarbamate, which is isolated as the bispiperazinium salt (252). Reaction of CS_2 with cysteine in the presence of ammonia gives the dithiocarbamate—thioxanthate, which can be isolated as the triammonium salt (253) (Fig. 2).

The utility of this trianion, as well as the bisdithiocarbamate ligands described previously, as multidentate ligands should be investigated.

A variety of other reactions of carbon disulfide with nucleophiles are presented in a review (252). In the last 10 years a number of new 1,1-dithio ligands have been synthesized, and the chemistry of others has been explored in detail.

The chemistry of the dithioformate anion was not explored until recently. A number of studies have appeared in the literature in which the formation of dithioformate complexes by the insertion of CS_2 into the M—H bond of metal

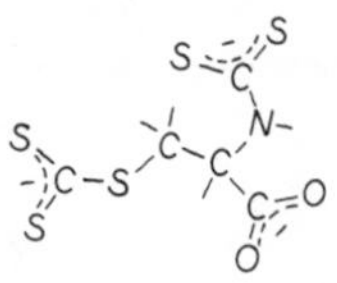

Fig. 2.   The dithiocarbamate—thioxanthate ligand derived from the reaction of cysteine with carbon disulfide (253).

hydrides is described. The properties of these dithioformate complexes are discussed in the appropriate sections, which are arranged according to element. The hydrides $MHX(CO)(PPh_3)_3$ ($M = Ru$, Os; $X = Cl$, Br), $RuH_2(PPh_3)_4$, $OsH_4(PPh_3)_3$, $IrHCl_2(PPh_3)_3$ (trans-chlorides), and mer-$[IrH_3(PPh_3)_3]$ undergo this insertion reaction in boiling benzene to yield $M(S_2CH)XCO(PPh_3)_2$ (two isomers), $[Ru(S_2CH)_2(PPh_3)_2]$, $Os(S_2CH)_2(PPh_3)_2$, $Ir(S_2CH)Cl_2(PPh_3)_2$, and $Ir(S_2CH)H_2(PPh_3)_2$, respectively, (556, 557).

A similar reaction is utilized for the synthesis of $Re(S_2CH)(CO)_2(PPh_3)_2$ (8, 256). The synthesis of the β-$Os(CO)Cl(PCy_3)_2HCS_2$ by the same route also has been described (464, 465). The x-ray structures of $Ru(S_2CH)_2(PPh_3)_2$ (370) and $Re(S_2CH)(CO)_2(PPh_3)_2$ (8) have been determined. The IR spectrum of the KS_2CH ligand shows bands at 1250, 980, and 786 cm^{-1}, which have been assigned to $\delta(HCS)$, $\nu(CS_2)_{asym}$ and $\nu(CS_2)_{sym}$, respectively (433). A kinetic study of the insertion reaction of CS_2 to trans-$PtH_2[P(C_6H_{11})_3]_2$ to form trans-$PtH(S_2CH)[P(C_6H_{11})_3]_2$ has been reported. The x-ray crystal structure of the product also has been determined (9). The structure of $Mn(HCS_2)(CO)_3$ (DPM) (DPM = diphenylphosphinomethane) shows the HCS_2 ligand chelated to Mn, however, the hydrogen atom was not located (210). For this and other HCS_2^- complexes, structural details can be found in the appropriate sections throughout this chapter.

The fluorine analogue of HCS_2^-, FCS_2^-, has been obtained by insertion of CS_2 into the Pt–F bond in $[PtF(PPh_3)_3]^+HF_2^-$. The structure of the $[(PPh_3)_2PtS_2CF]^+HF_2^-$ product shows the chelated ^-S_2CF ligand (225).

Numerous structural determinations have been reported on the alkali metal salt of various 1,1-dithio ligands. Structural details of these compounds can be found in Table I. Particularly interesting among these compounds are the dimeric $[M(n\text{-}Bu_2Dtc)]_2$, $M = K^+$, Rb^+ (618a). The metal atoms in these dimers are coordinated to six sulfur atoms from four different ligands (Fig. 3).

In the structure of $Cs(Me_2Dtc)$, chains of six-coordinate Cs atoms are linked by Cs–S interchain interactions such that the final Cs coordination is eight (617). An eight-coordinated Cs is also observed in the isostructural $M(H_2Dtc)$ salts ($M = Cs^+$, Rb^+, K^+, NH_4^+) (102, 290).

The synthesis and metal complexes of the pyrrole dithiocarbamate ligand were reported by Kellner et al. (386). At a later date two independent studies dealt with the coordination chemistry of this ligand (26, 211). An outstanding feature of this molecule is the pyrrole ring, which, by preserving aromaticity makes the contribution of resonance form B (Fig. 4) insignificant (26) and "creates a dominant π-accepting character at the sulfur atoms" (26).

A new interesting 1,1-dithio ligand was obtained by the addition of $KPPh_2$ to CS_2 in tetrahydrofuran (263) or dioxane (396). The KS_2CPPh_2 salt of the ligand shows strong absorptions in the infrared spectrum at 1000 and 853 cm^{-1} that have been attributed to the CS_2 moiety (263).

TABLE I
Structural Details of the 1,1-Dithioacid Ligand 'Salts'

Compound	M–S, Å	C–S, Å	$S_2C–X$, Å[a]	S–S, Å	S–C–S°	R
Na(Et$_2$Dtc)·3H$_2$O	3.052 (3)	1.720 (12)	1.344 (8)		120.4 (4)	1.
KHCS$_2$	3.576^b	1.643^b			131.3^b	2
	3.352^b					
KMeDta	3.31 (1)	1.67 (1)	1.40 (5)		123.5 (6)	6
	3.32 (1)					
	3.56 (1)					
K(n-Bu$_2$Dtc)^c	3.19 (3)	1.66 (3)	1.46 (12)	2.96 (3)	126 (6)	61
	3.28 (3)					
	3.37 (3)					
	3.49 (3)					
	3.27 (3)					
	3.40 (3)					
KEtXant	3.347 (6)					4:
	3.405 (6)					
	3.463 (5)					
	3.529 (5)					
	3.336 (5)					
	3.480 (5)					
Rb(n-Bu$_2$Dtc)^d		1.72 (2)	1.31 (2)	2.96 (1)	119 (1)	61
Cs(Me$_2$Dtc)^e	3.636 (1)	1.716 (2)	1.342 (5)	2.991 (2)	121.2 (2)	61
	3.713 (1)					
	4.099 (1)					
Cs(n-Bu$_2$Dtc)^f	3.532^g	1.739 (5)	1.340 (8)	3.000 (2)	119.1 (4)	1
	3.622					
	3.687					
	3.557					
	3.617					
	3.693					
Et$_4$N(NCCS$_2$)		1.684 (12)	1.453^b	128.9^b		21

^{a} X = N, C, O or CN.
^{b} Standard deviations of 0.005 Å and 0.3° were given for all the bond lengths and angles respectively. We believe that these values are too optimistic.
^{c} Closest K–K distance 3.87 (4) Å.
^{d} Closest Rb–Rb distance 3.998 (4) Å.
^{e} Closest Cs–Cs 4.773 (1) Å.
^{f} Closest Cs–Cs distance 4.294 (1) Å.
^{g} σ = 0.002 Å.

The crystal structure of the Ni[S$_2$CP(C$_6$H$_{11}$)$_2$]$_2$ complex shows the ligand coordinating through one sulfur and the phosphorus atom with the formation of a four-membered NiSCP ring (616).

The synthesis of the ferrocenedithiocarboxylic acid ligand and numerous, intractable metal complexes of this ligand have been reported (379).

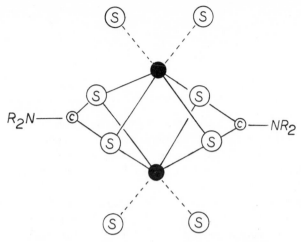

Fig. 3.   Schematic structure of the $[M(n\text{-}Bu_2 Dtc)]_2$ dimers, $M = K^+, Rb^+$ (618a).

Very recently the synthesis of the hitherto unknown aromatic xanthate ligands was reported by Fackler et al., who also determined several crystal structures of arylxanthate complexes (117, 238). The synthesis and reactions of air-stable piperidinium salts of aliphatic dithioacids have been reported (377).

The reaction of secondary diamines, carbon disulfide, and alkali metal hydroxides gives bisdithiocarbamate salts, which, upon reaction with bischloromethylated aromatics, afford soluble polydithiocarbamates (392). The coordination chemistry of these polymers should be investigated. They may prove to be interesting, and, perhaps, specific, metal-sequestering agents.

A few interesting reactions of the $R_2 Dtc^-$ ligands have been reported and should have bearing in the procedures used for the synthesis of metal complexes.

Heckley et al. reported that the reaction of $NaEt_2 Dtc$ with $CH_2 Cl_2$ heated under reflux affords the dithiocarbamate ester $Et_2 NC(S)SCH_2 S(S)CNEt_2$ in 86% yield. The same authors observed proton abstraction, by the $Et_2 Dtc$ ligand, from $CHCl_3$ or $CH_3 CN$, followed by partial decomposition of the protonated ligand and the eventual formation of $[Et_2 NH_2]^+ [Et_2 Dtc]^-$ (308).

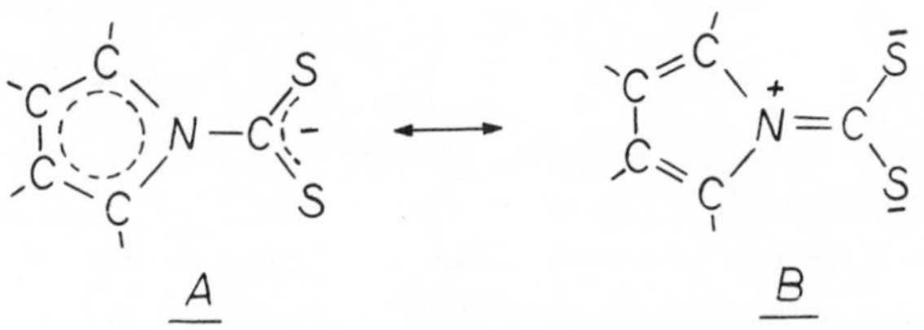

$$A \longleftrightarrow B$$

Fig. 4.   Resonance forms in the pyrrole dithiocarbamate ligand.

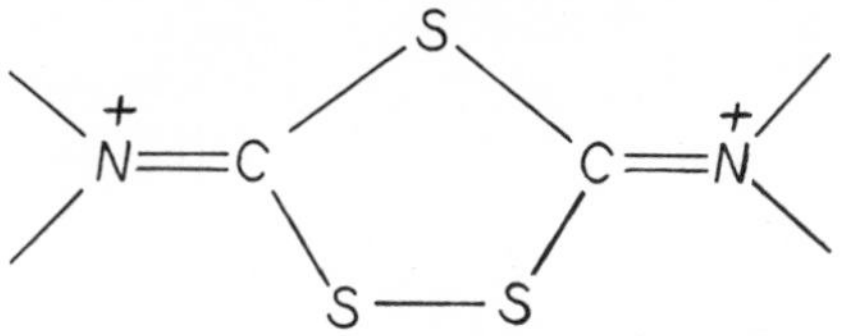

Fig. 5.   The structure of the 3,5-bis(N,N-dialkyliminium)trithiolane-1,2,4 cation (29).

Upon oxidation, certain dithiocarbamate complexes yield the 3,5-bis(N,N-dialkyliminium) trithiolane-1,2,4 cation (628). The structure of the $[Hg_2 I_6]^{2-}$ salt of this cation (R = Et) has been determined (29) (Fig. 5).

An electrochemical study of the oxidation of the $[R_2 Dtc]^-$ ligands has been carried out (109).

The π-electronic structure and electronic spectra of dithiocarbamic acid, $H_2 DtcH$, previously analyzed by Janssen (360) and Fabian (228), were re-examined by St. Nikolov and Tyutyulkov (585). In this later study calculations were carried out by the LCAO–MO–SCF–CI method. Excellent agreement between the calculated and observed energies of the two electronic transitions at ~34.6 and 39 kK was reported. By an analysis of the π-electronic charge densities, occurring on excitation, it was concluded that these two electronic transitions are located mainly on the CS_2 group.

B. Dithioacid Complexes with Nontransition Elements

1. Group II

Compounds of the alkaline earth metals with dithio ligands have been reported (153). However, these compounds, without exception, appear to be water-soluble ionic salts and rather uninteresting.

We would like to reiterate our initial suggestion (153) that studies of the structures of these compounds with highly hydrophobic ligands, in nonaqueous media, may reveal interesting aggregation characteristics. Such aggregations may persist in the solid state, particularly if these compounds are recrystallized from water-free, nonpolar solvents. Very little is known about the coordination properties of the alkaline earth–sulfur ligand compounds.

2. Group III

a. **Gallium, Indium, Thallium.**   While no structural information was available for the trivalent Group III 1,1-dithio complexes prior to 1969, several structural determinations have been reported in the last decade.

The crystal structures of the Ga(III) and In(III) tris-$Et_2 Dtc$ complexes have been determined by single-crystal x-ray diffraction studies (207). The

$M(Et_2Dtc)_3$ monomeric units in the two x-ray isomorphous complexes are situated on a crystallographic twofold axis. The symmetry of the MS_6 core in both molecules is closer to D_3 rather than O_h. A unique Et_2Dtc ligand is bisected by the twofold crystallographic axis and is symmetrically bound to the metal. The other two ligands are related by the C_2 axis and are described as "quasi-symmetrically" coordinated bidentate ligands. The statistically significant, unequal M–S bonds found with these two ligands are accompanied by unequal C–S bonds. The shorter C–S bonds are found for the sulfur atoms farther away from the metal ion. In the absence of stereochemically active lone pairs on the metal atoms, the observed structural asymmetry is attributed to crystal packing effects (207).

In the structure of the $In(PipDtc)_3$ complex (299), no appreciable difference between the In–S bonds is observed. The coordination geometry of the InS_6 core in this molecule approaches a trigonal prism ($\Phi = 25°$). A distorted octahedral InS_6 core also is found in the structure of the $In(BzDta)_3$ complex (46) (Table II).

A TlS_6 core of approximate D_3 symmetry was found in the structure of the $Tl(Me_2Dtc)_3$ complex (3). The average twist angle Φ in this complex was $33.2°$. A C_2 distortion superimposed on the trigonally twisted geometry of the TlS_6 core is similar to distortions of the same type already observed in the structures of the $In(BzDta)_3$ (46) and $In(PipDtc)_3$ (299) complexes and raises questions regarding the possible electronic–structural origin of these distortions.

The temperature-dependent PMR spectra of the $M(R_2Dtc)_3$ complexes [M = Ga(III), In(III)] show them to be stereochemically nonrigid. Kinetic parameters for the intramolecular metal-centered rearrangement (by a trigonal

TABLE II
Structural Details of the Group III, 1,1-Dithio Complexes

Complex	$\overline{M–S}$, Å	$\overline{C–S}$, Å	$\overline{C–N}$, Å	Ref.
$Ga(Et_2Dtc)_3$	2.436 (29)[a]	1.715 (4) 1.730 (6)[b]	1.325 (9)	207
$In(Et_2Dtc)_3$	2.597 (14)[c]	1.721 (4) 1.736 (5)[b]	1.328 (7)	207
$In(PipDtc)_3$	2.588 (4)	1.717 (11)	1.327 (12)	299
$In(BzDta)_3$	2.603 (15)[d]	1.691 (27)		46
$Ti(Me_2Dtc)_3$	2.659 (27)[e]	1.72 (4)	1.37 (4)	3

[a] Range: 2.408 (2) – 2.466 (1) Å.
[b] C–S distance of sulfur atom furthest away from the metal ion (2.466 and 2.611 Å for the Ga (III) and In (III), respectively.
[c] Range: 2.582 (2) – 2.611 (2) Å.
[d] Range: 2.577 (9) – 2.619 (5) Å.
[e] Range: 2.613 (10) – 2.677 (12) Å.

twist mechanism) for the $M(Bz_2 Dtc)_3$ complexes were determined by PMR line-broadening techniques (529).

In extensive crystallographic studies, Hesse, Jennische, and co-workers explored the coordination of Tl(I) in various $[Tl(R_2 Dtc)]_2$ complexes (13, 214, 361, 362, 486, 528). In all these dimeric structures, the thallium atoms are located on either side of a sulfur parallelepiped defined by the sulfur atoms of the two $R_2 Dtc$ ligands with a $Tl_2 S_4$ core similar to the one shown in Fig. 3.

The linkage of the dimers in the lattice is influenced by the size of the alkyl groups on the ligands and, as a result, the coordination number of thallium varies for different ligands. In general, the coordination number seems to decrease with increasing length of the ligands. Taking into consideration inter-dimer Tl–S interactions, the coordination numbers and geometries for the thallium atoms vary from six and a distorted trigonal prism in the $Bu_2 Dtc$ dimer (214), five and a distorted square pyramid in the i-$Pr_2 Dtc$ dimer (362), six and distorted trigonal prism in the $Et_2 Dtc$ dimer, and seven and a distorted capped trigonal prism in the $Me_2 Dtc$ dimer (361) (Table III). A coordination number five and a skewed tetragonal pyramid are found in the structure of the i-$Bu_2 Dtc$ dimer (13).

3. Group IV

a. Silicon, Germanium. Compounds of silicon or germanium with the 1,1-dithio ligands are scarce and, until recently (378), were limited to the $R_2 Dtc$ ligands.

Whereas trisilylamine or N-methyldisilylamine do not react with CS_2, N,N-dimethylsilylamine reacts with CS_2 to give silyl-N,N-dimethyldithio-carbamate (208). Evidence for hindered rotation around the $S_2 C–N$ bond in the $Me_2 Dtc$ ligand was found in the PMR spectrum of the $H_3 Si(Me_2 Dtc)$ "complex." Thus just below room temperature the broad methyl resonance of this compound splits into two peaks of equal intensity.

The $L_2 Si(R_2 Dtc)_2$ complexes (L = pip, R = pip; L = pyrrol, R = pyrrol) were obtained by the reaction of $SiCl_4$, pyrrolidine, or piperidine and CS_2 in CCl_4 (593). The IR, PMR, and mass spectra of these compounds were obtained. Two strong absorptions in the IR spectra at 995 and 955 cm^{-1} (in the pyrroli-dine derivative) and at 1005 and 975 cm^{-1} (in the piperidine derivative) were considered to be evidence for the presence of monodentate $R_2 Dtc$ ligands. Similar reaction conditions were employed for the synthesis of the corre-sponding germanium compounds from $GeCl_4$ (594). The infrared spectra of these compounds as well indicated the presence of monodentate $R_2 Dtc$ ligands.

Secondary or tertiary ammonium salts of dithioacids, $RCS_2^- XH^+$, (X = $Et_3 N$) react with chlorotrimethylsilane to give compounds of the type $(RDta)Si(Me)_3$ (R = Ph, p-Cl–Ph).

TABLE III
Structural Details of the $(TlR_2Dtc)_2$ Complexes

R	Tl–Tl	Tl–S	Tl–Tl[b]	Tl–S[b]	Ref.
i-Pr	3.584 (5)	2.98 3.03 3.04 3.05[a]	3.64	3.86	362
Et (dimer B)	3.602 (1)	3.047 (3) 3.104 (3) 3.020 (3) 3.069 (3)	3.47	3.80 3.89	528
Et (dimer A)	3.619 (1)	3.115 (3) 3.101 (3) 3.034 (3) 3.073 (3)	3.54	3.71 3.81	528
n-Bu	3.62 (1)	2.97 (3) 3.08 (3) 3.12 (3) 3.16 (3)	5.05	3.98 (4) 4.19 (4)	214
i-Bu	3.678 (2)	2.966 (6) 3.043 (5) 3.123 (6) 3.184 (5)	5.19	3.424 (6)	13
Me	3.85	2.99 3.03 3.28 3.44[c]	3.64	3.46 3.52 3.74	361
n-Pr	3.977 (4)	2.88[a] 2.91 3.02 3.11 3.12 3.12 3.29 4.37	4.002 (4)	3.38 3.52[a] 3.59 3.69	486

[a] Standard deviations 0.015 Å.
[b] Intermolecular distances.
[c] Standard deviations 0.01 Å.

The $RCS_2^-XH^+$ salts (X = piperidine) also react with chlorotrimethyl-germane to give the complexes $(RDta)Ge(Me)_3$ (R = tolyl, Ph, *p*-Cl–Ph). These complexes show two intense bands in the IR spectra at 1240 and 1050 cm^{-1} (silyl esters) and 1220 and 1050 cm^{-1} (germyl esters) that are assigned to the C=S vibrations of the monodentate RCS_2^- ligands (378). The $(Ph)_3Ge(EtXant)$ derivatives also show the C=S vibration at 1040 cm^{-1} (565). Undoubtedly these vibrations are coupled to other vibrational modes in either the RDta or the RXant ligands, and their occurrence can span a wide range.

The synthesis and the infrared and Raman spectra of the $(Me)_3M(MeDta)$ complexes (M = Si, Ge, Sn) and of $(Et)_3Pb(MeDta)$ have been reported (375). In all compounds a strong absorption that ranges from 1160 cm^{-1} in the $(Et)_3Pb(MeDta)$ to 1200 cm^{-1} in $(CH_3)_3C(MeDta)$ has been assigned to $\nu(C=S)_{asym}$. A medium-intensity band in the range from 862 to 880 cm^{-1} and a very strong band between 590 and 440 cm^{-1} have been assigned to the asymmetric and symmetric $\nu[C(S)-S]$ vibrations, respectively.

b. Divalent Tin and Lead Complexes. The crystal structures of several $M(L)_2$ complexes (M = Sn(II), Pb(II); L = Et_2Dtc, EtXant, n-BuXant) and of the $Pb(i\text{-}PrXant)_2Py$ complex have been determined (Table IV). A similar geometry for the MS_4 unit is observed (Fig. 6) in all these complexes.

The ligands are asymmetrically bound such that both a short and a long M–S bond arise from each ligand. The short bonds are positioned trans to each other in a PbS_4 polyhedron that resembles a distorted square pyramid with Pb at the apex. This structure, which appears to result from lone pair–bonding pair repulsions, is quite similar for the Pb(II) and Sn(II) complexes (Table IV). In the $Pb(i\text{-}PrXant)_2Py$ complex, the pyridine molecule is located above the lead atom with a Pb–N distance of 2.55(4) Å (288).

The synthesis (346) and crystal structure (471) of the $(Et_4N)Pb(EtXant)_3$ complex have been reported. The structure of this molecule has been described

TABLE IV

Structural Details of the Divalent Sn and Pb, 1,1-Dithio Complexes.[a]

Complex	M–S, Å[b]	C–S, Å[c]	S_1-M-S_3, °	S_2-M-S_4, °	Ref.
$Sn(Et_2Dtc)_2$	2.583 (12)	1.721 (9)	96.2 (2)	139.6 (2)	524
	2.792 (38)	1.703 (30)			
$Pb(E_2Dtc)_2$	2.765 (29)	1.73 (3)	96.2 (2)	133.2 (4)	359
	2.912 (38)	1.67 (5)			
$Pb(EtXant)_2$	2.76 (3)	1.74 (4)	98.2 (9)	137.2 (10)	287
	2.89 (8)	1.67 (1)			
$Pb(i\text{-}PrXant)_2 \cdot Py$	2.78 (8)	1.65 (11)	95.0 (6)	127.7 (5)	288, 289
	2.98 (11)	1.67 (11)			
$Pb(n\text{-}BuXant)_2$	2.80 (5)	1.76 (19)	91.7 (15)	141.2 (26)	286
	2.90 (18)	1.60 (28)			

[a] The numbering scheme is identical to that in Fig. 6.
[b] The mean values of the two short and the two long M–S bonds are reported assuming that they are chemically equivalent.
[c] The sulfurs in the C–S bond correspond to the sulfur in the M–S bonds in the same line of the adjacent column.

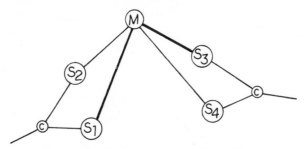

Fig. 6.   The MS_4 core structure in divalent Sn and Pb 1,1-dithio complexes. The heavy lines correspond with the "short" M-S distances (Table IV).

as a distorted pentagonal bipyramid with five of the sulfur atoms lying in an approximate plane and the sixth sulfur atom occupying one axial site. The other axial site is occupied by a stereochemically active lone pair (Fig. 7) (Table V).

c. **Quadrivalent Tin and Lead Complexes.**   The electrical conductance of the $Sn(Et_2Dtc)_4$ in polar solvents indicated a 1:1 electrolyte with the formula $Sn(Et_2Dtc)_3^+(Et_2Dtc)^-$ (55a). Evidence for this formulation is supplied by a report of the isolation of the $(Ph_4B)^-Sn(Et_2Dtc)_3^+$ complex, which shows Mössbauer spectra very similar to that of the $Sn(Et_2Dtc)_4$ complex (380). All evidence indicates that the tin in $Sn(Et_2Dtc)_4$ is six-coordinate. This could be accounted for by either the $Sn(Et_2Dtc)_3^+(Et_2Dtc)^-$ formulation and bidentate ligands or by the $Sn(Et_2Dtc)_4$ formulation, where two of the ligands are monodentate.

A structure determination of the complex showed the latter to be correct (Table V) (295). The tin atom is six-coordinate with distorted octahedral

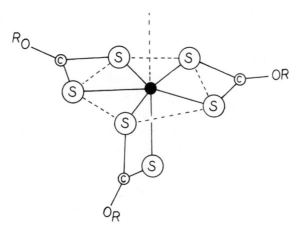

Fig. 7.   Schematic structure of the $Pb(EtXant)_3^-$ complex anion (471).

TABLE V
Structural Details of the Quadrivalent Sn, 1,1-Dithio Complexes
and $(Et_4 N)Pb(EtXant)_3$

Complex	$\overline{M-S}$, Å	$\overline{C-S}$, Å	$\overline{C-N}$, Å	Ref.
$Sn(Et_2 Dtc)_4$	$2.546 (7)^a$	$1.73 (2)^a$	$1.33 (2)$	295
	$2.504 (7)^b$	$1.67 (2)^b$		
		$1.77 (2)^b$		
$Me_3 Sn(Me_2 Dtc)^{c,d}$	$2.47 (1)$	$1.80 (2)^e$	$1.35 (3)$	570
	$3.33 (1)$	$1.70 (3)$		
$Me_3 Sn(Me_2 Dtc)^{f,g}$	$2.47 (1)$	$1.75 (1)^e$	$1.31 (2)$	570
	$3.16 (1)$	$1.72 (1)$		
$(Et_4 N)[(PbEtXant)_3]$	$3.00 (12)$			471

^{a} Bidentate Et_2 Dtc ligand.
^{b} Monodentate Et_2 Dtc ligand.
^{c} $\overline{Sn-C} = 2.22 (4)$ Å.
^{d} Orthorhombic modification.
^{e} Sulfur coordinated to the tin atom.
^{f} Monoclinic modification.
^{g} $\overline{Sn-C} = 2.19 (7)$ Å.

geometry. Two of the dithiocarbamate ligands are chelated, while the other two are monodentate and occupy cis positions in the octahedron. The possibility still exists that, in solution, the $Sn(Et_2 Dtc)_4$ complex dissociates with the formation of the six-coordinate $Sn(Et_2 Dtc)_3^+$ and $(Et_2 Dtc)$ ions. This process could be studied by variable-temperature PMR spectroscopy.

A variety of organometallic compounds of Sn(IV) and Pb(IV) with 1,1-dithio ligands have been reported. The interesting $MeSn(Et_2 Dtc)_3$ complex is cited (467) to have a pentagonal bipyramidal structure. The axial Sn—S distance opposite to the methyl group is 0.21 Å shorter than the equatorial Sn—S distances.

The synthesis of the $Cl_2 Sn(R_2 Dtc)_2$ and $(Me)_2 Sn(R_2 Dtc)_2$ complexes has been described (589). Line-shape analysis of the variable-temperature PMR spectra of the $X_2 Sn(i\text{-}Pr_2 Dtc)_2$ complexes in $CHCl_3$ was carried out to obtain the activation parameters of the hindered rotation around the $S_2 C \dot{=} N$ bond. Barriers to rotation of 18.3 kcal mole^{-1} for X = Cl and 16.4 kcal mole^{-1} for X = Me were reported.

The difference in these values (which may indicate more double-bond character in the $R_2 N–CS_2$ bonds in the $Cl_2 Sn(i\text{-}Pr_2 Dtc)_2$ complex) is supported (588) by the difference in the $C \dot{=} N$ vibrations in $Cl_2 Sn(Me_2 Dtc)_2$ when compared to $Me_2 Sn(Me_2 Dtc)_2$. These vibrations are found at 1546 and 1490 cm^{-1}, respectively (338).

The crystal and molecular structures of the orthorhombic and monoclinic modifications of $(Me)_3 Sn(Me_2 Dtc)$ have been determined (570). In both structures the Me_2Dtc ligand is bound to the tin atom in a monodentate fashion with the second sulfur atom of the ligand located 3.33(1) and 3.16(1) Å from the tin. The C_3SnS unit is distorted tetrahedral with one of the C–Sn–C angles at $119 \pm 1°$ (Table V).

Analogous crystalline compounds of the type $(Ph)_3Sn(RDta)$ and $Ph_2Sn(RDta)_2$ have been synthesized (376, 378) by the reaction of $ClSn(Ph)_3$ or $Cl_2Sn(Ph)_2$ with piperidinium salts of dithioaromatic ligand anions in CH_2Cl_2. Reactions of the "stannyl esters" $Ph_3Sn(RDta)$ with secondary amines result in cleavage of the Sn–S bond and formation of the R_2N–$Sn(Ph)_3$ and the free dithioacid HRDta. The $(Ph)_2Sn(RDta)_2$ compounds undergo an initial substitution reaction with R_2NH, and the proposed $R_2'NSn(Ph)_2(RDta)$ intermediates decompose to $(Ph_2SnS)_3$ and $RC(S)NR_2'$ compounds.

The use of $Me_3Sn(Me_2Dtc)$ in the synthesis of various transition metal carbonyldithiocarbamate complexes (Eq. 1) has been described (2).

$$M(CO)_nX + Me_3Sn(Me_2Dtc) \longrightarrow M(CO)_{n-1}(Me_2Dtc) + Me_3SnX + CO \qquad (1)$$

$$(M = Mn, n = 5; M = Rh, n = 3)$$

The syntheses of $Fe(CO)_3I(Me_2Dtc)$, $(\eta^5\text{-Cp})Fe(CO)(Me_2Dtc)$, $(\eta^5\text{-Cp})W(CO)_2(Me_2Dtc)$, and $(\pi\text{-allyl})Fe(CO)_2(Me_2Dtc)$ from $Fe(CO)_4I_2$, $(\eta^5\text{-Cp})Fe(CO)_2Cl$, $(\eta^5\text{-Cp})W(CO)_3Cl$, and $(\pi\text{-allyl})Fe(CO)_3I$, respectively, are accomplished by similar reactions with $Me_3Sn(Me_2Dtc)$ (2).

The $Ph_3Pb(RDta)$ compounds (R = aryl) also are known and yield thion-

esters $(R\overset{\overset{\text{S}}{\|}}{-}C{-}OR)$ upon treatment with alkoxides (374). The by-products in this reaction are NaRDta and $(Ph_3Pb)_2S$.

The Mössbauer spectra of the Sn(II) complexes (Table VI) show quadrupole splittings that are consistent with a sterically active pair of electrons. However, the magnitude of these splittings appears more sensitive to the nature of the ligand rather than the stereochemistry of the Sn(II) ion (342).

The Mössbauer parameters for a series of organotin(IV) Et_2Dtc complexes have been reported. A linear relationship between the isomer shift and the Mulliken-Jaffe electronegativity values is observed for a number of $XYSn(Et_2Dtc)_2$ complexes (Table VI). A similar plot of the Sn–S stretching frequency versus the isomer shift is also approximately linear and indicates low values for the isomer shifts in compounds with the higher frequency Sn–S vibrations (189). Both of these correlations are consistent with a greater positive charge on the metal atom with the more electronegative X and Y groups.

4. Group V

a. Arsenic, Antimony, Bismuth. The chemistry of the Group V elements, which was explored in a rather sporadic fashion prior to 1969, now

TABLE VI

Mossbauer Parameters for the Sn 1,1-Dithio Complexes

Compound	IS mm sec^{-1}	QS mm sec^{-1}	T,°K	Ref.
$Sn(Et_2Dtc)_2{}^a$	1.92	1.05	83	342
	1.92	1.05	4.2	342
$Sn(EtXant)_2{}^a$	1.93	0.75	83	342
$(Me_4N)Sn(EtXant)_3{}^a$	2.10	0.80	83	342
$R_2Sn(Et_2Dtc)_2{}^b$	$(1.51-1.56)^c$	$(2.82-3.04)^c$	81	189
$(Ph)_2Sn(Et_2Dtc)_2$	1.13	1.74	81	189
$I_2Sn(Et_2Dtc)_2{}^b$	1.12	0	81	189
$Br_2Sn(Et_2Dtc)_2{}^b$	0.95	0	81	189
$Cl_2Sn(Et_2Dtc)_2{}^b$	0.74	0	81	189
$F_2Sn(Et_2Dtc)_2{}^b$	0.19	0	81	189
$RSn(Et_2Dtc)_2Cl^{b,d}$	$(1.09-1.21)^d$	$(1.66-1.76)^d$	81	189
$(n\text{-}Bu)Sn(Et_2Dtc)_2Br^b$	1.27	1.80	81	189
$R_2Sn(Et_2Dtc)X^b$	$(1.43-1.53)^e$	$(2.80-2.84)^e$	81	189
$(Ph)_2Sn(Et_2Dtc)Cl^b$	1.28	2.21	81	189
$(Ph)_3Sn(Et_2Dtc)^b$	1.30	1.71	81	189
$Me_2Sn(R_2Dtc)Cl^f$	$(1.28-1.35)^g$	$(2.72-2.98)^g$	80	245
$Bu_2Sn(R_2Dtc)Cl^f$	$(1.31-1.45)^g$	$(2.76-3.14)^g$	80	245
$Ph_2Sn(R_2Dtc)Cl^f$	$(1.08-1.23)^g$	$(2.19-2.34)^g$	80	245

^{a} Relative to a Pd/^{119}Sn source.
^{b} Values of IS with respect to $BaSnO_3$ at room temperature.
^{c} Range of values for R = Me, Pr, i = Bu.
^{d} Range of values for R = Me, n = Bu, Ph.
^{e} Range of values for R = Me, i = Bu; X = Cl, Br.
^{f} Room temperature source of $Ba^{119}SnO_3$; IS relative to SnO_2.
^{g} Range of values for R_2 = Me_2, Et_2, Bz_2, pyrrol.

has been studied systematically for the R_2Dtc ligands. In addition, various x-ray structural studies have provided structural prototypes for the Group V dithio complexes (Table VII).

For the trivalent elements As, Sb, and Bi, all compounds with the composition $X_nM(R_2Dtc)_{3-n}$ (n = 0, 1, 2; X = Cl, Br, I) have been obtained. Furthermore, organoantimony(III) derivatives of the type $RSb(R_2'Dtc)_2$ and $R_2Sb(R_2'Dtc)$ also have been reported (450a).

The synthesis of $As(R_2Dtc)_3$ (419, 421) and $M(R_2Dtc)_3$ (M = Sb, Bi) complexes (418, 421) by direct reaction between MCl_3, CS_2, and R_2NH (R_2 = Pyrrol, Pip, i-Bu_2, Bz) in a 1:3:6 molar ratio in CCl_4 has been reported. It is suggested (421) that in the simultaneous mixing of the reagents the $M(NR_2)_3$ compounds first form and then undergo a CS_2 insertion reaction.

On the basis of the IR spectra (splitting of the C=S vibration), an asymmetric ligand bonding has been proposed, with one of the As—S or Sb—S

TABLE VII
Structural Details of the Group V 1,1-Dithio Complexes.

Compound	$\overline{M-S}$, Å	$\overline{C-N}$, Å	$\overline{C-S}$ (1), Å	$\overline{C-S}$ (2), Å	$\nu(\overline{C-N})$, cm^{-1}	Ref.
As(Et$_2$Dtc)$_3$	2.349 (5)		1.76 (1)	1.69 (2)		124, 537
	2.845 (51)					
Sb(Et$_2$Dtc)$_3$	2.581 (81)	1.34 (2)	1.71 (2)	1.73 (3)		542
	2.915 (43)					
Bi(Et$_2$Dtc)$_3$	2.700 (94)	1.32 (3)	1.71 (4)	1.75 (2)		542
	2.942 (30)					
[Sb(n–Bu$_2$Dtc)$_2$]$_2$Cd$_2$I$_6$	2.628 (37)^a	1.33 (3)	1.71 (2)	1.72 (2)	1511^c	597
	2.460 (5)^b					
Br$_2$As(Et$_2$Dtc)^d	2.270 (3)	1.31 (2)	1.729 (9)	1.712 (10)	1510^c (1542)^e	170
	2.374 (4)					
BrAs(Et$_2$Dtc)$_2^f$	2.294 (38)		1.796	1.684		22
	2.820, 2.475		1.698	1.727		

^{a} Mean axial Sb–S bond length.
^{b} Equatorial Sb–S bond length.
^{c} In CHCl$_3$.
^{d} As–Br 2.404 (2) (terminal bromine atom).
^{e} In Kbr.
^{f} As–Br 2.711.

bonds covalent and the others ionic in nature (421). The electronic and PMR spectra of these compounds also are discussed.

The crystal structures of the M(Et$_2$Dtc)$_3$ complexes with M = As (537a, 124), Sb, and Bi (542) have been determined. In the structure of the As(III) homologue, the molecular symmetry is close to C_3, with three short As–S distances occupying a triangular face in the AsS$_6$ core at an average distance of 2.349(5) Å. Three longer As–S bonds complete the coordination sphere at an average distance of 2.845(51) Å. The deviation from D_3 symmetry is attributed to a stereochemically active lone pair along the threefold axis which points away from the trigonal face defined by the covalently bound sulfurs.

While at first glance the structures of the Sb and Bi homologues appear similar to that of As(Et$_2$Dtc)$_3$, closer inspection shows a pseudododecahedral coordination arising from intermolecular M–S interactions at 3.389(4) and 3.210(5) Å in the Sb and Bi derivatives, respectively (542).

The PMR spectra of the M(R$_2$Dtc)$_3$ complexes (M = As, Sb, Bi) in benzene show solvent shifts induced by the ring current of benzene, which are interpreted in terms of 1:1 van der Waals benzene–M(R$_2$Dtc)$_3$ adducts. The association constants for these adducts appear to parallel the polarity of the C–N bond in the R$_2$Dtc ligands (422).

The mass spectra of the M(R$_2$Dtc)$_3$ (M = As, Sb, Bi) complexes have been obtained and the fragmentation patterns have been analyzed (420). The most prominent peak is due to the $M(R_2Dtc)_2^+$ cation, a species that has been obtained synthetically for M = Bi (423) and M = As, Sb (597). The mass spectra also show the presence of polynuclear ions of the type M_nS_m M = As, Sb, Bi; n = 1, 2, 3, 4; m = 0, 1, 2, 3, 4, 5. This interesting observation suggests that the pyrolysis of $M(R_2Dtc)_3$ complexes (M = As, Sb, Bi) under controlled conditions may lead to the synthesis of unusual M_nS_m cluster molecules.

The $Bi(R_2Dtc)_2^+$ cation has been obtained (R_2 = Pyrrol, Pip, Et_2) by the reaction of the $Bi(R_2Dtc)_3$ complexes with BF_3 in $CHCl_3$ under an inert (Ar) atmosphere (420). The corresponding cations of As(III) and Sb(III) also have been obtained as I_3^- "salts," either by the reactions of MI_3 with R_4Tds in a 1:1 molar ratio or by the reaction of I_2 with the $M(R_2Dtc)_3$ complexes. The apparent reduction of the thiuram disulfide in the first reaction is thought to occur by way of oxidative addition to MI_3 and formation of $M(V)(R_2Dtc)_2I_3$. A subsequent reductive elimination of I_2 gives the product and the I_3^- anion. This proposed mechanism is supported by the fact that the $Sb(CH_3)_3$ undergoes oxidative addition with Me_4Tds to give $Sb(V)(CH_3)_3(Me_2Dtc)_2$, which does not undergo reductive elimination (597).

The crystal structure of the $[Sb(n\text{-}Bu_2Dtc)_2]_2(Cd_2I_6)$ complex, obtained by the reaction of SbI_3 with $Cd(n\text{-}Bu_2Dtc)_2$ in $CHCl_3$, has been determined (597). Discrete $Sb(n\text{-}Bu_2Dtc)_2^+$ and $Cd_2I_6^{2-}$ ions are found in the structure, and the geometry of the cation is best described as a pseudotrigonal bipyramid with a stereochemically active lone pair in an equatorial position. The asymmetrically bound $n\text{-}Bu_2Dtc$ ligands form the longer Sb–S bonds in the axial sites of the trigonal bipyramid (597) (Table VII) (Fig. 8).

A large number of $IM(R_2Dtc)_2$ complexes, M = As(III), Sb(III), and Bi(III), have been obtained by the reaction of the $M(R_2Dtc)_3$ complexes with I_2 in a 2:1 molar ratio. The IR, electronic, and mass spectra have been obtained (423). Based on the analysis of these data, and the square pyramid structure reported for the $BrAs(Et_2Dtc)_2$ complex (22), a square pyramidal structure with the I^- at the apex is proposed for these complexes. The molecular weights of the $IM(R_2Dtc)_2$ complexes in $CHCl_3$ indicate monomeric *undissociated* species in solution. These results should be compared to conductivity studies reported for the $I_3M(R_2Dtc)_2$ complexes, which point to 1:1 electrolytic character for the latter complexes in nitromethane solution (597). The conclusions that must be drawn from the molecular-weight and conductivity studies are that the halide ligands in these molecules are only weakly bound and dissociation occurs in solvents of high dielectric character. In such solvents the existence of a trigonal bipyramidal structure (Fig. 8) should not be ruled out for the $IM(R_2Dtc)_2$ complexes. It is quite likely that a fluxional behavior may be observed for these species by variable-temperature PMR studies. A study of the analogous $BrM(R_2Dtc)_2$ complexes with various R_2Dtc ligands (M = As, Sb, Bi) has been reported (423).

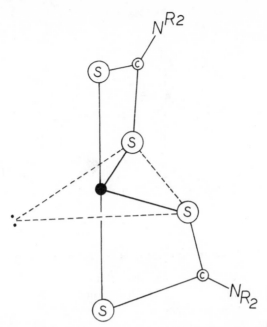

Fig. 8.   Schematic structure of the Sb(n-Bu$_2$ Dtc)$_2^+$ cation (597).

The synthesis of the $X_2 M(Et_2 Dtc)$ complexes $(X = Cl, Br, I; M = As, Sb, Bi)$ is accomplished by the reaction of the $M(Et_2 Dtc)_3$ complexes with MX_3 in a 1:2 molar ratio in $CHCl_3$ (170). The structure of the $Br_2 As(Et_2 Dtc)$ complex, which is monomeric and a nonelectrolyte in solution, in the solid state shows a loosely held centrosymmetric dimer with bridging bromide ions. The coordination geometry of the five-coordinate As(III) is intermediate between a square pyramid and a trigonal bipyramid, with a stereochemically active lone pair occupying a site of a distorted octahedral coordination sphere (Fig. 9).

The $RSb(R_2' Dtc)_2$ and $R_2 Sb(R_2' Dtc)$ complexes can be synthesized by the reaction of the $RSb(NR_2')_2$ and $R_2 Sb(NR_2')$ compounds with CS_2 in benzene (450a). The $R_2 Sb(R_2' Dtc)$ derivatives $(R = alky)$ are unstable and spontaneously disproportionate into a 1:1 mixture of trialkylstibine and $RSb(R_2' Dtc)_2$. The aryl derivatives $(R = Ph)$ are quite stable, and $Ph_2 Sb(Et_2 Dtc)$ is stable in refluxing benzene. Previous attempts to prepare the bismuth analogue resulted in a decomposition into $Ph_3 Bi$ and $PhBi(Et_2 Dtc)_2$ (398). The structures of the monoorganostibine bisdithiocarbamates most likely are distorted octahedra with bidentate $R_2 Dtc$ ligands and a stereochemically active lone pair at one of the sites. The possibility of asymmetrically bound $R_2 Dtc$ ligands in the $RSb(R_2' Dtc)_2$ complexes is not supported by the IR spectra. Only one strong absorption in the region 980 ± 70 cm^{-1}, generally associated with the C–S

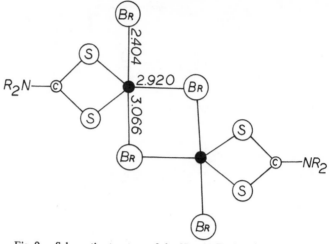

Fig. 9.   Schematic structure of the [Br$_2$As(Et$_2$Dtc)]$_2$ dimer (170).

vibrational frequency, is found. By contrast, solution spectra of the Ph$_2$Sb(Et$_2$Dtc) complex show two C–S absorptions at 986 and 1006 cm^{-1} that could be due to the presence of a monodentate R$_2$Dtc ligand and three-coordinate Sb(III).

The reaction of (CH$_3$)(CF$_3$)$_3$PN(CH$_3$)$_2$ with CS$_2$ proceeds with CS$_2$ insertion into the P–N bond and formation of the (CH$_3$)(CF$_3$)$_3$P(Me$_2$Dtc) compound (590). Very little is known about the chemistry of the xanthate or dithioacid complexes of the Group V elements. The P(EtXant)$_3$ has been reported (410). The IR spectrum of this molecule and those of the corresponding As(III) and Sb(III) compounds show a monotonic decrease of the C–O vibration around 1200 cm^{-1} (by 7 cm^{-1} intervals) from P to Bi (635).

The very interesting Sb(EtXant)$_4^-$ and Bi(EtXant)$_4^-$ complexes also have been reported (635). The presence of multiple C–S vibrations between 1020 and 1050 cm^{-1} suggests that these molecules may contain monodentate EtXant ligands. Six-coordinate central atoms with two monodentate and two bidentate chelates are very likely to be found in these compounds.

5. Group VI

a. Sulfur, Selenium, Tellurium.   Prior to 1969 the chemistry of the Group VI 1,1-dithio chelates, consisted primarily of synthetic studies (153). Since then a significant number of structural studies have appeared in the literature, mainly owing to the crystallographic work of Husebye and co-workers (347–350). These studies have systematized the structural aspects of compounds of divalent sulfur, selenium, and tellurium and of tetravalent tellurium.

b. Dithiocarbamates. The chemistry of sulfur with R_2Dtc ligands has been described (153). Although the products, $S(R_2Dtc)_2$, cannot be classified as inorganic complexes, their synthesis is very similar to that of other main element complexes. The structure of $S(MorphoDtc)_2$ has been determined (348) (Table VIII). The valency angle of the two-coordinated central sulfur atom $[108.8(4)°]$ is not significantly different than the average value of $107°54'$ found in ortho-rhombic sulfur (104). In each dithiocarbamate "ligand" both a short and a long C–S bond are found consistent with the monodentate mode of binding. The analogous $Se(MorphoDtc)_2$ complex is monomeric and the selenium atom is four-coordinate and planar. The two dithiocarbamate ligands are asymmetrically bound to the divalent selenium and form a trapezoid with two short and two long Se–S bonds (Table VIII) (14).

The same coordination geometry is found in the structure of the $Se(Et_2Dtc)_2$ complex (349) (Fig. 10). The trapezoidal coordination geometry also is observed in the structure of the $Te(MorphoDtc)_2$ complex (347), where individual $Te(MorphoDtc)_2$ molecules are linked into dimers by short intermolecular Te–S contacts in a manner not unlike that found for the structure of the $Cu(Et_2Dtc)_2$ complex (491). The synthesis of $Te(II)(Et_2Dtc)_2$ has been reported (583). The EPR spectrum of this molecule (C_6H_6, $103°K$) consists of a singlet $g_0 = 1.983$. A less intense doublet ($A = 17$ G) also was observed and was attributed to ^{123}Te and ^{125}Te ($I = \frac{1}{2}$). Based on the IR spectrum of this compound (a splitting of the C=S vibration at ~1000 cm^{-1}), a unidentate mode of binding was proposed for the Et_2Dtc ligands. This prediction was verified in the structure of the $Te(Et_2Dtc)_2$ "complex." In this structure an intermolecular Te–S contact $[3.579(5)$ Å$]$ is observed in the plane of the Te–S_4 trapezoid (229).

Bonding in the trapezoidal structures of Se(II) and Te(II) has been interpreted in terms of the three-center bond theory (229, 347). According to this theory (151), in the XS_4 group (X = Se, Te) there should be two linear three-center four-electron bonds, each formed by the overlap of a p orbital of the central atom and the σ orbitals of the two outer atoms. Another explanation that has been advanced for the geometry of the TeS_4S' core in $Te(Et_2Dtc)_2$ is based on a full hybridization theory (151). Seven pairs of electrons (4 electrons from tellurium and a total of 10 from the sulfur atoms) are arranged in such a fashion that five are bonding and two are lone pairs. An sp^3d^3 hybridization of metal orbitals is required in this case, as well as a pentagonal bipyramidal structure. The positions of the lone pairs are shown in Fig. 10.

The synthesis, IR, UV, and PMR spectra of $Te(Et_2Dtc)_4$ have been reported (583). A distorted octahedral environment was proposed for this molecule involving two bidentate and two monodentate ligands.

The crystal structure of $Te(Et_2Dtc)_4$ shows a distorted dodecahedral coordination around the eight-coordinate Te(IV) ion. The lone pair (ns^2) is stereochemically inert. The TeS_8 core can be envisioned as two interweaving planar TeS_4 trapezoids (350). A similar structure for the TeS_8 core is found in Te(MorphoDtc)$_4$ (222) (Table VIII). A comparison of the "short" interligand S–S distance in this compound (3.39 Å) with that observed in the structure of the Te(Et$_2$Dtc)$_4$ complex (350) (3.17 Å) has been made. It is suggested that the intramolecular electron transfer (Eq. 2) that leads to the decomposition of Te(Et$_2$Dtc)$_4$ may be associated with the short interligand S–S bond length (222). It should be noted that the Te(MorphoDtc)$_4$ is a considerably more stable molecule.

$$Te(IV)(Et_2Dtc)_4 \xrightarrow{\Delta} Te(II)(Et_2Dtc)_2 + Et_4Tds \qquad (2)$$

The crystal structure of Te(Et$_2$Dtc)$_3$Ph, originally synthesized by Foss (251), has been determined (220). The six sulfur atoms and the carbon atom of

TABLE VIII
Structural Details of the Group VI 1,1-Dithio Complexes.

Compound	$\overline{M-S}$, Å	$\overline{C-N(O)}$, Å	$\overline{S-M-S}$ °	$\overline{C-S}$, Å	Ref.
S(Morpho Dtc)$_2$	2.013 (9)	1.35 (3)	108.8 (3)^a	1.80 (2)	348
				1.63 (2)	
Se(Morpho Dtc)$_2$	2.298 (22)	1.35 (3)	70.5 (2)	1.78 (4)	14
			84.6 (2)	1.66 (5)	
			69.7 (2)		
			135.0 (2)		
Se(Et$_2$Dtc)$_2$	2.322 (14)	1.36 (3)	70.9 (2)	1.76 (2)	349
	2.75 (4)^b		87.7 (2)	1.66 (2)	
			70.9 (2)		
			131.5 (2)		
Te(Morpho Dtc)$_2$	2.514 (22)	1.33 (3)	66.8 (2)	1.75 (3)	347
	2.845 (15)		80.9 (2)	1.68 (2)	
			66.8 (2)		
			145.7 (2)		
Te(Et$_2$Dtc)$_4$	2.74 (7)^c	1.33 (2)^d	66.5 (28)^e	1.71 (3)^f	223
			159.8 (9)		350
Te(Et$_2$Dtc)$_3$Ph^g	2.72 (9)^h	1.35 (3)	63.4 (26)^i	1.71 (3)	220
					223
Te(Et$_2$Dtc)$_2$	2.519 (4)	1.36 (2)	66.7 (1)	1.74 (2)	229
	2.830 (3)		79.7	1.69 (2)	
	2.893 (4)		65.8 (1)		
			147.8		
Te(Morpho Dtc)$_4 \cdot (C_6H_6)_3$	2.688 (12)^j	1.34 (3)	64.9 (6)	1.70 (2)	221
	2.786 (44)		74.8 (14)^k		222
			64.9 (4)		
			155.3 (3)		
S(MeXant)$_2$	2.052 (9)	1.34 (2)	104.2 (3)	1.73 (2)	79, 80
				1.59 (3)	

TABLE VIII (continued)

Compound	$\overline{M-S}$, Å	$\overline{C-N(O)}$, Å	$\overline{S-M}-S°$	$\overline{C-S}$, Å	Ref.
Se(MeXant)₂	2.187 (25)	1.29 (2)	100.8 (3)	1.73 (2)	79, 80
				1.63 (3)	
Te(MeXant)₂	2.504 (7)	1.324 (15)	66.20 (4)	1.713 (5)	281
	2.843 (3)		85.18 (4)	1.644 (8)	
			66.33 (4)		
			142.29 (3)		
Et₄N[Te(EtXant)₃]	2.663 (22)	1.325 (19)	62.1 (1)	1.701 (19)	343
	3.055 (5)	1.312 (14)ⁱ	73.6 (1)	1.659 (24)	344
	2.503 (3)ⁱ		61.7 (1)	1.708 (12)ⁱ	411
			82.1 (1)	1.655 (12)ⁱ	
			80.8 (1)		

ᵃ Dihedral angles C–S–S/S–S–S, 92.4, 101.5°.
ᵇ Range: 2.719 (5) – 2.779 (5) Å.
ᶜ Range: 2.631 (8) – 2.845 (14) Å.
ᵈ Range: 1.32 (2) – 1.40 (2) Å.
ᵉ Range: 63.1 (3) – 70.8 (3) °.
ᶠ Range: 1.76 (2) – 1.65 (2) Å.
ᵍ Te–C, 2.124 (11) Å.
ʰ Range: 2.816 (3) – 2.606 (3) Å.
ⁱ Range of 65.04 (9) – 60.45 (11)° intraligand angles.
ʲ Mean values of the 'short' and 'long' bonds of the four asymmetrically bound ligands.
ᵏ Mean values of the four angles in the two interpenetrating trapezoids.
ˡ Mondentate ligand.

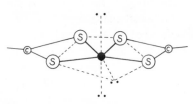

Fig. 10.   Idealized structure of the divalent Se and Te 1,1-dithio complexes (347, 349).

the phenyl ring define a pentagonal bipyramid configuration. The axial Te–C bond (Table VIII) is found at an angle of 144.6(2)° from the axial Te–S bond [3.228(4) Å]. The remaining equatorial Te–S bonds average to 2.72(9) Å.

c. Xanthates.   The synthesis of S(MeXant)₂ has been reported to occur by nucleophilic substitution on the central atom of the pentathionate ion (80). A similar procedure has been outlined by Foss (251) for the synthesis of the selenium analogue from the selenopentathionate ion. The structure of the S(MeXant)₂ molecule (Table VIII) shows the central sulfur atom bound to two sulfur atoms of the monodentate MeXant ligands. A trapezoid planar configuration is seen if two, rather long, intermolecular S—S contacts [3.625(6), 3.709(8) Å] are taken into consideration (80). The corresponding Se(MeXant)$_2$ is x-ray isomorphous and isostructural. In the structure of this compound, the intermolecular Se—S contacts are seen at 3.501(9) and 3.595(7) Å, on the average 0.3 Å shorter than the sum of the van der Waals radii.

The electronic and IR spectra of the Te(MeXant)$_2$ compound have been discussed. The structure of this molecule (281) has been determined to be quite similar to that of the corresponding Te(MorphoDtc)$_2$ (347).

The reaction of Te(EtXant)$_2$ with (Et$_4$N)EtXant in hot ethyl alcohol is reported to give the Te(EtXant)$_3^-$ ion. This ion can be isolated as the Et$_4$N$^+$ salt upon cooling. Molecular-weight studies on this molecule, in benzene solution, are indicative of dimer formation at high concentrations (411). A subsequent crystal-structure determination of Et$_4$NTe(EtXant)$_3$ showed (343, 344) a monomeric anion with a very interesting five-coordinate pentagonal planar geometry. This geometry arises from two bidentate and one monodentate EtXant ligands. The bonding in the molecule can be rationalized by the full hybridization theory (see above), with the lone pairs being stereochemically active and oriented along the axial sites of a pentagonal bipyramid.

C. Dithioacid Complexes with Transition Elements

1. Group I

a. **Copper.** The chemistry of the copper 1,1-dithio complexes is almost entirely limited to the dithiocarbamate ligands and the 1,1-dithiolate ligands (153). Major developments since 1969 are underscored by the synthesis of (1) oxidized Cu(R$_2$Dtc)$_2$ complexes and derivatives and (2) mixed-oxidation-state polynuclear complexes. Extensive studies of the magnetic properties of these complexes have been carried out by either EPR spectroscopy or magnetic susceptibility measurements.

The Cu(PyrrolDtc)$_2$ complex was found to be x-ray isomorphous to the square planar Ni(PyrrolDtc)$_2$ complex (473). A structure determination of the former complex revealed a monomeric Cu(PyrrolDtc)$_2$ unit with a square planar CuS$_4$ chromophore (Table IX). This result is particularly interesting considering that the Cu(Et$_2$Dtc)$_2$ (53) and Cu(n-Pr$_2$Dtc)$_2$ (507) complexes are dimeric species containing five-coordinate rectangular pyramidal Cu(II) ions.

It has been suggested that intermolecular copper—sulfur interactions in these complexes are of comparable magnitude to proton—sulfur and proton—metal interactions and should not be regarded as a general feature of the structural characteristics of the Cu(R$_2$Dtc)$_2$ complexes. It should be noted that in the structure of the Cu(PyrrolDtc)$_2$ complex, the crystal packing is domin-

TABLE IX
Structural Details of the Copper Dithiocarbamate Complexes.

Complex	$\overline{\text{Cu–S}}$, Å	$\overline{\text{C–N}}$, Å	ν(C–N), cm^{-1}	Ref.
Cu (n-Bu$_2$ Dtc) Br$_2^a$	2.193 (6)	1.30 (3)	1560	33
Cu (n-Bu$_2$ Dtc)$_2$ I$_3$	2.22 (2)	1.33 (8)	1570	270
				624
Cu$_3$ (n-Bu$_2$ Dtc)$_6$ (Cd$_2$ Br$_6$)	2.22 (1)^b	1.33 (4)^b	1512^b	626
	2.33 (6)^c	1.33 (3)^c	1548^{c,d}	
[Cu (PipDtc)$_2$ (CuBr)$_4$]	2.3163 (9)	1.320 (6)	1520	278
[Cu (PipDtc)$_2$ (CuBr)$_6$]	2.3059 (11)	1.319 (7)		278
Cu$_2$ (Et$_2$ Dtc)$_2$ Cl$_2^e$	2.274 (10)			315
Cu$_2$ (Et$_2$ Dtc)$_3$ Br$_2$	2.20 (4)^f		1535	598
	2.28 (2)^g		1515	
Cu (PyrrolDtc)$_2$	2.344 (2)	1.39 (1)		473
Cu (MePhDtc)$_2$	2.301 (39)	1.31 (2)		427
Cu (PPh$_3$)$_2$ EtSXant^h	2.379 (3)			18
	2.493 (3)			

^{a} Cu–Br 2.311 (4) Å.
^{b} From the B-type units containing the Cu (III) ions.
^{c} From the A=type unit containing the Cu (II) ion.
^{d} ν Cu–S for A unit at 362 cm^{-1}, for B units at 398 cm^{-1}.
^{e} $\overline{\text{Cu–Cl}}$ 2.321 (17) Å.
^{f} Distances in the Cu (Et$_2$ Dtc)$_2^+$ unit.
^{g} Distances in the [Cu (Et$_2$ Dtc) Br$_2$]$^-$ unit, Cu–Br, 2.40 (2) Å.
^{h} $\overline{\text{Cu–P}}$ = 2.256 (13) Å.

ated by intermolecular proton contacts from the ligand substituents to the metal and sulfur atoms (473). The crystal structure of the Cu(MePhDtc)$_2$ complex also shows monomeric units (Table IX). The dimeric arrangement found with other Cu(II) R$_2$ Dtc complexes is preempted in Cu(MePhDtc)$_2$ by the orientation of the phenyl rings. These rings are nearly normal to the plane of the rest of the molecule, an orientation that appears to be dictated by the steric interactions of the adjacent methyl substituents (427).

The very interesting Cu(R$_2$ Dtc)$_2$ complex (R$_2$ = pyrrole) has been synthesized. The tendency of the pyrrole ring to preserve aromaticity (Fig. 4), makes the resonance form R$_2$ $\overset{+}{\text{N}}$=CS$_2$ quite unimportant in the structure of the ligand. Complexes with this ligand, including the Cu(II) complex, show the C–N vibration between 1250 and 1350 cm^{-1} (26).

The EPR spectrum of the Cu(pyrroleDtc)$_2$ complex in a dilute Pd(II) matrix is indicative of a four-coordinate Cu(II) species. The parameters $A_{\parallel} = 144.8 \times 10^{-4}$ cm^{-1}, $A_{\perp} = 30.5 \times 10^{-4}$ cm^{-1}, $g_{\parallel} = 2.11$, and $g_{\perp} = 2.02$ have been used together with electronic spectra data for the calculation of MO coefficients that seem to indicate strong covalency in the in-plane σ bonding, and moderate covalency in the in-plane and out-of-plane π bonding. The moderate out-of-plane π-bonding is intriguing in view of the fact that the extent of ligand conjugation appears minimal for the pyrroleDtc ligand when compared to other R_2Dtc ligands.

Two types of copper(III) complexes have been isolated. Oxidation of $Cu(n\text{-}Bu_2Dtc)_2$ or $[Cu(n\text{-}Bu_2Dtc)]_4$ by halogens results in the formation of the diamagnetic square planar monomeric $Cu(n\text{-}Bu_2Dtc)X_2$ complexes (X = Cl, Br) (33). The crystal structure of $Cu(n\text{-}Bu_2Dtc)Br_2$ has been determined (33) (Table IX).

The $Cu(R_2Dtc)_2^+$ cations are obtained by the oxidation of the Cu(II) complexes with iodine (75) $FeCl_3$, $Fe(ClO_4)_3$, or $Cu(ClO_4)_2$ (270). The structure of the $Cu(n\text{-}Bu_2Dtc)_2I_3$ complex shows a Cu(III) ion in a four-coordinate planar coordination (624). A short Cu(III)–S bond [2.22(2) Å] compared to the Cu(II)–S bond in the $Cu(R_2Dtc)_2$ complexes can be rationalized in terms of a smaller ionic radius for the Cu(III) ion.

The increase in the Cu–S stretching frequency observed in the $Cu(R_2Dtc)_2^+$ complexes (385, 410 cm^{-1}) compared to that in the $Cu(R_2Dtc)_2$ complexes (~345, 370 cm^{-1}) has been attributed to a strengthening of the Cu–S bond as a result of removing an electron from an antibonding orbital composed primarily of metal and sulfur functions (75).

A detailed electrochemical study on 16 different $Cu(R_2Dtc)_2$ complexes has been reported (313). The oxidation of the $Cu(R_2Dtc)_2$ complexes is unambiguously defined as the one-electron reversible process shown by Eq. 3.

$$Cu(R_2Dtc)_2 \rightleftharpoons [Cu(R_2Dtc)_2]^+ + e^- \tag{3}$$

The reduction of the $Cu(R_2Dtc)_2$ complexes often appears quasi-reversible (based on voltametric parameters), presumably owing to a structural reorganization of the CuS_4 core of the $Cu(R_2Dtc)_2^-$ product. Coulometric reduction of $Cu(i\text{-}Pr_2Dtc)_2$ at 0.70 V gave an n value of 1.0 and was consistent with the simple reduction process shown by Eq. 4. Although the monoanion could not be

$$Cu(i\text{-}Pr_2Dtc)_2 + e^- \longrightarrow [Cu(i\text{-}Pr_2Dtc)_2]^- \tag{4}$$

isolated, the solutions containing this species show reversible two–one electron oxidation step by cyclic voltametry. A plot of oxidation potential versus reduction potential is nearly linear and suggests that those substituents that enhance ease of oxidation render reduction more difficult and vice versa (313). In general, for the reduction step given by Eq. 4 the potentials are more positive by about 1 V than the corresponding potentials in the reduction of the $Ni(R_2Dtc)_2$ complexes and illustrate convincingly the involvement of the metal ion in the reduction process.

A new series of mixed-valence-state complexes of the general formula $[Cu_3(n\text{-}Bu_2Dtc)_6][MBr_3]_2$ (M = Zn, Cd, Hg) with copper in the formal oxidation state of $+2\frac{2}{3}$ (M = Zn, Cd, Hg) have been described (172).

These compounds were obtained by the reaction of $Cu(n\text{-}Bu_2Dtc)_2$ with MBr_2 and Br_2 in stoichiometric amounts or by the reaction with the 3,5-bis(N,N-dialkyliminium 1,2,4-trithiolane (Fig. 5) salt of $Cu_2Br_6^{2-}$ (626).

Magnetic susceptibility measurements on $[Cu_3(n\text{-}Bu_2Dtc)_6][Cd_2Br_6]$ from 100 to $293°K$ show a normal Curie-Weiss behavior, and at $293°K$ they show a μ_{eff}^{corr} of 1.8 BM. Single-crystal EPR spectra of this complex are similar to those reported for the $Cu(Et_2Dtc)_2$ complex in the $Ni(Et_2Dtc)_2$ lattice (278). On the basis of the stoichiometry, infrared data, conductivity data, and magnetic characteristics, the presence of two Cu(III) and one Cu(II) ion in this trimer was implicated. The x-ray crystal structure of the $[Cu_3(n\text{-}Bu_2Dtc)_6][Cd_2Br_6]$ shows a loosely held centrosymmetric trimer not unlike that found in the structure of the Ni(II) and Pd(II) dithiobenzoate complexes (51) (Fig. 11).

The B-A-B order of stacking has the B-type units that contain the Cu(III) ions interacting with the A unit through Cu(III)–S and Cu(II)–S interactions of 2.88(1) and 3.19(1) Å, respectively (626). On the basis of the somewhat short Cu(III)–S axial distance and a slight deviation of the $Cu(III)S_4$ units from planarity, a square pyramidal structure is suggested for the Cu(III) ions in the B-type units.

The reaction between $[Cu(n\text{-}Bu_2Dtc)_2][HgBr_3]$ and $n\text{-}Bu_4Tds$ in a 4:1 molar ratio, and in the presence of 1 equiv of $[Au(n\text{-}Bu_2Dtc)]_2$, gives dark green crystals of the stoichiometry $[Cu_2Au(n\text{-}Bu_2Dtc)_6][HgBr_3]_2$ (264). This complex is x-ray isomorphous to the $[Cu_3(n\text{-}Bu_2Dtc)_6][CdBr_3]_2$ complex, for which the structure has been determined (see above). A crystal structure determination shows the B-type units occupied by both Au(III) and Cu(III) ions (0.5:0.5 occupancy) and the A-type units occupied by Cu(II) ions in the B-A-B order of stacking. The mean value of the Au(III)–Cu(III)–S bond (2.265 Å) was compared and was found to be similar to the mean value of Au(III)–S and

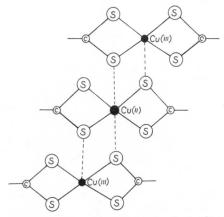

Fig. 11.   Schematic structure of the $[Cu_3(n\text{-}Bu_2Dtc)_6]^{2+}$ trimer (626).

Cu(III)–S bonds of 2.272 Å computed from distances found in various Au(III) and Cu(III) R_2Dtc complexes (264).

The reaction of Cu(II) halides with Cu(R_2Dtc)$_2$ complexes was reported to give polymeric materials (270). At a later date the exact nature of these products was probed and, by means of the reaction of Cu(PipDtc)$_2$ with CuBr$_2$, complexes of the type Cu(PipDtc)$_2$(CuBr)$_n$ ($n = 4$, and 6) were isolated. The structures of these molecules have been determined and consist of polymeric sheets of individual Cu(PipDtc)$_2$ molecules linked to polymeric CuBr chains by way of Cu–S bonds (Table IX). The Cu(PipDtc)$_2$(CuCl)$_4$ homologue is x-ray isomorphous and probably isostructural with the corresponding bromo compound.

The preparation and crystal structure of the dimeric chloro-bridged Cu$_2$(Et$_2$Dtc)$_2$Cl$_2$ complex has been reported (315). This compound was obtained by the reaction of Cu(Et$_2$Dtc)$_2$ with CuCl$_2$·2H$_2$O in a benzene–ethanol 3:1 mixture. A minor product of the composition Cu$_3$(Et$_2$Dtc)$_2$Cl$_3$ also was isolated in this reaction. The same type of complexes, Cu$_2$(R$_2$Dtc)$_2$Cl$_2$, have been reported to form in the reaction between Cu(II) chloride and mixed benzoic–dithiocarbamic anhydrides C$_6$H$_5$COS$_2$CNR$_2$ or C$_6$H$_5$COS$_2$CR′, where R = Me, Et; R′ = Pyrrol, Pip (472).

The structure of the Cu$_2$(Et$_2$Dtc)$_2$Cl$_2$ complex (315) consists of symmetrical chloro-bridged dimers weakly associated through intermolecular Cu–Cl [2.874(2) Å] and Cu–S [2.882(2) Å] interactions (Fig. 12). Each copper has a five-coordinate square pyramidal coordination not unlike that found in the structures of the Cu(R$_2$Dtc)$_2$ complexes (53, 507).

The structure of the mixed Cu(II), Cu(I) valence Cu$_3$(Et$_2$Dtc)$_2$Cl$_3$ complex shows a polymeric arrangement of centrosymmetric Cu$_2$(Et$_2$Dtc)$_2$Cl$_2$ dimers and centrosymmetric Cu$_4$(Et$_2$Dtc)$_2$Cl$_4$ units linked by Cu–S bonds in an alternating linear sequence (315).

In a subsequent study of the magnetic properties of the Cu$_2$(Et$_2$Dtc)$_2$Cl$_2$ dimer (65) (Fig. 12), the temperature dependence of the magnetic susceptibility

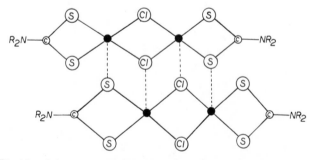

Fig. 12.   Schematic structure of the Cu$_2$(Et$_2$Dtc)$_2$Cl$_2$ dimer (315).

was interpreted in terms of three isotropic-exchange interactions acting between each of the four copper(II) ions. The magnetic moment of this molecule decreases from a value of 1.63 BM at $300°K$ to the value of 0.35 BM at $10°K$. A temperature-independent behavior is found from 10 to $4.2°K$ and is attributed to a paramagnetic impurity. A principal intradimer antiferromagnetic interaction ($2J = -104$ cm^{-1}), a smaller interdimer antiferromagnetic interaction ($2J = -55$ cm^{-1}), and an interdimer ferromagnetic interaction ($2J = 22$ cm^{-1}) were derived by a least-squares fit of experimental magnetic susceptibility data (65).

A similar study of the magnetic susceptibilities of the $Cu_2(R_2Dtc)_2Cl_2$ complexes was reported by Furneaux (259) ($R_2 = Bz$-n-Pr). In his analysis intradimer interactions ranging from $2J = -98$ to $2J = -140$ were reported for the different R groups and interdimer interactions were considered unimportant.

The remarkable variety in the permutations of the R_2Dtc ligands, various oxidation states of copper, and halide ions was again demonstrated in the synthesis of the $Cu_2(R_2Dtc)_3X_2$ complexes with copper in the oxidation states II and III. These paramagnetic compounds (1.77–1.86 BM per formula weight, R = Me, Et; X = Br, Cl) were obtained by the reaction of CuCl or CuBr and R_4Tds in CHCl$_3$ (598).

The crystal structure of the $Cu_2(Et_2Dtc)_3Br_2$ was determined and consists of alternating $Cu(Et_2Dtc)_2^+$ and $Cu(Et_2Dtc)Br_2^-$ units in chains. The rather short "interionic" distances in the chain that vary from 2.94(4) to 3.27(3) Å between copper atoms and ligand atoms in adjacent units have prompted van de Leemput et al. to describe the coordination of the copper atoms as being distorted octahedral (598).

Recently a number of polynuclear copper complexes were obtained by the reaction between CuBr and R_4Tds in CH_2Cl_2 in various molar ratios. In a 3:1 ratio, the dark blue microcrystalline $Cu_3(Me_2Dtc)_2Br_3$ complex was isolated. A 4:1 ratio gave $Cu_5(Et_2Dtc)_2Br_5$ and $Cu_5(n$-$Pr_2Dtc)_2Br_5$. A 6:1 ratio gave $Cu_7(Bu_2Dtc)_2Br_7$. On the basis of IR and ESCA data, these intractable complexes were considered to contain copper ions in the oxidation states I and III (599). It appears that all these complexes can be formulated as $[Cu(R_2Dtc)_2^+]$ - $[Cu_{n-1}Br_n^-]$ with n varying as a function of the R group. Values of n are 7, 5, 3 for n-Bu, n-Pr, and Et, respectively. With R = Pr the complex $[Cu(Pr_2Dtc)_2^+]_2$. $[Cu_2Br_4]^{2-}$ also has been isolated (599).

The interesting $XCu(Et_2Dtc)$ complexes obtained by the reaction of CuX_2 and $Cu(Et_2Dtc)_2$ (545), have been reported again (74) the crystal structure of the Cl complex also has been determined and is the same as that reported earlier (315). It consists of a dimeric chloride-bridged dimer with square Cu(II) ions (627). The mass spectrum fragmentation pattern of the $Cu(Et_2Dtc)_2$ complex differs from that of the $Zn(Et_2Dtc)_2$ complex. Thus, while the first loss from the former molecule is of C_2H_4 fragment, a C_2H_5 fragment is lost first from $Zn(Et_2Dtc)_2$.

These data are rationalized in terms of the Shannon-Swan rule and the low first ionization potential of the Cu(II) ion compared to the Zn(II) ion. This difference allows the Zn(II) complex to fragment by the route, $OE^{+\cdot} \to EE^+ +$ $OE \cdot$, (i.e., a radical loss from the molecular ion). The fragmentation scheme of the Cu(II) complex follows the route, $EE^+ \to EE^+ + EE$ (i.e., a neutral, stable molecule loss from the molecular ion), which suggests an internal electron transfer and cleavage of the Cu—S bond in the molecular ion (612).

The reaction of the $Cu(R_2 Dtc)_2$ complexes with alkyl hydroperoxides is reported to give alkoxy and alkyl peroxy radicals. The kinetics and a mechanism for this reaction have been reported (120).

The synthesis and structural characterization of the $(Ph_3 P)_2 Cu(EtSXant)$ complex has been reported (18). The Cu(I) ion is found in a highly distorted tetrahedral $P_2 CuS_2$ unit. The rather long Cu—S bonds (Table IX) and the acute S—Cu—S angle [73.6(1)] suggest that the RSXant ligand may occupy one coordination site of a pseudo-trigonal $P_2 CuL$ complex [P—Cu—P = 125.7(1)°].

b. Silver.   Since the pioneering synthetic studies by Akerström (6) and the structure determination of the $[Ag(n\text{-}Pr_2 Dtc)]_6$ hexamer by Hesse and Nilson (324) little work has been done with the silver 1,1-dithio chelates.

An interesting polynuclear complex of the stoichiometry $I_7 Ag_{11} (n\text{-}Bu_2 Dtc)_4$ has been isolated by iodine oxidation of $[Ag(n\text{-}Bu_2 Dtc)]_6$ in $CHCl_3$ (168).

The reaction of $[Ag(n\text{-}Bu_2 Dtc)]_6$ with $n\text{-}Bu_4 Tds$ in a 1:6 molar ratio afforded a dark red solution, which faded within 10 minutes to an EPR active blue solution. On the basis of the intensity of the EPR signal, generally attributed to the blue Ag(II) $R_2 Dtc$ complexes, it was suggested that the red solution contained an Ag(III) complex (27).

c. Gold.   Synthetic studies on gold 1,1-dithio chelates have placed emphasis on the reactions of the $[Au(R_2 Dtc)]_2$ dimer and the synthesis and structures of Au(III) complexes. The structure of the $[Au(n\text{-}Pr_2 Dtc)]_2$ dimer (321) has been published in detail (323) (Table X).

In a detailed study of the reactions of the $[Au(R_2 Dtc)]_2$ complex, Beurskens et al. reported (34) the reaction scheme shown in Fig. 13.

The $[Au(n\text{-}Bu_2 Dtc)_2] Br$ complex and the corresponding $AuBr_4^-$ salt obtained by a metathesis reaction (Fig. 13) are 1:1 electrolytes in nitrobenzene. Unusually large values for the C—N stretching frequencies in these complexes (1583 and 1560 cm^{-1}, respectively) have been attributed to "an increasing positive charge on the nitrogen atom induced by the anion." The Br^- anion has been observed to be close to the nitrogen atom of the $n\text{-}Bu_2 Dtc$ ligand in the structure of the $[Au(n\text{-}Bu_2 Dtc)_2] Br$ complex (34) (Table X).

A square planar Au(III) ion is found in this structure, which has structural

TABLE X
Structural Details of the Gold Dithiocarbamate Complexes.

Complex	$\overline{M-S}$, Å	$\overline{C-N}$, Å	$\nu(C-N)$, cm^{-1}	Ref.
[Au(n-Bu$_2$ Dtc)$_2$] Br	2.321 (17)	1.30 (2)	1583	34
[Au(n-Bu$_2$ Dtc)$_2$] [Au(MNT)$_2$]	2.335 (4)^d	1.29 (2)	1550	488
				603
Au(n-Bu$_2$ Dtc)(MNT)	2.326 (5)^b	1.34 (2)		489
	2.293 (13)^c			
[Au n-Pr$_2$ Dtc]$_2^d$	2.28 (2)	1.32 (5)		323
Au(Et$_2$ Dtc)$_3$	2.341 (10)^e	1.33 (2)^e		487
	2.343 (14)^f	1.28 (2)^f		

^{a} Distances in the cation.
^{b} S of the n-Bu$_2$ Dtc ligand.
^{c} S of the MNT ligand.
^{d} Au–Au, 2.76 (1) Å.
^{e} Monodentate ligand.
^{f} Bidentate ligand.

features similar to those of the Au(n-Bu$_2$ Dtc)$_2$ AuBr$_2$ complex (28) and of the Au(n-Bu$_2$ Dtc)$_2$ AgBr$_2$ complex (169). In a subsequent publication Van der Linden reported on the synthesis of a wide variety of [Au(R$_2$ Dtc)$_2^+$] X$^-$ complexes with R = H, Me, Et, n-Pr, n-Bu, and Ph; X = Br$^-$, ClO$_4^-$, PF$_6^-$, AgBr$_2^-$; AuBr$_2^-$; B(Ph)$_4^-$ [Au(S$_2$ C$_6$ H$_3$ CH$_3$)$_2$)], and [Au(S$_2$ C$_2$ (CN)$_2$)$_2$]$^-$. The increase in the C–N stretching frequency, detected previously, was noted again for certain anions (603).

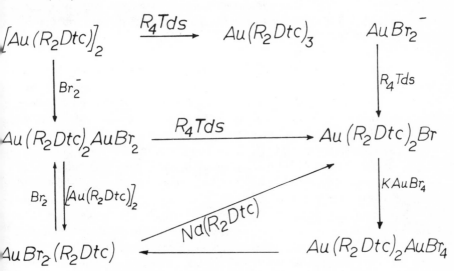

Fig. 13.   Reactions of the gold dithiocarbamate complexes (34).

The structure of the $[Au(n\text{-}Bu_2Dtc)_2]^+[Au(MNT)_2]^-$ complex (603) was determined (488) and, as expected, the structural parameters for the $Au(n\text{-}Bu_2Dtc)_2^+$ cation are similar to those reported previously for other structures containing the $Au(R_2Dtc)_2^+$ ion (28, 31, 34).

The mixed-ligand complex $Au(n\text{-}Bu_2Dtc)(MNT)$ has been synthesized (604), and its crystal structure was determined (489). A planar AuS_4 core was found, with the $n\text{-}Bu_2Dtc$ and MNT ligands involved as bidentate chelates on the same Au(III) center (Table X).

The structure of the $Au(Et_2Dtc)_3$ complex, originally reported by Åkerström (7), has been determined (487). One of the dithiocarbamate groups in this molecule functions as a bidentate ligand, while the other two function as monodentate ligands. The coordination of the Au(III) ion is square planar (Table X).

Recently a variety of $Au(R_2Dtc)Cl_2$ complexes were prepared and characterized ($R = n\text{-}Pr$, $n\text{-}Bu$, $n\text{-}pentyl$, $n\text{-}hexyl$; $R_2 = Pip$, Bz_2, MeBz) (201). These complexes are homologues of the previously reported $Au(n\text{-}Bu_2\text{-}Dtc)Cl_2$ (39).

The reaction of $Tl_2(7,8\text{-}C_2B_9H_{11})$ with $Br_2Au(Et_2Dtc)$ has been reported to give the $(1,2\text{-}C_2B_9H_{11})Au(Et_2Dtc)$ complex. An x-ray determination of the structure of this molecule shows the bidentate nature of the Et_2Dtc ligand. The 1,2-dicarbadodecaborane adopts a "slipped" structure with the metal situated over the B_3 fragment of the C_2B_3 face (133).

A Raman study of metal–metal bonding in $[Au(i\text{-}Bu_2Dtc)]_2$ was reported. The Au–Au bond order was estimated at about one-fourth of its value in Au_2^0. The metal–metal interaction was rationalized on the basis of charge transfer from the R_2Dtc ligands to the Au(I) ions and the partial formation of an Au_2^0 bond (243).

d. EPR Studies and Magnetic Properties.

Numerous studies on the EPR characteristics of the coinage metal 1,1-dithio complexes have appeared in the literature. The suitability of the $Cu(R_2Dtc)_2$ complexes for theoretical calculations and the relatively uncomplicated magnetic properties they exhibit allowed for extensive investigations of the electronic structures of these complexes.

EPR parameters for the $Cu(R_2Dtc)_2$ complexes in frozen solutions (515, 595) or in host lattices of $Ni(R_2Dtc)_2$ or $Zn(R_2Dtc)_2$ have been obtained for various guest/host ratios (167, 450b). Similar parameters have been obtained for the $Ag(R_2Dtc)_2$ complexes (605, 606).

Studies of the spectra of the homologous $Au(R_2Dtc)_2$ complexes have been reported; however, it appears that the analyses of the spectra are hampered by the large influence of the quadrupole moment of the gold nucleus (564, 606, 608). The interpretation of experimentally determined spin Hamiltonian parameters for the Ag(II) and Cu(II) complexes in terms of LCAO coefficients and

MO energies are problematic (381, 383, 384). A discussion of the problems associated with the extraction of information from LCAO–MO calculations has been presented (627) and will not be dealt with in detail here.

In summary, these studies and interpretations show that, in general, covalency in the M–S bonds increases strongly in going from copper to silver to gold, and that the metal $3d$ character of the MO of the unpaired electron decreases from 50% in $Cu(Et_2Dtc)_2$ to 26% in $Ag(Et_2Dtc)_2$ and 15% in the homologous Au(II) complex (627) (607).

Numerous studies of the EPR spectra of the $Cu(R_2Dtc)_2$ complexes in the presence of bases have been reported. The interaction of $Cu(n-Bu_2Dtc)_2$ with Pip, Py, and n-hexylamine (Hex) was studied by variable-temperature EPR measurements (139). Evidence for the formation of 1:1 adducts was presented and thermodynamic parameters were reported. For the Pip, Hex, and Py adducts, respectively, equilibrium constants of 3.9(1), 2.1(1), and 0.40(2) l mole^{-1} were determined. In the same order, ΔH^0 values of $-7.5(4)$, $-7.3(12)$, $-5(2)$ kcal mole^{-1} and ΔS^0 values of $-22(1)$, $-23(2)$, and $-19(3)$ eu were reported. The rate of adduct formation is primarily limited by the entropy of activation, while the rate of dissociation is limited by the enthalpy of activation.

In a similar study of the EPR and absorption spectra of the $Cu(PipDtc)_2$ and $Cu(EtBzDtc)_2$ complexes, Yordanov and Shopov (638) observed the formation of 1:2 adducts with Py and no evidence for the 1:1 adduct. The apparent contradiction with previous work (139) may well be associated with the influence of the R_2 group on the ability of the $Cu(R_2Dtc)_2$ complexes to form base adducts. A pronounced effect of the R_2 group in the base-adduct formation of the $Ni(R_2Dtc)_2$ complexes has been noted in the past (155). A calculation of the MO coefficients for $Cu/Zn(R_2Dtc)_2$ and $Cu(R_2Dtc)_2$ base adducts show that the contributions of d_{z^2} and $4s$ orbitals in the ground state of the unpaired electron are increased by comparison with the corresponding values for the $Cu/Ni(R_2Dtc)_2$ complex, where no rhombic distortion is present (639, 641). The interaction of $Cu(R_2Dtc)_2$ complexes with organic hydroperoxides has been studied by EPR spectroscopy (571).

Recently an EPR study of the base adduct formation of $Cu(Et_2Dtc)_2$ with Py, i-quinoline, quinoline, thiophane, and dioxane was reported (637). The formation of 1:1 adducts was detected and the equilibrium constants for the various bases follow the order Py $>$ i-quin $>$ quin $>$ thiophane $>$ dioxane.

The reactions

$$Ni(Et_2Dtc)_2 + 2CuX_2 \rightleftharpoons 2CuX(Et_2Dtc) + \ldots \tag{5a}$$

and

$$Ni(Et_2Dtc)_2 + 2Cu^{2+} \rightleftharpoons 2Cu(Et_2Dtc)^+ + \ldots \tag{5b}$$

were investigated by EPR spectroscopy. As expected, a dependency on solvent properties was found (642). Similar reactions between $Cu(Et_2Dtc)_2$ and CuX_2

or Cu^{2+} were reported to give the same products, which show superhyperfine splittings for X = Cl or Br (640).

The magnetic properties of the $Cu(Et_2 Dtc)_2$ complex have been investigated (292, 615). The magnetic susceptibility of this compound from 4.2 to $56°K$ (615) and magnetization studies from 1.5 to $10°K$ (449) indicate a triplet ground state for the dimeric (53) complex with a singlet—triplet separation ($2J$) of 24 cm^{-1}. A small interdimer antiferromagnetic interaction is observed also and is believed to occur by way of the copper atom in one dimer and the hydrogen atom of an ethyl group of an adjacent dimer at a distance of 2.86 Å.

The quadrupole coupling constant of the $Cu(Et_2 Dtc)_2$ complex doped in the $Ni(Et_2 Dtc)$ lattice was determined from single-crystal EPR data. The small value ($QD \sim 0.7 \times 10^{-4}$ cm^{-1}) was interpreted to imply an effective spherical symmetry, which was attributed mainly to the large covalent character of the Cu—S σ bond (619). Similar studies were reported for $Cu(Et_2 Dtc)_2$ in the $Zn(Et_2 Dtc)_2$ host lattice (382, 385). The value of QD (3.15 x 10^{-4} cm^{-1}) for the five-coordinate Cu(II) ion is in line with the previous study.

2. Group II

a. Zinc, Cadmium, Mercury. *(1) Dithiocarbamates.* The reactions and structures of $R_2 Dtc$ complexes of divalent zinc, cadmium, and mercury have been explored to a considerable extent. Oxidation of the $M(R_2 Dtc)_2$ [M = Zn(II), Cd(II), Hg(II)] complexes is centered on the ligands which form thiuram disulfides. With halogens as oxidants, the $X_2 M(R_4 Tds)$ complexes of the divalent elements are obtained (74, 77). The crystal structure of the HgI_2-($Me_4 Tds$) complex shows a four-coordinate Hg(II) ion with a thiuram disulfide bidentate ligand. A severe distortion from tetrahedral symmetry that brings the mercury atom very close to the plane of a triangle defined by the two I$^-$ and one of the sulfur atoms is observed. This distortion is such that the coordination polyhedron can best be described as a trigonal pyramid (32).

The PMR spectra of these complexes show that the hindered rotation about the partial $S_2 C–NR_2$ double bond in the thiuram disulfide ligands is affected by coordination to Zn(II), Cd(II), or Hg(II). Thus, while coalescence of the methyl proton resonances in $Me_4 Tds$ and $Bu_4 Tds$ occurs at 15 and $26°C$, respectively, coalescence of the same resonances in the Zn, Cd, and Hg complexes is seen at higher temperatures (ca. >30°C) (74). The increase in the rotational energy barrier that presumably arises from an increasing double-bond character of the $R_2 N{\doteq}C$ bond in the ligands also is reflected in the $C{\doteq}N$ stretching vibrations that occur at higher frequencies (78). It should be pointed out that the $M(X)_2 R_4 Tds$ complexes (M = Zn, X = Cl and M = Cd, X = Cl, Br) could not be obtained.

A series of interesting reactions of the $ZnI_2 (Me_4 Tds)$ complex have been reported by McCleverty and Morrison (440, 442). Reactions of this complex with 1, 2, or 3 equiv of PPh_3 give $ZnI_2S(SCNMe_2)_2$, $[PPh_3(CNMe_2)S]^+[ZnI_2(Me_2Dtc)]$, and $[PPh_3(CNMe_2)S]^+[ZnI_3PPh_3]^-$, respectively.

The reaction of $Zn(R_2Dtc)_2$ with R_2Dtc^- gives the tris ionic $Zn(R_2Dtc)_3^-$ complexes (R_2 = Me_2, MePh). By means of similar reactions the mixed $[Zn(R_2Dtc)_2(R_2'Dtc)]^-$ anions also were obtained, as was the very interesting $[Zn(O_2CR')(R_2Dtc)_2]^-$ complex anion. A single-crystal x-ray structure determination of $[NEt_4][Zn(Me_2Dtc)_3]$ shows two of the Me_2Dtc ligands asymmetrically bonded and one of the ligands bonded symmetrically as a bidentate chelate (17). The bonding has been described in terms of a distorted tetrahedral ZnS_4 unit, with the remaining sulfur atoms occupying positions over two faces of the ZnS_4 tetrahedron.

The $Hg(Et_2Dtc)_2$ complex can be obtained in two forms, one of which (β form) has been structurally elucidated by two independent x-ray studies (306, 358) (Table XI).

The mercury atom is surrounded by four sulfur donor atoms from two centrosymmetrically related Et_2Dtc ligands. A linear coordination is found for the Hg(II) ion with two short Hg–S bonds at 2.397(6) Å. The remaining sulfurs are found at 2.990(7) Å. As in the structure of the $As(Et_2Dtc)_3$ complex (124), the two M–S bond lengths are different for each of the two ligands and the C–S distance shows an inverse relation to the M–S distance.

The structure of the α form of $Hg(Et_2Dtc)_2$ is similar (358) to that of the dimeric Zn(II) and Cd(II) (153) analogues.

Polynuclear Hg(II) complexes are obtained according to the reaction:

$$2Et_2Dtc^- + mHgCl_2 \xrightarrow{H_2O} Hg(Et_2Dtc)_2 \cdot (HgCl_2)_{m-1} \qquad (6)$$

where the value of m can be as high as 5 (391). The interesting trinuclear $Hg_3Cl_2(Et_2Dtc)_4$ complex also has been reported (357). The propensity of Hg toward sulfur donors and the availability of lone pairs on the coordinated S atoms in the $Hg(R_2Dtc)_2$ complexes can explain the formation of these polynuclear complexes. The scope of these interactions in the synthesis of heteronuclear aggregates perhaps should be explored with other "soft" metal halides.

The reaction of Et_4Tds with MeHgCl gives both $Hg(Et_2Dtc)_2$ and $MeHg(Et_2Dtc)$. The structure of the latter has been determined (122). An asymmetrically bound Et_2Dtc ligand similar to the ligand in the β form of the $Hg(Et_2Dtc)_2$ complex is found in this molecule (306, 358). The coordination about the two-coordinate Hg(II) is slightly bent with a C–Hg–S angle of $171.2(8)°$.

Formation of the $Hg(R_2Dtc)_2$ complex is reported (60) to occur at the dropping mercury electrode in acetone according to reaction 7.

$$2Hg^0 + 2R_2Dtc^- \rightleftharpoons 2(HgR_2Dtc) + 2e$$

$$2(HgR_2Dtc) \overset{rapid}{\rightleftharpoons} Hg(R_2Dtc)_2 + Hg^0 \qquad (7)$$

TABLE XI
Structural Details of the Group II 1,1-Dithio Complexes

Complex	$\overline{M-S}$, Å	$\overline{S-M-S}$, °	Coordination geometry[a]	Ref.
PyZn(EtXant)$_2$	2.748 (3)[b] 2.294 (3)[c]	152.3 (1)[d]	A	544
PyZn(Et$_2$Dtc)$_2$	2.605 (10) 2.327 (4)	127 (3)[d]	B	255
Zn(PhDta)$_2$	2.347 (33)	76.6 (7)[e] 128 (4)[f]	C	49
Zn(Me$_2$Dtc)$_3$(Et$_4$N)	2.44[g] 2.31[h]		C	17
Cd(Et$_2$Dtc)$_2$	2.570[i] 2.536[j] 2.800[j]		D	203
Cd(EtXant)$_3$(Et$_4$N)	2.667 (32)[k] 2.508 (4)[l]	66.3[e] 113.9 (35)[m]	E	341
Cd(EtXant)$_2$-o-phen[n]	2.647 (3) 2.727 (3)	67.1 (1)[e]	F	535
α-Hg(Et$_2$Dtc)$_2$	2.520 (6); 2.663 (6) 2.418 (7); 2.698 (6)		E	358
β-Hg(Et$_2$Dtc)$_2$	2.397 (6); [2.398 (4)] 2.990 (7); [2.965 (4)]	66.4[e]	G	306 (358)
MeHg(Et$_2$Dtc)	2.418 (7)	67.1 (2)	G	122
I$_2$Hg$_2$(Et$_2$Dtc)$_2$[o]	2.644 (4) 3.422 (4) 3.042 (4)		H	121
HgI$_2$(Me$_2$Dtc)[p]	2.651 (7) 2.882 (7)		H	32

[a] A = distorted trigonal bipyramidal, B = irregular five-coordinate, C = distorted tetrahedral, D = intermediate between trigonal bipyramidal and square pyramidal, E = distorted tetragonal pyramidal, F = pseudooctahedral, G = linear, H = highly distorted tetrahedral.
[b] 'Axial' bonds.
[c] Equatorial bonds, Zn - N - 2.03 (1) Å
[d] $S_{ax}-M-S_{ax}$ angle.
[e] Intraligand sulfur angles.
[f] Interligand sulfur angles.
[g] Bidentate ligand sulfurs.
[h] Monodentate ligand sulfurs.
[i] Nonbridging ligand sulfurs.
[j] Bridging ligand sulfurs.
[k] 'Basal' Cd—S bonds.
[l] 'Axial' Cd—S bonds.
[m] 'Equatorial' angles.
[n] Monomeric molecule with crystallographic C$_2$ symmetry, Cd—N, 2.386 (8) Å.
[o] Hg—I distance 2.641 (1) Å.
[p] $\overline{Hg-I}$ distance, 2.657 (5) Å.

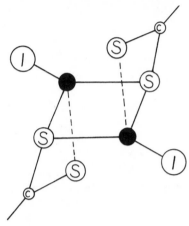

Fig. 14.   Schematic structure of the $[IHgEt_2Dtc]_2$ complex (121)

The following stability constants for the $Hg(II)(R_2Dtc)_x$ complexes have been determined from potentiometric studies in water (387):

$$Hg^{2+} + Et_2Dtc^- \;\rightleftharpoons\; Hg(Et_2Dtc)^+ \qquad K_1 = 2.1 \times 10^{22}$$

$$Hg(Et_2Dtc)^+ + Et_2Dtc^- \;\rightleftharpoons\; Hg(Et_2Dtc)_2 \qquad K_2 = 5.7 \times 10^{15} \qquad (8)$$

$$Hg(Et_2Dtc)_2 + Et_2Dtc^- \;\rightleftharpoons\; Hg(Et_2Dtc)_3^- \qquad K_3 = 1.7 \times 10^1$$

The reaction of HgI_2 with Et_4Tds has been reported (121) to give, in addition to $I_2Hg(Et_4Tds)$, the very interesting dimeric $[IHgEt_2Dtc]_2$ complex. The crystal structure of this molecule revealed the two Hg(II) ions bridged by unidentate Et_2Dtc ligands (Fig. 14) (Table XI).

Weak interactions of the mercury atoms with the nonbonded S atoms of the Et_2Dtc ligands (3.042 Å) define an eight-membered ring.

(2) Xanthates and Dithioacid Complexes.   Various facets of the chemistry of the $Zn(RXant)_2$ complexes have been explored (153). In recent years the interactions of these complexes with bases such as Py, DMSO, and *o*-phen have been studied to a considerable extent (331, 332, 346). The infrared spectra also have been reported (625).

The crystal structure of the $pyZn(EtXant)_2$ complex has been determined (544a). The coordination around the Zn(II) can be described as a distorted trigonal bipyramid. The zinc atom lies in the trigonal plane defined by the pyridine nitrogen and a pair of sulfur atoms. The remaining two sulfur atoms occupy the axial sites at a considerably longer $\overline{Zn-S}$ distance [2.748(3) Å] (Table XI). The analogous $pyZn(Et_2Dtc)_2$ complex shows (255) a very irregular

$NZnS_4$ core that could be envisioned as arising from the core found in the pyZn(EtXant)$_2$ complex by bending and shortening the two "axial" Zn–S bonds (Table XI).

A very interesting structure for the ZnS_4 unit is found in the Zn(PhDta)$_2$ complex. This molecule is four-coordinate and monomeric and shows a high degree of distortion from ideal tetrahedral symmetry imposed by the small "bite" size of the ligand (Table XI). The unusual aspect in this structure is that it remains a monomeric species and maintains chelation of the ligands in spite of the obvious strains evident in the ZnS_4 core (49). The structure of the Cd(EtXant)$_2$ complex has been determined (353).

The structure of the [Et$_4$N] Cd[EtXant]$_3$ complex (346) has been determined (341). The anion is five-coordinate, with two EtXant ligands functioning as bidentate chelates and the third bonded to the Cd(II) center through one of its sulfur atoms. A tetragonal pyramidal coordination is seen for the CdS$_5$ core, with the Cd(II) ion above the square base by 1.079(1) Å (Table XI).

A pseudooctahedral, six-coordinate Cd(II) is found in the structure of the (o-phen)Cd(EtXant)$_2$ complex (535) (Table XI).

It has been pointed out (293) that asymmetric M–S bonds in xanthate complexes are observed when the central element has fewer vacant valence orbitals of low energy than the number of electron pairs offered by the sulfur atoms. An examination of the structures of the pyZn(EtXant)$_2$ and o-phen Cd(EtXant)$_2$ complexes shows asymmetrically bound EtXant ligands. It has been suggested that the introduction of the Lewis bases generates a surplus of ligand electron pairs. A consequence of this is the contribution of more than one pair of ligand electrons to the same metal coordination site, which results in longer M–L bonds and asymmetric ligand binding (294).

Evidence for the existence of the Hg(RXant)$_3^-$ complex is reported in a polarographic study of the RXant$^-$ anions. The results of this study (56) indicate the formation of Hg(RXant)$_2$ by means of the following reaction sequence :

$$2Hg + 6RXant^- \rightleftharpoons 2Hg(RXant)_3^- + 4e$$

$$2Hg(RXant)_3^- + Hg^0 \rightleftharpoons 3Hg(RXant)_2 + 2e$$

$$(9)$$

3. Group III and the Lanthanides

To our knowledge dithio complexes with either scandium or yttrium still have not been prepared. Since 1968 two reports have appeared that deal with the lanthanide dithiocarbamate complexes.

A detailed procedure for the synthesis of pure crystalline M(Et$_2$Dtc)$_3$ complexes (M = La to Lu inclusive, except Pm) has been described (87). Anhydrous MBr_3 and the stoichiometric amount of $NaEt_2Dtc$ in anhydrous EtOH afford the crude complexes, which can be recrystallized from $CH_3CN/ether$ mixtures. X-ray powder diffraction results indicate a series of isomorphous complexes from La to Nd inclusive. Another series of isomorphous $M(Et_2Dtc)_3$ complexes were observed from Sm to Lu inclusive.

Reaction of the $M(Et_2Dtc)_3$ complexes with $NaEt_2Dtc$ in the presence of NEt_4Br in anhydrous alcohol, in 1:1:1 molar ratios, affords the crystalline $(Et_4N)M(Et_2Dtc)_4$ complexes.

In a recent paper Hill et al. reported on a study of the temperature-dependent PMR spectra of the $(Et_4N)M(Et_2Dtc)_4$ complexes (M = Pr, Nd, Tb, Dy, Ho, Er, Tm, and Yb). The temperature dependence of the paramagnetic shifts was interpreted in terms of both a T^{-1} and a T^{-2} contribution (326).

4. Group IV

a. Titanium, Zirconium, Hafnium. With a few exceptions, the synthetic and structural chemistry of the Group IV 1,1-dithio complexes is limited to the R_2Dtc ligands. The hafnium chemistry with 1,1-dithio ligands is still unknown.

The crystal structure of $Ti(Et_2Dtc)_4$ has been determined (128). Two independent molecules are found in the asymmetric unit and both contain eight-coordinate Ti(IV) and chelating Et_2Dtc ligands. The coordination geometry of the TiS_8 core in both molecules is very close to dodecahedral. Chelation is observed along the *m* edges [Hoard and Silverton notation (327)] of the dodecahedron (Fig. 15).

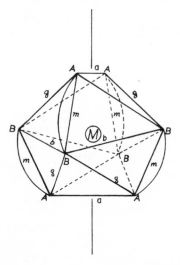

Fig. 15.   The idealized dodecahedron. The notation is after Hoard and Silverton (327).

TABLE XII
Selected Structural Parameters of the Group IV and Actinide Complexes.

Complex	$\overline{M-S}$, A	$\overline{C-N}$, A	S–S(bite), A	Geometry	Ref.
Ti(Et$_2$Dtc)$_4$	2.605 (11)^a	1.34 (3)	2.84 (3)	Dodecahedral^b	128
	2.521 (13)^c				
Ti(Me$_2$Dtc)$_3$Cl	2.477 (3)^d	1.32 (1)	2.898 (4)^e	Pentagonal bipyramidal	35, 404
	2.519 (39)^f		2.840 (6)^g		
(η^5–Cp)Zr(Me$_2$Dtc)$_3^h$	2.681 (1)^d	1.334 (7)	2.910 (2)^e	Pentagonal bipyramidal	89
	2.692 (25)^f		2.882 (5)^g		
Th(Et$_2$Dtc)$_4$	2.87 (2)	1.31 (6)		Distorted dodecahedral	85, 86
[Np(Et$_2$Dtc)$_4$] $^-$Et$_4$N$^+$	2.86 (3)			Distorted dodecahedral	85, 87
UO$_2$(MeDta)$_2$Ph$_3$PO^i	2.85 (1)^j			Pentagonal bipyramidal	41b
UO$_2$(Et$_2$Dtc)$_2$Ph$_3$PO	2.85 (1)			Pentagonal bipyramidal	282
UO$_2$(Et$_2$Dtc)$_2$Ph$_3$AsO	2.85 (1)			Pentagonal bipyramidal	282

^{a} Mean value for the Ti–S$_A$ bond lengths in both molecules of the asymmetric unit.
^{b} a edges, 2.999 (36) Å; b edges, 3.64 (15) Å; g edges, 3.30 (7) Å.
^{c} Mean value for the Ti–S$_B$ bond lengths in both molecules of the asymmetric unit.
^{d} Axial Ti–S bond.
^{e} Axial–equatorial ligand.
^{f} Mean value of the five equatorial Ti–S bond lengths.
^{g} Equatorial ligands.
^{h} C$_6$H$_5$Cl solvate.
^{i} U–O = 1.66 (2) Å (uranyl group).
^{j} U–O (PPh$_3$O), 2.34 (1) Å.

The dihedral angles between the two interlocking trapezoids in the two independent units are very close to the 90° required for D_{2d} symmetry (89.8, 87.8°). Slightly shorter Ti–S$_B$ bond lengths are observed compared to the Ti–S$_A$ bond lengths (Table XII) (Fig. 15).

Seven-coordinate Ti(IV) complexes of the type XTi(R$_2$Dtc)$_3$ were obtained (R = Me, i-Pr, i-Bu; X = Cl and R = Me, Et; X = Br) by the reaction of TiX$_4$ with NaR$_2$Dtc in CH$_2$Cl$_2$ (11a, 35). The compounds were found to be monomeric in benzene and nonconducting in CH$_3$NO$_2$. The nonrigidity of these molecules is apparent in the PMR spectra, which show equivalent alkyl protons even at −80°C. The IR spectra show C–S bands consistent with bidentate R$_2$Dtc groups. The structure of the ClTi(Me$_2$Dtc)$_3$ has been determined and establishes the monomeric nature of this complex and a seven-coordinate distorted pentagonal bipyramidal ClTiS$_6$ moiety (35, 404a). The chlorine atom occupies one axial position [Ti–Cl = 2.305(3) Å]. The remaining coordination sites are occupied by two bidentate Me$_2$Dtc ligands in the pentagonal plane and one ligand spanning both an equatorial and an axial site (Fig. 16).

The structure shows a distortion that can be ascribed to the short S–S intraligand distance. Although the S–S bite of the ligand that spans the

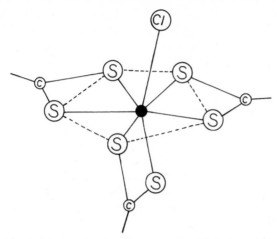

Fig. 16.   Schematic structure of the CITi(Me$_2$Dtc)$_3$ complex (404a).

equatorial and axial sites is significantly larger than the bites of the equatorially positioned ligands, it cannot span the two sites without strain. The result is a Cl–Ti–S$_{ax}$ angle of 165.03(12)° and an appreciable deviation of the TiS$_5$ unit from planarity. The very short interligand S–S distances indicate a crowded TiS$_5$ unit and may also account for the observed distortion.

Under the appropriate stoichiometric conditions, six-coordinate Ti(IV) R$_2$Dtc complexes of the type Cl$_2$Ti(R$_2$Dtc)$_2$ can be obtained (11a, 35) by means of a reaction similar to the one employed for the synthesis of the XTi(R$_2$Dtc)$_3$ complexes. The dipole moments of these complexes (~9 D) indicate a *cis* arrangement of the Cl atoms.

Variable-temperature PMR studies on the stereochemically nonrigid, (Cl$_{4-n}$)Ti(R$_2$Dtc)$_n$ (n = 2, 3, or 4) complexes show the metal-centered rearrangements to be fast on the PMR time scale at temperatures higher than −90°. Hindered rotation about the C–N bonds was observed for R = i-Pr, and activation parameters were determined for this process.

By means of a variable-temperature ^{1}H and ^{13}C NMR study, Muetterties has shown (468) that in the M(R$_2$Dtc)$_4$ complexes (M = Ti, Zr, Nb) there is no evidence of nonequivalence in the alkyl groups of the symmetrical ligands from +30 to −160°.

Similar studies on the Zr(RC$_6$F$_5$Dtc)$_4$ complexes did not produce evidence of stereoisomers from +30 to −130° (469).

The (η^5-Cp)Zr(Me$_2$Dtc)$_3$ complex was obtained by the reaction of (η^5-Cp)$_2$ZrCl$_2$ with NaMe$_2$Dtc in dry CH$_2$Cl$_2$ under N$_2$ (89). Evidence for stereochemical rigidity (four Me resonances of relative intensity 2:1:2:1 at 37°C), and the obvious contrast with the stereochemically nonrigid CITi(Me$_2$Dtc)$_3$ complex prompted Bruder et al. (89) to undertake a structural determination of this molecule. The pentagonal bipyramidal arrangement (consistent with the PMR spectra) was found with the η^5-Cp group occupying one axial site. The other positions are occupied as described for the Ti(Me$_2$Dtc)$_3$Cl complex (see above). An umbrella bending of the five equatorial sulfur atoms away from the η^5-Cp ring is observed with the Zr(IV) ion situated 0.60 Å above the equatorial plane in the direction of the η^5-Cp ring. In the Zr structure, the axial Zr–S bond length is not significantly shorter than the equatorial bonds, contrary to the situation in the structure of the Ti(Me$_2$Dtc)$_3$Cl complex.

b. Uranium and the Actinides.   Following the successful synthesis of the quadrivalent M(Et$_2$Dtc)$_4$ (M = Th, U, Np, and Pu) actinides (21), Brown and Holah reported on the synthesis of the trivalent M(Et$_2$Dtc)$_3$ (M = Np, Pu) and [M(Et$_2$Dtc)$_4$]$^-$ (M = Np, Pu) complexes (84). The IR spectra of these complexes indicate bidentate R$_2$Dtc ligands.

The crystal structures of the Th(Et$_2$Dtc)$_4$ and (Et$_4$N)Np(Et$_2$Dtc)$_4$ complexes have been determined (85, 86). The thorium atom is eight-coordinated by four bidentate Et$_2$Dtc ligands in a geometry that has been described as intermediate between the ideal dodecahedral and square antiprismatic arrangements (86).

A large distortion from dodecahedral symmetry also is observed in the structure of the (Et$_4$N)Np(Et$_2$Dtc)$_4$ complex. Here, however, the distortion does not tend towards the alternative square antiprismatic coordination. Instead the arrangement of the sulfur atoms in the NpS$_8$ planar fragment has one S atom above and two S atoms below the plane (87).

The synthesis of UO$_2$(Et$_2$Dtc)$_2$Ph$_3$XO (X = P, As) from K[UO$_2$(Et$_2$Dtc)$_3$]·H$_2$O and Ph$_3$XO has been described. The brilliant red crystals of this complex show the P–O and As–O vibrations at 1130, 1117, and 880 cm^{-1}, respectively (282). The structures for X = P and As have been determined. The U atom is seven-coordinate with a pentagonal bipyramidal arrangement of the ligand atoms. A linear UO$_2$ group is nearly perpendicular to the plane containing the four sulfur atoms and the oxygen atom of the Ph$_3$XO ligand in an irregular pentagon (282) (Table XII). The synthesis of the corresponding complexes with MeDta and PhDta ligands has been described (416) from the reaction between uranyl nitrate hexahydrate and the appropriate dithioacid in methanol under basic conditions and followed by the addition of Ph$_3$XO. The crystal structure of the UO$_2$(MeDta)$_2$Ph$_3$PO is very similar to that of the corresponding dithiocarbamate complexes (416).

5. Group V

Numerous synthetic and crystallographic studies of the Group V 1,1-dithio complexes have been reported in the literature since 1970.

a. **Vanadium.**   A series of V(III), V(IV), and V(V) complexes were reported with substituted pyrazolinedithiocarbamate ligands, $R-C_4H_4N_2S_2^-$ (R = 3-phenyl, 5-phenyl, or 3,5-diphenyl) (98). The stoichiometries of these complexes were $VO(R_2Dtc) \cdot H_2O$, $VO(R_2Dtc)_2$, and $VO_2(R_2Dtc)$ containing V(III), V(IV), and V(V), respectively. Only the 3,5-diphenylpyrazoline dithiocarbamate gives the $VO_2(R_2Dtc)$ complex.

The EPR spectra of the V(IV) complexes have been recorded, as have the IR spectra of all the complexes. The V–O vibration is found between 1020 and 1050 cm^{-1}.

A series of $V(R_2Dtc)_3$ complexes were reported recently (372). These compounds were obtained by the direct reaction between VCl_3 and anhydrous NaR_2Dtc ligands (R_2 = Morpho, Pip, Pyrrol, Me$_2$). The $V(R_2Dtc)_2Cl$ complexes (R_2 = Morpho, Pyrrol) also were obtained. The electronic and infrared spectra of the complexes were obtained and analyzed. The synthesis of $V(Et_2Dtc)_3$ also is accomplished by CS_2 insertion into the V–N bond in $V[NEt_2]_4$ (70).

A considerable amount of research has been devoted to the chemistry of V(IV) 1,1-dithio complexes. The 1:1 base adducts of the $VO(R_2Dtc)_2$ complexes (445, 446, 610) have been isolated and characterized (R_2 = Me$_2$, Pyrrol with pyridine and 4-methylpyridine, respectively (444).

Characteristic perturbations of the C$\doteq$N and V=O stretching vibrations were detected following base-adduct formation. These perturbations (bathochromic shifts of the V=O stretch and C$\doteq$N vibrations) also were detected in DMSO solutions of the $VO(R_2Dtc)_2$ complexes.

The EPR spectrum of the $VO(Et_2Dtc)_2$ complex in toluene solution at room temperature and at 77°K shows the typical eight-line spectrum [$g_\| = 1.956(3)$, $g_\perp = 1.982(3)$, $A = 160(2)$ Oe, $B = 55(2)$ Oe]. The coefficients in the molecular orbitals of the $VO(Et_2Dtc)_2$ complex were calculated from the parameters of the EPR spectra and show a high degree of covalency in the V-ligand σ bond (402).

The reaction of vanadyl salts with dithiocarboxylate ligands occurs with cleavage of the V–O bond and gives the $V(RDta)_4$ complexes (518), R = Ph, p-Me–Ph, Me, Bz. The magnetic moments are in the expected range (1.7–1.8 BM) for a $3d^1$ system and are almost temperature independent. These complexes are monomeric in C_6H_6 and CH_2Cl_2 solutions and show IR spectra characteristics of only bidentate ligands. The solution EPR spectra (CH_2Cl_2 or C_6H_6) show the typical eight-line signals [$g_\| = 1.969, g_\perp = 1.977$ for $V(MeDta)_4$ and $g_\| = 1.966$, $g_\perp = 1.981$ for $V(p$-Me–PhDta)$_4$]. On the basis of the EPR spectra, a dodecahedral (D_{2d} symmetry) coordination was proposed for these complexes (517, 518).

The crystal structure of the $V(BzDta)_4$ complex verified the proposed coordination geometry (45). The ligand bite, as in the structure of the $Th(Et_2Dtc)_4$ complex (86), coincides with the m edge of the dodecahedron.

A similar situation prevails with the $Ti(Et_2Dtc)_4$ (127) and $Np(Et_2Dtc)_4^-$ (87) complexes where the ligand bite also is coincident with the m edge of the dodecahedron (Fig. 15) (581c).

At a later date the structure of the $V(MeDta)_4$ also was determined (242). Two independent molecules in the asymmetric unit show geometries consistent with the two different stereoisomers belonging to the dodecahedral subclasses $I_d(D_{2d} - \overline{42}_m)$ and $V_d(C_2 - 2)$.

Here again the ligand bite is coincident with the m dodecahedral edges in one of the two units. In the second unit chelation is found along two m edges and two g edges (Fig. 15).

The first xanthates of V(IV) to be isolated were reported by Casey and Thackeray (106). They obtained the $[(\eta^5\text{-}Cp)_2V(RXant)]^+$ complexes as BF_4^- "salts" by means of the reaction of $(\eta^5\text{-}Cp)_2VCl_2$ with NaRXant in aqueous media (R = Me, Et, i-Pr, n-Bu, and Pip). The magnetic moments (1.65 BM, room temp.) and magnetic susceptibilities follow the Curie law. The EPR, IR, and electronic spectra for these molecules are tabulated and discussed (106).

The synthesis of the corresponding R_2Dtc complexes also has been reported (107) and their electrochemical properties have been investigated (57). Two one-electron polarographic reduction waves between +0.75 and −2.2 V were observed on a DME in acetone (versus Ag|AgCl), the first of which was reversible. The following sequence of electrode reactions was suggested (57).

$$(Cp)_2V(IV)(R_2Dtc)^+ + e \; \rightleftharpoons \; (Cp)_2V(III)(R_2Dtc)$$

$$(Cp)_2V(R_2Dtc) \; \overset{kf}{\rightleftharpoons} \; (Cp)_2V^+ + R_2Dtc^-$$

$$R_2Dtc^- + Hg^0 \; \rightleftharpoons \; Hg(R_2Dtc) + e^-$$

$$2Hg(R_2Dtc) \; \rightleftharpoons \; Hg(II)(R_2Dtc)_2 + Hg^0$$

(10)

A similar sequence of electrode reactions is proposed for the corresponding xanthate complexes (58, 59).

The electrochemistry of $V(Et_2Dtc)_3$, $VO(Et_2Dtc)_2$, and $VO(Et_2Dtc)_3$ has been studied by cyclic voltametry (555). One-electron reversible reduction is found for the $V(Et_2Dtc)_3$ complex (-1.075V versus SCE) and one-electron irreversible oxidation (+0.35 V) also has been found with the formation of both V(IV) and V(V) products. The $VO(Et_2Dtc)_3$ is reduced to $VO(Et_2Dtc)_2$ at −0.90 V. The product is further reduced at −1.35 V to an electroinactive V(III) species (555).

The structure of the $V(PhDta)_4$ complex (50) is very similar to that of the corresponding $V(BzDta)_4$ (45) complex, and a dodecahedral coordination geometry with chelation along the m edges of the dodecahedron was reported.

The crystal structure of the $VO(Et_2Dtc)_2$ complex (319) shows a five-

TABLE XIII

Selected Structural Parameters of the Group V Transition Element 1,1-Dithio Complexes.

Complex	V=O, Å	M–S, Å	g edge, Å	m edge, Å	Geometry	Ref.
V(MeDta)$_4$		2.50 (2)^a 2.46 (2)^b	3.15	2.75	Dodecahedral	242
V(BzDta)$_4$		2.524 (7)^a 2.470 (7)^b	3.197 (10)	2.752 (10)	Dodecahedral	45
V(PhDta)$_4$		2.56 (2)^a 2.45 (1)^b	3.15 (4)	2.79 (2)	Dodecahedral	50
VO(Et$_2$Dtc)$_2$	1.591 (4)	2.401 (10)			Rectangular pyramidal	319
VO(Et$_2$Dtc)$_3$	1.65 (2)	2.478 (20)^c 2.629 (10)^d			Distorted pentagonal bipyramidal^e	193
NbO(Et$_2$Dtc)$_3$	1.74 (1)	2.574 (22)^c 2.753 (4)^d			Distorted pentagonal bipyramidal	193

^{a} Mean value of the V–S bond.
^{b} Mean value of the V–S$_B$ bond.
^{c} Equatorial $\overline{V–S}$ bonds.
^{d} Axial $\overline{V–S}$ bond.
^{e} O–V–S$_{ax}$ = 166.2 (8)°.

coordinate rectangular pyramidal vanadium coordination sphere. The oxygen is at the apex of the pyramid [V–O, 1.591(4) Å] and the vanadium atom is 0.75 Å above the basal plane toward the oxygen atom (Table XIII).

The EPR spectrum of VO(Et$_2$Dtc)$_2$, enriched with ^{13}C at the CS$_2$ group of the E$_2$Dtc ligand, shows ^{13}C superhyperfine interactions.

The isotropic g value, hyperfine coupling constant, and ^{13}C super-hyperfine coupling constant are found to be 1.981, 88 G, and 6.6 G, respectively.

The ^{13}C superhyperfine splitting is interpreted in terms of a *trans*-annular interaction involving the vanadium $d_{x^2-y^2}$ orbital and the carbon 2s orbital (581b) (586).

The synthesis and characterization of V(V) complexes of the type VO(R$_2$Dtc)$_3$ has been accomplished by the reaction of VO(SO$_4$) with NaR$_2$Dtc in aqueous media and in the presence of H$_2$O$_2$ (R$_2$ = Me$_2$, Et$_2$, Pyrrol) (105). The infrared spectra of these molecules have been obtained and discussed. The V=O stretching frequency occurs at 950 cm^{-1}. The crystal structure of the VO(Et$_2$Dtc)$_3$ has been determined (193). A pentagonal bipyramidal VOS$_6$ core shows two equatorially bound Et$_2$Dtc ligands and an axially–equatorially bound ligand. Some of the distortions of the pyramid seem to arise from the short bite of the axially–equatorially bound ligand. This ligand shows the largest S–S bite

[2.894(13) Å] compared to the other two ligands [S—S; 2.791(19)]. However, it appears that this unique ligand cannot span the sites, as shown by the bent O—V—S_{ax} group [166.2(8)°].

b. Niobium, Tantalum. 1,1-dithio complexes of quadrivalent and pentavalent Nb and pentavalent Ta are known. The $NbO(R_2Dtc)_3$ complexes are obtained by the reaction of $NbCl_5$ and $Na(R_2Dtc)$ in anhydrous methanol (69). The crystal structure of the ethyl derivative has been determined and is very similar to that of the corresponding V(V) analogue (193) (see above and Table XIII). The O—Nb—S_{ax} angle is 161.3°.

The synthesis of the dithiocarbamates of Nb(IV) and Nb(V) and of Ta(V) is accomplished by reactions of the $Nb(NR_2)_5$ and $Ta(NR_2)_5$ complexes with CS_2 (69), which produce the $M(R_2Dtc)_5$ complexes (R = Me, M = Nb, Ta). Subsequent reduction of the $Nb(R_2Dtc)_5$ results in the formation of $Nb(R_2Dtc)_4$. Reactions of $TaCl_5$ (496, 557), $NbCl_4$ (412, 577), and $NbCl_5$ (496) with NaR_2Dtc salts under anhydrous conditions also produce $Ta(R_2Dtc)_5$, $Nb(R_2Dtc)_4$, and $Nb(R_2Dtc)_4X$, respectively. Reactions of Nb(V) and Ta(V) chlorides and bromides with NaR_2Dtc in alcohols produce mixed-metal alkoxide—R_2Dtc complexes (496).

The $M(Et_2Dtc)_4X$ complexes (M = Nb, Ta; X = Cl, Br) and $Ta(Et_2Dtc)_4I$ react with MX_5 in CH_2Cl_2 or C_6H_6 to produce the new $M(Et_2Dtc)_2X_3$ complexes. The infrared spectra have been recorded and band assignments have been made. The reactions of the $M(Et_2Dtc)_2X_3$ complexes with $NaEt_2Dtc$ in a 1 : 2 molar ratio in benzene afford the $M(Et_2Dtc)_4X$ complexes and the very interesting $M(Et_2Dtc)_3S$ complexes. The $C \overset{..}{-} N$ absorptions are observed in the IR spectra of the $M(Et_2Dtc)_3S$ complexes (307). The nature of the "extra" sulfur in these molecules is not entirely clear. On the basis of the lack of a metal-X stretching frequency it is proposed that the $M(Et_2Dtc)_4X$ complexes are probably the ionic $M(Et_2Dtc)_4^+X^-$ species (307). The crystal structure of the $Ta(Me_2Dtc)_4Cl \cdot CH_2Cl_2$ complex shows the $Ta(Me_2Dtc)_4^+$ cation with the ligands spanning the m edges of an idealized dodecahedron (404b). Molecular-weight determinations on the $M(Et_2Dtc)_2Cl_3$ complexes in CH_2Cl_2 suggest the formulation $[M(Et_2Dtc)_2Cl_3]_n$, where $n = 1.5$. The observed molecular weights were nearly constant over the limited concentration range (1×10^{-2} to 4×10^{-2} M) (308). The synthesis of the interesting $Nb(Et_2Dtc)_3Cl_2$ also has been reported (308). This molecule appears to be a 1 : 1 electrolyte, possibly of the type $Nb(Et_2Dtc)_3Cl^+Cl^-$.

Conflicting reports have appeared concerning the room-temperature magnetic moments of the $Nb(R_2Dtc)_4$ complexes. Thus values as low as 0.5 BM (69, 412) and as high as 1.57 BM (88, 577) have been reported.

A rather interesting dimeric (412) complex, $Nb_2Br_3(Et_2Dtc)_5$, is obtained upon reaction of $NbBr_4$ and $Na(Et_2Dtc)$ in a 1 : 2 molar ratio. Only bidentate

R_2Dtc ligands are present in this monomeric complex, which shows a field-dependent magnetic susceptibility. On the basis of the equivalent conductance in CH_3NO_2 (which increases at low concentrations), the compound is formulated as $[Nb_2(Et_2Dtc)_5Br_2]Br$ (412). It would not be surprising if either the $Nb(Et_2Dtc)_3Cl_2$ or the $[M(Et_2Dtc)_2Cl_3]_n$ complexes (308) (see above) are somehow related to the $[Nb_2(Et_2Dtc)_5Br_2]Br$ complex. Obviously these complexes should be reinvestigated.

The reaction of NbX_5 and TaX_5 with NaR_2Dtc in methanol gives a large number of crystalline monomeric complexes of the form $XM(OCH_3)_2(R_2Dtc)_2$ (X = Cl, Br, NCS; R = Me, Et, Bz). The nitrogen—M linkage of SCN in the appropriate complexes is proposed on the basis of the IR spectra (496). Preliminary x-ray data indicate trans axial OCH_3 groups in a pentagonal bipyramid containing four sulfurs and the halogen atom in the basal pentagon (496).

6. Group VI

a. Chromium.   Unlike the case for other transition metal ions, since 1969 a relatively limited amount of synthetic work has been carried out with chromium and 1,1-dithio chelates.

Dark brown crystalline perfluoroalkyl derivatives of the type $R_fCr(R_2Dtc)_2$ py have been synthesized for $R_f = C_2F_5$, C_3F_7 and R = Et, for $R_f = C_3F_7$ and R = Me, i-PR, Bz; and for $R_f = C_4F_9$ and R = Et. The synthesis was accomplished by the reaction of a "mixture" of $CrCl_2R_f$ and $CrCl_xI_{3-x}$ in acetonitrile, with NaR_2Dtc ligands (425).

The x-ray crystal structure of the $C_3F_7Cr(Me_2Dtc)_2$ py complex has been determined and shows a cis arrangement of the pyridine and R_f groups. The Cr—S bond trans to the C_3F_7 group is considerably longer (2.457 Å) than the mean of the other Cr-S bonds (2.392 Å).

The structures of the benzene (96) and CH_2Cl_2 (94, 302) solvates of the $Cr(MorphoDtc)_3$ complex and of $Cr(Et_2Dtc)_3$ (542a) have been determined (Table XIV).

The coordination spheres for these solvates are similar and show a trigonal distortion from octahedral symmetry. Such a distortion is dictated by the small bite of the R_2Dtc ligand and ligand—ligand repulsions (388). The Cr—S stretching vibrations occur at 361 and 364 cm^{-1} for the benzene and CH_2Cl_2 solvates respectively (96). Similar structural features are observed in the CrS_6 core of the $Cr(PyrrolDtc)_3$ hemibenzene solvate complex (575) (Table XIV).

Very recently the interesting reaction between NaR_2Dtc (R = Me, Et) and $K_2Cr_2O_7$ was reported (339). The product of this reaction, $Cr(R_2Dtc)_2$-(R_2DtcO), shows new bands in the IR spectrum at 1009 and 488 cm^{-1}, which have been assigned to the S—O and Cr—O vibrations, respectively. A structure determination of this molecule (339) confirmed the conclusion that

TABLE XIV
Selected Structural Parameters of the Cr (III) 1,1-Dithio Complexes.

Complex	M–S, Å	C–N, Å	ν(C–N), cm^{-1}	μ_{eff}, BM	Ref.
$C_3F_7Cr(Me_2Dtc)_2py$	2.457^a 2.392^c	1.38	1480–1590^b	4.02	425
Cr(Morpho Dtc)$_3$^{d}	2.396 (10)	1.32 (1)			96
Cr(Morpho Dtc)$_3$^{e}	2.406 (12)	1.348 (9)			94
					302
Cr(Pyrrol Dtc)$_3$^{f}	2.404 (9)	1.333 (9)		3.80–3.75^g	244
					575
Cr(Et$_2$Dtc)$_2$(Et$_2$DtcO)^h	2.406			3.68	339
Cr(EtXant)$_3$^{i}	2.393 (8)				453
Cr(EtSXant)$_3$	2.398 (8)^{j,k}				613

^{a} Trans to C_3F_7 group.
^{b} Range given for all homologues of the $R_fCr(R_2Dtc)_2$ py series.
^{c} Mean of all other Cr–S bond lengths; no deviation from the mean or individual Cr–S bond lengths are given.
^{d} Dibenzene solvate.
^{e} CH$_2$Cl$_2$ solvate.
^{f} Hemibenzene solvate.
^{g} Values reported for the unsolvated (?) complex (244).
^{h} Cr–O, 2.01 Å; S–O, 1.26 Å.
^{i} S$_2$C–O bond, 1.297 (11) Å; C–OEt bond, 1.471 (11) Å.
^{j} Mean value for 12 independent Cr–S bonds from the two complex molecules in the asymmetric unit.
^{k} S$_2$C–S bond = 1.723 (6) Å; S–C$_2$H$_5$ bond = 1.816 (6) Å.

the oxygen atom was inserted into a Cr–S bond with concomitant ring expansion (Fig. 17). This reaction, which is reminiscent of the sulfer insertion reaction giving 1,1-dithiolates (156), must occur with a different mechanism, since in the 1,1-dithiolates, sulfur is inserted into the C–S bond (20, 232).

Various studies probing the electronic structure of Cr(III) 1,1-dithio

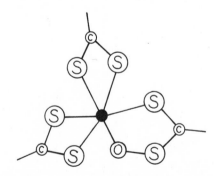

Fig. 17.   Idealized structure of the Cr(Et$_2$Dtc)$_2$(Et$_2$Dtc)O complex (339).

chelates have appeared. The $^2E \leftrightarrow {}^4A_2$ emission and absorption maxima have been measured in solution, single crystals (461), and polymer solvent media (185). Measurements of lifetimes and quantum yields for the Cr(EtXant)$_3$, Cr(Et$_2$Dtc)$_3$, and Cr(Me$_2$Dtc)$_3$ complexes reveal low quantum yields and short lifetimes (461). Assignments of the three lowest energy spin-forbidden transitions in O_h symmetry have been made for the Cr(EtXant)$_3$ and Cr(Me$_2$Dtc)$_3$ complexes. These assignments were derived from single-crystal high-resolution spectral data (462).

Ambient-temperature magnetic circular dichroism (MCD) studies for the low-energy region $(^2T_{2g}, {}^4T_{1g})$ for Cr(Et$_2$Dtc)$_3$ and Cr(EtXant)$_3$ established the ordering: $^2A_1(^2T_{2g}) < {}^2E(^2T_{2g}) < {}^4E(^4T_{1g})$ (301, 567). The low-temperature MCD measurements for the Cr(Et$_2$Dtc)$_3$ complex also have been reported, and energies for the trigonal d–d states have been obtained (300).

The magnetic susceptibility of Cr(PyrrolDtc)$_3$ was measured to 2.5°K. The magnetic moment was found to be virtually independent of temperature from room temperature to 40°K (3.80–3.75 BM). A decrease in the magnetic moment was found from 40 to 2.5°K. At 2.5°K moments ranging from 3.61 to 3.54 BM were found (244).

The temperature dependence of the PMR spectra of some Cr(R$_2$Dtc)$_3$ complexes was studied in CDCl$_3$ over the range −60 to +60°C. Both Δ and Λ enantiomers were detected, and the broad signals in the spectra were ascribed to the lack of an appropriate relaxation mechanism (271).

The EPR spectrum of magnetically dilute Cr(EtXant)$_3$ in an In(EtXant)$_3$ lattice is axially symmetric. The values $g_x = g_y = 1.9985(5)$, $g_z = 1.9947(2)$, $E < 10^{-3}$ cm^{-1}, and $D = 0.169(3)$ cm^{-1} were obtained (390).

The crystal structure of the Cr(EtXant)$_3$ complex has been determined. The Cr(III) is found in a distorted octahedral environment provided by the six sulfur atoms of the bidentate ligands. The rather short S$_2$C–O bond of 1.297(11) Å is thought to be an indication of considerable double-bond character of this bond. The contribution of the resonance form ^{2} $^-$S$_2$ C=ÖR to the structure of the ligand is considered to be ~30%. Concomitant with this short bond is the unusually long bond between the oxygen atom and the first carbon atom of the ethyl group [1.471(11) Å]. (453).

This lengthening of the O–C$_2$H$_5$ bond, which also is found in the structures of the Cd(EtXant)$_2$ o-phen (535) and Fe(EtXant)$_3$ (340, 618b) complexes may account for the ease with which the OR$^-$ group is removed from certain xanthate complexes. The conversion of coordinated xanthate ligands to dithiocarbonates (S$_2$CO^{2-}) has been described (141, 144). The crystal structure of the "Co(EtSXant)$_3$" has been determined (613). As pointed out by Li and Lippard (407) the similarity of the M–S bond lengths in this structure to the Cr–S bond lengths in other complexes suggests strongly that the alleged "Co(EtSXant)$_3$" crystals were, in fact, Cr(EtSXant)$_3$. The results for the Co(EtSXant)$_3$ complex are shown in Table XIV in the column labeled Cr(EtSXant)$_3$. In this structure the CrS$_6$ core' is six-coordinate and distorted octahedral. The relatively short S$_2$C–S bond [1.723(6) Å], when compared to the S–C$_2$H$_5$ bond [1.816(6) Å], underscores the importance of the resonance form $^2{}^-$S$_2$C–̈SR in describing the structure of the ligand. The very long S–C$_2$H$_5$ bond also is consistent with the reactivity of the coordinated thioxanthate ligand and the nucleophilic substitution of the SR$^-$ group by NR$_2^-$ (239, 634) in certain thioxanthate complexes.

The Cr(EtSXant)$_3$ complex (μ_{eff} = 3.94 BM, room temp.) (504) should be an ideal candidate for a study of the nucleophilic substitution reactions on the coordinated EtSXant ligand.

The very interesting [Cr(L)$_2$Cl$_2$]Cl complexes have been described (138). (L = tetraalkyl thiuram monosulfide or disulfide). The infrared and electronic spectra demonstrate convincingly that these molecules cannot be described as oxidative addition products. The thiuram ligands are placed between R$_2$Dtc$^-$ and RXant$^-$ in the spectrochemical series. The electrochemistry of these complexes and, particularly, their reduction characteristics should be examined.

b. Molybdenum.   Interest in the inorganic biochemistry of molybdenum and the implication of Mo–S bonding in the active site of nitrogenase have contributed substantially to intense studies on molybdenum complexes with 1,1-dithio ligands.

(1) Simple Monomeric Molybdenum Complexes.   The first report on the synthesis of simple Mo dithio complexes appeared in 1971 when Mo(IV)(R$_2$Dtc)$_4$ complexes were obtained by CS$_2$ insertion into Mo(NR$_2$)$_4$ complexes (68). Shortly after this report, the synthesis of these complexes from MoCl$_4$ and NaR$_2$Dtc in acetonitrile was reported (88). A variety of tetrakis R$_2$Dtc complexes of Mo(IV) also were obtained by the oxidative decarbonylation of Mo(CO)$_6$ with tetraalkyl thiuram disulfide (481, 482, 609).

Nieuwpoort et al. have obtained Mo(IV)(R$_2$Dtc)$_4$ complexes with R = Me, Et, i-Pr, Bz, and Ph with this method. Careful studies of the magnetic properties of these complexes (481) led the authors to conclude that, contrary to previous reports (88, 609), the pure Mo(IV)(R$_2$Dtc)$_4$ complexes are diamagnetic.

Extensive tabulations of the visible spectra of these compounds and extended Hückel calculations have been reported (485). Assignments of the electronic transitions have been made. Typically the absorption of 35–38 kK is assigned to internal ligand transitions. Six absorptions between 30 and 17 kK are assigned to charge transfer (CT) bands on the basis of their high extinction coefficients. Absorptions of low intensity at 10.8 and 14.5 kK are assigned to d–d transitions (485).

Electrochemical studies on the Mo(R$_2$Dtc)$_4$ complexes revealed the electrochemical behavior shown in Eq. 11. Substituent effects on the $E_{1/2}$ values for the

$$[Mo(III)(R_2Dtc)_4]^{1-} \xleftarrow{\quad E_{1/2}(0/-1) \quad} [Mo(IV)(R_2Dtc)_4]^0 \xrightleftharpoons{\quad E_{1/2}(0/1) \quad} \qquad (11)$$

$$[Mo(V)(R_2Dtc)_4]^+ \xrightarrow{\quad E_{1/2}(1/2) \quad} [Mo(VI)(R_2Dtc)_4]^{2+}$$

different processes shown in Eq. 11 were examined by a plot of $E_{1/2}$ versus $\Sigma\sigma^*$ (Taft substituent constants). A linear correlation was found for each plot (482). The observed potentials, recorded in CH_2Cl_2 versus SCE, for different R groups range as follows (Eq. 11): $E_{1/2}(0/1)$, -1.22 to -1.13 V; $E_{1/2}(0/1)$, -0.58 to -0.28 V; $E_{1/2}(1/2)$, 1.28 to 1.10 V (482).

Simple Mo xanthate complexes were obtained by the reaction between Mo(II) acetate and the xanthate ligands in inert atmosphere (550, 581a). The red crystalline compounds had the stoichiometry $Mo_2(EtXant)_4$ and reacted readily with donor molecules such as tetrahydrofuran (550) pyridine, 4-picoline, and Et_3As (581a). A Mo(II)-acetate type of dimeric structure was proposed (581a) for this molecule and was confirmed by a subsequent x-ray study (550).

Thioxanthate complexes of Mo(IV) were obtained by Hyde and Zubieta (351) by the reaction of $MoCl_3(THF)_3$ and $Bu_4NRSXant$. The $Mo(IV)(RSXant)_4$ complexes (R = i-Pr, t-Bu, Et) are diamagnetic as expected for D_{2d} or D_{4d} symmetry with the $d_{x^2-y^2}$ orbital lying lowest in energy and not participating in σ bonding (351).

Electrochemical studies by cyclic voltametry (351) established a scheme for these compounds not unlike the one observed with the dithiocarbamate analogues (Eq. 12). Not unexpectedly, and in accord with the ability of R_2Dtc ligands to stabilize higher oxidation states, the two-electron oxidation found with the $Mo(IV)(R_2Dtc)_4$ complexes is not observed in the $Mo(IV)(RSXant)_4$ compounds. Instead, a two-electron reduction is observed in the latter, with the formation of a formally Mo(II)-containing monomer (Eq. 12). The synthesis and

$$Mo(RSXant)_4^+ \xrightleftharpoons{\quad E_{1/2}+0.37V \quad} Mo(RSXant)_4 \xrightleftharpoons{\quad E_{1/2}-0.31V \quad}$$

$$Mo(RSXant)_4^{1-} \xrightleftharpoons{\quad E_{1/2}-1.28V \quad} Mo(RSXant)_4^{-2} \qquad (12)$$

characterization of Mo(IV) diamagnetic dithiocarboxylate complexes was reported by Piovesana and Sestili (520). The $Mo(RDta)_4$ complexes (R = Ph, Bz, p-CH_3O–Ph) were obtained by the reaction of $MoCl_6^{3-}$ and various $RDta^-$ ligands.

By contrast to the R_2Dtc complexes, which show low-intensity d–d absorption bands, the dithioacid complexes show six bands between 11 and 25 kK (ϵ, 2000–14,000) that have been assigned to charge-transfer absorptions (520). Electrochemical studies on the $Mo(RDta)_4$ complexes for R = tolyl and naphthoyl groups have been reported (482). More recent electrochemical studies

(351) on the dithiobenzoate Mo(IV) complexes established the following electrochemical scheme:

$$[Mo(PhDta)_4]^+ \underset{+0.6}{\overset{E_{1/2}(0/1)}{\longleftarrow}} [Mo(PhDta)_4] \underset{-0.79\,V}{\overset{E(0/-2)}{\longrightarrow}} Mo(PhDta)_4^{2-} \tag{13}$$

The potentials seen in Eq. 13 (obtained in CH_2Cl_2 versus SCE) are similar to those reported for the tolyl and naphthoyl derivatives (482).

Differences in $E_{1/2}$ values between thioxanthate complexes and dithioacid complexes for equivalent processes have been related to the superior electron-

releasing character of the RSXant ligand by virtue of the $R-\overset{+}{S}=C\overset{\overset{\displaystyle -\bar{S}}{\diagup}}{\diagdown_{S}}$ form.

Predominance of this form, in fact, is expected to result in low oxidation potentials. A comparison of the $E_{1/2}(0/1)$ values for the R_2Dtc, RSXant, and RDta complexes in Eqs. 11–13 shows a difference of 1 V between the R_2Dtc and RDta complexes.

The spectra of the thioxanthate complexes (351) are qualitatively similar to those observed for the dithioacid derivatives (520), where efficient mixing of ligand π^* and the metal $d_{x^2-y^2}$ orbital has been suggested to account for the absence of transitions of pure $d-d$ origin.

A ligand field model has been used for the interpretation of the magnetic behavior of the $Mo(R_2Dtc)_4$ complexes (485). Extended Hückel calculations compare the electron-donating characteristics of the dithiocarbamate ligand to other 1,1-dithio chelates. This calculation concluded that: (1) there is more metal character in the lowest unoccupied MO in the dithiocarbamate complexes, and (2) there is more ligand character in the lowest unoccupied MO in the dithioacid complexes (485). This seems to be in agreement with the electronic spectra.

The dimeric complex $Mo_2(i\text{-PrSXant})_6$ has been synthesized (351). The electrochemical behavior of this molecule is shown in Eq. 14.

$$[Mo_2(i\text{-PrSXant})_6]^+ \underset{+0.11}{\overset{(0/1)}{\rightleftharpoons}} Mo_2(i\text{-PrSXant})_6 \underset{-0.18\,V}{\overset{(0/-1)}{\rightleftharpoons}}$$
$$[Mo_2(i\text{-PrSXant})_6]^{1-} \underset{-1.01\,V}{\overset{(-1/-2)}{\rightleftharpoons}} [Mo_2(i\text{-PrSXant})_6]^{-2} \tag{14}$$

Complexes of Mo(III) with analogous stoichiometry have been obtained with the R_2Dtc ligands. Thus the reaction between $Mo_2O_2Cl_2(H_2O)_6$ and $NaEt_2Dtc$ in water at pH 4 gives (458) the red $Mo_2(Et_2Dtc)_6$.

The IR spectrum of this compound is consistent with both mono and bidentate R_2Dtc ligands and the structure shown in Fig. 18 has been proposed for this molecule (458). Apparently the same compound was made previously (83, 366). An interesting fact about this molecule is that it gives an EPR spectrum. In solution $g_{av} = 1.985$ and $\langle A \rangle(Mo) = 37.5\,G$. A six-coordinate

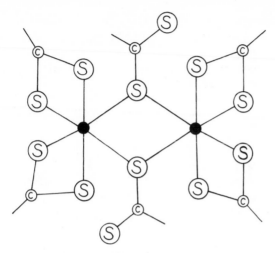

Fig. 18.   Proposed structure for the $Mo_2(Et_2Dtc)_6$ complex (458).

diamagnetic Mo(IV) complex is obtained with the cyclopentadienedithio-carboxylate ligand (CpD^{2-}) (368). The dark brown diamagnetic $(Et_4N)_2Mo(CpD)_3$ complex shows no EPR spectrum and it is suggested (368) that an "orbitally nondegenerate level is lowest in energy."

An interesting series of $(Cl)_{5-n}Mo(R_2Dtc)_n$ complexes ($R_2 = Et_2$, Morpho, Pip, Pyrrol; $n = 1, 2,$ or 3 and $R_2 = Et_2$; $n = 4$) have been synthesized by the reaction of $MoCl_5$ with the appropriate stoichiometric amounts of NaR_2Dtc in CH_2Cl_2 (373). These complexes, which are nonconducting in solution, should have interesting structures. For most of these species, multiple bands in the C–S region of the IR spectrum indicate asymmetrically bound or monodentate R_2Dtc ligands.

(2) Structures of Molybdenum Dithio Complexes.   The structure of the $Mo(Et_2Dtc)_4$ complex has been determined by x-ray crystallography (600) (Table XV). The geometry is described best as square antiprismatic or triangular dodecahedral. The structure of the $Mo(PhDta)_4$ also has been determined (52) (Table XV). The Mo(IV) ion is eight-coordinate and lies on a crystallographic $\bar{4}$ axis of symmetry. The coordination geometry is dodecahedral. The $Mo–S_A/$ $Mo–S_B$ ratio is 1.03 and θ_A and θ_B are 37.90(3) and 75.26(5)$^\circ$, respectively. Values for the a, b, g, and m edges (Fig. 15) have been reported as 3.124(1), 3.612(1), 3.176(1), and 2.764(1) Å, respectively (52).

The structure of the $Mo_2(EtXant)_4 \cdot 2C_4H_8O$ contains a dimeric core similar to the one found in transition metal carboxylate dimers (550). The observed Mo–Mo distance of 2.125(1) Å is very similar to that reported for the

$Mo_2(O_2CCF_3)_4 \cdot 2C_5H_5N$ complex (150). The Mo—O bond length is rather long at 2.795(1) Å and indicates a weak THF—dimer bonding interaction.

(3) Reactions of the Molybdenum Dithio Complexes.

The extensive electrochemical properties of the Mo dithio complexes are well documented. A variety of derivatives of the MoL_4 complexes, suggested by these properties, have been synthesized by chemical or electrochemical means. The chemical (I_2 oxidation in benzene) and electrochemical syntheses of $Mo(Et_2Dtc)_4^+$ have been described (558). Numerous other $Mo(R_2Dtc)_4^+$ complexes have been reported (481). The EPR spectra of these Mo(V) complexes in solution, at room temperature, show strong main lines that are flanked by 6 or 2 satellite lines arising from hyperfine interactions with ^{95}Mo, ^{97}Mo (natural abundance 25.2%; $I = 5/2$). Typical values of solution EPR parameters are: $g = 1.980$ and $A = 34.3$ G. An

TABLE XV
Selected Structural Parameters of the Molybdenum Dithio Complexes.

Compound	$\overline{Mo-S}^a$	$\overline{Mo-O}_t^b$	C—N	Ref.
$Mo(Et_2Dtc)_4$	2.529 (8)		1.33 (1)	600
$Mo(Phdta)_4^d$	2.543 (1)			52
	2.475 (1)^c			
$Mo_2(EtXant)_4 \cdot 2C_4H_8O$	2.477 (2)			550
$Mo(NO)(n\text{-}Bu_2Dtc)_3^e$	2.52 (2)			71
$Mo(MNT)_2(Et_2Dtc)$	2.455 (1)			63
	2.357 (4)			
$Mo_2(SCNnPr)_2(n\text{-}Pr_2Dtc)_2S_2^f$	2.51 (3)		1.323 (16)^g	547
$MoO(TCNE)(n\text{-}Pr_2Dtc)_2$	2.576 (2)^h	1.682 (4)		549
	2.419–2.486 (2)^i			
$MoO_2(n\text{-}Pr_2Dtc)_2$	2.413 (3)	1.664 (8)	1.312 (9)	549
$MoO(i\text{-}PrSXant)_2$	2.370 (7)^j	1.66 (1)		643
	2.446 (12)^k			
$MoO(S_2)(n\text{-}Pr_2Dtc)_2$	2.466 (29)^l	1.695 (5)^m	1.302 (7)	549
	2.651 (42)^n			
$MoO(Et_2Dtc)_3^+$	2.486 (25)	1.684 (6)	1.298 (18)	199
	2.630 (2)^n			
$MoO(S_2)(n\text{=}Pr_2Dtc)_2$	2.372 (25)^o	1.667 (13)		200
	2.503 (51)			
	2.663 (6)^n			
$MoOCl_2(Et_2Dtc)_2$	2.482 (1)^p	1.701 (4)	1.306 (6)	198
	2.515 (4)^q			
$MoOBr_2(Et_2Dtc)_2$	2.485 (11)	1.648 (5)	1.319 (2)	198
$Mo_2O_4(E_2Dtc)_2$	2.454 (5)	1.678 (2)	1.304 (3)	551
$Mo_2O_3S(n\text{-}Pr_2Dtc)_2$	2.305 (5)^r	1.665 (1)		197
$Mo_2S_4(n\text{-}Bu_2Dtc)_2^r$	2.446 (4)		1.36 (2)	579
$[Mo(S_2O)(Et_2Dtc)_2]_2$	2.381 (19)^s			200
	2.504 (27)			

TABLE XV   *continued*

Compound	$\overline{\text{Mo}-\text{S}}^a$	$\overline{\text{Mo}-\text{O}}_t^b$	$\overline{\text{C}-\text{N}}$	Ref.
Mo_2O_3 (n-Pr_2Dtc)$_4$	2.678 (12)	1.671 (7)	1.330 (15)	549
	2.485 (39)			
Mo_2O_3 (i-PrSXant)$_4$	2.694 (6)	1.687 (12)		645
	2.514 (48)			
Mo_2O_3(SPh)$_2$(Et$_2$Dtc)$_2$, ^{t}	2.483 (25)^u	1.683 (9)	2.010 (5)	636
	2.446 (8)^v			
	2.496 (12)^w			

^{a} Average distances: numbers in parentheses are maximum deviations from the mean.
^{b} t = terminal.
^{c} Mo$-$S$_B$
^{d} The individual Mo$-$S$_A$ and Mo$-$S$_B$ distances are given.
^{e} The Mo$-$N bond length is 1.731 (8) Å.
^{f} The Mo$-$Mo distance is 2.705 (2) Å; the Mo$-$C distance is 2.069 (7) Å.
^{g} C$-$N distance in the intact ligand.
^{h} Axial.
^{i} Equatorial.
^{j} Tridentate ligand (see text).
^{k} 'Normal' ligand.
^{l} Sulfur cis to 0.
^{m} Mean of two values.
^{n} Sulfur trans to 0.
^{o} Sulfurs in the S$_2$ group.
^{p} S cis to the equatorial chlorine.
^{q} S trans to the equatorial chlorine.
^{r} Mo$-$S bridge 2.307 (4) Å; Mo$-$S terminal 1.937 (3) Å.
^{s} Sulfur of the S$_2$O molecules.
^{t} Mo$-$Mo distance 2.683 (2) Å.
^{u} Et$_2$Dtc sulfur atoms trans to the bridging oxo group.
^{v} Et$_2$Dtc sulfur atoms trans to the SPh bridging group (Fig. 28).
^{w} SPh bridge sulfur cis to the terminal oxo groups.

interesting type of Mo(R$_2$Dtc)$_4^+$X$^-$ complex was obtained by halogen (Br$_2$ or I$_2$) oxidation of the Mo(R$_2$Dtc)$_4$ complexes. The oxidation of Mo(CO)$_6$ with R$_4$Tds afforded Mo(R$_2$Dtc)$_5$ which was shown by IR spectroscopy and conductivity studies to contain the nonbonded ionic R$_2$Dtc$^-$ anion and Mo(R$_2$Dtc)$_4^+$ (482). The Mo(RDta)$_4$ complexes do not react with I$_2$ and decompose upon treatment with Br$_2$ (482). An extensive study of the EPR and electronic spectra of the Mo(R$_2$Dtc)$_4^+$ complexes has been reported (483). An extended Hückel MO calculation has been carried out for the [Mo(H$_2$Dtc)$_4$]$^+$ cation (485). The electronic absorptions of low intensity between 13 and 20 kK were assigned to CT bands. The magnetic moments of these compounds ~1.7 BM are indicative of one unpaired electron. The Curie-Weiss law is followed for these compounds (485).

The $Mo(SRXant)_4^+$ cation was generated by electrolysis of the $Mo(SRXant)_4$ complexes ($R = i$-Pr and t-Bu) (351). The magnetic properties and EPR parameters of this cation are comparable to those of the R_2Dtc analogues. In general the synthesis of the $Mo(L)_4^+$ species and its stability follow the trends expected by the ability of the ligands to stabilize high oxidation states.

A number of "mixed"-ligand Mo 1,1-dithio complexes have been reported. The nitrosyl $(NO)Mo(R_2Dtc)_3$ complex originally was proposed (364) to contain seven-coordinate Mo. This structure was verified for the n-Bu derivative by a crystallographic study (71). In this molecule (Table XV) the MoS_6N core is found as a distorted pentagonal bipyramid similar to the structure of the $ClTi(Me_2Dtc)_3$ complex (Fig. 16). All three of the ligands are bidentate. The $Mo-S_{ax}$ bond is slightly longer than the $Mo-S_{eq}$ and may reflect the trans effect of the coordinated NO^+ group.

The synthesis of the $Mo(NO)_2(Me_2Dtc)_2$ complex has been reported (364). A cis arrangement of the NO groups was suggested. The temperature dependence of the PMR spectrum of this compound is consistent with a rigid cis octahedral structure at $25°C$ and rapid interconversion of the two types of the N-methyl groups at higher temperatures ($80–140°C$). A proposed mechanism postulated for this interconversion is partial dissociation of a R_2Dtc ligand together with a simultaneous rotation of the N-methyl groups about the $C–N$ bond ($\Delta G^* \sim 21$ kcal mole^{-1}).

The $Mo^{IV}(MNT)_2R_2Dtc$ complex has been obtained by the oxidative decarbonylation of $Mo(CO)_5I^-$ (63). The crystal structure of this diamagnetic molecule has been determined, and a trigonal prismatic coordination of the Mo atom has been found. The very short $Mo-S$ distances (Table XV) found with the MNT ligand may reflect an indeterminant oxidation state for the Mo atom. It is possible that this complex can be described formally as a Mo(II) complex with the two MNT ligands oxidized by two electrons. The redox properties of this molecule should be investigated. The reaction of Mo(III) acetate with the n-Pr$_2$Dtc ligand does not take place with the formation of a stable Mo_2L_4 complex as it does with the EtXant$^-$ ligand (550). Instead, an oxidative addition reaction occurs and the anticipated $Mo_2(R_2Dtc)_4$ complex is converted to a dimer of Mo(IV) ions (547, 548). The four electrons lost by the two Mo(II) ions formally were gained by the two R_2Dtc ligands, which then are converted to S^{2-} and $-S-C-NR_1$ units. The S^{2-} ions bridge the two Mo(IV) ions (Fig. 19), while the $-S-C-NR_2$ groups are bound to the Mo atoms by a sulfur and a carbene type bond. The large standard deviation shown in Table XV for the $\overline{Mo-S}$ bond reflects averaging of the two different $Mo-S$ bond lengths, 2.242(2) and 2.340(2) Å in the disymmetric bridge. The intrabridge angles $S–Mo–S$ and $Mo–S–Mo$ are $106.5(1)°$ and $72.3(1)°$, respectively.

The reaction of Mo(II) complexes such as the acetate dimer with dithioligands seems to preserve the dimeric Mo(II) structure with ligands that are

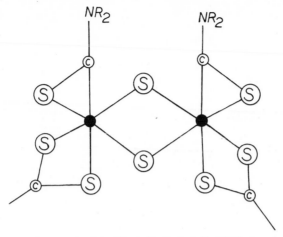

Fig. 19.   Idealized structure of the $Mo_2(n\text{-}Pr_2Dtc)_2(n\text{-}Pr_2NCS)_2S_2$ complex (548).

expected to stabilize lower rather than higher oxidation states, that is, $RXant^-$ rather than R_2Dtc^-. It is expected that dithioacid ligands very likely will give Mo(II) dimers, with the Mo(II)-acetate type of structure.

(4) Oxo and Sulfido Monomeric 1,1-Dithio Molybdenum Complexes. Oxo and sulfido derivatives of the 1,1-dithio complexes of molybdenum are very common and are found with a variety in stoichiometries and structures. The synthesis of $MoO(R_2Dtc)_2$ is accomplished by the reduction of aqueous sodium molybdate in the presence of the R_2Dtc ligand (366). The IR and visible spectra of this molecule have been recorded. The same molecule is obtained by the abstraction of oxygen from the $MoO_2(R_2Dtc)_2$ complexes (118) according to reaction 15. In DMSO solutions the $MoO(R_2Dtc)_2$ complexes dimerize (187) to

$$MoO_2L_2 + PR_3 \longrightarrow OMoL_2 + OPR_3 \qquad (15)$$

$Mo_2O_4(R_2Dtc)_4$. The $MoO(R_2Dtc)_2$ complexes catalyze the reaction of diethylazodicarboxylate with thiophenol (447) (Eq. 16). The reduction of

$$\left.\begin{array}{l} EtO_2CN = NCO_2Et + PhSH \longrightarrow EtO_2CN(SPh) - NHCO_2Et \\ EtO_2CN(SPh) - NHCO_2Et + PhSH \longrightarrow EtO_2CNHNHCO_2Et + PhS - SPh \end{array}\right\} \qquad (16)$$

azobenzene is catalyzed in a similar manner by the same complex. It has been demonstrated (566) that the $MoO(R_2Dtc)_2$ complex reacts with diazenes and activated acetylenes to yield 1:1 adducts. The adducts with diazenes hydrolyze to give $cis\text{-}MoO_2(IV)(R_2Dtc)_2$ (Eq. 17). The so-called adduct actually is an

$$\text{MoO(R}_2\text{Dtc)}_2 + \text{R'N=NR'} \longrightarrow \text{MoO(R}_2\text{Dtc)}_2 \cdot (\text{R'N}_2\text{R'}) \xrightarrow{\text{H}_2\text{O}} \qquad (17)$$

$$\text{MoO}_2(\text{R}_2\text{Dtc)}_2 + \text{R'NHNHR'}$$

oxidative addition product, with the Mo atom in the 6+ oxidation state. Similar "adducts" have been reported with dimethyl acetylene dicarboxylate and tetracyanoethylene (566). The structure of the tetracyanoethylene adduct, $\text{TCNE} \cdot \text{MoO}(n\text{-Pr}_2\text{Dtc})_2$, has been determined (552). The geometry around the Mo(VI) is a deformed pentagonal bipyramid if the TCNE is considered as a bidentate ligand (Fig. 20). A slightly asymmetric bonding of the TCNE is observed [Mo—C(1) = 2.263(6) Å; Mo—C(2) = 2.306(6) Å; C(1)—C(2) = 1.473(9) Å]. A trans effect of the oxygen is detected in the unequal lengths of the axial and equatorial Mo—S bonds, the axial bond trans to the oxygen being longer (Table XV). The reaction of $\text{MoO(Et}_2\text{Dtc})_2$ with O_2, pyridine N-oxide, dimethylsulfoxide, Ph_3PO, and nitrates proceeds with an abstraction of oxygen from these molecules and the formation of $\text{MoO}_2(\text{Et}_2\text{Dtc})_2$. The reaction of $\text{MoO(Et}_2\text{Dtc})_2$ with N_2O is reported to give $\text{Mo}_2\text{O}_3(\text{Et}_2\text{Dtc})_4$ (459). The reactions of $\text{MoO(Et}_2\text{Dtc})_2$ with Ph_3PO and N_2O have been questioned by McDonald et al. (118), who instead claim that the equilibrium lies very much to

$$\text{MoO}_2(\text{R}_2\text{Dtc})_2 + \text{PPh}_3 \rightleftharpoons \text{MoO(R}_2\text{Dtc})_2 + \text{Ph}_3\text{PO} \qquad (18)$$

the right. The same indication is given by Ricard et al. (549), who report the synthesis of $\text{MoO(R}_2\text{Dtc})_2$ by the abstraction of O from $\text{MoO}_2(\text{R}_2\text{Dtc})_2$ by Ph_3P. The same authors report on the structure of the $\text{MoO}(n\text{-Pr}_2\text{Dtc})_2$ complex. A square pyramidal geometry is found with the Mo(IV) 0.83 Å above the plane defined by the four sulfur atoms and toward the oxygen atom. The synthesis of the analogous thioxanthate compound, $\text{MoO}(i\text{-PrSXant})_2$, has been

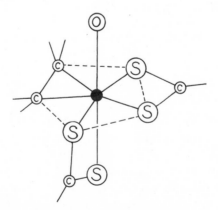

Fig. 20.   Schematic structure of the TCNE · $\text{OMo}(n\text{-Pr}_2\text{Dtc})_2$ complex (552).

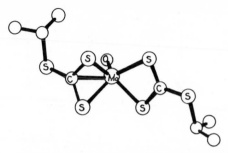

Fig. 21.   The structure of the MoO(*i*-PrSXant)$_2$ complex (643).

reported (643). This complex was obtained by the addition of PPh$_3$ to the Mo$_2$O$_3$(*i*-PrSXant)$_4$ complex (645). The air-stable compound shows a very unusual thioxanthate coordination (Fig. 21). The basic geometry is square pyramidal with the molybdenum 0.86 Å out of the S(1)S(2)S(3)S(4) plane and toward the oxygen atom. There is a "tridentate" type of coordination for one of the thioxanthate ligands such that both sulfurs and the CS$_2$ carbon are involved in bonding. The Mo–S bond lengths for the "tridentate" ligand are considerably shorter than those observed with the normal ligand. The Mo–C bond length is 2.25(1) Å (643). The five-coordinate red-brown (Et$_4$N)MoO(L)$_2$ complex, with L = cyclopentadienedithiocarboxylate, has been reported (368). This compound, which presumably contains Mo(V), is diamagnetic in the solid state, however, in solution it shows an EPR spectrum. Spectral evidence for MoO(RXant)$_2$ has been presented (448), but no report of the corresponding dithioacid complex has appeared in the literature.

The Mo(VI) dithiocarbamate dioxo complexes were first reported by Malatesta (414). At a later date other syntheses by Moore and Larson (466) and by Pilipenko and Gridchina (516) were reported.

The structure of the MoO$_2$(Et$_2$Dtc)$_2$ was first reported in 1972 (394). A more accurate determination was reported for the structure of the *n*-propyl analogue (549). Both structures show a deformed octahedral coordination about the Mo atom with the oxygen atoms in the *cis* position. In both structures (Table XV) the Mo–S bond trans to an oxygen atom is considerably longer than the Mo–S bond cis to an oxygen atom. In the *n*-Pr$_2$Dtc analogue the O–Mo–O angle is 105.7(1)°. A dimeric version of the MoO$_2$(R$_2$Dtc)$_2$ complexes (β form) is obtained (135) when these molecules are synthesized by oxidation of the Mo(CO)$_2$MPh$_3$(R$_2$Dtc)$_2$ complexes (135) (M = P, As, Sb) in CCl$_4$ solution. The dimeric nature of the yellow crystalline product, [MoO$_2$(R$_2$Dtc)$_2$]$_2$, was established by mass spectrometry. The mass spectra of the "normal" MoO$_2$(R$_2$Dtc)$_2$ (α form) also were examined (135) and their monomeric nature was verified. On the basis of fragmentation patterns of the β form and also of the Mo$_2$O$_3$(R$_2$Dtc)$_4$ complexes (135), either of the two structures shown in Fig. 22

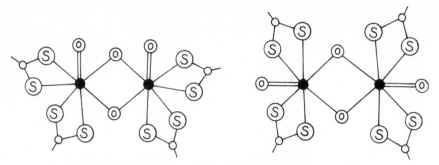

Fig. 22. Possible structures of the β-form of $[MoO_2(R_2Dtc)_2]_2$ (135).

was proposed for the β form. The IR spectra of the β form show the Mo=O stretch at 949 cm^{-1}. Two Mo–O vibrations are found in the cis α isomer (466).

The reactive nature of the $MoO_2(R_2Dtc)_2$ complex has been demonstrated in various reactions. The catalytic oxidation of phosphines has been reported (23) and K_R and K_0 have been determined (Eq. 19). The oxidative

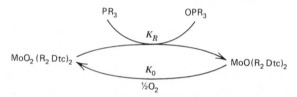

addition product of $OMo(R_2Dtc)_2$ with dimethylacetylenedicarboxylate hydrolyzes to give $MoO_2(R_2Dtc)_2$ and 1,2-bisethoxycarbonylhydrazine (566).

The reactions of $MoO_2(R_2Dtc)_2$ (R = CH$_3$, Et, Ph) with phenylhydrazine or 1,1-dimethylhydrazine in refluxing CH$_3$OH and in the presence of NaR$_2$Dtc have been described (36). Both azo(1–) and N,N-disubstituted hydrazido-$N(2–)$ complexes of Mo were obtained. With phenylhydrazine high yields of brown Ph–N=N–Mo(IV)(R$_2$Dtc)$_3$ complexes were obtained. With 1,1-dimethylhydrazine, the Mo(IV)ONNMe$_2$(R$_2$Dtc)$_2$ was isolated [ν(Mo=O) at 890 cm^{-1}]. The azo complexes react with alkylating agents, MeI, Et$_3$O$^+$, and so on, to give the Mo(N$_2$RR″)(R$_2$Dtc)$_3^+$ complexes. The R groups in these complexes are equivalent by NMR, which suggests that they are located on the terminal N atom. The azo complexes react with HCl to give the nonconducting MoClN$_2$HR(R$_2$Dtc)$_3$ complexes and with HBF$_4$ to give the [Mo(N$_2$HR)R$_2$ (Dtc)$_3$]$^+$BF$_4^-$ "salts" (36).

In electrochemical studies of the $MoO_2(R_2Dtc)_2$ complexes the dimerization of these molecules to $Mo_2O_4(R_2Dtc)_4$ was observed (187). Decomposition of $MoO_2(R_2Dtc)_2$ solutions in the presence of (CH$_3$O)$_2$SO$_2$ gives the [MoO(R$_2$Dtc)$_3^+$]$_2$(Mo$_6$O$_{19}$)$^{2-}$ salt, which contains the MoO(R$_2$Dtc)$_3^+$ cation

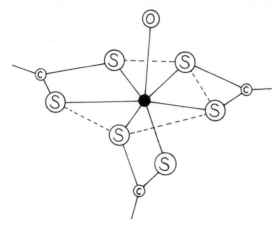

Fig. 23.   Schematic structure of the MoO(Et$_2$Dtc)$_3^+$ complex (46).

(546). The same cation is obtained by the spontaneous decomposition of MoOF$_2$(R$_2$Dtc)$_2$ (199). The structure of the cation (R = Et) has been determined (46) and the OMoS$_6$ core is a distorted pentagonal bipyramid (Fig. 23) (Table XV). The reactions of MoO$_2$(R$_2$Dtc)$_2$ with H$_2$S give (196, 200) MoO(S$_2$)(R$_2$Dtc)$_2$, Mo$_2$S$_4$(R$_2$Dtc)$_2$, Mo$_2$S$_3$O(R$_2$Dtc)$_2$, and Mo$_2$O$_2$S$_2$-(R$_2$Dtc)$_2$. Reaction with P$_4$S$_{10}$, in addition to the above products, gives Mo$_2$O$_2\mu$-O-μ-S(R$_2$Dtc)$_2$. The reaction of Mo(CO)$_2$(R$_2$Dtc)$_2$ with S$_8$, in addition to MoO(S$_2$)(R$_2$Dtc)$_2$, gives the green [Mo(S$_2$O)(R$_2$Dtc)$_2$]$_2$ complex (200).

The structure of the blue MoO(S$_2$)(n-Pr$_2$Dtc)$_2$ complex has been determined (200) (Table XV). The S$_2$ group occupies two coordination sites in the equatorial plane of a deformed pentagonal bipyramid. The oxygen and one of the ligand sulfur atoms occupy the axial positions. Again the trans effect of the oxygen is demonstrated in the long Mo—S bond trans to the oxygen atom. The structure is similar to that reported for the TCNE "adduct" of OMo(n-Pr$_2$Dtc)$_2$ (549), with the two equatorial sites occupied by two sulfur atoms instead of TCNE (Fig. 20).

The reactivity of the MoO$_2^{2+}$ toward hydrohalic acids leads to the formation of MoOX$_2$(R$_2$Dtc)$_2$ compounds, X = F, Cl, Br; R = Me, Et, n-Pr (198). The structures of the chloro and bromo derivatives with R = Et have been determined (Table XV). The geometry is the same in both structures, and the Mo(VI) ion is seven-coordinate and pentagonal bipyramidal. The axial sites are occupied by an X$^-$ and an O^{2-} ligand. A large trans effect is obvious in the distances of the Mo—Cl and Mo—Br bonds, which are trans to the Mo=O bonds. Distances of 2.504(1) and 2.729(1) Å are observed for these bonds in the dichloro and dibromo compounds, respectively. The corresponding distances for the cis Mo—S

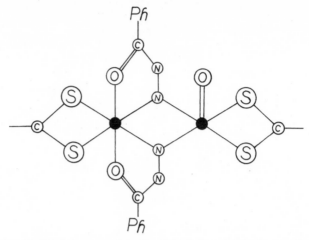

Fig. 24.   Idealized structure of the $Mo_2O(PhCON_2)_2(Et_2Dtc)_2$ complex (37).

bonds are 2.417(1) and 2.550(1) Å. The rather short Mo—O bond in the dibromo derivative is reflected in the high (960 cm^{-1}) frequency of the Mo—O stretching vibration (198).

The product of the reaction between $MoO_2(Et_2Dtc)_2$ and benzoylhydrazine hydrochloride is shown by x-ray crystallography to be an asymmetric dinuclear complex $[Mo_2O(PhCON_2)_2(Et_2Dtc)_2]$ with bridging benzoyl diazenido ligands (Fig. 24). One Mo atom has approximate octahedral symmetry, while the other Mo atom is essentially square pyramidal and is displaced from the basal plane by 0.72 Å toward the apical oxygen atom (37). The Mo—Mo distance is 2.661 Å and the Mo—N—Mo angles are 90.9 and 87.9°.

The corresponding compound with thiobenzoyl diazenido ligands also has been prepared (37).

(5) Dimeric μ-Bridged Molybdenum Dithio Complexes.

Several syntheses of $Mo_2O_3(R_2Dtc)_4$ have been reported (105, 136, 414, 466). The most satisfactory synthesis was reported by Newton et al. (477). With their procedure the dimer is obtained by the reaction of $MoCl_5$ with NaR_2Dtc in a 1:9 molar ratio in cold water (0°C). The same compound is obtained by the addition of NaR_2Dtc to a solution of $Mo_2O_4(R_2Dtc)_2$ in a 1:1 mixture of $CHCl_3/CH_3OH$. The reaction of this dimer with diethylazodicarboxylate has been reported (477) to give $cis\text{-}MoO_2(R_2Dtc)_2$ and $MoO(R_2Dtc)_2 \cdot EtO_2CN{=}NCO_2Et$. A detailed discussion of the IR and visible spectra of $Mo_2O_3(R_2Dtc)_4$ also is given (477). The synthesis of $Mo_2O_3(R_2Dtc)_4$ by the reaction of $MoO_2(R_2Dtc)_2$ with Ph_3P has been described (549). The structure of the $Mo_2O_3(n\text{-}Pr_2Dtc)_4$ has been determined and is very similar to that reported previously for $Mo_2O_3(EtXant)_4$ (41a). The distances observed (549) in the $Mo_2O_3S_8$ core are nearly identical for both compounds (Table XV) (Fig. 25).

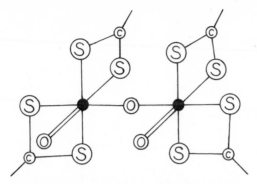

Fig. 25.   Schematic structure of the Mo_2O_3 (n-Pr_2Dtc)$_4$ complex (549).

The interesting $Mo_2O_4(Et_2Dtc)_2$ complex is a decomposition product of the 1:1 adduct of $MoO(Et_2Dtc)_2$ with dimethyl acetylenedicarboxylate (551). The structure of this molecule has been determined (Fig. 26). The Mo(V) ions in the dimer are five-coordinate and show square pyramidal coordination (197).

The synthesis and structure of the analogous $Mo_2S_4(n$-$Bu_2Dtc)_2$ have been described (579) (Table XVI). The structure of this molecule has a geometry very similar to that of the μ-dioxo analogue (551) (Fig. 26). The preparation and structure of the $Mo_2O_2\mu$-O–μ-$S(R_2Dtc)_2$ also has been reported (197). It has been pointed out that the alleged $Mo_2OS(\mu$-$O)_2(n$-$Bu_2Dtc)_2$ complex (560) has the same IR spectrum as, and probably is, the Mo_2O_2-μ-O μ-$S(R_2Dtc)_2$ complex. The structure of the interesting $[Mo(S_2O)(Et_2Dtc)_2]_2$ has been reported (200). Two $Mo(R_2Dtc)_2$ moieties are bridged by two S_2O molecules (Fig. 27) (Table XV). The Mo—Mo distance in the molecule is 2.754(1) Å. An electrochemical study of $MoO_2(R_2Dtc)_2$, $MoO_3(R_2Dtc)_4$, $MoO(R_2Dtc)_2$, $Mo_2O_4(R_2Dtc)_2$, $MoS_2(R_2Dtc)_2$, and $Mo_2O_2S_2(R_2Dtc)_2$ has appeared (187). In this study the very interesting $[Mo_3^VMo^{IV}O_4S_4(R_2Dtc)_4]^-$ tetramer is reported to form following reduction of the $Mo_2O_2S_2(R_2Dtc)_2$ complexes. This tetramer and its one-electron reduction product (tetramer^{2-}) show interesting EPR spectra.

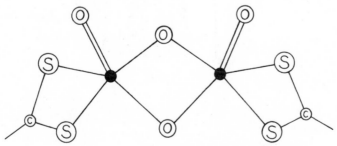

Fig. 26.   Idealized structure of the $Mo_2O_4(Et_2Dtc)_2$ complex (551).

TABLE XVI

Structural Parameters in the Cores of Di-μ-Bridged Molybdenum Dithio Complexes.[c]

Compound	Bond Lengths, Å				Bond angles, °			Ref.
	Mo–O$_t$[a]	Mo–O$_B$	Mo–S$_b$	Mo–Mo	Mo–Ob–Mo	Mo–S$_b$Mo	Xb–Mo–Xb	
Mo$_2$O$_4$(Et$_2$Dtc)$_2$	1.678 (2)	1.940 (2)		2.580 (1)	83.3 (1)		91.9 (1)	551
Mo$_2$O$_3$S(nPr$_2$Dtc)$_2$	1.665 (1)	1.927 (11)	2.305 (5)	2.673 (3)	87.8 (4)	70.9 (2)	97.0 (1)	197
Mo$_2$S$_4$ (nBu$_2$Dtc)$_2$[b]			2.307 (4)	2.801 (2)		74.7 (1)	101.8 (2)	579

[a] Average distances and angles; numbers in parentheses are maximum deviations from the mean.
[b] Mo–S$_t$ = 1.937 (3) Å.
[c] The Inpared spectra for some of these complexes have been assigned (148).

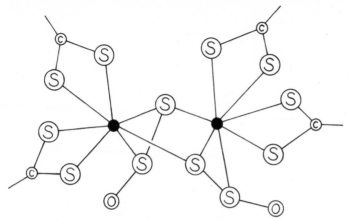

Fig. 27.   Idealized structure of the $[Mo(S_2O)(Et_2Dtc)_2]_2$ dimer (200).

An extensive detailed account of the reactions of $Mo_2O_4(R_2Dtc)_2$ was recently presented by Newton et al. (476). Under various conditions this molecule reacts with NaR_2Dtc to give $Mo_2O_3(R_2Dtc)_4$; with P_4S_{10} to give $Mo_2S_4(R_2Dtc)_2$, with H_2S to give $Mo_2O_3S(R_2Dtc)_2$, with RSH to give $Mo_2O_3(SR)_2(R_2Dtc)_2$, and with $o\text{-}C_6H_4XHSH$ ($X = O$, NCH_3) to give $Mo_2O_3(C_6H_4SX)(R_2Dtc)_2$ or $MoR_2Dtc(C_6H_4SX)_2$.

The EPR spectra of the $Mo(Et_2Dtc)(HNSC_6H_4)_2$ have been reported (262, 497). Hydrogen-1, hydrogen-2, and nitrogen-14 superhyperfine splittings were observed (497).

The crystal structure of the $Mo_2O_3(SPh)_2(Et_2Dtc)_2$ complex has been determined (636) and shows a triply bridged complex with two different bridging thiophenolate groups, and an oxo bridge. One of the bridging SPh⁻ groups is cis to the terminal oxo group of each Mo atom, while the other bridging SPh⁻ is trans to the terminal oxo group of each Mo atom (Table XV). Two different types of bridging Mo–S distances are found, and they differ by ~0.2 Å. The Mo–S bonds trans to the terminal oxo group are the longer bonds in the bridge [2.702(3) and 2.673(3) Å]. The Mo(1), S(2), O(3), Mo(2) unit is nearly planar (Fig. 28). A distorted octahedral coordination is observed for the Mo(V) ions in the dimer.

A simple method for the synthesis of $Mo_2O_3(RXant)_4$ complexes has been reported (478) from Na_2MoO_4, KRXant, and $6M$ HCl with R = Me, Et, i-Pr, n-Bu, and i-Bu. Also the synthesis of $Mo_2O_2S_2(RXant)_2$ has been achieved by the reaction of $Mo_2O_3(RXant)_4$ in $CHCl_3$ with H_2S. The reaction of $Mo_2O(RXant)_4$ and $2Et_4N^+RXant^-$ gives $2RRXant$ plus $(Et_4N)_2Mo_2O_2S_2\text{-}(OCS_2)_2$ in what appears to be a transalkylation reaction. The reactions of $Mo_2O_3(RXant)_4$ with ROH and Me_2NH (Eq. 20) also have been reported (478). The $Mo_2O_3(i\text{-}PrSXant)_4$ has been synthesized by the reaction of

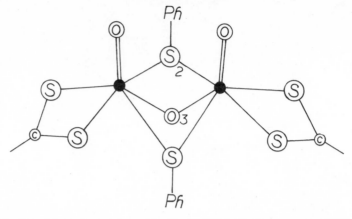

Fig. 28.   Schematic structure of the $Mo_2O_3(SPh)_2(Et_2Dtc)_2$ complex (636).

$$Mo_2O_3(RXant)_4 + 2ROH \longrightarrow Mo_2O_2S_2(RXant)_2 + 2(RO)_2CS + H_2O$$

(20)

$$Mo_2O_3(RXant)_4 + 4Me_2NH \longrightarrow ROCSNMe_2 + \text{"}Mo_2O_2S_4H_2\text{"}$$

$(NH_4)_2MoOCl_5$ and i-PrSXant$^-$ in water (645). The structure of this molecule (645) is similar to that reported for the $Mo_2O_3(EtXant)_4$ (41) and $Mo_2O_3(R_2Dtc)_4$ (549) complexes (Table XV). The electrochemical behavior of the $Mo_2O_3(RSXant)_4$ complexes was interpreted (645) as:

$$Mo_2O_3(RSXant)_4 + 2e \rightleftharpoons [Mo_2O_3(RSXant)_4]^{2-}$$

$\downarrow$ slow $\downarrow$ (21)

X   Y

X and Y were not identified.

c. **Tungsten.**   Considerable progress has been made in the synthesis and characterization of tungsten 1,1-dithio chelates. These complexes were virtually unknown prior to 1970. The $W(R_2Dtc)_4$ complexes were made by the reaction of WCl_4 and NaR_2Dtc in acetonitrile (577). For the dark brown $W(PyrrolDtc)_4$ complex, the IR spectra, electronic spectra, and magnetic moment (μ_{eff}, room temp. 0.98 BM) were reported (88). On the basis of IR data, an eight-coordinate structure was proposed for this molecule. At a later date the synthesis of $W(R_2Dtc)_4$ and $W(RDta)_4$ complexes was accomplished by the oxidative decarbonylation of $W(CO)_6$ with appropriate disulfides (482). Synthesis of the $W(R_2Dtc)_4X$ complexes was accomplished by the same reaction in air and in the presence of Bu_4NX ($X = Cl, Br, I$) (482). An alternate procedure for the synthesis of the W(V) complexes involves the oxidative decarbonylation of the

$Bu_4N[W(CO)_5X]$ compounds. The EPR spectrum of the $W(Ph_2Dtc)_4I$ in $CHCl_3$ solution (481) is reported to show $g = 1.9002$ and $A = 56.8 \times 10^{-4}$ cm^{-1} (hyperfine splitting due to ^{183}W, $I = \frac{1}{2}$, natural abundance 14.4%).

Electronic and EPR spectra also have been reported for the $W(R_2Dtc)_4X$ complexes (X = halide; R = Me, Et, i-Pr, Ph, Bz; R_2 = Pyrrol, Pip). The d–d transitions are fitted into a ligand field model for W(V) to yield 10 Dq = 33.8 kK and $d = 1610$ cm^{-1} (480, 484). The electronic spectra and magnetic susceptibilities of the $W(RDta)_4$ complexes (R = tolyl, naphthoyl) have been reported (485). The crystal structure of the $W(Et_2Dtc)_4Br$ complex has been determined (623). The tungsten atom is bonded to eight sulfur atoms from four chelating R_2Dtc ligands. The coordination geometry approximates very closely the triangular dodecahedron (Table XVII).

The synthesis of the $WO_2(R_2Dtc)_2$ complexes is accomplished by an "oxo transfer" reaction from $Mo_2O_3(L)_4$ (L = diethyldithiophosphate) to the W(II) complex $W(CO)_2(PPh_3)(R_2Dtc)_2$. This latter complex was obtained (119) according to the reaction 22

$$W(CO)_3(PPh_3)_2Cl_2 + 2NaR_2Dtc \longrightarrow W(CO)_2(PPh_3)(R_2Dtc)_2 + 2NaCl + CO + PPh_3 \quad (22)$$

It is suggested that oxidation of W(II) to W(IV) in $W(CO)_2(PPh_3)(R_2Dtc)_2$ is affected by the $MoO(L)_2$ complex, which exists in equilibrium (Eq. 23) with the reactant $Mo_2O_3(L)_4$ complex (see above).

$$Mo_2O_3(L)_4 \rightleftharpoons MoO(L)_2 + MoO_2(L)_2 \quad (23)$$

The W=O stretching frequencies in the $WO_2(R_2Dtc)_2$ complexes are found in the range 890–935 cm^{-1} (R = Me, Et, n-Pr). The mixed-ligand complex $W(MNT)_2(Et_2Dtc)$ is obtained in the same way that the Mo(IV) analogue is obtained, by aerobic oxidative decarbonylation of $(Bu_4N)(W(CO)_5I$ (63) in the presence of Na_2MNT and NH_4Et_2Dtc.

The temperature-dependent PMR spectra of the $W(NO)_2(Me_2Dtc)_2$ complex (364) have been obtained (182) and, just as in the Mo analogue, an interconversion of the nonequivalent N-methyl groups is observed at high temperatures (80–140°C).

7. Group VII

Prior to 1969 complexes of the heavier Group VII elements with 1,1 dithio ligands were scarce and were limited to the R_2Dtc ligands. Only complexes of the type $Cl_2Re(R_2Dtc)$ had been described (134). The originally proposed tetrahedral stereochemistry for these diamagnetic complexes has been modified. The reasonable suggestion, that these molecules may be structurally related to the Re_3Cl_{12} cluster, has been advanced (559).

TABLE XVII
Properties of the Tungsten 1,1-Dithio Complexes.

Complex	Color	ν(C–N), cm^{-1}	ν(M–S), cm^{-1}	μeff, BM	$E^{0/1}_{1/2}$, V^a	$E^{1/2}_{1/2}$, V^b	$\overline{\text{M–S}}$, Å	Ref.
W(Et₂Dtc)₄Br	Dark green	1519	356		−0.73	+ 0.81	2.509 (18)	481, 623
W(Me₂Dtc)₄⁺	Dark green	1552	356		−0.71	+ 0.84		482, 623
W(i-Pr₂Dtc)₄⁺	Dark green	1500	379		−0.81	+ 0.74		482, 623
W(Pyrrol Dtc)₄⁺	Dark green	1516	333		−0.73	+ 0.81		482, 623
W(PipDtc)₄⁺	Dark green	1526	346		−0.76	+ 0.73		482, 623
W(Bz₂Dtc)₄⁺	Dark green	1507	382		−0.57			482, 623
W(Ph₂Dtc)₄I	Dark green	1398	331		−0.50	+ 1.26		481, 482
W(naphthoyl Dta)₄	—				+ 0.63			481, 482
W(Pyrrol Dtc)₄	Dark brown	1520	346	0.98				88
W(Et₂Dtc)₄	Black	1525	365	1.17^c				577

^{a} Reversible reduction in CH₂Cl₂ versus SCE.
^{b} Irreversible oxidation in CH₂Cl₂ versus SCE.
^{c} 298 °K.

a. **Manganese.** *(1) Manganese(III) Complexes.* Three structures of the $Mn(R_2 Dtc)_3$ complexes have been determined. In $Mn(Et_2 Dtc)_3$ the distorted MS_6 core can be described as arising from a tetragonal distortion of an ideal O_h symmetry superimposed on the approximate overall D_3 molecular symmetry (305). The molecule seems to display a distortion that conforms to the requirements of the Jahn-Teller effect expected for the Mn(III) ion with a d^4 configuration. The far IR spectrum of this compound shows a very broad band centered at 370 cm^{-1}, which is attributed to the Mn–S stretching vibration. The abnormally large width of this band is considered as another indication of the asymmetry in the MnS_6 core. A tetragonal distortion also is evident in the structure of the $Mn(MorphoDtc)_3 \cdot CHCl_3$ complex (Table XVIII) (96a). In this structure, as well as in the structure of the corresponding $CH_2 Cl_2$ solvate (537), the solvent of crystallization interacts with the sulfur donors. In both complexes the elongation of the Mn–S bonds is observed for the sulfur atoms near the solvent. It is not clear whether interaction causes this elongation or simply assists a distortion that was bound to occur because of the Jahn-Teller effect. The magnetic moment of the $Mn(MorphoDtc)_3 \cdot CHCl_3$ complex is relatively tem-

TABLE XVIII

Selected Structural Parameters of the Group VII 1,1-Dithio Complexes.

Compound	$\overline{M-S}$, Å	$\overline{C-N}$, Å	$\nu(C-N)$, cm^{-1}	μ_{eff}^{corr}, BM	Ref.
$Mn(Et_2 Dtc)_3$	2.549 (11)^a	1.35 (3)	1490	5.31	272,
	2.403 (30)^b		1510	room temp.	305
$Mn(Morpho Dtc)_3 \cdot CHCl_3$	2.555 (40)^c	1.322 (7)		5.5 (294°K)	96
	2.458 (35)				
	2.354 (15)				
$Mn(PipDtc)_3 ClO_4 \cdot CHCl_3$	2.325 (3)	1.31 (1)	1530		82
$Mn(CO)_3 (DPM\ CHS_2)^d$	2.40 (1)				210
$Re_2 O_3 (Et_2 Dtc)_4{}^e$	2.438 (20)	1.313 (12)	1515		248,
					559
$NRe(Et_2 Dtc)_2{}^f$	2.388 (5)	1.323 (12)	1535		247,
					249,
					559
$Re(CO)_2 (HCS_2)(PPh_3)_2{}^g$	2.516 (22)				8
$(Ph_3 P)Mn(CO)_3 (Me_2 Dtc)^h$	2.377 (6)^i	1.36 (5)	1480		183

^{a} Mean value of the two axial Mn–S bonds in tetragonally distorted MnS_6 octahedron.
^{b} Mean value of the four equatorial Mn–S bonds.
^{c} Mean value of three pairs of apparently unequal Mn–S bonds.
^{d} DMP = diphenylphosphinomethane.
^{e} Mean terminal Re–O bond, 1.722 (7) Å.
^{f} Re–N bond, 1.656 (8) Å.
^{g} $\overline{Re-C}$, 1.91 (2); $\overline{Re-P}$, 2.419 (10).
^{h} $\overline{Mn-P}$, 2.352 (7) Å; $\overline{Mn-C}$, 1.777 (33) Å.
^{i} The mean value of the Mn–S distances in the two molecules of the asymmetric unit.

perature independent and is considered normal. A decrease of the moment at very low temperatures is observed and has been attributed to antiferromagnetic interactions (96).

The PMR spectra and magnetic susceptibilities of a series of $Mn(R_2Dtc)_3$ complexes have been studied (272). The conclusion reached on the basis of these studies is that the electronic ground state is the ^5E state and the $^3T_1$ state is the lowest-lying excited state with δ much greater than the spin—orbit coupling constant. Similar conclusions were reached in a theoretical study (280).

The EPR studies of a series of $Mn(R_2Dtc)_3$ complexes diluted in the Co(III) analogues were rather interesting (279). It was concluded that the signal observed, showing neither hyperfine nor superhyperfine splittings, arises from an electron that resides at or near the sulfur atom. The EPR parameters obtained from a single-crystal study [$Mn(PyrrolDtc)_3$ in the $Co(PyrrolDtc)_3$ lattice] are $g_1 = 2.140$, $g_2 = 2.126$, and $g_3 = 2.030$ (279). These values are similar to those obtained for $Fe(Me_2Dtc)_3$ in the corresponding Co(III) lattice (553) ($g_1 = 2.111$, $g_2 = 2.076$, and $g_3 = 2.015$) and suggest that in the latter case, the EPR spectra may arise from a sulfur radical.

The temperature-dependent PMR spectra of the $Mn(MePhDtc)_3$ complex demonstrate that cis—trans and optical isomerization are slow on the PMR time scale below $\sim-60°C$. For this complex the metal-centered inversion in CD_2Cl_2 solution is characterized by a $\Delta H^{\ddagger}$ of 9.8 ± 1.0 kcal mole^{-1} and a $\Delta G^{\ddagger}$ of 9.1 ± 0.5 kcal mole^{-1} ($-50°C$). For the $Mn(Bz_2Dtc)_3$ complex, $\Delta H^{\ddagger}$ is 11.0 ± 1.0 kcal mole^{-1}, $\Delta S^{\ddagger}$ is 1.5 ± 5.0 eu and $\Delta G^{\ddagger}$ is 10.6 ± 0.2 kcal mole^{-1} at $-35°C$. The $\Delta S^{\ddagger}$ value is consistent with the trigonal twist mechanism for the inversion, since a larger, positive $\Delta S^{\ddagger}$ would be expected for a bond-rupture mechanism (531).

The synthesis of the mixed-ligand $Mn(Et_2Dtc)_2(TFD)$ complex [TFD = bis(perfluoromethyl)-1,2 dithietene] has been described (512).

(2) Manganese(II) and Manganese(IV) Complexes. Following the studies on $Mn(R_2Dtc)_2$ complexes by Fackler and Holah (236), who reported that these very air-sensitive compounds were x-ray isomorphous to the corresponding $Cu(R_2Dtc)_2$ complexes, a contradicting report appeared in the literature (400). In this later study, the $Mn(Et_2Dtc)_2$ complex was reported to be isomorphous with the square planar monomeric $Ni(Et_2Dtc)_2$ complex. In the same study, EPR data ($g_{\parallel} + 1.92$, $g_{\perp} = 4.11$) and magnetic susceptibility data [$\mu_{eff}^{corr} = 4.1$ BM (room temp.)] were considered consistent with a Mn(II) complex with the uncommon quartet ground state (400). These results were later disputed by Hill et al. (325), who attributed the low value of the magnetic moment to contamination and reported a sextet ground state for this molecule. The same authors confirmed the x-ray isomorphism of $Mn(Et_2Dtc)_2$ to $Cu(Et_2Dtc)_2$ as was reported previously (236).

The Mn(II)-containing complexes $Mn(EtXant)_2$ and $Mn(EtXant)_3^-$ are obtained in aqueous solution under nitrogen by the reaction between $MnCl_2$ and KXant in the appropriate molar ratios (330). The $Mn(EtXant)_3^-$ is precipitated as the Et_4N^+ salt. The $Mn(EtXant)_2$-o-phen and two crystalline modifications of the $Mn(EtXant)_2$bipy form under the same synthetic conditions in the appropriate molar ratios (330).

The $Mn(EtXant)_2$ complex (μ_{eff} = 4.8 BM, room temp.) shows reflectance spectra that suggest the presence of six-coordinated metal ions. The low magnetic moment of this complex may reflect solid-state antiferromagnetic interactions in a polymeric structure. The $Mn(EtXant)_3^-$ ion shows magnetic properties (μ_{eff} = 6.0 BM, room temp.) and visible spectra ($^6A_{1g} \rightarrow {}^4T_{1g}$, 14,700 cm^{-1}) as expected for an octahedral d^5 Mn(II) complex. The magnetic properties and electronic spectra of the $Mn(EtXant)_2$-o-phen, $Mn(EtXant)_2$-bipy, $Mn(Et_2Dtc)_2$-o-phen, and $Mn(Et_2Dtc)_2$-bipy are as expected for octahedral high-spin Mn(II) complexes.

The Mn(IV) dithiocarbamate complexes, $Mn(R_2Dtc)_3^+$, are readily obtained by either the reaction of BF_3 with $Mn(R_2Dtc)_3$ complexes in CH_2Cl_2 (500, 562) or by the reaction of the $Mn(R_2Dtc)_3$ complexes in benzene with acetone solutions of $Mn(H_2O)_6(ClO_4)_2$ or $Mn(H_2O)_6(BF_4)_2$ (82, 270). These dark purple complexes, which are obtained as the ClO_4^-, or BF_4^- salts, are 1:1 electrolytes in nitrobenzene and have magnetic moments (3.70–4.25 BM) consistent with the Mn(IV) d^3 formulation (82). The EPR spectra of the $Mn(R_2Dtc)_3ClO_4$ complexes (R_2 = Pip, Et_2) in dilute frozen $CHCl_3$ solution at $113°K$ are characterized by isotropic g values (1.99 ± 0.005) and small values for the hyperfine interaction ($\langle A \rangle$ = 95 ± 1) (82).

The structure of the $Mn(PipDtc)_3ClO_4$ complex contains a $Mn(IV)S_6$ core of approximate D_3 symmetry (82) (Table XVIII). The deviation from octahedral symmetry is as expected for the short bite of the ligand (388). The $\overline{Mn-S}$ bond [2.325(3) Å] is shorter than that found in the Mn(III) complexes (Table XVIII). A comparison of the Mn–S and C–N stretching frequencies of the $Mn(R_2Dtc)_3^+$ and $Mn(R_2Dtc)_3$ complexes show that in the former, the Mn–S and C$\doteq$N vibrations occur at higher energies (82). These data are in accord with the shorter Mn–S bond and greater positive charge of the Mn(IV) ion in the $Mn(R_2Dtc)_3^+$ complexes.

An electrochemical study of a large number of manganese–R_2Dtc complexes shows that, in aprotic solvents, the $Mn(R_2Dtc)_3$ complexes undergo single one-electron oxidation and one-electron reduction with the formation of the $Mn(R_2Dtc)_3^+$ and $Mn(R_2Dtc)_3^-$ species, respectively (311). Electrochemical studies were carried out on a Pt electrode, in acetone or CH_2Cl_2 with Et_4NClO_4 or n-Bu_4NBF_4 as supporting electrolytes, and versus a Ag|AgCl reference electrode.

For the reversible process:

$$Mn(R_2Dtc)_3 \;\rightleftharpoons\; Mn(R_2Dtc)_3^+ + e^- \tag{24}$$

potentials ranged from +0.25 V for $Mn(Cx_2 Dtc)_3$ (Cx = cyclohexyl) to +0.53 V for $Mn(Bz_2 Dtc)_3$ in a series of 16 different $Mn(R_2 Dtc)_3$ complexes.

The reversible reduction on the same series of complexes (Eq. 25) showed a range of potentials, from −0.23 V for the $Mn(Cx_2 Dtc)_3$ to +0.07 V for the $Mn(Bz_2 Dtc)_3$ complexes.

$$Mn(R_2 Dtc)_3 + e^- \rightleftharpoons Mn(R_2 Dtc)_3^- \qquad (25)$$

As the data for the above two complexes indicate, there exists a definite correlation of the redox potentials with substituent inductive effects. This is further illustrated in a plot of the potentials for all the $Mn(III)(R_2 Dtc)_3/Mn(IV)(R_2 Dtc)_3$ couples versus the potentials of the $Mn(II)(R_2 Dtc)_3/Mn(III)(R_2 Dtc)_3$ couples. The plot is very nearly linear, with a slope $\Delta E_{red}/\Delta E_{ox} = 1.1$, and clearly indicates that substituents that stabilize Mn(IV) destabilize Mn(II) and vice versa.

A titration of $PyrrolDtc^-$ solution with Mn^{2+} ion was monitored electrochemically. The formation of $[Mn(PyrrolDtc)_3]^-$ was first detected, followed by the formation of $Mn(PyrrolDtc)_2$, which was found to undergo irreversible oxidation to the $Mn(PyrrolDtc)_2^+$ cation (311).

Attempts to isolate the $Mn(R_2 Dtc)_3^-$ complexes have been unsuccessful, (331), perhaps because the anionic tris complex reacts with excess Mn(II) ions to give the insoluble $Mn(R_2 Dtc)_2$ complex (311).

b. Technetium, Rhenium. As late as 1977 the chemistry of technetium with 1,1-dithio ligands remained unexplored. A considerable number of new rhenium complexes have been reported.

The simple $Re(III)(R_2 Dtc)_3$ compounds were obtained by the reaction of $ReCl_3(CH_3 CN)(PPh_3)_2$ with $NaR_2 Dtc$ in acetone. With R = Et the monomeric tris chelate is obtained. However, with R = Ph, the deep purple crystalline $Re(Ph_2 Dtc)_3 PPh_3$ is isolated. The IR spectrum of this latter complex indicates the presence of a unidentate $Ph_2 Dtc$ ligand (559), and it is suggested to contain six-coordinate Re(III). In two papers Rowbottom and Wilkinson uncovered the rich coordination chemistry of Re(III) and Re(V) ions with dithiocarbamate ligands (558, 559).

The reaction of $ReOCl_3(PPh_3)_2$ with $NaR_2 Dtc$ (R = Me, Et, Ph) in acetone affords brown, crystalline, diamagnetic $Re_2 O_3(R_2 Dtc)_4$ complexes. A strong band in the IR spectrum between 660 and 670 cm^{-1} is assigned to the Re—O—Re group, while a band at 955 cm^{-1} is assigned to the Re=O vibration. No other data are available for these interesting complexes; however, the x-ray crystal structure of the $Re_2 O_3(Et_2 Dtc)_4$ complex has been determined. In this molecule the linear O—Re—O—Re—O array is observed with two octahedrally coordinated Re(V) ions sharing a bridging oxygen (Fig. 29).

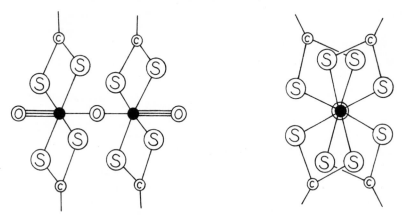

Fig. 29.   Two views of the idealized structure of the $Re_2O_3(Et_2Dtc)_4$ complex (248).

The dithiocarbamate ligands are bidentate, and normal Re–S distances are observed (Table XVIII). The very short terminal Re–O bonds [1.722(7) Å] are considered to be "normal" triple bonds. The formation of the triple bond is accounted for by overlap between Re empty d_{yz} and d_{xz} orbitals with P_y and P_x orbitals on the terminal oxygen, in addition to the σ bond formed by a d^2sp^3 hybrid ($5d_{z^2}$, $5d_{x^2-y^2}$, $6s$, $6p_x$, $6p_y$, $6p_z$) orbital on the Re atom and the p_x orbital on the oxygen atom (248). The $Re_2O_3(Et_2Dtc)_4$ complex also can be obtained from the reactions between $NaEt_2Dtc$ and $ReOBr(PPh_3)_2$, Re_2O_3-$(py)_4Cl_4$, $ReCl_4(MeCN)_2$, or $ReCl_4(PPh_3)_2$.

The reaction of R_4Tds with $ReOX_3(PPh_3)_2$ (X = Cl, Br) results in the formation of the $ReOX(R_2Dtc)_2$ complexes (R = Me, Et, or R_2 = Pip). These compounds (magnetic properties?) react with H_2O in acetone in the presence of Na_2CO_3 to give the $Re_2O_3(R_2Dtc)_4$ complexes quantitatively (559). The $ReOCl_2(Et_2Dtc)PPh_3$ is obtained as blue crystals by the reaction of $ReOCl_3$-$(PPh_3)_2$ with $Tl(Et_2Dtc)$ in dry acetone in a 1:1 molar ratio (559). The reaction of aniline with the $Re_2O_3(Et_2Dtc)_4$ complex in refluxing benzene affords red-brown crystals of $[Re(NPh)(Et_2Dtc)_2]_2O$, which is thought to contain the PhN=Re–O–Re=NPh array. The dark green crystalline $Re(NPh)Cl(Et_2Dtc)_2$ complex is obtained in high yield by the reactions of $Re(NPh)Cl_3(PPh_3)_2$ with Et_4Tds in acetone. The Re(III) complex $ReCl_2(Et_2Dtc)(PPh_3)_2$ is obtained by refluxing a solution of $ReOCl_2(Et_2Dtc)(PPh_3)_2$ in alcohol in the presence of PPh_3 (559).

The interaction between $ReNCl_2(PPh_3)_2$ and $NaEt_2Dtc \cdot 3H_2O$ in acetone affords yellow crystals of $ReN(Et_2Dtc)_2$ (559). The crystal structure of this diamagnetic molecule has been determined (247, 249) and reveals a monomeric five-coordinate Re(V) in a distorted square pyramidal coordination with two bidentate Et_2Dtc ligands at the base and the nitride at the apex of the pyramid.

The short Re—N bond [1.656(8) Å] is considered to be a triple bond on the basis of arguments previously presented for the Re—O terminal triple bond in the $Re_2O_3(Et_2Dtc)_4$ complex (see above).

The oxidative addition of Et_4Tds to $Re(CO)_5Cl$ in benzene solution affords the $Re(CO)_2(Et_2Dtc)_3$ complex, which is diamagnetic and monomeric in solution. The IR spectra suggest bidentate chelation of the R_2Dtc ligands, and, consequently, a rather unusual eight-coordinate Re(III) (558). In the same reaction the $[Re(Et_2Dtc)_4][Re(CO)_3Cl(Et_2Dtc)]$ complex was isolated and was found to be a diamagnetic 1:1 electrolyte. The cation and anion in this salt have been separated, and the $(n$-$Bu_4N)[Re^{(I)}(CO)_3I(Et_2Dtc)]$ and $BF_4[Re^{(V)}(Et_2Dtc)_4]$ have been isolated (558). A red crystalline paramagnetic $ReCl_2(Et_2Dtc)$ can be isolated from the reaction of $ReCl_4(PPh_3)_2$ with Et_4Tds in benzene, which also affords the $Re(V)(Et_2Dtc)_4^+$ complex cation (558). The stoichiometry of the $ReCl_2(Et_2Dtc)$ complex is rather unusual and this compound should be reinvestigated.

c. **Organometallic Derivatives of Manganese and Rhenium.** Organometallic derivatives of 1,1-dithio Mn and Re complexes are obtained by the CS_2 insertion into the M—H bond of various carbonyl hydride complexes of Mn and Re.

The synthesis of the $M(S_2CH)(L)(CO)_3$ complexes [M = Mn, Re; L = diphenylphosphinomethane (DPM) or bisdiphenylphosphinoethane (DPE)] is feasible by the reaction of the cis-$HM(L)(CO)_3$ complexes with CS_2 (210). The $M(CO)_3(DPE)(HCS_2)$ complexes appear to contain a monodentate dithioformate ligand, and strong C=S vibrations are found in the IR and Raman spectra of these complexes at about 1000 cm^{-1}. With the $HM(CO)_3DPM$ complexes, however, the CS_2 insertion reaction is followed by nucleophilic attack of one of the P atoms of the DMP ligand at the HCS_2 carbon (following scission of the M—P bond) and the formation of the novel $R_2PCH_2PR_2CHSS$ tridentate ligand (210). The C—S vibrations in these complexes occur at about 818 cm^{-1}. It would be interesting to investigate the possibility that deprotonation of the ligand takes place to give the $R_3P=CS_2^{2-}$ ylid, which may be stabilized by coordination to the metal ion. Analogous nucleophilic attack by a phosphorus on other coordinated dithioacid ligands (MeDta, PhDta, etc.) may also occur with the formation of the corresponding zwitterionic R_3P^+—$RCSS^{2-}$ ligands.

The crystal structure of the $Mn(CO)_3(DPMCHS_2)$ complex shows this intriguing ligand with a HCS_2 sp^3 hybridized carbon atom and a normal P—C carbon bond of $1.79(3)$Å in the HS_2C—P moiety (210) (Fig. 30).

The CS_2 insertion also occurs in the reaction of the $HRe(PPh_3)_3(CO)$ complex with CS_2 (256). The crystal structure of the $Re(CO)_2(HCS_2)(PPh_3)_2$ product has been determined (8). The molecule shows a distorted octahedral

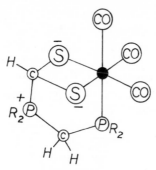

Fig. 30.   Schematic structure of the Mn(CO)$_3$ (DPMCHS$_2$) complex (210).

coordination, and two cis positions are occupied by the sulfur atoms of the CS$_2$ group. The two CO groups occupy positions trans to the sulfur atoms.

The chelated HCS$_2^-$ group in the structure shows C–S distances of 1.64(2) and 1.68(1) Å and a S–C–S angle of 116.7(1)°. Although the hydrogen atom on the HCS$_2^-$ unit was not located in the structure, a peak in the PMR spectrum at 4.85τ is probably due to the hydrogen atom of the SSCH group (256).

The insertion of CS$_2$ into transition metal–carbon σ bonds also has been reported (408). The complexes (LCS$_2$)M(CO)$_4$ (L = Ph, p-CH$_3$–Ph, Me; M = Mn, Re) are obtained by the reaction of CS$_2$ and the LM(CO)$_4$ complexes in CS$_2$. The structure and bonding of these new dithiocarboxylate complexes are discussed on the basis of their mass, PMR, and IR spectra (408).

The (Ph$_3$P)Mn(CO)$_3$(Me$_2$Dtc) complex was obtained as a by-product in the synthesis of the thiocarboxamido complex (Ph$_3$P)Mn(CO)$_3$CSN(Me)$_2$ (183, 184).

The structure of this complex has been determined (183). It is very similar to that of Re(CO)$_2$(HCS$_2$)(PPh$_3$)$_2$ (see above), if the HCS$_2^-$ ligand is replaced by Me$_2$Dtc$^-$ and one of the PPh$_3$ ligands is replaced by CO (Table XVIII).

The formation of this complex may arise from sulfur insertion into the M–C bond of the corresponding thiocarboxamido complex (183).

8. Group VIII

a. **Iron, Ruthenium, Osmium.** *(1) Iron.*   Since 1970 research in the chemistry of the iron 1,1-dithio chelates has produced new compounds and has probed in detail phenomena reported prior to 1970. Outstanding among the new synthetic contributions are complexes that contain iron in either low [Fe(II)] or high [Fe(IV)] formal oxidation states and mixed-ligand complexes. Physical studies on the 1,1-dithio chelates of iron concentrated on the spin-crossover behavior of the R$_2$Dtc complexes, which was probed by crystallographic, magnetic resonance, and Mössbauer techniques. Other areas that were studied in considerable detail are the chemistry and electronic structures of the five-coordinate $XFe(R_2Dtc)_2$ complexes.

(a) Synthesis and Reactions. Complexes of iron with remarkable properties were obtained (512) in the anaerobic reaction between $Fe(R_2Dtc)_2$ and bis(perfluoromethyl-1,2-dithietene) (TFD) in THF solution. The complexes, $Fe(R_2Dtc)_2$ (TFD), are considered oxidative addition products with the iron "oxidized above the divalent state and the dithietene reduced below the α-dithione level." Outstanding properties of these complexes are: (1) a singlet–triplet spin-state equilibrium and isotropically shifted PMR spectra and (2) at least two intramolecular rearrangement reactions that differ by $\sim10^3$ in rate at $25°C$.

Of the rearrangement reactions the low-temperature one has been identified by NMR to be an inversion of molecular configuration. The high-temperature process is assigned to a C–N bond rotation. For the first process, activation energies for different R groups vary from 9.6 ± 2.1 to 17.2 ± 2.4 kcal mole^{-1}. Evidence is presented for the operation of a twist mechanism in the rearrangement reactions of chelates. The physical basis for this mechanism is proposed to be a distortion toward a trigonal prismatic configuration. These compounds also show redox properties indicative of a three-membered electron transfer series with the $0 \rightleftharpoons -1$ interconversion electrochemically reversible.

Similar mixed-ligand complexes of the type $(R_1R_2Dtc)_2(MNT)Fe$ have been synthesized. The complexes were obtained initially as dianions, $[(R_1R_2-Dtc)_2(MNT)Fe]^{2-}$, and were subsequently oxidized either by air or $Cu(II)$ ions in acetonitrile (510). They also exhibit the singlet–triplet equilibrium; however, they show a higher population of the triplet state than is found for the TFD analogues. The complexes are stereochemically nonrigid and display the same type of kinetic processes as their TFD counterparts. Thermodynamic activation parameters for inversion of the two complexes (TFD versus MNT) do not differ within experimental error.

An x-ray study of $Fe(Et_2Dtc)_2$ TFD has shown that the molecule is chiral (crystallographic C_2 symmetry) and that there is considerable double-bond character in the C–N bond of the R_2Dtc groups (365). The coordination of the six sulfur atoms (Table XIX) about the iron atom can be described as an octahedron distorted toward trigonal prismatic geometry. On the basis of the TFD structural parameters, an effective $1-$ oxidation state for the ligand was proposed, and consequently an effective $3+$ oxidation level was suggested for the iron. This description obviously is not in accord with the classical properties of Fe(III) complexes, where an odd number of unpaired electrons are expected for the ground state.

The first report on the synthesis of $Fe(IV)R_2Dtc$ complexes was given by Pasek and Straub (500). The $[Fe(IV)(R_2Dtc)_3]^+$ complexes were obtained by the oxidation of the $Fe(R_2Dtc)_3$ complexes with BF_3 in the presence of air.

The magnetic moments (3.37–3.15 BM) and Mössbauer spectra (at room temperature, Isomer Shift, $IS = 0.46 \pm 0.02$ mm sec^{-1} and Quadrupole Splitting, $QS = 2.0$–2.3 mm sec^{-1}) are consistent with a Fe(IV) ion and a low-spin d^4 configuration. A C–N vibration higher by 30–40 cm^{-1} than that found in the Fe(III) complexes is also consistent with the increased contribution of the $R_2N^+{=}CS_2^{2-}$ resonance form, expected to occur in the ligands of the Fe(R_2Dtc)$_3^+$ complex (500). The structure of the ClO$_4^-$ salt of the [Fe(PyrrolDtc)$_3$]$^+$ cation has been determined (428). The coordination geometry (Table XIX) is very similar to that observed in the structure of the only other 1,1-dithio Fe(IV) complex reported, [Fe(DED)$_3$]$^{2-}$ (DED^{2-} = 1,1 dicarboethoxy-2,2-ethylene dithiolate) (158, 337), and conforms neither to an ideal octahedral nor trigonal prismatic arrangement. The geometry of the MS$_6$ moiety, in fact, has been described (158) as originating from a trigonal prism that suffers individual rotations of the chelating ligands around the C_2 axis (Fig. 55) (see also Section III.C). The synthesis of Fe(IV)(R$_2$Dtc)$_3^+$FeCl$_4^-$ and Fe(R$_2$Dtc)$_3^+$ClO$_4$ has been achieved by the addition of FeCl$_3$ or Fe(ClO$_4$)$_3$·6H$_2$O, respectively, to Fe(R$_2$Dtc)$_3$ in

TABLE XIX
Magnetic and Structural Characteristics of the Iron 1,1-Dithio Complexes.

Complex	$\overline{\text{Fe–S}}$, Å[a]	$\overline{\text{C–N}}$, Å[a]	ν(C–N), cm^{-1}	μ_{eff}, BM	Ref.
Fe(Me$_2$Dtc)$_2$TFD			1533	1.01[b]; 1.13, 1.17[c]	512
Fe(Et$_2$Dtc)$_2$TFD	2.310 (3)[d] 2.195 (3)[e]	1.302	1507	2.25[b]; 1.34[c]	365, 512
Fe(MeBzDtc)$_2$TFD			1508	1.75[b]; 1.20[c]	365, 512
Fe(PyrrolDtc)$_2$TFD			1501	1.13[b]; 1.50[c]	365, 512
Fe(MePhDtc)$_2$TFD			1472	1.27[b]; 0.9 [c]	365, 512
Fe(PipDtc)$_2$TFD			1503	1.18[b]; 1.42[c]	
(Et$_4$N)[Fe(Et$_2$Dtc)$_2$MNT]			1490	2.07[f]	510
Fe(Et$_2$Dtc)$_2$MNT			1501	2.43[f]; 2.46[c]	510
Fe(MePhDtc)$_2$MNT				1.80[c] (24°C)	
Fe(Et$_2$Dtc)$_3$BF$_4$			1520	3.28[g]	500
Fe(Me$_2$Dtc)$_3$BF$_4$			1560		500
Fe(i-Pr$_2$Dtc)$_3$BF$_4$			1500	3.29[g]	500
Fe(Cx$_2$Dtc)$_3$BF$_4$			1490	3.22	500
Fe(PyrrolDtc)$_3$BF$_4$[h]			1470	3.37	500
Fe(PyrrolDtc)$_3$ClO$_4$	2.300 (2)	1.32 (2)			428
Fe(PyrrolDtc)$_3$	2.41 (1)	1.31 (4)		5.9	303
Fe(Et$_2$Dtc)$_3$[i]	2.357 (3)	1.33 (1)	1483	4.3[i]	226, 403
Fe(Et$_2$Dtc)$_3$[j]	2.306 (1)	1.323 (6)		2.2[j]	226, 403

TABLE XIX *continued*

Complex	$\overline{\text{Fe–S}}$, Å[a]	$\overline{\text{C–N}}$, Å[a]	$\nu(\text{C–N})$, cm^{-1}	μ_{eff}, BM	Ref.
Fe(MePhDtc)$_3$	2.32 (1)	1.37 (3)		2.9	303
Fe(MorphoDtc)$_3$[k]	2.317 (5)	1.319 (11)		3.5	96[b]
Fe(MorphoDtc)$_3$[l]	2.353 (12)	1.324 (10)			96
Fe(Bu$_2$Dtc)$_3$[m]	2.341 (4)	1.33 (2)		3.6	463
Fe(MorphoDtc)$_3$[n]	2.435 (14)	1.31 (2)		5.5	302
Fe(MorphoDtc)$_3$[o]	2.416 (12)	1.323 (8)		5.1	96, 174
Fe(pyrrolDtc)$_3$[p]	2.432 (10)	1.317 (5)		5.6	575
Fe(MorphoDtc)$_3$[q]	2.442 (10)	1.318 (3)		'5.7'[r]	96
[Fe(Et$_2$Dtc)$_2$]$_2$	2.425 (24)	1.330 (6)			354
Fe(EtXant)$_2$				4.4[f]	330
Fe(EtXant)$_3^-$(Et$_4$N)$^+$				5.5[f]	330
Fe(EtXant)$_3$	2.316			2.7	340, 618[b]
α-Fe(EtXant)$_2$bipy				4.8	330
Fe(EtXant)$_2$phen				5.1	330
Fe(t-BuSXant)$_3$	2.297 (7)			2.46	406
ClFe(Et$_2$Dtc)$_2$[s]	2.30 (1)	1.36 (3)		4.00	345, 429
BrFe(Et$_2$Dtc)$_2$	2.42				114
IFe(Et$_2$Dtc)$_2$[t]	2.27 (2)	1.37 (3)			304
Fe(Salen)(PyrrolDtc)	2.452 (2) 2.566 (1)	1.312 (6)		5.68	159
Fe(p–Cl–PhNC)$_2$(Et$_2$Dtc)$_2$ ZnI$_2$[u]	2.374 (5)[v] 2.305 (6)	1.33 (2)	1516	Diamagnetic	439

[a] Average values; numbers in parentheses are deviations from the mean.
[b] Solid-state measurement, 22°C.
[c] Solution measurement in CH$_2$Cl$_2$ at 30°C.
[d] For the R$_2$Dtc ligand.
[e] For the TFD ligand.
[f] Solid-state measurement, 24°C.
[g] Solid-state measurement, 31°C.
[h] Pyrrol = pyrrolidyl.
[i] Room temperature data.
[j] 79°K data.
[k] Dibenzene solvate.
[l] Nitrobenzene solvate.
[m] Hemibenzene solvate.
[n] CH$_2$Cl$_2$ solvate.
[o] CHCl$_3$ solvate.
[p] Hemibenzene solvate.
[q] Water solvate.
[r] Value estimated from Fig. 6, Ref. 96a.
[s] Fe–Cl, 2.26 (1) Å.
[t] Fe–I, 2.59 (1) Å.
[u] Zn–S, 2.422 (5) Å; Zn–I, 2.552 (3) Å; Fe–C, 1.83 (2) Å.
[v] Bridging S atoms.

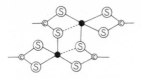

Fig. 31.   Idealized structure of the $[Fe(Et_2 Dtc)_2]_2$ complex (354).

acetone or alcohol. IR visible and Mössbauer spectra for these complexes have been reported (270).

The structure of the $[Fe(Et_2 Dtc)_2]_2$ complex that was synthesized previously (236, 512) has been determined (354). The iron atoms in the dimer are surrounded by a distorted trigonal bipyramid arrangement of the sulfur donors (Fig. 31).

Four of the Fe–S bond lengths are relatively similar. However, the fifth bond is rather long (2.613(2) Å) (Table XIX) (Fig. 31). The dimer actually can be "viewed as one formed from two bridging monodentate dithiocarbamate ligands" (354). The experimental data (Mössbauer and electronic spectrum) for this compound have been interpreted in terms of a high-spin ground state for the Fe(II) ion (192). A molecular orbital calculation, on which the above interpretation was based, was carried out assuming square pyramidal coordination (190). In view of the trigonal bipyramidal structure of the iron atoms in the $[Fe(Et_2 Dtc)_2]_2$ dimer, these calculations and interpretations should be reevaluated.

The isolation of "salts" of the $Fe(R_2 Dtc)_3^-$ complexes has been reported (192). The Mössbauer spectra of these complexes indicate the presence of high-spin Fe(II). The $10 Dq$ value of $R_2 Dtc^-$ was found to be $8000\ cm^{-1}$, considerably smaller than the value observed in the Fe(III) complexes.

In contrast to the previous work (192) Holah and Murphy report that all attempts to make complexes of the type $Fe(R_2 Dtc)_3^-$, in various solvents and employing a multitude of synthetic procedures, failed (330). An examination of the analytical data reported for the $Fe(R_2 Dtc)_3^-$ complexes (192) leaves some doubt as to their analytical purity. However, the reported equilibria (Eq. 26) may account for difficulties in synthesizing and purifying these complexes.

$$(R_4 N)Fe(Et_2 Dtc)_3 \underset{\substack{excess \\ R_4 N^+}}{\overset{H_2 O}{\rightleftharpoons}} Fe(Et_2 Dtc)_3^- + R_4 N \underset{\substack{excess \\ Et_2 Dtc^- \\ + R_4 N^+}}{\overset{H_2 O}{\rightleftharpoons}} Fe(Et_2 Dtc)_2 + Et_2 Dtc^- \qquad (26)$$

The synthesis of the blue $Fe(R_2 Dtc)_2(o$-phen$)$ and $Fe(R_2 Dtc)_2(bipy)$ complexes has been reported (330). The visible spectra of these complexes are characteristic of octahedral coordination.

The syntheses and properties of the $Fe(RXant)_2$ and $Fe(RXant)_3$ complexes have been described. The powder pattern of the $Fe(EtXant)_2$ complex is identical to the Mn(II) analogue but different than the corresponding Zn(II) or

Ni(II) complexes. The magnetic moments and reflectance spectra suggest six-co-ordinate Fe(II), presumably possible through polymerization (330). The Fe-$(EtXant)_3^-$ complex shows a powder pattern identical to that of the Ni$(EtXant)_3^-$ complex for which the octahedral coordination has been established by x-ray crystallography (177). The compound is high spin (Table XIX). The high-spin base adducts Fe$(EtXant)_2$(bipy) (α and β form) and the Fe$(EtXant)_2$ o-phen also have been reported (330).

Reactions of Fe(III) xanthates with pyridine have been reported by Saleh and Straub (561). In these reactions bright yellow complexes of the type Fe(II)$(RXant)_2$$(py)_2$ have been isolated. The magnetic moments (4.9–5.0 BM), isomer shifts (1.2 mm sec^{-1}), and large quadrupole splittings ($\sim$3 mm sec^{-1}, room temp.) of these complexes support the apparent reduction of the Fe$(RXant)_3$ complexes (Eq. 27).

$$2Fe(RXant)_3 + 4py \longrightarrow 2Fe(RXant)_2py_2 + ROC(S)SSC(S)OR \qquad (27)$$

Few Fe(III) complexes have been reported with dithioaliphatic or dithio-aromatic ligands. The deeply colored tris-complexes with the PhDta$^-$ and BzDta$^-$ ligands have been reported to contain a pseudooctahedral Fe(III)S$_6$ chromophore (111). They all are low spin as expected from the high field strength of the ligands. The Mössbauer spectra of these compounds were record-ed (111). Attempts to prepare the Fe(PhDta)$_2$Cl complex by the reaction of the tris complex with HCl failed.

The preparation and properties of Fe(t-BuSXant)$_3$ have been reported (163, 227). Unlike the analogous complexes with R = Et, n-Pr, n-Bu, and Bz, this complex does not eliminate CS$_2$ to form the dimeric [Fe(RSXant)$_2$(SR)]$_2$ complexes (163). The Fe(t-BuSXant)$_3$ complex undergoes reversible one-elec-tron oxidation (+0.73 V) and one-electron reduction (−0.36) (in CH$_2$Cl$_2$ versus an Ag|AgI reference electrode).

A detailed study of the PMR spectra, IR spectra, and electrochemical behavior of the dimeric [Fe(RSXant)$_2$SR]$_2$ complexes is available (163). The synthesis and electronic properties of the unstable precursors Fe(RSXant)$_3$ have been reported (227, 504).

The structure of the monomeric low-spin (2.46 BM, room temp.) Fe(t-BuSXant)$_3$ complex has been determined (406) (Table XIX). The coordination geometry of the iron atom is a distorted octahedron of sulfur atoms contributed by the three chelating thioxanthate ligands. The structure of the low-spin (μ_{eff} = 2.7 BM) Fe(EtXant)$_3$ complex has been found to contain a distorted octahedral FeS$_6$ core (340).

The reactions of the Fe(Et$_2$Dtc)$_3^{+,0}$, IFe(Et$_2$Dtc)$_2$ and [Fe(Et$_2$Dtc)$_2$]$_2$ complexes with DPPE, CNR, and PPh$_3$ have been described (439) (DPPE = Ph$_2$-PCH$_2$CH$_2$PPh$_2$).

The Fe(IV) complexes upon reaction with p-Cl–PhNC give the yellow

$[Fe(p\text{-Cl}-PhNC)_4(Et_2Dtc)]^+$ and Et_4Tds. The reaction with PPh_3 affords Ph_3-PS, $[Ph_3P(CNEt_2S)]^+$, and $Fe(Et_2Dtc)_3$. In the presence of an excess of PPh_3, a red insoluble solid was obtained but was not characterized. With DPPE the $[Fe(DPPE)(Et_2Dtc)_2]^+$ complex was isolated.

$Fe(Et_2Dtc)_3$ reacts with RNC, under anaerobic conditions and in the absence of light, to give the Fe(II) complexes $Fe(CNR)_2(Et_2Dtc)_2$ (R = i-Pr, t-Bu, or p-ClPh). The reduction of the Fe(III) ion is accompanied by oxidation of the Et_2Dtc ligands to Et_4Tds. The reaction of $Fe(Et_2Dtc)_3$ with DPPE in refluxing acetone affords the $Fe(DPPE)(Et_2Dtc)_2$ complexes. The reaction of $IFe(Et_2Dtc)_2$ with either p-Cl–PhNC or DPPE in the presence of PF_6^- gives the $[(p\text{-Cl}-PhNC)_2Fe(Et_2Dtc)_2]PF_6$ and $[Fe(DPPE)(Et_2Dtc)_2][PF_6]$ complexes, respectively.

The electrochemistry and spectral properties of these complexes have been examined. The IR spectra of the $Fe(L)_2(Et_2Dtc)_2$ complexes suggest a cis geometry.

The interesting reaction of the $Fe(Et_2Dtc)_3$ with ZnI_2 in the presence of p-Cl–PhNC results in the formation of the binuclear $[Fe(p\text{-Cl}-PhNC)_2(Et_2Dtc)_2][ZnI_2]$ complex.

The crystal structure of this molecule has been determined. The iron is found in a distorted octahedral coordination, and the zinc is tetrahedrally coordinated. The ZnI_2 unit shares the Et_2Dtc sulfur atoms trans to the RNC groups with the iron (Fig. 32). Attempts to use the $Fe(Et_2Dtc)_3$ complexes as ligands for ZnI_2 have not been successful (439).

The fungicidal activity of the $Fe(R_2Dtc)_3$ complexes have been found to correlate with the frequency of the $S_2C\text{-}NR_2$ vibration. Thus the highest activity was observed for the complexes with a high frequency for this vibration.

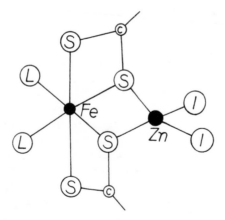

Fig. 32. Schematic structure of the $[Fe(L)_2(Et_2Dtc)_2][ZnI_2]$ complex; L = p-Cl–PhNC (439).

A model has been presented that relates the biological activity of the $Fe(R_2-Dtc)_3$ complexes to their ability to fit into a receptor site and undergo reversible redox reactions at suitable potentials (276). It seems quite likely that the relative solubility properties of these complexes in water are determined to some extent by the contributions of the $^{2-}S_2C={}^+NR_2$ resonance form to the structure of the ligands. The observed correlations, therefore, could be related to the solubility of these complexes in water.

The synthesis of the $(\eta^5\text{-Cp})Fe(CO)_2(Me_2Dtc)$ was accomplished by the reaction of $NaMe_2Dtc$ with $(\eta^5\text{-Cp})Fe(CO)_2Cl$ in acetone (490). A unidentate Me_2Dtc ligand is present in this complex, as evidenced by the IR spectrum.

(b) Five-Coordinate 1,1-Dithio Complexes. Since the initial synthesis of the $XFe(R_2Dtc)$ complexes $(X = Cl^-, Br^-, I^-)$ (429) considerable emphasis has been placed on the structural and physical properties of these compounds. The structures of the monomeric chloro (345) and iodo (304) derivatives have been determined. The configuration about the iron atom in both structures is approximately square pyramidal. The iron atom lies above the plane defined by the four sulfur atoms and toward the halogen that occupies the apex of the pyramid. In the chloride complex the Fe(III)-to-plane distance is 0.62 Å and the Fe—Cl bond is 2.26(1) Å. In the iodo derivative the same Fe(III)-to-plane distance is observed (0.6 Å) and the Fe—I bond length is 2.59(1) Å. Other data on these complexes can be found in Table XIX.

The structure of the bromo analogue also was determined and was found to be very similar to that of the chloro complex (Fe—Br, 2.42 Å) (114). The $XFe(R_2Dtc)_2$ complexes were "rediscovered" in 1975 (527) and a series of these molecules were obtained with the piperidine, thiomorpholine, and morpholine dithiocarbamates.

The $ClFe(R_2Dtc)_2$ complex also was obtained (456) by the interesting photolysis reaction of $Fe(R_2Dtc)_3$ complexes in $CHCl_3$ or C_6H_5Cl. The proposed mechanism for this reaction is shown in Eq. 28.

Recently the reaction of $IFe(R_2Dtc)_2$ and $Fe(R_2Dtc)_3$ with iodine was reported to give $I_2Fe(R_2Dtc)_2$ complexes $(R = Me, i\text{-Pr}, R_2 = Pyrrol) I_3Fe(R_2-Dtc)_2$, $(R = Et)$ and $I_3Fe(R_2Dtc)_3$ $(R = i\text{-Pr}, Cx, Bz)$. Mössbauer spectroscopy was used to show that all the above complexes contained Fe(III) with the exception of the $I_3Fe(R_2Dtc)_3$, which contained Fe(IV) (498).

The $I_3Fe(Et_2Dtc)_2$ was assumed to contain coordinated I_3^-. The magnetic moment of this complex was not reported. It would be interesting to know if this complex has a quartet ground state, since the quadrupole splitting is considerably larger (3.65 mm sec^{-1}) than the $IFe(Et_2Dtc)_2$ complex [2.87 mm sec^{-1} (219)]. Unambiguous structures could not be formulated from the Mössbauer data for the $I_2Fe(R_2Dtc)_2$ complexes. It is possible that the molecular formula is a multiple of the empirical formula. Molecular-weight and conductivity studies may shed some light on the nature of these interesting molecules.

Another series of $XFe(R_2Dtc)_2$ molecules were obtained (499) when the $Fe(R_2Dtc)_3$ complexes were reacted with AgX (X = NCO$^-$, NCS$^-$, NCSe$^-$), RHgX (X = Cl), RX (R = CH$_3$, X = I; R = allyl, X = Br; R = totyl, X = Cl), and strong organic acids XH (X=CF$_3$COO$^-$, picrate). The average isomer shift for these complexes is 0.64 ± 0.02 mm sec^{-1}. This constancy was interpreted to indicate that the total π bonding in these compounds is invariant. Although the quadrupole splittings are independent of R, the values depend on X and vary from 2.2 to 3.0 mm sec^{-1}. The splitting increases in the order NCO$^-$ < NCS$^-$ < Cl$^-$ < Br$^-$ < NCSe$^-$ ∼ CF$_3$COO$^-$ < I$^-$ ∼ picrate, an expected order that is inversely related to the spectrochemical series (499). The implied spin-quartet ground state for these molecules ought to be verified by magnetic measurements.

Various low-temperature magnetic susceptibility (114, 429), Mössbauer (114, 190, 620, 622), and EPR (114, 621) studies show that in the $XFe(R_2-Dtc)_2$ complexes the ferric ion is in a spin-quartet orbital singlet ground state. Far infrared spectroscopic studies have given an estimate of the zero field splitting (ZFS) for some of these molecules (67a, 67b). Paramagnetic anisotropy measurements in the 80–300°K temperature range on single crystals of the $XFe(Et_2Dtc)_2$ series (X = Cl, Br, I) have been reported (554). The ZFS values have been calculated using a spin-Hamiltonian formalism. For the $ClFe(Et_2Dtc)_2$ complex the values of the axial and rhombic ZFS parameters were D = 1.73 cm^{-1} and E = −1.19 cm^{-1} (266, 554).

These values are in reasonable agreement with the values reported on the basis of far IR work (67a, 67b). In general the anisotropy measurements establish that the ligand field in the $XFe(R_2Dtc)_2$ complexes is rhombic and provide a method for estimating spin-Hamiltonian parameters.

Although the molecular structures of the $XFe(Et_2Dtc)_2$ complexes are quite similar for different X groups, the magnetic properties of the ground states in the solid state are quite different. The I$^-$ compound is antiferromagnetic (T_N = 1.9°K), the SCN$^-$ and Br$^-$ homologues are paramagnets (to 1.4°K), and the Cl$^-$ homologue is a ferromagnet. These differences have been discussed in terms of exchange and uniaxial crystal field interactions (114). For the $ClFe(Et_2Dtc)_2$ complex Mössbauer studies between 2.5 and 300°K show a well-defined doublet with a practically temperature-independent quadrupole splitting of

2.68 mm sec^{-1} (114, 190, 620), magnetic ordering, and a well-defined sextet below 2.50°K. These and EPR results, which indicate a spin-quartet ground state (622), have been interpreted (5) as being indicative that the ground electronic state of the Fe(III) (C_{4v} symmetry) ion is an orbital singlet, $^4A_2$. The spin—orbit interaction with the excited states splits $^4A_2$ into two spin multiplets $M_s = \pm \frac{1}{2}$ and $M_s = \pm \frac{3}{2}$. In this crystal field investigation (5) a negative value for D and $E = 0$ were assumed. The apparent discrepancy between these values and those obtained from single-crystal studies (266, 554) suggests that a reexamination of the theory (5) is necessary.

In synopsis, the main features of the electronic properties of the ground states in the XFe(R_2Dtc)$_2$ complexes (67a, 395, 554) are: (1) a dependence of the sign of the ZFS on the X group and/or the alkyl groups R and (2) the existence of intermolecular exchange interactions of comparable magnitude to the crystal field splitting, which in some cases lead to magnetic ordering at temperatures below 4.2°K. The relation between the occurrence of magnetic ordering and the character of the ground-state Krammers doublet at present is not well understood (5, 114).

PMR spectra between 60 and −60°C on the XFe(R_2Dtc)$_2$ complexes allowed for the measurement of paramagnetic anisotropies and an estimation of the pseudocontact shift from these measurements (194). A large effect of the T^{-2} term in the iodo derivative was rationalized in terms of a large ground-state zero field splitting of the Fe(III) ion. The temperature-variation study revealed a conformational change, which was ascribed to hindered rotation about the C$\dot{=}$N bond in the ligand.

Several other studies concerning the proton magnetic resonance spectra of the XFe(R_2Dtc)$_2$ complexes have been reported (114, 202, 284, 401) and the results have been interpreted in terms of current relaxation theory. In a number of these studies (114, 202, 498) the contact contribution to the isotropic contact shift of the proton resonances was considered to be controlled by the transmission of unpaired spin density through the π framework of the ligand.

A new series of Fe(III) complexes with the Salen (N,N'-ethylenebissalicylaldiminate) anion and the dithiocarbamate ligands have been synthesized and their magnetic properties have been determined (505, 576). Four different derivatives were obtained with different R groups of the type Fe(Salen)(R_2Dtc) (R_2 = Et$_2$, n-Bu$_2$, Pyrrol, Pip). Mössbauer studies in the temperature region 1.4–300°K show that the iron ion is in the $S = \frac{5}{2}$ spin state. ZFS are positive for R = Et, n-Bu and negative ($\sim$−1 cm^{-1}) for the pyrrolidyl and piperidyl derivatives. The pyrrolidyl derivative orders magnetically at 2.0°K. The x-ray isomorphous piperidyl homologue does not show ordering down to 1.4°K; however, magnetic susceptibility data suggest that ordering is likely at lower temperatures. The crystal structures of the Fe(Salen)(PyrrolDtc) and Fe(Salen)(PipDtc) complexes have been determined. Both structures are extremely similar (159) and the coordination about the iron, to a first approximation, can be described as highly distorted octahedral. An examination of the rather long Fe–S bond lengths (Table XIX) (Fig. 33) suggests an alternate description for the coordination geometry. If the R_2Dtc^- ligand is assumed to occupy one coordination site, the coordination geometry can be envisioned as distorted trigonal bipyramidal with the R_2Dtc ligand O(1) and N(2) occupying the equatorial triangle and N(1) and O(2) on the pyramid axis (Fig. 33).

(c) Electrochemical Properties of Fe(III) Dithio Complexes.   An examination of the electrochemical properties of the Fe(III) dithio chelates was not made until recently (110, 116, 443). A comprehensive study of the electron transfer behavior of 20 iron(III) dithiocarbamates has been conducted in acetone at a platinum electrode (115).

The $Fe(R_2Dtc)_3$ complexes undergo a simple reversible one-electron oxidation consistent with the equation

$$Fe(R_2Dtc)_3 \rightleftharpoons Fe(R_2Dtc)_3^+ + e \qquad (29)$$

The redox potentials of the Fe(III)–Fe(IV) couples (0.42–0.66 V) are independent of whether the data are measured on $Fe(R_2Dtc)_3$ or $Fe(R_2Dtc)_3^+$ solutions. A one-electron reversible reduction is also observed with reduction potentials spanning the region -0.25 to 0.55 V. The redox couple involves the anionic complexes $Fe(R_2Dtc)_3^-$:

$$Fe(R_2Dtc)_3 + e \rightleftharpoons Fe(R_2Dtc)_3^- \qquad (30)$$

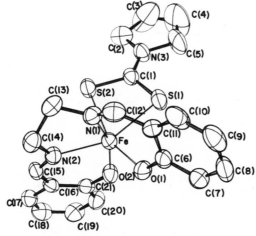

Fig. 33.   The molecular structure of the Fe(Salen)(PyrrolDtc) complex (159).

The reaction between $Fe(R_2Dtc)_3^+$ and R_2Dtc^- was followed by electrochemical means and was found to result in the formation of thiuram disulfides. While $HClO_4$ does not affect the Fe(IV) complexes, HCl destroys them.

The reaction of mercury with $Fe(R_2Dtc)_3$ complexes in the presence of O_2 has been reported to form $Hg(R_2Dtc)_2$. This may explain the complex electrochemistry on mercury electrodes found for $Fe(R_2Dtc)_3$, as well as for other complexes (115). The order for ease of reduction relative to substituents (R) is

$$\underset{\substack{\text{aryl substituents}}}{Ph_2 > Ph, Me} \; \underset{\substack{\text{satd. chain}}}{> Me_2 > Et_2} \; \underset{\substack{\alpha\text{ branched, satd. chains}}}{> \;\;\;(i\text{-Pr})_2 > (Cx)_2} \qquad (31)$$

In addition, there appears to be a steric effect depending on the expected R–N–R angle at the nitrogen atom. Ease of reduction of the $Fe(R_2Dtc)_3$ complex is accompanied by difficulty in oxidation of the same complex.

For the $Fe(RXant)_3$ complexes reductions at about +0.1 V and oxidations at about +1.2 V are observed. The ease of reduction decreases in the order: R = benzyl > Me > Et > i-Pr. The reduction involves only one electron and is reversible. The oxidation of $Fe(EtXant)_3$ is not strictly reversible. Mixed $Fe(R_2$-$Dtc)_3$ complexes show oxidation potentials that reflect the nature of the ligands in an additive fashion. The existence of the interesting $Fe(III)(R_2Dtc)_2^+$ observed in the titration of $Fe(i\text{-Pr}_2Dtc)_3$ with Fe(III) ions is claimed on the basis of an end point that corresponds to the $Fe(III)(R_2Dtc)_2^+$ cation (115). In solution $Fe(BzXant)_3$ and $Fe(i\text{-Pr}_2Dtc)_3$ undergo ligand exchange, as shown by their electrochemical behavior. The redox properties expected for a system in spin-state equilibrium with $K_{eq} < 10^{-7}$ sec have been discussed (115). $Fe(R_2Dtc)_3$ complexes with the stronger electron-releasing R groups exhibit magnetic moments in which the low-spin state predominates, as evidenced by a plot of the oxidation potential versus μ_{eff}^2 (115).

Determinations of reduction potentials for a series of Fe(III) and Mn(III) tris-dithiocarbamate complexes by voltametry in the presence of various concentrations of polar molecules were conducted. The $E_{1/2}$ values varied linearly with the mole fraction of the particular polar molecule present (274). A theoretical model that was consistent with the experimentally derived $E_{1/2}$ values was advanced.

Substituent effects have been interpreted in another way (213), namely, by considering the aqueous uncorrected pK_a values of the parent secondary amine HNR_1R_2 as a measure of the inductive effects of the R_1 and R_2 substituents. It has been argued that the greater the pK_a, the more favored the limiting form $^{2-}S_2C=\overset{+}{N}R_2$ is. This form has been associated with diminished covalency in the Fe–S bond and high-spin behavior (213).

In a study of the isotropic proton hyperfine interaction constants for various $Fe(R_2Dtc)_3$ complexes and their redox behavior, a simple relationship between $E_{1/2}$ and the NCH_2 proton hyperfine interaction constant was established (275). It was argued that the reversible reduction of the $Fe(R_2Dtc)_3$ complexes occurs through the ligand nitrogen atom rather than at the iron center. Furthermore, the more positive the nitrogen atom, the easier the reduction. Chant et al. (115) argue that since the $^{2-}S_2C=\overset{+}{N}R_2$ form is known to stabilize the high oxidation states, the exact opposite is true, as is evident in their electrochemical studies.

(d) Kinetic Studies on the Iron 1,1-Dithio Complexes. The progress in understanding the dynamic behavior of transition metal 1,1-dithio complexes has been reviewed (509). It is through the efforts of Pignolet et al. (205, 206a, 493, 495, 511, 531) that a considerable amount of information concerning the dynamic properties of Fe(III) dithiochelates is available. In view of the existing review, only a brief account of this area is presented here.

The stereochemical nonrigidity of the $Fe(MePhDtc)_3$ was the first to be recognized (493, 495). At $-108°C$ the complex is "frozen out" and two isomers are present (cis and trans). The cis isomer is unaffected upon warming. However, two of the three sites of the trans isomer are interchanged. No geometric cis–trans isomerization is observed, and at higher temperatures S_2C-N bond rotation occurs.

Similar studies were reported for the $Fe(Bz_2Dtc)_3$ and $Fe(MeBzDtc)_3$ complexes (493) and the $Fe(IV)(R_2Dtc)_3$ complexes (205). The rates of electron transfer between $Fe(R_2Dtc)_3$ and $Fe(R_2Dtc)_3^+$ complexes were studied by PMR, and rate constants (of the order of 10^{-8} and mole^{-1} sec^{-1}) were given (494).

Barriers to $S_2C\dot{=}N$ bond rotation in complexes of the type $Fe(R_2Dtc)_2$-o-phen, $Fe(R_2Dtc)_2Bipy$, $Fe(R_2Dtc)_3$, and $[Fe(R_2Dtc)_3]BF_4$ have been measured by PMR spectra line-shape analysis.

The kinetic parameters depend on the formal oxidation state of the iron, and $\Delta F^{\ddagger}$ (298°C) values of ~8.6, 12.0, and 15.0 kcal mole^{-1} were reported for the Fe(II), Fe(III), and Fe(IV) complexes, respectively (209).

(e) Mössbauer Studies. The use of Mössbauer spectroscopy has contributed significantly to the understanding of the electronic and structural properties of the iron 1,1-dithio complexes. Solvation effects were observed for the $XFe(III)(Et_2Dtc)_2$ complexes ($X = Cl^-$, SCN^-) (191) by Mössbauer spectroscopy in frozen solutions. A large reduction of the quadrupole splitting from solid state to frozen solution was explained in terms of a change in the iron coordination from five- to six-coordinate. A similar study on $XFe(R_2Dtc)_2$ complexes (415) reports the formation of a highly covalent six-coordinate complex when $XFe(R_2Dtc)_2$ complexes are dissolved in THF. Studies of the paramagnetic hyperfine structure and temperature dependence of the isomer shift and quadrupole splitting conclusively prove the formation of this high-spin ($S = \frac{5}{2}$) complex.

The Mössbauer spectra of various $Fe(R_2Dtc)_3$ complexes were studied from 100 to 300°K (454). The temperature dependence of the isomer shift was found to be nearly identical for a number of complexes. The temperature dependence of the quadrupole splitting was considered to fall "short of that expected on the basis of a $^2T_{2g} \rightarrow {}^6T_{1g}$ equilibrium." It was suggested (454) that these compounds could be described in terms of a mixed-spin ground state, a concept proposed by G. Harris on the basis of crystal field calculations (296, 297). The variable magnetic moment was proposed to arise as a result of a change in character of the ground state with temperature.

An extended Hückel calculation of the electric field gradient at the nucleus of the iron in the $XFe(R_2Dtc)_2$ complexes has been reported. It was shown that the abnormally large electric field gradient in the $ClFe(R_2Dtc)_2$ and $[Fe(R_2Dtc)_2]_2$ complexes is caused by covalency effects (190).

The Mössbauer spectra of Fe(III) doped in $M(PyrrolDtc)_3$ complexes [M = Co(III), Ga(III), and In(III)] at 298, 78, and 4.2°K show the usual six-line pattern due to magnetic hyperfine splitting. With M = Cr(III) a five-line spectrum is observed that is interpreted to be a result of spin–spin relaxation between Fe(III) and Cr(III) ions (397). The zero-field Mössbauer spectra of the $Fe(R_2Dtc)_2(MNT)$ and $Fe(R_2Dtc)_2(TFD)$ complexes and the $[Fe(R_2Dtc)_2 (MNT)]^-$ anion have been measured (514). For the neutral species the spectral parameters are nearly identical at 298 and 4.2°K. Since the magnetic properties of these complexes are consistent with a singlet–triplet spin equilibrium, the failure to detect the two spin-state isomers is suggested to imply an interconversion rate in excess of the reciprocal lifetime ($\sim 10^7$ sec^{-1}) of the excited ^{57}Fe nucleus.

(f) Magnetic Properties of the Fe(R$_2$Dtc)$_3$ Complexes. Outstanding among the magnetic properties of the 1,1-dithio complexes is the $^6A_1 - {}^2T_2$ spin-state equilibrium observed with the $Fe(R_2Dtc)_3$ complexes. This equilibrium was first observed by Cambi et al. (101) in 1931. This phenomenon was reinvestigated by Martin et al., and their studies appeared in a series of papers starting in 1964. These early studies have been reviewed (153). More recent reviews on this subject have appeared (431, 574) and the reader is referred to these reviews for details concerning the theory for the representation of the magnetic behavior of $Fe(R_2Dtc)_3$ complexes. An overview of the literature since these reviews were written is presented here.

The ligand field strength (Δ) and Racah parameters (B) of the dithiocarbamate ligands place the $Fe(R_2Dtc)_3$ complexes very near the $^6A_1 - {}^2T_2$ electronic crossover. As a result the magnetic and spectroscopic properties are very sensitive to changes of temperature and pressure. Furthermore, changes of the ligand substituents (R) are often sufficient to change the position of the $^6A_1 \rightleftharpoons {}^2T_2$ equilibrium through inductive, mesomeric, and steric effects.

A detailed study of the magnetic behavior of 18 $Fe(R_2Dtc)_3$ complexes with various R groups was carried out as a function of temperature and pressure (226). The magnetic susceptibility data as a function of temperature were fit by an equation that optimized the parameters g, C, and E (226), where g is the g value of the $^2T_{2g}$ state and E is the energy separation between the zero-point levels of the $^2T_{2g}$ and $^6A_{1g}$ states. The parameter C accounts for the ratio of the vibrational partition coefficients for the $^6A_{1g}$ and $^2T_{2g}$ levels. The physical meaning of C has been questioned, however (393). Various refinements of the theory of the spin crossover system have appeared, including configuration interaction and spin–orbit interaction of the $^6A_{1g}$ with the $^4T_{1g}$ state and of the $^2T_{2g}$ state with the $^4T_{1g}$ state (188). Figgis and Toogood (244) fit the data on Fe(Et$_2$Dtc)$_3$ from 4.2 to 100°K with the equations for a $^2T_{2g}$ level including spin–orbit interaction and trigonal distortion (D). A satisfactory fit was obtained only at very low temperatures. In their study of the "high-spin" Fe(PyrrolDtc)$_3$ between 295.5 and 2.26°K the authors were unable to totally account for their data (244) in terms of zero-field splitting of the $^6A_{1g}$ state. Higher order magnetic field effects and fourth-order ligand field terms have been included in a recent theoretical analysis (424). This theory was used to successfully fit the magnetic data for the Fe(PyrrolDtc)$_3$ complex obtained by Hall and Hendrickson (291).

The $^2T_2 \rightleftharpoons {}^6A_1$ equilibrium is pressure dependent and the $^2T_2$ state, having the smaller volume is favored by increasing pressure. The difference in volume ΔV is given by $\Delta V = RT[d \ln k/dP]$ where K is the population ratio of molecules in the $^6A_1$ and $^2T_2$ states (431). Studies under pressure (226) for the Fe(R$_2$Dtc)$_3$ complexes allowed Ewald et al. to evaluate ΔV, which is about 5–6 cm^3 mole^{-1}. This change, which can be attributed to the contraction of the FeS$_6$ core of a mole of the complex going from $^6A_1$ to $^2T_2$, represents a shortening of the Fe–S bond length in the $^2T_2$ state molecules by ~0.1 Å. The electronic spectra of the Fe(R$_2$Dtc)$_3$ complexes were recorded, and bands assigned to the $^6A_1 \rightarrow {}^4T_1$ and $^2T_2 \rightarrow {}^4T_1$ transitions were found to change as a function of pressure (226).

On the basis of the magnetic moments at room temperature and 1 atm pressure, the Fe(R$_2$Dtc)$_3$ complexes have been classified as follows (226):

1. The "high-spin," $\mu_{eff} = 5.8 \pm 0.1$ BM, complexes Fe(PyrrolDtc)$_3$.
2. The $\mu_{eff} = 4.3 \pm 0.2$ BM group (R = R' = n-alkyl groups or nonstrained rings).
3. The $\mu_{eff} = 3.5 \pm 0.2$ BM group (R = n-alkyl, R' = N-aryl groups).
4. The $\mu_{eff} = 2.5 \pm 0.2$ BM group (R = R'-di-sec-alkyl groups.

It has been suggested that the R$_1$NR$_2$ angle in the Fe(R$_1$R$_2$Dtc)$_3$ complexes might be expected to increase toward 120° with the bulkier R groups. The concomitant increase in the C–N bond order then could modify Δ, μ_{eff}, and δ. It was concluded that, as the electron density at the sulfur atoms is enhanced, stronger S $\rightarrow$ M σ bonding should generate larger values of Δ favoring the low-spin $Fe(R_2 Dtc)_3$ isomers (226). The implication that a larger $R_1 NR_2$ angle may result in lower magnetic moments is contradicted by the fact that the room-temperature moment of the recently synthesized (26, 211) trispyrrole dithiocarbamate iron (III) (2.2 BM) is lower than that of the corresponding dicyclohexyl derivative. A larger $R_1 NR_2$ angle is expected for the latter. On the basis of redox potentials, the pyrrole homologue is found to dramatically stabilize the lower oxidation state [Fe(II)] and destabilize the high oxidation state [Fe(IV)] (211). The very low C–N stretching frequency observed in the IR spectrum of the bis-Zn complex (1330 cm^{-1}) also indicates that the pyrrole ring, preserving aromaticity, does not allow for a significant contribution of the $S_2 C=N^+$ canonical form to the electronic structure of the ligand, and the π-accepting character of the sulfur atom is enhanced (211).

On the basis of these observations, it is proposed (211) that the low-spin state and larger quadrupole splitting in the $Fe(PyrroleDtc)_3$ complex "must be attributed to effects *other than* a stronger σ-bonding interaction of iron with a more negatively charged sulfur donor orbital." It is also proposed that the results are consistent with earlier suggestions of extensive metal-to-ligand π back bonding (149, 213). Other amines in which the nitrogen atom is a poor electron donor should also give $R_2 Dtc$ ligands in which the $S_2 C=\overset{+}{N}$ resonance form is not favored and the π-acceptor character is enhanced. Of the suggested $R_2 Dtc$ ligands derived from amines such as indole, carbazole, imidazole, and indoline (213), the indoline derivative has been reported (212) to give an $Fe(R_2 Dtc)_3$ complex that appears to be low spin on the basis of the Mössbauer isomer shift.

An important crystallographic study (403) of the $Fe(Et_2 Dtc)_3$ complex at 297 and 79°K has been reported. The contraction of the FeS_6 core of the isomer in the "low-spin" state at 79°K $(\mu_{eff} = 2.2 \text{ BM})$ has been demonstrated. The hypothesis of Ewald et al. (226) that the $^2 ^- S_2 C=\overset{+}{N}$ resonance form is more important in the low-spin complexes is marginally supported. In the high-temperature, "high-spin" structure the $\overline{C-N}$ is 1.337(6) Å and the $\overline{C-S}$ is 1.708(4) Å. In the low-temperature, "low-spin" structure, the corresponding distances are 1.323(4) and 1.721(2) Å. The average Fe–S distances of known structures plotted versus μ_{eff} give a smooth curve, and an apparent and reasonable correlation of the FeS_6 core (226) also is observed in the structure of the low-spin $Fe(MePhDtc)_3$ complex when compared to the structure of the high-spin $Fe(PyrrolDtc)_3$ complex (303) (Table XIX).

In a recent publication, Manoussakis et al. (596) reported on the synthesis of a series of mixed-ligand $Fe(R_2 Dtc)(R_2' Dtc)_2$ complexes (R_2 = Pyrrol, R' = Et or i-Pr; R_2' = Pyrrol, R = Et or i-Pr; R = Et, R' = i-Pr; R = i-Pr, R' = Et). Values of the ligand field parameters Δ, B, and β_{35} were calculated from a combination of the spectral and magnetic properties of the complexes. The effective positive charge on the iron was calculated as +1.13 for the low-spin complexes and +0.55 for the high-spin complexes. On the basis of these results and covalency arguments for the low-spin complexes the authors suggested that the $S_2C=\overset{+}{N}R_2$ form of the ligand is not favored in the low-spin complexes, in agreement with a previous suggestion (213) but contrary to others (115, 226). The possibility of higher covalency in the low-spin iron(III) dithiocarbamates has been raised previously (111).

Whereas there is little doubt that the FeS_6 core contraction takes place following the high spin to low spin conversion, the change in the ligand electronic structure is still not demonstrated beyond doubt. This is particularly obvious in recent structural results. The predominantly low-spin (3.5 BM, room temp.) $Fe(MorphoDtc)_3$ benzene solvate is characterized by mean $\overline{C-N}$ and $\overline{C-S}$ bond lengths of 1.319(11) and 1.715(3) Å, respectively. The corresponding distances in the $Fe(MorphoDtc)_3$, chloroform, and H_2O solvates, respectively, which are predominantly high spin (5.1, 5.7 BM, room temp.), are 1.323(8), 1.318(3) and 1.720(6) Å; and 1.721(7) Å (Table XIX).

The infrared spectra and, particularly, the C–N frequency should be examined in detail for the *same* complex as a function of temperature (317). The resolution of IR spectroscopy compared to the anticipated shifts (30–40 cm^{-1}) is more appropriate for an examination of the C–N bond than x-ray studies. In addition, a structure determination of the $Fe(PyrroleDtc)_3$ complex would be necessary.

Thus far studies on the magnetic properties of the $Fe(R_2Dtc)_3$ complexes that implicitly consider the $^6A_1 \rightleftharpoons ^2T_2$ equilibrium of magnetic states have been reviewed. In this model the observed magnetic, Mössbauer, solution PMR (277), and metal–ligand bond length data are the weighted means of the separate high-spin and low-spin forms of the dithiocarbamates. As indicated by Mössbauer spectroscopy, the spin-state interconversion must be faster than 1.5×10^{-7} sec.

An alternate model has been proposed (296–298) that attempts to interpret the data in terms of a single mixed state. For this model S is no longer a good quantum number. The unpaired electron spin density at the metal and the wave function that represents the single state are functions of temperature and pressure. This latter model was considered as a possibility by Leipoldt and Coppens. In their structure of the $Fe(Et_2Dtc)_3$ complex at 297 and 79°K they attempted an analysis of the temperature parameters at 297°K. The analysis could not distinguish between a mixed-spin state and a mixture of two different spin states (403).

In the Mössbauer spectra of the $Fe(R_2Dtc)_3$ complexes, simple quadrupole split doublets are observed that are interpreted as the weighted means of the high-spin and low-spin forms (153). It has been pointed out (454) that a rapid transition ($>10^{-7}$ sec) is not characteristic for complexes with different spin states, and indeed two doublets are observed in the Mössbauer spectra of iron(II) compounds with temperature-dependent magnetic moments. The same authors (454) point out that the temperature dependence of the quadrupole splitting in the $Fe(R_2Dtc)_3$ complexes "falls short" of a $^2T_{2g} \rightleftharpoons {}^6T_{1g}$ equilibrium and perhaps could best be explained in terms of a mixed-spin state. The PMR spectra (277) similarly can be accounted for by either the spin-state equilibrium model or by the single mixed-spin state model.

Two recent papers deal with the problem of choosing a model for the magnetic behavior of the $Fe(R_2Dtc)_3$ complexes by studies of the infrared and EPR spectra of these molecules.

The far-infrared spectrum of $Fe(EtPhDtc)_3$ as a function of pressure shows that the intensity of the band assigned to the $^2T_2$ state increases relative to that assigned to the $^6A_1$ state upon increasing pressure (93b) (38). For the $Fe(n\text{-}Pr_2\text{-}Dtc)_3$ complex, the Fe–S vibration at 367 cm^{-1} was assigned to the low-spin-state ($^6A_1$) isomer. On the basis of these results, the spin-state equilibrium was adopted as the true model with a spin-state interconversion rate lower than the vibrational time scale ($\sim 10^{-13}$ sec).

Recently, in a comprehensive study of the magnetic, IR, and EPR, properties of the $Fe(R_2Dtc)_3$ complexes, Hall and Hendrickson also observed a change in the far IR spectra as a function of temperature (291). Unfortunately the absorption previously assigned (93b) to the high-spin form of the $Fe(n\text{-}Pr_2Dtc)_3$ complex (307–305 cm^{-1}) was assigned to the low-spin form and was found to *gain in intensity* at the expense of the 360 cm^{-1} absorption as the temperature was lowered. Clearly the high-pressure and low-temperature IR results are in direct contradiction. On the basis of the Fe–S bond length data, it would appear reasonable that the high-energy band (360 cm^{-1}) is associated with the low-spin form. It is obvious that a reexamination of both experiments is necessary.

Several studies of the EPR spectra of the $Fe(R_2Dtc)_3$ complexes have been reported (153, 250, 553). For the $Fe(Me_2Dtc)_3$ complex doped into the Co(III) analogue relatively isotropic spectra were reported (553) with $g_1 = 2.111$, $g_2 = 2.076$, and $g_3 = 2.015$. Similarly room-temperature spectra on magnetic concentrated solids were reported for the $Fe(R_2Dtc)_3$ complexes (250). In their EPR studies Hall and Hendrickson (291) could not detect spectra for any of the $Fe(Dtc)_3$ complexes at room temperature in magnetically concentrated solids. They attributed previous results (250, 553) to possible contamination of the samples with Cu(II) ions. In $CHCl_3$ glasses at 85°K the X-band spectrum shows two signals at $g = 6.5$ and 4.3 and a weak signal at $g \simeq 2.0$, which was dismissed as an impurity. Upon lowering the temperature the $g = 4.3$ signal decreased in intensity and new signals at $g = 3.27$ and 1.83 were observed and were attributed to the $^2T_2$ state. The $g = 4.3$–4.6 signals were assigned to the population of the $^6A_1$ state. On the basis of these results, the rate of the spin-state change was positioned between 10^7 and 10^{10} sec^{-1}.

In a series of crystallographic and cryomagnetic studies, Sinn et al. uncovered yet new variations in the magnetic properties of the $Fe(R_2Dtc)_3$ complexes (93a).

In the Fe(MorphoDtc)$_3$ complex, crystallographic and variable-temperature magnetic data show marked changes in magnetic moments and average Fe–S bond lengths due to incorporation of solvent molecules in the lattice. Increases in the Fe–S bond lengths and magnetic moments are observed with hydrogen bonding solvents such as CH$_2$Cl$_2$ (95, 302), CHCl$_3$, and H$_2$O (96). In the lattice these molecules seem to interact with the sulfur atoms. With non-hydrogen bonding solvents such as benzene, a low magnetic moment and shorter $\overline{\text{Fe–S}}$ bond lengths are observed (96b). The magnetic moments of the hydrogen-bonded lattices are near the $S = \frac{3}{2}$ value at 4°K and for the CH$_2$Cl$_2$ solvate an $S = \frac{3}{2}$ ground state is observed.

Dramatic spin-state perturbations are observed for solvated lattices of the Fe(PyrrolDtc)$_3$ complex as well. It is reported that 7% benzene in the lattice of the normally high-spin Fe(PyrrolDtc)$_3$ complex results in a situation where a high spin–low spin equilibrium prevails (175). It is perhaps remarkable that with nitrobenzene in the lattice, the Fe(PyrrolDtc)$_3$ complex shows short Fe–S bond lengths (2.353 Å) (Table XIX) and is principally low spin at room temperature (175). A spin-state equilibrium also is induced in Fe(PyrrolDtc)$_3$ by adding large amounts of the Co(III) analogue in the lattice (175). In this magnetically dilute lattice the iron complex shows a sextet–quartet spin-state equilibrium (176).

The high-spin form of unsolvated Fe(PyrrolDtc)$_3$ shows antiferromagnetic interactions that diminish with increasing separation of the Fe(III) complex molecules (175).

In view of these recent studies, it appears that (1) incorporation of solvent molecules into the lattice of the Fe(R$_2$Dtc)$_3$ complexes can be as effective as temperature or pressure changes in shifting the spin-state equilibria, and (2) some of the magnetic data in the literature on the Fe(R$_2$Dtc)$_3$ complexes may well be misleading owing to the presence of different types and quantities of solvents in the lattice.

(g) Nitrosyl Dithio Complexes of Iron.   Since 1970 Feltham et al. have investigated the chemistry of five- and six-coordinate nitrosyl derivatives of various Fe(R$_2$Dtc)$_2$ complexes.

High yields of pure NOFe(R$_2$Dtc)$_2$ complexes were reported for the reaction among FeSO$_4$·7H$_2$O, NO$^+$, and NaR$_2$Dtc in an inert atmosphere (97). The structures of the NOFe(Me$_2$Dtc)$_2$ and NOFe(Et$_2$Dtc)$_2$ complexes have been determined (125, 180, 181). Both have square pyramidal geometry with a nearly linear NO group at the axial site. Oxidation of various of these complexes by X$_2$ (X = I, Br, SCN) or NO$_2$ proceeds readily, and six-coordinate products of the type NOFe(R$_2$Dtc)$_2$X can be isolated (355). For these complexes the order of decreasing stability with X is:

$$\text{CH}_3\text{CN} > \text{CH}_3\text{NC} > \text{I} > \text{SCN} > \text{Br} > \text{NO}_2 > \text{Cl} > \text{N}_3$$

It is proposed that these complexes decompose by homolytic cleavage of the Fe–X bond (355).

The cis-NOFe(R$_2$Dtc)$_2$NO$_2$ complexes were found to be identical to the ones described previously (417) as (NO)$_2$Fe(R$_2$Dtc)$_2$. The trans isomers of these complexes can be isolated by a low-temperature ($-78°$C) synthetic procedure. These "kinetic" isomers transform on heating to the "thermodynamic" cis form. The crystal structure of the cis-NOFe(Et$_2$Dtc)$_2$NO$_2$ has been determined (355). A six-coordinate iron is found in the monomeric complex. The NO and NO$_2$ ligands both are attached to the iron through the nitrogen atoms. The Fe—N—O angle is 174.9(5)$°$ and the (ON)—Fe—(NO$_2$) angle is 91.3$°$. The Fe—S bond length trans to the NO group is 2.308(2) Å, while the one trans to the NO$_2$ group is 2.321(2) Å.

Exchange studies of ^{14}NO with ^{15}NO monitored by IR spectroscopy show fast exchange with the NOFe(R$_2$Dtc)$_2$ complexes and slow exchange with the NOFe(R$_2$Dtc)$_2$X complexes (355).

A direct, slow intramolecular exchange reaction between coordinated NO and NO$_2$ groups in cis-NOFe(Me$_2$Dtc)$_2$NO$_2$ has been reported (356). The proposed rate-determining step for this exchange, followed by IR spectroscopy, is shown in Fig. 34. Consistent with the slow exchange properties of the NOFe(R$_2$Dtc)$_2$X complexes is the observation that above 5$°$C the trans-NOFe(Me$_2$Dtc)$_2$NO$_2$ complex is converted to the cis isomer with retention of the ^{15}NO label.

EPR (285) studies indicate that the unpaired electron in NOFe(R$_2$Dtc)$_2$ complexes occupies an orbital mainly comprised of d_{z^2} (Fe) and σ^* (NO) orbitals.

In a recent review (215), a qualitative MO scheme is presented for the iron nitrosyl complexes. The two major features in this scheme are that (1) the Fe—NO group is linear because the $3b$ and $3b_2$ orbitals are not occupied and (2) the highest occupied (and slightly antibonding) orbital is $5a_1$.

The diamagnetic cis-Fe(Et$_2$Dtc)$_2$(CO)$_2$ has been prepared by the reaction of NaEt$_2$Dtc, CO, and FeSO$_4 \cdot$7H$_2$O in water solution (97). The dimeric [Fe(CO)$_2$SMe(Et$_2$Dtc)]$_2$ complex was obtained from the reaction between Fe(CO)$_3$SMe and NaEt$_2$Dtc in alcohol under oxygen-free conditions. The very complicated PMR spectrum of this diamagnetic molecule did not allow for a choice of structure from a multitude of possible isomers (97).

(2) Ruthenium.   Since the original studies by Malatesta (413), very little attention has been paid to the chemistry of ruthenium with 1,1-dithio chelates.

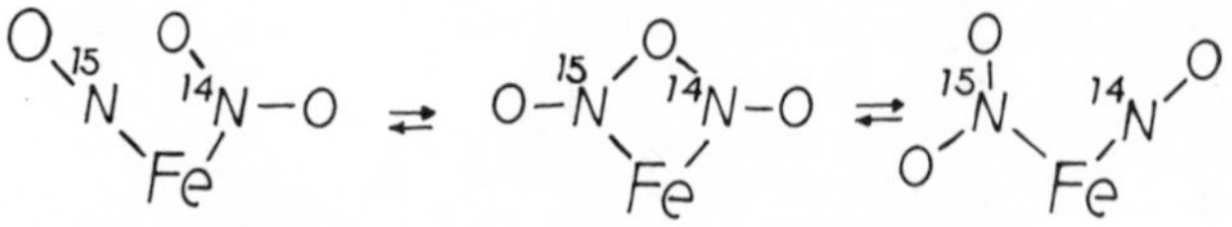

Fig. 34.   Proposed rate-determining step in the NO—NO$_2$ intramolecular exchange in NOFe(Me$_2$Dtc)$_2$NO$_2$ (356).

Recently, however, the yellow crystalline $Ru(R_2Dtc)_2CO$ (R = Et, Me) and $Ru(R_2Dtc)_2(CO)_2$ (R = Me, Bz) complexes and light brown $[Ru(R_2Dtc)_2$-$(CO)_2]Cl$ (R = Bz) were obtained (389) by the reaction of "carbonylated" ethanolic solutions of $RuCl_3$ with NaR_2Dtc. A magnetic moment of 1.8 BM was reported for the $[Ru(R_2Dtc)_2(CO)_2]Cl$ complex.

At a later date Pignolet et al. (512) used the $Ru(R_2Dtc)_2CO$ complexes in oxidative addition reactions with oxidized bis(perfluoromethyl-1,2-dithietene)(TFD). The diamagnetic $Ru(R_2Dtc)_2(TFD)$ (R_2 = Et_2, Me_2, MePh) complexes were obtained. By a PMR temperature study of the $Ru(MeBzDtc)_3$ complex (511) the stereochemical nonrigidity of this compound was established. A $\Lambda \rightleftharpoons \Delta$ isomer conversion was observed in the low-temperature coalescence ($\sim$–13 to +43°C) of the PMR resonances and a S_2C–N bond rotation was detected in high-temperature coalescence (+43 to +64°C). Cyclic voltametry of $Ru(Et_2Dtc)_3$ in DMF/Et_4NClO_4 versus SCE shows (501) a one-electron "nearly reversible" reduction at 0.741 V and a one-electron reversible (?) oxidation (0.382 V). It was suggested that $Ru(Et_2Dtc)_3^+$ should be synthetically accessible.

The temperature-dependent PMR spectra of $Ru(R_2Dtc)_3$ complexes have been examined in detail (206a). As in the studies of the Fe(III) complexes (see preceding section) intramolecular metal-centered inversion and ligand-centered geometric isomerization were reported. $\Delta G^{\ddagger}$ values of $\sim$13 and 15.6 kcal mole^{-1} were found for these two processes and were attributed to a trigonal twist and S_2C–N bond rotation, respectively (206). To assess the influence of solid-state geometry on the stereochemical nonrigidity of the $Ru(R_2Dtc)_3$ complexes, the detailed structure of the $Ru(Et_2Dtc)_3$ complex – which had been communicated earlier (204), was reported (508) (Table XX). The RuS_6 core has approximate D_3 symmetry and a twist toward the trigonal prismatic geometry.

An average twist angle ϕ of 38° is observed. The rather short S_2C–NR_2 bond (Table XX) compared to values reported for other complexes is consistent with the comparatively high barrier to S_2C–NR_2 bond rotation found in the Ru(III) complex. It has been suggested that the distortion of the RuS_6 core toward trigonal prismatic coordination may be a reason for the ease of metal-centered inversion, which has been shown to involve the trigonal twist pathway (206). Essentially the same RuS_6 core geometry is found in the structure of the $CHCl_3$ solvate of the $Ru(MorphoDtc)_3$ complex (Table XX). The hydrogen atom of each chloroform molecule is found to hydrogen bond to the sulfur atoms of various ligands (537).

The reaction of $Ru(R_2Dtc)_3$ with BF_3 under aerobic conditions was reported to give diamagnetic products (261). The exact nature of these products was clarified by Pignolet et al., who by utilizing the same reaction apparently isolated the same diamagnetic compounds (434, 513). The stoichiometry

TABLE XX

Selected Structural Parameters and Electrochemical Properties of the Ruthenium 1,1-Dithio Complexes.

Complex	$\overline{Ru-S}$, Å	$\overline{C-N}$, Å	$E_{1/2}^{n\|n+1}$, V	$E_{1/2}^{n\|n-1}$, V	μ_{eff}, BM	Ref.
Ru(Et₂Dtc)₃	2.376 (4)	1.30 (3)	+0.38[a]	−0.75[b]		434, 501, 508
Ru(MorphoDtc)₃·2.5 CHCl₃	2.383 (12)	1.35 (3)			1.75–1.95	537
β-Ru₃(Et₂Dtc)₅BF₄[c]	2.417 (13)[d] 2.309 (17)[e] 2.380 (15)[f]	1.37 (5)[d] 1.30 (5)[e] 1.40 (5)[f]	−0.46[b]	−1.05[a]	Diamagnetic	434
Ru(Me₂Dtc)₃			+0.48[a]	−0.72[b]		434
Ru(Bz₂Dtc)₃			+0.51[a]	−0.59[a]		434
Ru(PyrrolDtc)₃			+0.40[a]	−0.74[a]		434
β-Ru₂(Me₂Dtc)₅Cl[g]			−0.45[b]	−0.96[a]		434, 538
α-Ru₂(i-Pr₂Dtc)₅Cl[g]	2.296 (25)[i] 2.396 (10)[k]	1.34 (4)			0.7–1.8[j]	434, 538
[α-Ru₂(i-Pr₂Dtc)₅]₂Ru₂Cl₆[h,l]	2.299 (8)[l] 2.400 (12)[k]	1.34 (4)			0.7–1.0	434, 538
Ru₂(Et₂Dtc)₄(CO)₂	2.399 (7)[m]	1.32 (1)				539
Ru₃(Et₂Dtc)₄(CO)₃Cl₂[h]	2.37 (2)[e] 2.40 (2)[d]	1.41 (4)[e] 1.34 (5)[d]				540
Ru(Et₂Dtc)₃Cl[o]	2.416 (13)[p] 2.352 (7)[q]	1.35 (4)				268
Ru(Me₂Dtc)I₃	2.41 (4)[p]	1.42 (3)				435
Ru(Ph₃P)₂(S₂CH)₂	2.337 (8)[q] 2.387 (6)[r] 2.449 (6)[s]	1.33 (4)			Diamagnetic	371

[a] Irreversible.
[b] Reversible.
[c] Ru–Ru, 2.743 (3) Å, acetone solvate.
[d] Non-bridging ligands.
[e] Monobridging ligands.
[f] Dibridging ligand.
[g] Contains 2.5 C₆H₆ of solvation.
[h] Contains 2CHCl₃ of solvation.
[i] Bridging sulfur atoms.
[j] T.I.P. (Temperature Independent Paramagnetism).
[k] Non-bridging sulfur atoms.
[l] Ru–Ru 2.789 (4) Å.
[m] Ru–C, 1.79 (5).
[n] Ru–Ru, 3.70 (4) Å; Ru–Cl, 2.52 (7) Å; Ru–C, 1.67 (3) Å.
[o] Ru–Cl, 2.448 (7) Å.
[p] Equatorial sulfurs.
[q] Axial sulfur.
[r] Sulfur atoms cis to Ph₃P; Ru–P = 2.340 (6) Å.
[s] Sulfur atoms trans to Ph₃P.

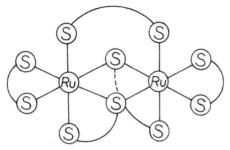

Fig. 35.   Idealized structure of the β-Ru$_2$(Et$_2$Dtc)$_5^+$ compled (434).

Ru$_2$(R$_2$Dtc)$_5$BF$_4$ was verified in the crystal structure determination of Ru$_2$(Et$_2$Dtc)$_5$BF$_4$ (434). The two ruthenium(III) ions are triply bridged by three Et$_2$Dtc ligands (Fig. 35).

Two of the bridging ligands adopt a bridging mode similar to that found in the Cu$_4$(Et$_2$Dtc)$_4$ tetramer (322), while the third bridges in a fashion similar to that observed for the bridging ligands in other dimeric complexes (162, 550). The Ru–Ru distance is short [2.743(3) Å]. The diamagnetism of the dimer coupled with the short Ru–Ru distance has led to the conclusion that Ru–Ru bonding is present in the dimer. Other structural parameters for this interesting molecule are shown in Table XX. An electrochemical study of the Ru(R$_2$Dtc)$_3$ and Ru$_2$(R$_2$Dtc)$_5$BF$_4$ complexes has been carried out.

For the dimer, polographic results in CH$_3$CN reveal a four-membered series (Eq. 32), and the one-electron reduction product was isolated (434).

$$[Ru_2(R_2Dtc)_5]^{2+} \longleftarrow [Ru_2(R_2Dtc)_5^+] \Longleftrightarrow [Ru_2(R_2Dtc)_5]^0 \longrightarrow [Ru_2(R_2Dtc)_5]^- \quad (32)$$

The polarographic results for the Ru(R$_2$Dtc)$_3$ complexes reveal a three-member series:

$$[Ru(R_2Dtc)_3]^+ \longleftarrow [Ru(R_2Dtc)_3]^0 \Longleftrightarrow [Ru(R_2Dtc)_3]^- \quad (33)$$

The one-electron oxidation observed at +0.38 V is an irreversible process.

Following the characterization of the [Ru$_2$(R$_2$Dtc)$_5$]$^+$ dimers, the isolation and structural characterization of two derivatives of the [Ru$_2$(i-Pr$_2$Dtc)$_5$]$^+$ cation was reported (538). These complexes, which were obtained during the chromatographic purification of the Ru(R$_2$Dtc)$_3$ complexes, were found to be isomers of the previously reported dimer. The structure of the [Ru$_2$(R$_2$Dtc)$_5$]$^+$ cation in both derivatives (Cl$^-$ and Ru$_2$Cl$_6^{2-}$) was described as a combination of Ru(R$_2$Dtc)$_3$ with cis-[Ru(R$_2$Dtc)$_2$]$^+$. The Ru(R$_2$Dtc)$_3$ molecule acts as a ligand for the [Ru(R$_2$Dtc)$_2$]$^+$ cation, and of the two possible isomers, the one shown in Fig. 36A (α isomer) was observed.

The Ru–Ru distance [2.789(4) Å] is similar to that reported for the other

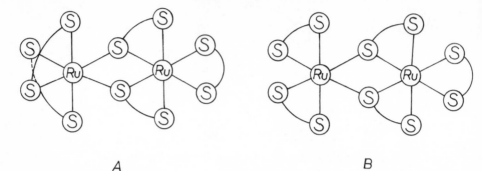

Fig. 36. Possible isomers of the α-Ru$_2$(Et$_2$Dtc)$_5^+$ complex; A is the isomer actually observed (538).

isomer (henceforth the β isomer). A low (temperature-independent) magnetic moment for the α isomer (0.7—1.0 BM) was attributed to temperature-independent paramagnetism. Differences in the synthesis and/or ligand substituents were suggested to be dominant factors in the type of isomer (α versus β) obtained.

Electrochemical studies on Ru(R$_2$Dtc)$_3$ derivatives have been reexamined (309). The reversible reduction (Eq. 33) previously reported (434) was verified. The oxidation of Ru(R$_2$Dtc)$_3$ also was described as "generally not reversible"; however, a substituent dependence on the degree of reversibility was noted. It was suggested that the Ru(R$_2$Dtc)$_3^+$ exhibits various degrees of instability toward decomposition depending on the substituents.

The α-[Ru$_2$(Me$_2$Dtc)$_5$]$^+$ (Fig. 36A) was converted to the β-[Ru$_2$(Me$_2$Dtc)$_5$]$^+$ isomer (Fig. 35) on heating in CHCl$_3$ solution (Cl$^-$, BF$_4^-$, or PF$_6^-$ salts). This conversion was monitored by PMR spectroscopy (309). Both isomers showed five methyl singlets. Heating the β isomer to 120°C in (CD$_3$)$_2$SO showed no coalescence of the methyl proton resonances, which indicates a high barrier to rotation around the S$_2$C—N bond. Both isomers are electroactive. The reduction potentials for the process

$$[\text{Ru}_2(\text{Me}_2\text{Dtc})_5]^+ \underset{-e}{\overset{+e}{\rightleftharpoons}} [\text{Ru}_2(\text{Me}_2\text{Dtc})_5] \tag{34}$$

show that the α isomer is slightly easier to reduce (35 mV) than the β isomer. This trend was more pronounced for the second reduction process:

$$[\text{Ru}_2(\text{Me}_2\text{Dtc})_5)] \underset{-e}{\overset{+e}{\rightleftharpoons}} [\text{Ru}_2(\text{Me}_2\text{Dtc})_5]^- \tag{35}$$

Following reduction of the β isomer to the monoanion, a rapid conversion to the anion of the α isomer was observed. The equilibrium position for the two

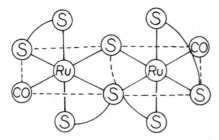

Fig. 37.   Idealized structure of the $[Ru(Et_2 Dtc)_2 CO]_2$ complex (539).

isomers and their reduction products were determined electrochemically. Thus $K(\beta^0/\alpha^0) = 5.6$, $K(\beta^+/\alpha^+) = 22$, and $K(\beta^-/\alpha^-) = 0.09$. The neutral dimer $Ru_2(Me_2 Dtc)_5$ was isolated.

The possibility that the previously reported $Ru(R_2 Dtc)_2 CO$ complexes (389) may be dimers prompted Raston and White to determine the structure of the "$Ru(Et_2 Dtc)_2 CO$" complex (539). The structure indeed reveals a $[Ru(Et_2-Dtc)_2 CO]_2$ dimer. Each Ru(II) is six-coordinate and is appreciably distorted from octahedral geometry. The two dithiocarbamate ligands on each Ru(II) ion are opposed to each other and bridge axial and equatorial positions (Fig. 37). Selected structural parameters are recorded in Table XX.

Another interesting Ru(II) trinuclear complex was isolated following the reaction between $R_2 Dtc$ and "carbonylated" $RuCl_3$ in methanol (540). The structure of the $Ru_3(Et_2 Dtc)_4(CO)_3 Cl_2$ has been determined and can be described as an equilateral triangular arrangement of pseudo-octahedrally coordinated ruthenium atoms (Fig. 38) (Table XX).

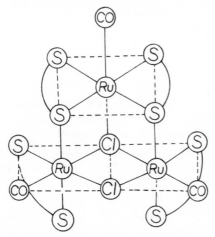

Fig. 38.   Idealized structure of the $Ru_3 (Et_2 Dtc)_4 (CO)_3 Cl_2$ trimer (540).

Seven-coordinate $Ru(R_2 Dtc)_3 Cl$ complexes (R = Et, Me) were obtained by the photolysis of $Ru(R_2 Dtc)_3$ complexes in $CHCl_3$ or $CH_2 Cl_2$ or by the reaction with gaseous HCl in benzene (268). Addition of I_2 to $Ru(Me_2 Dtc)_3$ in $CHCl_3$ solution gives microcrystalline $Ru(Me_2 Dtc)_3 I_3$. The $Ru(R_2 Dtc)_3 X$ complexes are stereochemically nonrigid at $-95°C$ in $CD_2 Cl_2$ solution. The $Ru(Et_2 Dtc)_3 Cl$ reacts with $AgBF_4$ in acetone solution. $AgCl$, $[Ru_2 (Et_2 Dtc)_5]-BF_4$, $(R_2 Dtc)_2$, and $Ru(BF_4)_3$ (?) were obtained. It appears that, as stated earlier (309), the $Ru(Et_2 Dtc)_3^+$ complex is unstable toward formation of the dimeric Ru(III) complex.

The crystal structures of the $Ru(Et_2 Dtc)_3 Cl$ (268) and $Ru(Me_2 Dtc)_3 I_3$ (435) complexes have been determined. On both, the Ru(IV) coordination geometry is pentagonal bipyramidal (Fig. 39) (Table XX). The ruthenium atoms in the ethyl derivative are pendant on infinite chains of iodine atoms in the lattice. This observation and the unusual golden color of the complex suggest that the crystals of this compound may possess interesting electrical properties.

Detailed photochemical studies of $Ru(R_2 Dtc)_3$ complexes show (267) that upon irradiation in $CHCl_3$, $CH_2 Cl_2$, or PhCl at $\lambda = 265$ nm, these compounds yield $ClRu(R_2 Dtc)_3$ and α-$Ru_2 (R_2 Dtc)_5 Cl$ and small amounts of $R'R_2 Dtc$ (R = Ph). A possible mechanism was proposed for this reaction:

$$Ru(R_2 Dtc)_3 \; \underset{}{\overset{h\nu}{\rightleftharpoons}} \; \left[(R_2 Dtc)_2 Ru(II) \underset{S}{\overset{S}{\diagdown}} C-N \underset{R}{\overset{R}{\diagup}} \right]^*$$

$$Ru(R_2 Dtc)_3^* + R'Cl - \left[\begin{array}{l} ClRu(R_2 Dtc)_3 + R' \\ \\ \text{"}ClRu(R_2 Dtc)_2\text{"} + R'R_2 Dtc \end{array} \right. \tag{36}$$

$$\text{"}ClRu(R_2 Dtc)_2\text{"} + Ru(R_2 Dtc)_3 \longrightarrow \alpha[Ru_2 (R_2 Dtc)_5]^+ Cl$$

A large number of complexes of the general formula $Ru(Me_2 Dtc)_2 (L)_2$ have been synthesized [L = PPh_3, $PMe_2 Ph$, $PMePh_2$, $P(OPh)_3$, etc.] by the reaction of various Ru(II) and Ru(III) tertiary phosphine and phosphite complexes with $NaMe_2 Dtc$ (129). PMR spectroscopy indicates a cis configuration for the ligands (L). Restricted rotation about the $S_2 C-N$ bond in the *cis*-$Rh(R_2 Dtc)_2 (L)_2$ complexes also was detected (130). The $Ru(R_2 Dtc)_2 (PPh_3)_2$ complexes were reported previously by O'Connor et al. (R = Me, Et, Ph), who obtained these compounds by the reaction of *mer*-$[RuCl_3 (PPh_3)_3]$ with an excess of $NaR_2 Dtc$ (490). The same reaction with $NaRXant$ ligands gives the corresponding xanthate complexes (490). The $Ru(Et_2 Dtc)_2 (Me_2 SO)_2$ complex has been reported (224).

The synthesis of the dithioformate complexes $Ru(Ph_3 P)_2 (S_2 CH)_2$ and

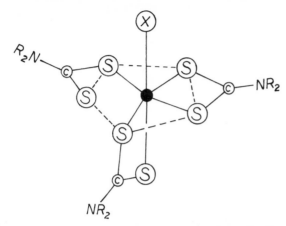

Fig. 39. Schematic structure of the $Ru(R_2Dtc)_3X$ complexes $(R = Et, X = Cl; R = Me,$ $X = I_3^-)$ (268, 435).

$RuCl(CO)(PCy_3)_2(S_2CH)$ has been accomplished by CS_2 insertion into the Ru–H bond of $H_2Ru(PPh_3)_4$ or $HRu(O_2CH)(PPh_3)_3$ for the first complex (371) and $Ru(HCl)(CO)(PCy_3)_3$ for the second complex (465) $(PCy_3 = $ tricyclohexyl phosphine). The structure of $Ru(Ph_3P)_2(S_2CH)_2$ has been determined (371). The Ru(II) ion is octahedrally coordinated by two Ph_3P ligands and four sulfur atoms of the two bidentate HCS_2^- ligands, which are situated cis to each other in the coordination polyhedron (Table XX). The S_2CH ligand is asymmetrically bonded and two long Ru–S bonds are found with the S atoms trans to Ph_3P. This asymmetry is not reflected in the C–S bond lengths, which are equal within the accuracy of the structural determination.

(3) Osmium.   The chemistry of osmium 1,1-dithio complexes was not explored until 1972 when the synthesis of $Os(R_2Dtc)_3$ complexes was claimed from the reaction between $(NH_4)OsCl_6$ and NaR_2Dtc (66). The reaction of $OsCl_2(bipy)_2$ with KPipDtc also was reported to give $OsCl(bipy)_2(PipDtc)$ (66). A brief mention of the $OsO_2(R_2Dtc)_2$ was made (572). Recently a wide variety of Os(II) mixed ligand complexes were reported (173). $Os(R_2Dtc)_2(PPh_3)_2$ complexes $(R = Me, Et)$ were obtained by the reaction between $OsH(MeCOO)$-$(PPh_3)_3$ and NaR_2Dtc. The corresponding xanthates are obtained in a similar fashion. On the basis of PMR studies, the cis configuration was proposed for the PPh_3 ligands. The reaction between $OsCl(CF_3COO)(CO)(PPh_3)_3$ and $NaMe_2Dtc$ was found to give $Os(Me_2Dtc)_2(CO)PPh_3$. The corresponding Et analogue could not be obtained and, instead, the $(Cl)(CO)Os(Et_2Dtc)PPh_3$ was isolated. The structure proposed for this molecule on the basis of PMR spectra is shown in Fig. 40. Analogous complexes with xanthate ligands could not be obtained.

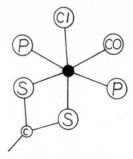

Fig. 40.   Proposed structure for the $(Cl)(CO)Os(Et_2 Dtc)(PPh_3)_2$ complex (173).

A similar structure (Fig. 40) was proposed for the $OsH(R_2 Dtc)CO(PPh_3)_2$ complexes and the RXant analogues (R = Me, Et). These compounds are obtained by the reaction of $OsHCl(CO)(PPh_3)_3$ with $NaR_2 Dtc$ and NaRXant ligands (173). The C=O stretching vibrations occur between 1930 and 1905 cm^{-1} and the M–H frequencies occur between 2020 and 2040 cm^{-1}.

Similar studies have been reported by Cole-Hamilton and Stephenson (132). The reaction of mer-[$OsCl_3(PMe_2 Ph)_3$] with $NaMe_2 Dtc$ in EtOH gives $Os(PMe_2 Ph)_2(Me_2 Dtc)_2$. A new complex, $OsCl(PMe_2 Ph)_3(Me_2 Dtc)$, was obtained from the same reaction. This compound reacts with KEtXant in EtOH to produce several compounds among which the new $Os(PMe_2 Ph)_2(Me_2 Dtc)$-(EtXant) and $Os(OEt)(PMe_2 Ph)_3(Me_2 Dtc)$ were identified. Another xanthate complex also was synthesized by the reaction of mer-[$OsCl_3(PMe_2 Ph)_3$] and KEtXant. The $OsCl(PMePh)_3(EtXant)$ and the bromo analogue have the halide trans to a phosphine group. The mass spectra of these complexes are reported (132).

Very recently the synthesis and structural characterization of the interesting compound $(Me_2 Dtc)_2 Os$-μ-N-μ-$(Me_2 Dtc)Os(Me_2 Dtc)_2$ was reported (269). The two osmium(IV) atoms have distorted octahedral $OsS_5 N$ coordination and very short Os–N bonds (1.75 Å). A rather large S–S "bite" is found for the bridging $Me_2 Dtc$ ligand (3.15 Å) (Fig. 41).

The Os–S bonds vary in length from 2.36 to 2.46 Å. It is suggested that the nitrido N atom comes from the excess $R_2 Dtc^-$ ligand used in the synthesis.

The dithioformates of Os(II) and Ru(II) $MX(L)(CO)(PPh_3)_2$ (X = Cl, Br, L = $S_2 CH$) have been reported. Their synthesis is accomplished by the CS_2 insertion reaction into the Os–H or Ru–H bond of the $HMX(CO)(PPh_3)_3$ complexes. Both the cis and trans isomers of these complexes were obtained (556).

b.   Cobalt, Rhodium, Iridium.   *(1) Cobalt.*   *(a) Dithiocarbamate Complexes.*   The divalent cobalt 1,1-dithio chelates are extremely unstable toward

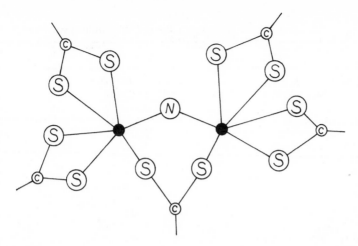

Fig. 41.   Schematic   structure   of   the $(Me_2Dtc)_2Os$-μ-N-μ-$(Me_2Dtc)Os(Me_2Dtc)_2$ complex (269).

oxidation, and in the case of the $Co(Et_2Dtc)_2$ complex, oxidation to $Co(Et_2Dtc)_3$ is observed even under rigorously oxygen-free conditions (236). Following these initial observations, a report appeared in the literature on the preparation and properties of $NaCo(Et_2Dtc)_3$ (179). An independent investigation of this complex by Holah and Murphy convincingly demonstrated that the alleged $NaCo(Et_2Dtc)_3$ complex was, in fact, the well-known Co(III) complex $Co(Et_2Dtc)_3$ (333).

An improved stability toward oxidation is found for the $Co(R_2Dtc)_2$ complexes with large R groups (426). The green paramagnetic complexes with R_2 = piperidine, thiomorpholine-4-, and n-methylpiperazine-4- (μ_{eff} = 2.00, 2.03, and 1.60 BM, respectively) have been obtained and are thought to contain four-coordinate planar Co(II) (426).

Several structures of $Co(R_2Dtc)_3$ complexes have been reported. The structure of the $Co(MorphoDtc)_3$ complex has been determined both as a CH_2Cl_2 (302) and as dibenzene (96b) solvate. The similarity of the CoS_6 cores in the two structures suggests that the differences in lattice forces arising from different solvents in the lattice are not significant.

In the structure of the $Co(H_2Dtc)_3$ complex (543) the symmetry of the CoS_6 seems to be lowered from D_3 to approximate C_3 and two threesomes of Co—S bond lengths are observed. The difference between the mean values of these bond lengths, however, is marginally significant at the 3σ level. Intermolecular S—H interactions have been advanced as a probable cause for this distortion. The structure of the $Co(Et_2Dtc)_3$ complex also has been determined, and is quite normal (72, 451) (Table XXI).

TABLE XXI

Structural Parameters in the Co, Rh, and Ir 1,1-Dithio Complexes.

Complex	M–S, Å	Ref.
Co(MorphoDtc)$_3$ · CH$_2$Cl$_2$	2.274 (11)	302
Co(MorphoDtc)$_3$ · (C$_6$H$_6$)$_2$	2.273 (4)	96b
Co(H$_2$Dtc)$_3$[a]	2.276 (12)	543
Co(Et$_2$Dtc)$_3$	2.267 (4)	72
Co(Et$_2$Dtc)$_3$	2.258 (2)	451
Co(EtSXant)$_3$	2.266 (7)	407
Co(EtXant)$_3$	2.277	452
[Co$_2$(Et$_2$Dtc)$_5$]BF$_4$[b]	2.315 (21)[c]	310, 316
	2.266 (8)[d]	
	2.447 (11)[e,f]	
[Co(SEt)(EtSXant)$_2$]$_2$	2.246 (6)[g]	405
	2.271 (15)[h]	
Co$_2$(SEt)$_2$(EtSXant)(Et$_2$Dtc)$_3$	2.251 (3)[g]	634
	2.278 (13)[h]	
Rh(Et$_2$Dtc)$_3$	2.364 (9)	540
Rh$_2$(Me$_2$Dtc)$_5$BF$_4$	2.386 (8)[c]	314
	2.372 (33)[d]	
	2.349 (14)[e]	
Rh(MorphoDtc)$_3$ · (C$_6$H$_6$)$_2$	2.364 (5)	96b
ClRh(Me$_2$Dtc)(Me$_2$NCS)PPh$_3$[f]	2.355 (47)[i]	62, 265
	2.432 (5)[j]	
Ir(Et$_2$Dtc)$_3$	2.367 (3)	541
Ir(MorphoDtc)$_3$ · (C$_6$H$_6$)$_2$	2.371 (7)	96b

[a] The mean value of six Co–S bond lengths is reported; however, a systematic distortion in the structure lowers the symmetry to approximate C_3. The mean values of the two three-somes of Co–S bonds are 2.267 (9) and 2.285 (7) Å. A similar, but less pronounced, distortion is seen in the structure of Co(MorphoDtc)$_3$ (96b).
[b] Co–Co distance, 3.37 Å.
[c] Bridging sulfurs.
[d] Terminal sulfurs at Co (1).
[e] Terminal sulfurs at Co (2).
[f] CHCl$_3$ solvate.
[g] Bridging mercaptide sulfur.
[h] Chelate sulfurs.
[i] Me$_2$Dtc sulfurs.
[j] Thiocarboxamido sulfur.

For the Co(R$_2$Dtc)$_3$ complexes, oxidation of the $3d^6$ ion is expected to be difficult. This becomes apparent in the electrochemical studies of Chant et al., who show that for a given R$_2$Dtc ligand, the oxidation potential for the processes:

$$Co(R_2Dtc)_3 \rightleftharpoons Co(R_2Dtc)_3^+ + e^- \tag{37}$$

is higher than any other oxidation of $R_2 Dtc$ complexes of first-row elements (116).

Following the successful synthesis of the $Fe(IV)–R_2 Dtc$ and $Mn(IV)–R_2 Dtc$ complexes, Saleh and Straub attempted the oxidation of the $Co(R_2 Dtc)_3$ complexes with BF_3 (562). Complexes formulated as $Co(Et_2 Dtc)_3^+$ and $Co(C_6 H_{11})_2 Dtc)_3^+$ were isolated as BF_4^- salts and were reported to be *paramagnetic*. Using a similar method $[Co(R_2 Dtc)_3$ and $BF_3]$, Gahan and O'Connor obtained the *diamagnetic* oligomeric "$Co(R_2 Dtc)_3 BF_4$" complexes (261) (260).

A reexamination of the same reaction at a later date indicated that the BF_3 oxidation of the $Co(R_2 Dtc)_3$ complexes affords the diamagnetic Co(III) dimers $[Co_2 (R_2 Dtc)_5] BF_4$ (310, 316).

The existence of paramagnetic $Co(R_2 Dtc)_3^+$ complexes at present appears problematic, particularly since it is pointed out that the reported (562) Co(IV) complexes have spectra "virtually identical to that found for $Co_2 (R_2 Dtc)_5 BF_4$ species" (316).

The stoichiometry $Co_2 (R_2 Dtc)_5 BF_4$ has been verified by an x-ray crystallographic study of the $[Co_2 (Et_2 Dtc)_5] BF_4$ complex (316). The cation, $[Co_2 (Et_2 Dtc)_5]^+$, can be considered as a $Co(Et_2 Dtc)_3$ unit coordinated in the cis positions of an octahedral $Co(Et_2 Dtc)_2^+$ species of opposite chirality. The geometry of the CoL_3 unit resembles that found for $Co(Et_2 Dtc)_3$. The mean Co–S bond lengths in the CoL_2^+ unit are marginally shorter than those in the CoL_3 unit (Table XXI) (Fig. 42). A rather long Co–Co distance (3.37 Å) precludes any M–M bonding. These observations and the rather long Co–S bond lengths in the bridge prompted Hendrickson et al. to attempt the interaction of the CoL_2^+ unit with other ligands (316).

The reaction

$$Co_2 (R_2 Dtc)_5^+ + L^- \longrightarrow Co(R_2 Dtc)_3 + Co(R_2 Dtc)_2 L \qquad (38)$$

proceeds with L = acac, sacsac, $Bz_2 Dtc$, *n*-PrXant, and PipDtc. The mixed-ligand complexes are inert to substitution and can be separated from the reaction mixture by chromatographic techniques. Mixed 1,1-dithio ligand Co(III) com-

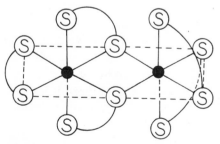

Fig. 42.   Idealized structure of the $[Co_2 (Et_2 Dtc)_5]^+$ dimer (316).

plexes also have been obtained from the reaction between $KCo(bi)_2$ (bi = biuret) and the R_2Dtc^- or $RXant^-$ ligands (73). Both the $[Co(bi)_2R_2Dtc]^{2-}$ and $[Co(bi)(R_2Dtc)_2]^-$ can be isolated depending on the stoichiometry of the reaction and the temperature. The former of the two is only stable at $0°C$, and at higher temperatures it is converted to the latter. The analogous complexes with the $RXant^-$ ligand also can be obtained.

The synthesis of the interesting $Co(EtXant)(Et_2Dtc)_2$ complex by the reaction of $Co(EtXant)_3$ and $Co(Et_2Dtc)_3$ in $CHCl_3$ (refluxing for 10 min) (73) is remarkable in view of the kinetic inertness of the starting complexes.

The above report is even more interesting since intermolecular exchange of dithiocarbamate ligands was not found upon refluxing nitrobenzene-d_5 solutions of $Co(R_2Dtc)_3$ and $Co(R_2'Dtc)_3$ complexes for several hours at $195°C$ (493).

The temperature dependence of the PMR spectra of the $Co(III)(R_2Dtc)_3$ complexes has been studied and certain unusual aspects of the spectra are accounted for by a total line-shape analysis in terms of a general two-proton-exchange model. In similar studies Siddall (573) assigned certain of the complexities of the PMR spectra of $Co(RR'Dtc)_3$ complexes to rotation about the C—N bond.

The temperature-dependent PMR spectra of $Co(R_2Dtc)_3$ complexes have been examined (493). The molecules are stereochemically nonrigid; however, the mechanism that results in optical inversion could not be determined. By contrast, the $Rh(R_2Dtc)_3$ complexes were rigid up to $200°C$ in $NO_2C_6D_5$. The kinetic parameters derived for the metal-centered inversion in $Co(Bz_2Dtc)_3$ are: $\Delta H^{\ddagger}$, 25.5 ± 1.0 kcal mole^{-1}; $\Delta S^{\ddagger}$, 4.1 ± 5.0 eu, and $\Delta G^{\ddagger}$, 23.6 ± 0.2 kcal mole^{-1} ($+168°C$). The values of $\Delta H^{\ddagger}$ and $\Delta G^{\ddagger}$ are much larger than the corresponding parameters reported for the $Fe(R_2Dtc)_3$ complexes and have been related to the large difference in the crystal field stabilization energy (CFSE) between the trigonal prismatic and trigonal antiprismatic geometries of the $Co(R_2Dtc)_3$ complexes. As a result of this correlation, a trigonal twist mechanism for the metal-centered inversion is favored.

The cobalt-59 NMR spectra of various $Co(R_2Dtc)_3$ complexes and of $Co(RXant)_3$ have been reported. For $B_0 = 450$ cm^{-1} a linear correlation between the resonance frequency (ν_j) and the $^1T_{1g} \leftarrow {}^1A_{1g}$ absorption maximum is observed (430). In an electrochemical study, Toropova et al. (592) reported on the catalytic liberation of hydrogen in solutions of $Co(R_2Dtc)_2$ complexes R_2 = (hydroxyethyl)$_2$, H-hydroxyethyl, carboxymethyl, etc.). The following mechanism was proposed for this very interesting reaction:

$$Co^{II}(R_2Dtc)_2 + 2e^- \longrightarrow Co^0(R_2Dtc)_2$$
$$Co^0(R_2Dtc)_2 + 2H^+ \longrightarrow Co^0(R_2DtcH)_2 \qquad (39)$$
$$Co^0(R_2DtcH)_2 + 2e^- \longrightarrow Co^0(R_2Dtc)_2 + H_2\uparrow$$

$$\downarrow$$

$$Co^0 + 2R_2Dtc^-$$

The structure of the $NoCo(Me_2Dtc)_2$ has been determined (216). The coordination geometry about the cobalt is a tetragonal pyramid with the NO group at the apex. The Co atom is found 0.52 Å above the basal plane and toward the NO group, which is disordered and bent [Co–N–O, 135.1(17)°].

(b). *Xanthate and Thioxanthate Complexes.* By comparison to other Co(III) 1,1-dithio complexes, abnormally long Co–S bond lengths were reported for the $Co(EtSXant)_3$ complex (613) [2.398(1) Å]. A reinvestigation of the structure of $Co(EtSXant)_3$ showed a mean Co–S bond length of 2.266(7) Å (407), which compares favorably with corresponding distances in other Co(III) structures (Table XXI).

The apparent discrepancy between the two determinations has led Li and Lippard (407) to suggest that the alleged $Co(EtSXant)_3$ crystals used in the first structural analysis (613) were, in fact, $Cr(EtSXant)_3$. This suggestion seems reasonable, since the reported "Co–S" bond length is very close to the Cr–S bond lengths in the $Cr(PhDta)_3$ (42) and $Cr(EtXant)_3$ (453) complexes.

Just as the $Fe(RSXant)_3$ complexes undergo a CS_2 elimination reaction to give mercaptide-bridged dimers of the type $[Fe(RS)(RSXant)_2]_2$, the $Co(RSXant)_3$ complexes undergo the same reaction (405):

$$2Co(RSXant)_3 \longrightarrow [Co(SR)(RSXant)_2]_2 + 2CS_2 \qquad (40)$$

A kinetic study of the reaction (R = Et) in chloroform shows first-order kinetics. The rate-determining step in the CS_2 elimination process is thought to be cleavage of the MS_2C–SR bond (405). The crystal structure of the $[Co(SR)(RSXant)_2]_2$ dimer (R = Et) has been determined. The EtSXant ligands are terminally coordinated and the ethyl groups of the bridging EtS⁻ ligands are restricted to the anti configuration (Fig. 43A). The structure should be compared to that of the isomeric Fe(III) analogue (162), in which two terminal and two bridging EtSXant ligands are found (Fig. 43B).

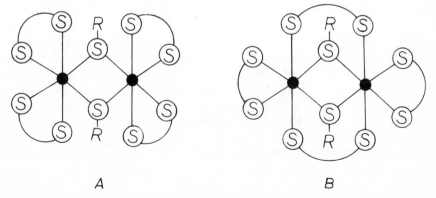

A   B

Fig. 43. Schematic structures of the $[M(RS)(RSXant)_2]_2$ dimers for (*A*) M = Co(III) and (*B*) M = Fe(III) (162, 405).

Whereas there is evidence for M—M bonding in the Fe(III) dimer, the long Co—Co distance [3.321(2) Å] precludes metal—metal bonding in the Co(III) dimer. The Co(III) dimers undergo irreversible one-electron oxidation, unlike the corresponding monomeric Co(RSXant)$_3$ complexes, which show reversible or nearly reversible one-electron reductions at potentials ranging from −0.80 to −1.00 V (405).

The [Co(SEt)(EtSXant)$_2$]$_2$ complexes react with diethylamine to give Co$_2$(SEt)$_2$(EtSXant)$_n$(Et$_2$Dtc)$_{2-n}$ (n = 1−3) complexes (634). This reaction (Eq. 41) has been described previously for the Ni(II) system (239) and is

$$[Ni(SR)(RSXant)]_2 + 2NH(Et)_2 \longrightarrow [Ni(SR)(Et_2Dtc)]_2 + 2RSH \qquad (41)$$

thought to occur by nucleophilic attack of diethylamine on the carbon atom of the CS$_2$ moiety in each RSXant ligand. The structure of one homologue in the mixed RSXant—R$_2$Dtc ligand dimers, Co$_2$(SEt)$_2$(EtSXant)(Et$_2$Dtc)$_3$, has been determined (634). It resembles that of the [Co(SEt)(EtSXant)$_2$]$_2$ dimer (Fig. 43A) (405) with a nonbonded Co—Co distance of 3.350(4) Å. The EtSXant ligand in this dimer is equally disordered among the four terminal chelating ligand sites.

(2) Rhodium, Iridium. Considerable advances in the chemistry of rhodium 1,1-dithio chelates have taken place in the last 10 years. The structural prototypes for the Rh(R$_2$Dtc)$_3$ and Ir(R$_2$Dtc)$_3$ complexes are now available in x-ray structural determinations of these complexes with R$_2$ = Et$_2$ and Morpho. In the Rh(Et$_2$Dtc)$_3$ structure (540, 537a) the RhS$_6$ core shows only trivial deviations from D_3 symmetry (Table XXI). The same distorted octahedral structure is found for the RhS$_6$ core in the bis benzene solvate of the Rh(MorphoDtc)$_3$ complex (96b). The mean Rh—S bond lengths in the two structures are identical (Table XXI).

The Ir(Et$_2$Dtc)$_3$ complex is x-ray isomorphous to the Co(III) analogue and also contains an IrS$_6$ core of distorted D_3 symmetry (541). A similar core is found in the structure of the Ir(MorphoDtc)$_3$·(C$_6$H$_6$)$_2$ complex (96b). Here again the mean values of the Ir—S bond lengths in the two structures are very similar (Table XXI). While there is a significant increase from the ($3d^6$) Co—S bond length to the ($4d^6$) Rh—S bond length, an insignificant change is found in the ($5d^6$) Ir—S bond length when compared to the Rh—S bond (Table XXI).

The cationic Rh(IV) complexes Rh(R$_2$Dtc)$_3$BF$_4$, ostensibly obtained by the BF$_3$ oxidation of Rh(R$_2$Dtc)$_3$ complexes (261), are now known to be the binuclear Rh(III) complexes of the stoichiometry Rh$_2$(R$_2$Dtc)$_5$BF$_4$. A structural study of the [Rh$_2$(Me$_2$Dtc)$_5$]BF$_4$·0.5CH$_2$Cl$_2$ complex shows the dimer closely related to the Co(III) analogue (Fig. 42) with a Rh—Rh distance of 3.556(1) Å (314).

The nonexistence of the Rh(R$_2$Dtc)$_3^+$ cation is further supported by the fact that the $Rh(Me_2Dtc)_3$ complex is oxidized *irreversibly* (cyclic voltametry on a platinum electrode in acetone/$0.1M$ Et_4NClO_4) at a potential of +1.24 V versus Ag|AgCl (314).

Numerous Rh(III) mixed-ligand complexes have been reported. In the presence of an excess of PPh_3, methanolic solutions of Rh_2^{4+} react with $NaEt_2Dtc$, and the yellow crystalline $[Rh(Et_2Dtc)_2(PPh_3)_2]BF_4$ is obtained (460). The complex is a 1:1 electrolyte in nitromethane, and a single C–N absorption in the IR spectrum, at 1518 cm^{-1}, suggests the presence of bidentate dithiocarbamate ligands. A different product is obtained following the reaction between $[Rh(I)(CO)(PPh_3)_3]BF_4$ and $NaEt_2Dtc$ in methanol. The orange crystals of $Rh(Et_2Dtc)_3PPh_3$ thus obtained have been reported previously (490).

A PMR temperature-dependence study on the $Rh(Me_2Dtc)_3PPh_3$ complex revealed a six-coordinate Rh(III) with one unidentate Me_2Dtc ligand cis to the PPh_3 site. At elevated temperatures (72°C) the PMR spectra coalesce to a single resonance by a dynamic process suggested to be an exchange of the ligand Me_2Dtc sites (182).

An extensive series of Rh(III) mixed-ligand complexes were reported recently by Cole-Hamilton and Stephenson (131). In these studies the reactions between *mer*-$RhCl_3(PMe_2Ph)_3$ and R_2Dtc or RXant ligands proceeded in a stepwise manner. Complexes of the type *trans*-$[RhCl_2(L)(PMe_2Ph)_2]$ (L = Me_2Dtc, EtXant), *mer*-$RhCl_2L(PMe_2Ph)_3$ (L = Me_2Dtc, EtXant), *mer*-$[RhClL(PMe_2Ph)_3]^+$ (L = Me_2Dtc, EtXant), *cis*-$[Rh(L)_2(PMe_2Ph)_2]^+$ (L = Me_2Dtc), and *trans*-$[Rh(L)_2(PMe_2Ph)_2]^+$ (L = Me_2Dtc) have been synthesized.

While the reaction of KEtXant with *mer*-$RhCl_3(PMe_2Ph)_3$ in acetone or EtOH|CHCl$_3$ proceeds with the formation of $RhCl_2(EtXant)(PMe_2Ph)_2$, in pure EtOH the products were $RhCl(S_2CO)(PMe_2Ph)_3$, $K[RhCl_2(S_2CO)(PMe_2Ph)_2]$, and two isomers of $Rh(S_2CO)(EtXant)(PMe_2Ph)_2$ (131). The dithiocarbonate ligand apparently is obtained by attack on a coordinated xanthate ligand by a nucleophile (10).

In all these complexes, the Rh(III) ion is presumed to be six-coordinate, and possible structures have been assigned on the basis of PMR and IR data.

Until recently 1,1-dithio complexes of univalent Rh or Ir were virtually unknown. The metathesis of the bridging chloride by Et_2Dtc in the $[MCl(C_8H_{14})_2]_2$ or $[MCl(C_8H_{12})]_2$ complexes [M = Rh(I), Ir(I)] proceeds under N_2 in toluene or acetone with the formation of $M(Et_2Dtc)(C_8H_{14})_2$ and $M(Et_2Dtc)(C_8H_{12})$, respectively (186). These complexes undergo oxidative addition with Et_4Tds to give the $M(Et_2Dtc)_3$ complexes (M = Rh). The very interesting $Ir(Et_2Dtc)_3(C_8H_{12})$ is obtained from $Ir(Et_2Dtc)(C_8H_{12})$. This compound shows PMR and IR spectra indicative of monodentate as well as bidentate, Et_2Dtc ligands, and probably is a six-coordinate Ir(III) complex containing one bidentate and two monodentate Et_2Dtc ligands. The oxidative

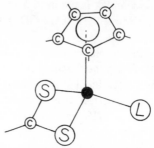

Fig. 44.   Proposed structure for the $(\eta^5 C_5 Me_5)Rh(Me_2 Dtc)L$ complex; L = NCBPh$_3$ (143)

addition of Et$_4$Tds to the [MCl(C$_8$H$_{14}$)$_2$]$_2$ complexes results in the formation of the dimeric [M(Et$_2$ Dtc)$_2$ Cl]$_2$ compounds (186).

The oxidative addition of Me$_3$Dtc to the RhCl(PPh$_3$)$_3$ complex proceeds with cleavage of a C–S bond and the formation of the RhCl(SMe)(SCNMe$_2$)-(PPh$_3$)$_2$·C$_6$H$_6$ complex (265). The SMe ligand in this compound reacts with CS$_2$ and the RhCl(MeSXant)(SCNMe$_2$)PPh$_3$ can be isolated. The C–S bond cleavage, as a result of an oxidative addition reaction, also is observed in the reaction of Me$_4$Tms with RhCl(PPh$_3$)$_3$, and the RhCl(Me$_2$Dtc)(SCNMe$_2$)PPh$_3$ complex is obtained (265).

The crystal structure of the CHCl$_3$ solvate of this molecule has been determined (62, 265). The coordination of the Rh(III) ion is distorted octahedral and the thiocarboxamido group is coordinated by way of a short Rh–C bond [1.895(16) Å] and a rather long Rh–S bond [2.432(5) Å] (Table XXI).

Recently the synthesis and characterization of the η^5-(C$_5$Me$_5$)-Rh(Me$_2$Dtc)(NCBPh$_3$) was reported (143). The proposed structure for this molecule is shown in Fig. 44 and is similar to the structure of the corresponding dimethyldithiophosphinate complex, which has been determined by x-ray crystallography (143). The [NORh(Me$_2$Dtc)$_3$]PF$_6$ complex is obtained by the reaction of [Rh(NO)MeCN)$_4$]$^{2+}$ with excess NaMe$_2$ Dtc ($\nu_{(NO)}$ = 1545 cm^{-1}). A bent nitrosyl group, bidentate Me$_2$Dtc ligands, and seven-coordinate Rh have been proposed for this complex on the basis of IR data (137).

c. **Nickel, Palladium, Platinum.** *(1) Nickel.* Several advances in the chemistry of nickel with 1,1-dithio chelates have been made since 1969. Outstanding among these are the synthesis and characterization of Ni(III) and Ni(IV) dithiocarbamate complexes, first reports on aryl xanthate complexes, the synthesis of a variety of mixed-ligand complexes, and the structural characterization of perthio complexes.

(a). Ni(IV) Complexes.   In a number of papers (75, 77, 632) the Nijmegen group, including J. J. Steggerda, J. A. Cras, H. C. Brinkhoff, and J.

Willemse, reported on the reaction of $Ni(R_2Dtc)_2$ complexes with Cl_2, Br_2, or I_2 and characterized the products as $Ni(IV)(R_2Dtc)_3X$ ($X = Cl, Br, I$). The structure of the $Ni(n$-$Bu_2Dtc)_3Br$ was determined by Fackler et al. (19, 231).

The coordination geometry in the cation is distorted octahedral with the NiS_6 core twisted by $45.5°$ from trigonal prismatic geometry. The molecular parameters (Table XXII) are consistent with a ligand-stabilized nickel(IV) description. The shift in the C—N stretching vibration from 1505 cm^{-1} in the $Ni(n$-$Bu_2Dtc)_2$ complex to 1545 cm^{-1} in the Ni(IV) complex and the short C—N bond $1.318(8)$ Å suggest that the resonance form $^{2-}S_2$ C=N$^+$ of the ligand plays an important role in stabilizing the Ni(IV) ions. The bulk magnetic susceptibility for this complex, μ_{eff}^{corr}, is 0.7 BM (231).

The brown color of the $Ni(n$-$Bu_2Dtc)_3^+$ cation is photochemically and thermally bleached reversibly in CH_3CN solution. Thiuram disulfide and the Ni(II) complex were identified as products of this bleaching process. The kinetics of the reappearance of color were reported to depend on the square of the concentration of the bleached nickel and were inversely proportional to the concentrations of thiuram disulfide and Br$^-$ (231).

The reaction of $[Ni(n$-$Bu_2Dtc)_3]Br$ with $Ni(n$-$Bu_2Dtc)_2$ results in a solution with an EPR spectrum that has been attributed to a reactive asymmetric Ni(III) complex (231). This complex may be similar to that reported by Willemse et al. (631), who obtained $Ni(III)(n$-$Bu_2Dtc)_2I$ by the low-temperature ($-30°C$) oxidation of $Ni(n$-$Bu_2Dtc)_2$ with I_2. This complex is characterized by a C—N stretching frequency of 1518 cm^{-1} and a M—S vibration at 377 cm^{-1}. The μ_{eff} of 1.33 BM is low for a five-coordinate or six-coordinate low-spin complex. Axial EPR spectrum was observed with $g_1 = 2.260$, $g_2 = 2.215$, and $g_3 = 2.027$.

The $Ni(R_2Dtc)_3X$ complexes also were obtained by the oxidation of the $Ni(R_2Dtc)_2$ complexes with the novel oxidant 3,5-bis(N,N-dialkyliminium)-1,2,4-trithiolane ion, for which the structure is known (29) (Fig. 5).

The reactions of the $Ni(R_2Dtc)_3^+$ complexes with Lewis bases [L = CNR$'$ (R$' = i$-Pr, t-Bu, or p-Cl—Ph),] are reported to give $[NiL_2(R_2Dtc)]^+$. With L = CNR$'$ the R_4Tds by-products have been isolated (441). The PMR, electronic, and IR spectra of these complexes have been reported.

The synthesis of $Ni(R_2Dtc)_2Br_2$ from the reaction of $Ni(R_2Dtc)_2$ with Br_2 reported by Nigo et al. (479) could not be reproduced by Wilemse et al. (629), who instead could only obtain the $Ni(R_2Dtc)_3Br$ complexes.

A two-proton exchange model using a density matrix formalism has been used to analyze the unusual structure of the PMR spectra in the diamagnetic $Ni(R_2Dtc)_2$ complexes (R = n-Pr, Et, i-Pr, i-Bu, and Bz) (273).

(b). Structures of $Ni(R_2Dtc)_2$ Complexes. A series of crystal structures for the $Ni(R_2Dtc)_2$ complexes were determined to delineate the effects of change of ligand substituent on the ligand geometry and electron distribution (Table XXII). As expected, in all these structures the Ni(II) ions are four-

TABLE XXII
Selected Structural Parameters and Redox Potentials for Nickel 1,1 dithio complexes.

Complex	$\overline{\text{Ni–S}}$, Å	$\overline{\text{C–N}}$, Å	$E_{1/2}$, V[a]	$E_{1/2}$, V[b]	Ref.
Ni(i-Pr$_2$Dtc)$_2$	2.181 (3)	1.329 (15)	+ 0.88	−1.48	475
Ni(MePhDtc)$_2$	2.203 (7)	1.30 (1)			427
Ni(PyrrolDtc)$_2$	2.209 (11)	1.33 (1)	+ 0.88	−1.34	473
Ni(i-Bu$_2$Dtc)$_2$	2.193 (3)	1.31 (1)	+ 0.90	−1.41	534
Ni(MeHDtc)$_2$	2.199 (9)	1.30 (2)	+ 0.89	−1.35	475
Ni(n-Bu$_2$Dtc)$_3$Br	2.261 (1)	1.318 (8)		0.369[c]	19, 231,
				0.337[d]	312
Ni(o-t-BuPhXant)$_2$	2.200 (5)				117
[Ni(Et$_2$Dtc)(TFD)]$_2$	2.234 (2)[e]	1.32 (1)		+ 0.29[g]	320
	2.154 (2)[f]				
[Ni(n-Bu$_2$Dtc)(MNT)]$_2$[h]	2.237 (4)[e]	1.31			318
	2.169 (4)[f]				
[Ni(SBz)BzSXant]$_2$[i]	2.209 (4)[j]				241
	2.189 (4)[k]				
[Ni(PhDta)$_2$]$_3$[l]	2.228 (4)				51
[Ni(BzDta)$_2$]$_2$[m]	2.215 (10)				43
[Ni(t-BuDta)$_2$]$_2$[n]	2.217 (12)				47
Ni$_3$(MeDta)$_3$(MeCS$_3$)[o]	2.20 (2)[p]				43
	2.19 (3)[q]				
cis-Ni(BzDta)$_2$(Py)$_2$[r]	2.449 (3)				44
Ni(PhDtaS)$_2$	2.161 (6)[s]				48
Ni(p-DtcmS)(p-Dtcm)[t]	2.212 (6)				233, 257
	2.134 (6)[u]				

[a] One-electron oxidation in acetone versus Ag|AgCl at a platinum electrode.
[b] One-electron reduction as in footnote a.
[c] [NiL$_3$]$^+$ generated by oxidation of NiL$_2$, cathodic scan.
[d] As for footnote c anodic scan.
[e] R$_2$Dtc ligand.
[f] TFD or MNT ligand.
[g] Two-electron reduction in acetone or CH$_2$Cl$_2$ at a gold electrode versus SCE.
[h] Ni–S intermolecular distance 2.415 (4) Å.
[i] Ni–Ni intermolecular distance 2.795 (3) Å.
[j] RSXant ligand.
[k] RS bridging ligand.
[l] Ni–S intermolecular distance 3.108 (5) Å.
[m] Ni–Ni intermolecular distance 2.56 Å.
[n] Ni–S intermolecular distance 2.7 Å.
[o] Ni–Ni intermolecular distance 3.0 Å.
[p] MeDta ligand.
[q] Trithio orthoacetate ligand.
[r] Ni–N, 2.10 (1) Å.
[s] S–S bite 3.173 Å.
[t] p-Dtcm-p-isopropylphenyldithiobenzoate.
[u] Sulfur-rich ligand, S–S bite 3.052 Å.

coordinate and planar. Intramolecular hydrogen bonding effects with the sulfur atoms were considered instrumental in determining molecular packing and out-of-plane distortions in the Ni(HMeDtc)$_2$ complex (474).

Both intra- and intermolecular hydrogen bonding also were observed in the structure of the complex with R$_2$ = i-Pr$_2$ (475). With R$_2$ = MePh, intermolecular hydrogen bonding was considered to be the cause for the nearly orthogonal (82°) tilt of the phenyl ring relative to the conjugated ligand (427). With R$_2$ = Pyrrol the structure also is subject to hydrogen bonding interactions (473). It is interesting that this molecule is x-ray isomorphous to the Cu(II) analogue, for which a monomeric structure also was found. The structure of the Ni(H-i-PrDtc)$_2$ is the first example of a Ni(RR′Dtc)$_2$ complex with different ligand substituents that adopts the cis configuration in the solid. This conformation is due to strong intermolecular –NH– hydrogen–sulfur hydrogen bonding (536). Hydrogen bonding interactions also are observed in the structure of the Ni(i-Bu$_2$Dtc)$_2$ complex (534). The Ni(PipDtc)$_2$·2H$_2$O complex has been reported (526).

(c). Mixed-Ligand Complexes.   The reaction of (n-Bu$_4$N)$_2$Ni(MNT)$_2$ with Ni(n-Bu$_2$Dtc)$_2$ in boiling acetonitrile was employed to obtain (n-Bu$_4$N)-Ni(MNT)(n-Bu$_2$Dtc) as golden green crystals (604).

The [Ni(MNT)(n-Bu$_2$Dtc)]⁻ anion shows a reversible one-electron oxidation at +0.33 versus SCE (CH$_2$Cl$_2$ solution). This value is intermediate between $E_{1/2}$ values for the dithiolene and the dithiocarbamate complex. A continuation of this work appeared (601) and a series of Ni(MNT)(R$_2$Dtc)$^{1-}$ and Ni(MNT)-(R$_2$Dtc)^0 complexes were isolated. The Ni(MNT)(RXant)⁻ also was reported. Exchange reactions between Ni(R$_2$Dtc)$_2$ complexes and Ni(RXant)$_2$ complexes proceeded very slowly and no mixed products could be isolated. The exchange of ligands between the dithiolene, Ni(TFD)$_2$, and Ni(R$_2$Dtc)$_2$ or Ni(RXant)$_2$ complexes is rapid. The xanthate complexes were found to react more slowly than the dithiocarbamate complexes (320). Kinetic measurements along with spectroscopic and conductivity measurements at −60°C indicate the rapid and reversible formation of a polarized 1:1 intermediate between Ni(TFD)$_2$ and Ni(L)$_2$, where L = R$_2$Dtc or RXant.

The dimers [Ni(TFD)(L)]$_2$ (L = R$_2$Dtc⁻ or RXant⁻) can be reduced chemically by o-phenylene diamine to give yellow solutions of the dianions. The crystal structure of the [Ni(TFD)(Et$_2$Dtc)]$_2$ complex was determined (320). It consists of well-separated centrosymmetric sulfur-bridged dimers (Table XXII). The nickel ion has a square pyramidal coordination and lies 0.36 Å out of the equatorial plane in the direction of the apical bridge sulfur. The structure resembles that of the dimeric Cu(R$_2$Dtc)$_2$ complexes.

A similar dimeric structure was found in the [Ni(n-Bu$_2$Dtc)(MNT)]$_2$ complex and was considered to contain Ni(III) (318). Unsymmetrical planar complexes of the type Ni(R$_2$Dtc)(PR′$_3$)X (R = i-Bu, Et; R′ = alkyl, aryl; X = Cl, Br, I,

SCN, SR) were reported by Maxfield (436) and later by Fackler and coworkers. Fast halide exchange was considered responsible for the magnetically equivalent R groups observed in the PMR spectra of the complexes in the presence of free phosphine (237).

The equilibria

$$Ni(R_2 Dtc)_2 + 2Py \; \rightleftharpoons \; Ni(R_2 Dtc)_2 Py_2 \qquad (42)$$

were studied and K_{eq}, ΔH^0, and ΔS^0 were reported for $R_2 = Et_2$, n-Pr_2, and n-Bu_2 (611). The stability order given by ΔH^0 was Et $< n$-Pr $< n$-Bu; however, K_{eq} at 25°C was Et $> n$-Pr $> n$-Bu, indicating the importance of the entropy factor in the ΔG^0 values.

A very interesting reaction occurs between Ni(n-Bu_2Dtc)$_2$ and excess $ZnCl_2$ in diethyl ether under *anaerobic* conditions to give Ni(n-Bu_2Dtc)$_3$Cl. This reaction is quite mysterious since no ligand reduction has been observed (171).

Hendrickson et al. (312) made a detailed investigation of the electro-chemical behavior of the Ni(R$_2$Dtc)$_2$ complexes with 16 different R$_2$ groups in acetone at a platinum electrode. A quasi-reversible reduction was found spanning the range -1.24 to -1.49 V (versus Ag|AgCl). The reduction of the Ni(R$_2$Dtc)$_3^+$ complexes is characterized by two facile one-electron reduction steps, both couples being more positive than -0.35 V versus Ag|AgCl. The first reduction wave is quasi-reversible, as is the second reduction step:

$$Ni(R_2 Dtc)_3^+ \xrightarrow{\;+e\;} Ni(R_2 Dtc)_3 \xrightarrow{\;+e\;} Ni(R_2 Dtc)_3^- \qquad (43)$$

It appears that both of the reduced species are unstable toward decomposition. The limiting current ratio of the anodic versus the cathodic process in these couples approaches unity at $-70°C$.

The electrochemical oxidation of the Ni(R$_2$Dtc)$_2$ complexes appears near $+0.9$ V and is irreversible. The overall process has been shown to be

$$3Ni(R_2 Dtc)_2 \longrightarrow 2Ni(R_2 Dtc)_3^+ + Ni^{2+} + 4e^- \qquad (44)$$

The Ni(EtXant)$_3^-$ complex (155) undergoes a reversible oxidation at $+0.92$ V (312). The substituent effects in the electrochemical characteristics of the Ni(R$_2$Dtc)$_2$ and Ni(R$_2$Dtc)$_3^+$ complexes are related to the inductive and meso-meric effects of the R groups.

(d). Xanthate Complexes.   Following the successful synthesis of the aryl xanthate ligands (see also Section II.A), Fackler et al. reported on the synthesis and characterization of various substituted aryl xanthate nickel complexes (238). The structure of the Ni(RXant)$_2$ complex (R = 4-t-Bu—phenyl) shows a planar Ni(II) ion with the phenyl ring nearly perpendicular to the NiS$_4$ plane (117).

The reaction between tetraethylthiuram disulfide with the Ni(EtXant)$_2$ complex gives a red-brown oxidation product with an EPR spectrum consisting of one line and has an isotropic g value of 2.106 (at low temperatures). At room temperature this product is rapidly converted into Ni(Et$_2$Dtc)$_2$. The composition Ni(III)(EtXant)(Et$_2$Dtc)$^+$ was suggested for this product (75).

The equilibria

$$\text{Ni(EtXant)}_2 + \text{B} \xrightleftharpoons{K_1} \text{Ni(EtXant)}_2\text{B}$$

$$\text{Ni(EtXant)}_2\text{B} + \text{B} \xrightleftharpoons{K_2} \text{Ni(EtXant)}_2\text{B}_2 \tag{45}$$

were studied with various bases. For the reactions with 2-picoline, 2,6-lutidine, and quinoline K_1 values of 30, 15, and 2, and K_2 values of 2, 1, and <0.1 were reported for the three amines, respectively. The enthalpies were quite insensitive to methyl substitution in the 3, 4, and 5 position of the pyridine ring even though substitution effects due to increased basicity were evident in the equilibrium constants (103).

(e). Thioxanthates.   The chemistry of nickel(II) with thioxanthate ligands is dominated by the CS$_2$ elimination reaction (see also Section IV) (40, 152):

$$2\text{Ni(RSXant)}_2 \longrightarrow 2\text{CS}_2 + [\text{Ni(SR)(RSXant)}]_2 \tag{46}$$

The structure of the [Ni(EtS)(EtSXant)]$_2$ complex was determined by Villa et al. (614) (Table XXII), who considered the Ni–Ni distance of 2.76 Å as an indication of Ni–Ni bonding. Soon thereafter Fackler et al. reported the structure of the coresponding [Ni(BzS)(BzSXant)]$_2$ dimer (241) (Fig. 45). A similar Ni–Ni distance of 2.795(3) Å and a nonplanar syn–endo Ni$_2$S$_2$ rhombus was found. It was pointed out that the Ni–Ni distance was "not consistent with a

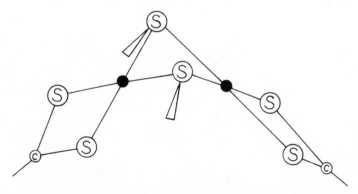

Fig. 45.   Idealized structure of the [Ni(BzS)(BzSXant)]$_2$ dimer (241).

metal–metal bond alone being the feature controlling the gross stereochemistry of the complexes" (241) and that S–S interactions also were important. A discussion of the Ni–Ni bonding and S–S interactions in this dimer and the palladium analogue has been presented (230). Kinetic studies and a proposed mechanism for the CS_2 elimination reaction are discussed in Section IV.

(f). *Dithioacid Complexes.*   In recent years major emphasis has been placed on the structural properties and sulfur addition reactions of the Ni(RDta)$_2$ complexes. The crystal structures of the Ni(PhDta)$_2$ (51) and Ni(*t*-BuDta)$_2$ (47) complexes have been determined. In both, weak intermolecular interactions were observed. In Ni(PhDta)$_2$, trimers were observed such that, in a "skewed sandwich" fashion, one molecule of type A is linked centrosymmetrically to two molecules of type B by means of metal–sulfur bridges. The structure is reminiscent of the mixed-valence Cu(II), Cu(III) *n*-Bu$_2$Dtc trimer (Fig. 11). Ni–S bridging distances of 3.11 and 2.77 Å were found. The other Ni–S bond lengths are typically in the range 2.21–2.23 Å. A dimeric arrangement similar to that observed in the structure of the Cu(R$_2$Dtc)$_2$ complex (491) was found in the Ni(*t*-BuDta)$_2$ complex (47). Here again the dimers are only loosely held with Ni–S bridge distances of 2.7 Å.

The structure of the [Ni(BzDta)$_2$]$_2$ complex was determined by Bonamico et al. (43). Four bridging (BzDta) ligands hold the planar NiS$_4$ units at a Ni–Ni distance of 2.56 Å. The two NiS$_4$ squares are staggered by 26° from an ideal N$_2$S$_8$ core of D_{4h} symmetry. The Ni atoms are displaced toward each other and out of the S$_4$ planes by 0.13 Å (Fig. 46) (Table XXII). The apparent Ni–Ni interaction suggested by this structure is also evident in the single-crystal polarized spectrum of [Ni(BzDta)$_2$]$_2$ (230). Spectra not resembling those of typical quadratic diamagnetic complexes with the NiS$_4$ core presumably reflect the perturbation between the two NiS$_4$ units.

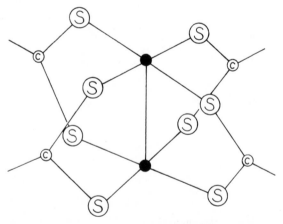

Fig. 46.   Schematic structure of the [Ni(BzDta)$_2$]$_2$ dimer (43).

The structure of the Ni(BzDta)$_2$py$_2$ was reported (44). A distorted octahedral coordination with a cis arrangement of the pyridine molecules was found. The rather long $\overline{\text{Ni-S}}$ bond length of 2.449(2) Å is perhaps a result of the high-spin nature of Ni(II) ion.

Treatment of Ni(MeDta)$_2$ with CS$_2$–EtOH (230) or of NiCl$_2$ with NaMeDta in water (43) results in the formation of two crystalline forms of the $\mu^{I},\mu^{II},\mu^{III}$-tris(MeDta)-=3-trithio-orthoacetato-triangulo-trinickel(II)   complex. The structures of both forms were determined and are similar. Three nickel atoms form an equilateral triangle with Ni—Ni distances of about 3 Å. The three MeDta ligands bridge the Ni$_3$ triangle pairwise on one side of the triangle. On the other side of the triangle the MeCS$_3^{3-}$ ligand acts in a tridentate triply bridging fashion to complete the coordination spheres of the square planar Ni(II) ions (Fig. 47) (Table XXII).

The formation of the R–CS$_3^{3-}$ anion from RDta and CS$_2$ is very interesting and poses the question of the mechanism and generality of this reaction.

Sulfur addition to 1,1-dithio complexes of Ni(II) was first reported by Coucouvanis and Fackler (156). The initial proposal that sulfur addition was accompanied by ring expansion was verified by the crystal structure determination of the FeL$_3$S complex (160) (L = p-dithiotoluate). Since then, several studies that dealt with the synthesis, structures, and reactivities of perthio complexes have appeared.

The perthiopivalate ligand (t-BuDtaS) was obtained (246) essentially by the original synthetic procedure used for the synthesis of aryl dithioacids complexes (156). The electronic spectra of Ni(t-BuDtaS)$_2$, Ni(PhDtaS)$_2$, and Ni(p-MePhDtaS)$_2$ complexes have been discussed (246) (see also Section IV).

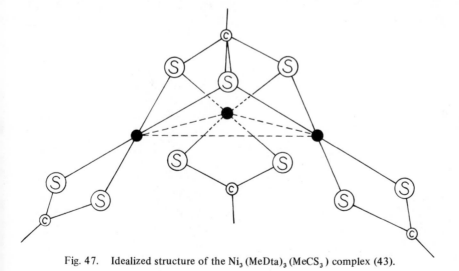

Fig. 47.   Idealized structure of the Ni$_3$ (MeDta)$_3$ (MeCS$_3$) complex (43).

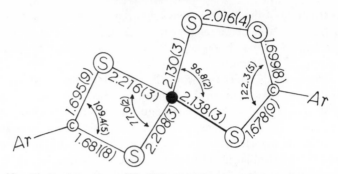

Fig. 48.   Idealized structure of the Ni(p-i-PrPhDta)(p-i-PrPhDtaS) complex (233).

The structure of the Ni(PhDtaS)$_2$ complex was reported (48) and verified the ring expansion once more. The molecule possesses a crystallographic center of symmetry. The observed $\overline{\text{Ni—S}}$ bond length [2.161(6) Å] is short when compared to other 1,1-dithio complexes of Ni(II). It was suggested that in five-membered rings there are better conditions for sp^2 sulfur hybridization because of the large dimension of the bite. The shortening of the Ni—S bond was rationalized in terms of the ability of the highest filled $-CS_3^-$ π orbital to give M_π bonds with the Ni(II) M_π orbitals (258). The S—Ni—S ($\sim$90°) and Ni—S—X (closer to 120°) angles seems to support this argument.

The significant changes imposed on the dithioaromatic ligands and complexes upon sulfur addition are illustrated in the structure of the Ni(p-i-PrPhDta)(p-i-PrPhDtaS) complex (Fig. 48) (Table XXII), determined by Fackler et al. (233, 257). The same workers explored the sulfur addition and abstraction reaction in depth (232) (see also Section IV). The rates and mechanisms of substitution reactions of square planar nickel(II) 1,1-dithiolate complexes (502) is discussed in Section IV.

(2). Palladium, Platinum.   (a). Synthetic and Structural Studies.   In the last 10 years, primary emphasis has been placed on the reaction chemistry of the 1,1-dithio Pd and Pt complexes. The crystal and molecular structure of the Pd(Et$_2$Dtc)$_2$ complex has been reported (30). The compound is x-ray isomorphous to the Pt(II) analogue, for which the crystal structure was reported previously (12). The expected four-coordinate planar coordination of the Pd(II) ion is characterized by a $\overline{\text{Pd—S}}$ bond length of 2.316(3) Å.

The M(n-Bu$_2$Dtc)$_3$X complexes (M = Pd, Pt; X = Br, I) were obtained by the oxidation of the M(n-Bu$_2$Dtc)$_2$ complexes with Bu$_4$Tds (629). The reaction of M(n-Bu$_2$Dtc)$_2$ complexes with halogens failed to produce the tris-M(IV) complexes, and instead M(Bu$_2$Dtc)$_2$X$_2$ complexes were isolated (M = Pd, Pt; X = Br, I). Both the cis and trans isomers of the Pt(n-Bu$_2$Dtc)$_2$I$_2$ complex were isolated and the crystal structure of the cis isomer was determined (629) (Table

XXIII). The infrared, Raman, PMR, and ESCA spectra of these molecules were reported. The molar magnetic susceptibilities for the $M(n\text{-}Bu_2Dtc)_3X$ complexes ($M = Pd$, Pt; $X = Cl^-$, I^-, I_3^-) ranged from 860×10^{-6} to 1847×10^{-6} an unusually high temperature independent paramagnetism. The $M(n\text{-}Bu_2Dtc)_2I_2$ and $M(n\text{-}Bu_2Dtc)_3I_3$ complexes undergo exchange and reduction reactions.

A detailed PMR study of the $cis\text{-}Pt(n\text{-}Bu_2Dtc)_2I_2$ complex in $CDCl_3$ as a function of temperature was carried out. An intramolecular process between

TABLE XXIII

Selected Structural Parameters of the Palladium and Platinum 1,1-Dithio Complexes.

Complex	$\overline{\text{M–S}}$, Å	$\overline{\text{M–M}}$, Å	$\overline{\text{S–M–S}}$,° [a]	$\overline{\text{M–S–M}}$,°	Ref.
$Pd(Et_2Dtc)_2$ [b]	2.316 (3)		75.5 (1)		30
$cis\text{-}Pt(n\text{-}Bu_2Dtc)_2I_2$ [c,d]	2.35 (2)		73.4 (4) 90.6 (4)[e]		629
$Pd(2,4,6\text{-}Me_3PhXant)_2$ [f]	2.330 (4)		75.2 (2)		117
$Pd(PhDta)_2$	2.328 (3)		74.5 (2)		51
$[Pd(t\text{-}BuS)(t\text{-}BuSXant)]_2$	2.330 (7)[g] 2.321 (12)[h]	3.162 (1)	74.3 (1) 84.4 (1)[i]	85.8 (4)	241
$[Pd(EtS)(EtSXant)]_3$	2.324 (4)[g] 2.338 (8)[h]	3.655 (2) 3.307 (2) 3.303 (2)	87.2 (2)[i] 87.6 (3)[i] 88.6 (3)[i] 74.4 (4)	103.8 (2) 90.4 (3) 90.8 (2)	241
$Pt(EtXant)_3(Ph_4As)$	2.307 (32)		76.5 (10) 86.3 (9)		141,144
$[Pt(p\text{-}i\text{-}PrPhDta)_2]_2$	2.303 (25)[g] 2.314 (8)[h]	2.870 (2)	73.6 (5)		92
$Pd(MeDta)_2$ [k]	2.331 (6)		73.7 (2)		522, 521
$[Pd(MeDta)_2]_2$ [k]	2.320 (13)	2.755 (2)	90.0 (1)		522, 521
$trans\text{-}PtH(S_2CH)(PCy_3)_2$ [l]	2.368 (6)[m] 3.748 (7)				9
$Pt(S_2CF)(PPh_3)[HF_2]$	2.331 (13)[n]		74.7		225

[a] Intraligand.
[b] $\overline{\text{C–N}}$, 1.32 (1) Å.
[c] $\overline{\text{Pt–I}}$, 2.651 (3) Å.
[d] $\overline{\text{C–N}}$, 1.29 (6) Å.
[e] Interligand.
[f] S_2C–O, 1.330 (4) Å.
[g] Bridging ligands.
[h] Nonbridging ligands.
[i] Mercaptide bridge sulfurs.
[j] Monodentate ligand sulfurs.
[k] Both units are found in the crystal structure of one of the "Pt(MeDta)₂" complexes and alternate in a stacking fashion.
[l] S–C–S, 129.0 (5)°.
[m] $\overline{\text{Pt–P}}$, 2.276 (5) Å; C–S, 1.73 (2), 1.68 (3) Å.
[n] Pt–P, 2.282 (19) Å; S–C–S, 108°.

-100 and $100°C$ and a $\Delta H^{\ddagger}$ of 7.2 kcal mole^{-1} and $\Delta S^{\ddagger}$ of -37 eu were observed. The experimental data were explained in terms of a slow exchange (up to $100°C$) and a fast racemization process involving a heteropolar breaking of a Pt–ligand bond (630). The isomerization of *trans*- into *cis*-Pt(n-Bu$_2$Dtc)$_2$I$_2$ between -10 and $+35°C$ by PMR proceeds with $\Delta H^{\ddagger} = 17.2$ kcal mole^{-1} and $\Delta S^{\ddagger} = -17$ eu and is suggestive of an ionic intermediate with rupture of a Pt–I bond.

Following the successful synthesis of the aryl xanthate complexes, Chen and Fackler reported on the synthesis and structural characterization of the diamagnetic, monomeric bis(O-2,4,6-trimethylphenylxanthato)palladium(II) complex (117). A typical PdS$_4$ unit was found in this complex in which the shorter S$_2$C–O bond length [1.330(12) Å] compared to the Ph–O bond length [1.439(16)] indicates a certain importance of the S$_2$C=Ö resonance form in the structure of the xanthate ligand.

As a result of the CS$_2$ elimination reaction from the Pd(RSXant)$_2$ complexes, the [Pd(t-BuSXant)(t-BuS)]$_2$ and [Pd(EtSXant)(EtS)]$_3$ complexes were isolated and structurally characterized (241). The former has a dimeric structure, not unlike that observed in [Ni(BzSXant)(BzS)]$_2$ (see preceding section). A distance of 3.162(1) Å separates the two Pd(II) ions in the Pd$_2$S$_2$ rhombus. A folded anti disposition of the bridging mercaptido groups was observed. In the structure of the trimer, a triangular array of Pd atoms is bridged by the three mercaptide ligands (Fig. 49).

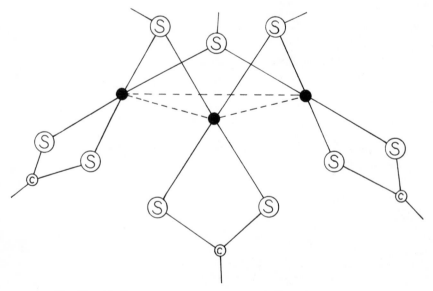

Fig. 49.   Idealized structure of the [Pd(EtS)(EtSXant)]$_3$ trimer (241).

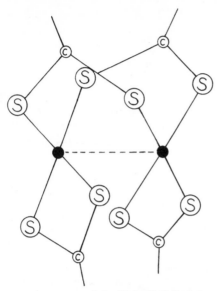

Fig. 50.   Schematic structure of the [Pt(p-i-PrPhDta)$_2$]$_2$ complex (92).

The Pd(II) ions are four-coordinate and planar. An examination of this molecule reveals a triangular array of exposed mercaptido sulfur donors. It is conceivable that this molecule may serve as a donor for other metal ions with the formation of polynuclear complexes.

The structure of the bisdithiocumatoplatinum(II), [Pt(p-i-PrPhDta)$_2$], complex was determined to be dimeric, with two bridging and two terminal ligands (Fig. 50). A Pt–Pt distance of 2.870(2) Å is observed, as well as a distorted square antiprismatic arrangement of the sulfur atoms. On the basis of infrared data, a similar type of structure is predicted for the Pt(p-MePhDta)$_2$ and Pt(PhDta)$_2$ complexes (92).

Recently we were informed (522) of the very interesting synthesis and structural characterization of two distinct Pd(MeDta)$_2$ complexes. Both of these complexes contain square planar chromophores stacking in columns with the Pd atoms directly above each other. In one of the structures, mononuclear and binuclear units alternate along a crystallographic axis with metal separations of 2.755 (in the dimer) and 3.398 Å (monomer–dimer). In another structure, of an apparent homologue of the same compound, dimers only are observed stacking along a crystallographic fourfold axis. The intra- and interdimer Pd–Pd distances are 2.739 and 3.257 Å respectively. Molecules of CS$_2$ are incorporated between different columns in this structure.

The solid-state electrical conductance properties of these materials perhaps should be investigated.

(b). Reaction Chemistry.   The reaction of tertiary phosphines with the

$M(L)_2$ complexes (M = Pd, Pt; L = RXant, R_2Dtc, RDta) gave the complexes $M(L_2)PR_3$. These compounds were formulated as five-coordinate species on the basis of PMR data (234, 239, 240). With excess phosphine the xanthate adducts underwent a further reaction and the interesting $(PR_3)_2M(S_2CO)$ complexes were isolated (239). The $Pt(R_2Dtc)_2PR_3'$ complexes are nonconducting and diamagnetic. The PMR spectra of these complexes are temperature dependent and at low temperatures two nonequivalent R groups of the R_2Dtc ligand are observed. This temperature dependence is not considered a result of hindered rotation about the C–N bond of the ligand but instead is explained in terms of the equilibrium between two PMR-equivalent, four-coordinate complexes:

$$\tag{47}$$

At lower temperatures a "slowdown" of this intramolecular process accounts for the nonequivalence of the R groups on the ligand (10). This interpretation is at variance and, perhaps, equally as likely as the one suggested earlier and based on a trigonal bipyramidal structure of a five-coordinate complex.

A four-coordinate Pt(II) with a "dangling" R_2Dtc ligand also is assumed to be the low-temperature form of the $(\pi$-allyl)Pd(Me$_2$PhP)(Me$_2$Dtc) and $(\pi$-allyl)Pd(Me$_2$PhP)(MeXant) complexes (525). An x-ray structure is reported (10) to show that in the solid state the $Pt(Et_2Dtc)_2(PMePh_2)$ contains four-coordinated Pt(II) and both uni- and bidentate R_2Dtc ligands.

A similar structure is observed for the Pd(p-i-PrPhDta)$_2$PPh$_3$ complex, where a PdS$_3$P core is attained and both uni- and bidentate dithiocumate ligands are found (587). A mechanism involving nucleophilic attack on the OR group of the xanthate ligand by free xanthate ion is proposed (10) for the reaction of $M(RXant)_2$ complexes with tertiary phosphines:

$$M(RXant)_2 + PR_3' \longrightarrow PR_3'M(RXant)_2$$

$$Et_2O \parallel PR_3' \tag{48}$$

The synthesis of the Pt(RXant)$_3^-$(Ph$_4$As)$^+$ salt (R = Et, i-Pr) was accomplished by the reaction between Pt(RXant)$_2$ and Ph$_4$AsRXant in CH$_2$Cl$_2$ (141, 144). The x-ray crystal structure determination of this molecule (R = Et) shows a planar PtS_4 unit with one bidentate and two monodentate RXant ligands (141) (Table XXIII). The variable-temperature PMR spectra of these complexes indicate rapid unidentate–bidentate exchange at room temperature (141).

The corresponding $Pd(RXant)_3^-$ complex anion appears to be unstable and is converted to the $Pd(RXant)(S_2CO)$ complex (141). Nucleophilic attack by RXant$^-$ at the OR group of a coordinated RXant ligand may be responsible for this conversion.

The reaction of the halide-bridged dimers $[MX_2PR'_3]_2$ (M = Pd; X = Cl, Br, I; PR'_3 = PMe_2Ph; M = Pt; X = Cl; PR'_3 = PMe_2Ph, $PMePh_2$) with R_2Dtc^- and RXant$^-$ alkali metal salts has been described (145, 147). The products, R_2-$DtcM(PR'_3)X$, are monomeric and can be used effectively for the synthesis of "mixed" 1,1-dithio complexes by metathesis of the X group by 1,1-dithio ligands (145). The $(R_2Dtc)Pd(PR'_3)Cl$ complexes have been obtained by the reaction of $[PdCl_2(PR'_3)]_2$ with $Me_2SnCl(R_2Dtc)$ (578).

The reaction of $Pd(C_8H_{12})Cl_2$ with $NaEt_2Dtc$ gives an orange solid of empirical formula $[PdCl(Et_2Dtc)]_n$. This compound is dimeric by molecular weight measurements and is considered to have a chloride-bridged structure from the presence of the bridging Pd–Cl band at 300 cm^{-1} in the far IR spectrum. The analogous EtXant complex also has obtained and was found to be unstable toward decomposition to intractable products (140, 146).

The Cl$^-$ bridges in the Et_2Dtc complex could be cleaved by various Lewis bases to give monomeric $Pd(Et_2Dtc)(Cl)(L)$ (L = PPh_3, $AsPh_3$, $SbPh_3$, or C_5H_5N).

The $[PdCl(Et_2Dtc)]_2$ complex, in a metathetical reaction with SPh$^-$ gave $[Pd(SPh)(Et_2Dtc)]_2$, which upon reaction with PPh_3 gave the red crystalline monomeric $Pd(SPh)(PPh_3)(Et_2Dtc)$ complex (140). The $[M(S$-t-$Bu)(Et_2Dtc)]_2$ complexes (M = Pd or Pt) were previously synthesized by the reaction of $[M(S$-t-$Bu)(t$-$BuSXant)]_2$ complexes with NEt_2H (15).

The reaction of Pd(II) and Pt(II) 1,1-dithio complexes with tertiary phosphinites PPh_2OR has led to the isolation of $[M(L)(PPh_2O)(PPh_2OH)]$ (M = Pd, L = Et_2Dtc; M = Pt, L = Et_2Dtc). A careful study of the $Pt(Et_2Dtc)_2PPh_2OMe$, reaction has resulted in the isolation of the intermediate $[Pt(Et_2Dtc)(PPh_2(OMe)]_2X$ (X = BPh_4^- or Cl$^-$) and the formulation of a mechanism for the formation of the $[Pt(Et_2Dtc)(PPh_2O)(PPh_2OH)]$ complex (142).

The synthesis of the $trans$-$PtH(S_2CH)(PCy_3)_2$ complex has been attained by CS_2 insertion into the Pt–H bond of $trans$-$PtH_2(PCy_3)_2$. The x-ray crystal structure of this molecule shows a planar Pt(II) ion. The S_2CH^- ligand is bonded to the metal through a sulfur atom as a monodentate anion (Table XXIII). A kinetic study of the CS_2 insertion reaction has been conducted. The data are consistent with a mechanism that involves CS_2 addition to $trans$-$PtH_2(PCy_3)_2$, to give a five-coordinate intermediate, which collapses to $trans$-$PtH(S_2CH)(PCy_3)_2$ (9).

The fluoro–thioformate complex $[Pt(S_2CF)(PPh_3)_2][HF_2]$ is synthesized by the reaction of CS_2 with $[PtF(PPh_3)][HF_2]$. The structure of the product of this interesting CS_2 insertion reaction has been determined (Table XXIII). A distorted square planar coordination around the Pt atom is found, as well as a bidentate S_2CF ligand (225).

D. Infrared Spectra of Dithio Complexes

1. Dithiocarbamates

Three regions in the IR spectra of the $M(R_2Dtc)_n$ complexes are of considerable interest and have proven valuable in arguments concerning the electronic and structural characteristics of these compounds.

The 1450–1550 cm^{-1} region is associated primarily with the "thioureide" vibration and is attributed to the ν(C–N) vibration of the S_2C–NR_2 bond. An increase in the double-bond character of the C–N bond (Fig. 51C) results in higher frequencies for this vibration (153).

A second region between 950 and 1050 cm^{-1} is associated with the ν(CSS) vibrations and has been used effectively in differentiating between monodentate and bidentate R_2Dtc ligands. Bonati and Ugo were the first to recognize the presence of two absorptions in the region of 1050–950 cm^{-1} as a diagnostic criterion for asymmetrically bound R_2Dtc groups (55b). Finally, the 300–400 cm^{-1} region is associated with M–S vibrations.

A number of theoretical studies have appeared in the past decade, and vibrational analyses of various $M(R_2Dtc)_n$ complexes provided both a theoretical justification and a "fine tuning" of certain semiempirical criteria used in the interpretation of the IR spectra of these molecules.

Normal coordinate analyses of a series of $Cr(R_2Dtc)_3$ complexes have been reported (81). In these analyses the complexes were treated as planar 1:1 metal–ligand systems of C_{2v} symmetry and only the 11 in-plane vibrations $(6A_1 + 5B_2)$ were considered (81). The alkyl groups were treated as point masses. One interesting result of these studies was that the observed decrease in ν(C–N) in the sequence R = Me > Et > Pr $\sim$ Bu, is paralleled by the calculated sequence obtained by increasing the point mass of the alkyl group

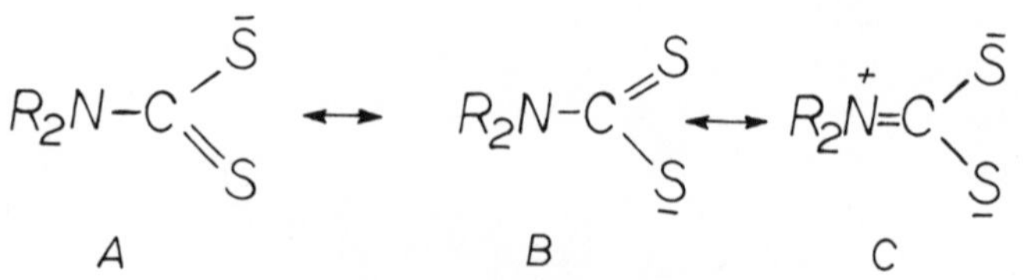

$$R_2N-C{\overset{\bar{S}}{\underset{S}{\diagup}}} \longleftrightarrow R_2N-C{\overset{S}{\underset{\underset{S}{\diagdown}}{\diagup}}} \longleftrightarrow R_2\overset{+}{N}=C{\overset{\bar{S}}{\underset{\underset{S}{\diagdown}}{\diagup}}}$$

$$A \qquad\qquad\qquad B \qquad\qquad\qquad C$$

Fig. 51. Resonance forms in the R_2Dtc ligands.

only. To reproduce the observed sequence (within 0.5%), however, it was found necessary to decrease the C–N force constant in the sequence: $f_{Me} > f_{Et} > f_{pr} \sim f_{Bu}$.

These results indicate that both electronic and kinematic effects are important in determining the frequency of the C–N vibration. The masses of the alkyl substituents also were found to affect the mixing of the asymmetric N-alkyl and the symmetric C–S modes. The mixing was found to increase with increasing substituent mass.

The Ugo-Bonati criterion (55b) that distinguishes monodentate from bidentate binding by the number of bands in the 950–1050 cm^{-1} region was shown to be valid, provided comparison is made between complexes containing the same alkyl group. It appears that the splitting of the ν(C–S) vibration also should occur, with unsymmetrical *bidentate* bonding. It is suggested that monodentate bonding should be assumed only if the splitting exceeds 20 cm^{-1} (81).

A similar suggestion was advanced by Alyea and Ramaswamy (11b). Another interesting result of the above calculations (81) was that small splittings ($\sim$15 cm^{-1}) of the C–N vibration also should be observed for unsymmetrical bidentate, or monodentate, binding of the R_2Dtc ligands. Unfortunately, this contention cannot be verified experimentally because of the inherently large width of the C–N band.

The Ugo-Bonati criterion has been verified a number of times (11b, 69, 76). A detailed compilation of the absorptions in the 900–1050 cm^{-1} region for a number of complexes with both symmetrically bound bidentate R_2Dtc ligands and complexes with monodentate R_2Dtc ligands has appeared (76). An observation of the data (Table XXIV) conclusively shows that, while two bands are observed in that region for the $M(R_2Dtc)_n$ complexes with symmetrically bound bidentate R_2Dtc ligands, three bands are observed for complexes with asymmetrically bound monodentate ligands.

An analysis of the region between 400 and 60 cm^{-1} for the $M(Me_2Dtc)_2$ (M = Ni, Pd, Cu), $FeCl(Me_2Dtc)_2$, and $Co(Me_2Dtc)_3$ complexes has been presented (492). For the $M(Me_2Dtc)_2$ complexes the three metal translation vibrations expected in this region are: (1) the symmetric Ni–S stretch B_{1u} (D_{2h} symmetry); (2) the antisymmetric Ni–S stretch B_{3u}; and (3) the out-of-plane vibration of the ligands relative to the metal. These vibrations have been assigned to frequencies observed at 387, 300 and 179 cm^{-1} in the Ni(II) complex; 361, 270, and 163 cm^{-1} in the Pd(II) complex, and 351, 250, and 120 cm^{-1} in the Cu(II) complex (492). While they agree with the assignment for the symmetric Ni–S stretch (B_{1u}), Jensen et al. (363) argue that, in species B_{3u}, the asymmetric Ni–S stretch should be the band observed at 376 cm^{-1} (nearly coincident with the symmetric stretch!). In this later study (363) the IR spectra in the range 40–4000 cm^{-1} of the $Ni(Me_2Dtc)_2$, the selenocarbamate analogue, and the perdeuterated forms of these complexes were recorded. A normal coordinate

TABLE XXIV
IR Data for R_2 Dtc Derivatives Measured in $CHCl_3$ (76).

Compound	Band (I, II) ν(C=S), cm^{-1}		Band (III) ν(CSS), cm^{-1}
ASYMMETRIC			
Et(Et$_2$Dtc)	1007 (m)	990 (m)	920 (m)
Me(Et$_2$Dtc)	1012 (m)	990 (m)	919 (m)
HgI$_2$Me(Et$_2$Dtc)^a	1002 (m–w)	989 (m)	909 (m)
CdI$_2$(Et$_4$Tds)^a	992 (m–w)	971 (m)	907 (m)
HgI$_2$(Et$_4$Tds)	1000 (m–w)	972 (m)	910 (m)
Et$_4$Tds	1008 (m)	972 (m)	918 (m)
ZnI$_2$(Et$_4$Tds)^a	991 (m–w)	971 (m)	907 (m)
As(Et$_2$Dtc)$_3$	1004 (m–w)	988 (m)	915 (m)
Sb(Et$_2$Dtc)$_3$	1000 (sh)	988 (m)	915 (m)
Bi(Et$_2$Dtc)$_3$	1000 (sh)	985 (m)	913 (m)
Se(Et$_2$Dtc)$_2$	1005 (m)	983 (m)	915 (m)
Te(Et$_2$Dtc)$_2$	1002 (m)	986 (m)	916 (m)
Sn(Et$_2$Dtc)$_4$	1000 (sh)	991 (m)	917 (m)
SYMMETRIC			
Co(Et$_2$Dtc)$_3$	1005 (m–w)		918 (mw)
Zn(Et$_2$Dtc)$_2$	991 (m)		914 (m)
Hg(Et$_2$Dtc)$_2$	989 (m)		913 (m)
Ni(Et$_2$Dtc)$_2$^{a}	993 (m)		914 (m)
Ni(Et$_2$Dtc)$_3^+$I$_3^-$ ^{a}	998 (m–w)		913 (w)
Cu(Et$_2$Dtc)$_2$^{a}	998 (m)		915 (m)

^{a} Spectra recorded in KBr.

analysis for these compounds was carried out using a 30-parameter force field in the generalized force field (GVFF) approximation. A full 2:1 ligand-to-metal model (D_{2h} symmetry) was used. The "thioureide" band was described as an "out-of-phase skeletal stretching" with varying contributions from internal modes of the methyl groups. A similar interpretation has been offered previously for this band, that is, a weakly coupled ν(C–N) vibration to the ν_s(–CNC) and ν_s(–CSS) vibrations (206b). The strong broad band in the region between 950 and 1000 cm^{-1} that undergoes a bathochromatic shift in the selenocarbamate analogue has been described as mainly a ν_{as}(–CSS) vibration. This vibration is coupled in-phase with ν_{as}(–CNC) in various proportions.

A nearly linear correlation is found in a plot of the ν(C–N) vibrational frequency versus the methylene proton resonance from PMR data (τ, ppm) on symmetric Et$_2$Dtc complexes. The correlation is interpreted in terms of the argument that an increase of the partial positive charge on the nitrogen (and higher C–N vibrational frequencies) results in a deshielding of the N-bonded methylene protons (582).

The $\tau/\nu(C-N)$ correlation for asymmetric $E_2 Dtc$ complexes shows that $\tau(CH_2)$ and $\nu(C-N)$ are not interdependent. The use of PMR data as a diagnostic criterion for symmetric versus asymmetric $R_2 Dtc$ bonding has been proposed (582).

2. Xanthates

Normal coordinate analyses of the $M(MeXant)_2$ complexes and the corresponding perdeuterated analogues have been reported (M = Ni, Pd, Pt) (432, 544b). The force constant of the $O-CS_2$ bond (6.4–6.8 mdyne $Å^{-1}$) is considered to be evidence for a significant contribution of the canonical form $CH_3-\overset{+}{O}=CS_2^{2-}$ to the structure of the ligand (432). A normal coordinate analysis of the $Ni(EtXant)_2$ complex shows that many of the vibrations of the ligand are strongly coupled. The $RO-CS_2$ group is characterized by four bands near 1250, 1100, 1020, and 550 cm^{-1}. Each of these "xanthate" bands arises primarily from combinations of the $\nu(C=S)$, $\nu(C-O)$, $\nu(C-S)$, and $\nu(R-O)$ modes (4).

3. Dithioacids

The infrared spectra of several $M(ArDta)_2$ complexes (M = Ni, Pd, Pt, Zn; Ar = Ph, Tol) and their sulfur-rich products $M(ArDtaS)_2$ and $M(ArDta)(ArDtaS)$ have been obtained from 200 to 4000 cm^{-1} (92).

Three regions of importance in the spectra are: (1) the region between 1240 and 1290 cm^{-1}, (2) the region between 900 and 1100 cm^{-1}, and (3) the region between 200 and 400 cm^{-1}.

Vibrations in the first region are associated with the stretching frequency of the $Ph-CS_2$ bond and occur between 1245 and 1251 cm^{-1} in the Zn(II) complexes and between 1265 and 1289 cm^{-1} in the Ni(II) and Pd(II) complexes. In the mixed-ligand perthiodithioarylate complexes this band shows a shoulder on the low-frequency side. In the bisperthioarylate species, $M(ArDtaS)_2$, the band shifts to lower frequencies by 25 cm^{-1}. The position of this band is higher in energy than expected for a C–C single bond and may indicate a partial double-bond character for the Ph–C bond.

The 900–1100 cm^{-1} region shows bands that are associated with the C–S vibrations. Two bands are found for the $Ni(ArDta)_2$ and $Ni(ArDtaS)_2$ complexes in this region, at 985 and 951 cm^{-1}, and at 1035 and 927 cm^{-1}, respectively. The existence of a four-membered ring and a five-membered ring in the $Ni(ArDta)(ArDtaS)$ complex (Fig. 60) is reflected by the presence of C–S bands at 1021, 954, and 927 cm^{-1} (Ar = cumyl).

The appreciable separation ($\sim$100 cm^{-1}) of the two vibrations and the lower frequency of the Ph–C vibration in the bisperthioarylato complexes may indicate the loss of conjugation and appreciable localization of the negative charge on the perthio moiety of the ligand. A band of medium-to-strong intensity at 480 cm^{-1} in the Raman spectra of the $Zn(ArDta)_2$ complexes has been assigned to the S–S vibration (92).

An analysis of the IR spectra of various $M(PhDta)_n$ complexes (M = Ni, Pd, Pt; n = 2 and M = Cr, Fe, Rh, In; n = 3) has been reported (416). The results of this analysis and the assignments of the Ph–C and C–S vibrations are in good agreement with the previous study (92).

A similar IR study was reported (519) for the dithio and perthiocarboxylate complexes. The results of this study and the assignments made are also in good agreement with the study by Burke and Fackler (92). In this later study a comparison of the CSS vibrations is made between the free acids and the dithioaromatic and dithioaliphatic free acids, and between the dithioaromatic and dithioaliphatic complexes. It appears that the energy separation is considerably smaller between the ν_{as} and ν_{sym} CSS vibrations in the dithioaromatic than in the dithioaliphatic complexes. Thus in $Ni(PhDta)_2$ these vibrations occur at 980 and 945 cm^{-1}. By comparison the same vibrations occur at 1173 and 1147 cm^{-1} and at 899 and 987 cm^{-1} in the spectrum of the $Ni(MeDta)_2$ complex.

E. Electronic Spectra of Dithio Complexes

An extensive compilation of the electronic spectra of the 1,1-dithio ligands and complexes was presented in an earlier review (153). Since then many new complexes and 1,1-dithio ligands have been discovered and their electronic spectra have been recorded. The basic ideas concerning the nature of the electronic transitions in these complexes still are based on semiempirical calculations.

Theoretical studies for the 1,1-dithio complexes of the extended Hückel or LCAO–MO–SCF type are few, and accounts of such studies are briefly given in the appropriate sections, which are arranged according to element.

Studies that attempted to probe the electronic structure of the dithio complexes include the correlation of the electronic spectra in a series of $R_2 Dtc$ complexes of closed-shell configuration (584). In this study three major electronic absorptions in the spectra were found to be essentially unaffected by the central metal atom of the complexes and were considered intraligand transitions.

Small perturbations of these transitions were detected (584) as described previously (153), and for the high-energy absorption (ca. 39 kK) it was suggested that these perturbations may reflect the type of coordination adopted by the ligand.

An extended Hückel calculation of the quadrupole splitting in $Fe(R_2$-$Dtc)_2 Cl$ and $[Fe(R_2 Dtc)_2]_2$ has been reported (384). The abnormally large electric field gradients in these complexes were attributed to covalency effects. Polarized single-crystal spectra of the $Ni(Et_2Dtc)_2$ complex were recorded and analyzed (195). On the basis of polarization and energy, the four weak absorptions in the visible region of the spectrum were assigned as d–d transitions, vibronically induced by a b_{2u} vibrational mode. The d orbital ordering $d_{xy} > d_{x^2-y^2} > d_{yz} > d_{xy} > d_{z^2}$ was suggested.

A LCAO–MO–SCF calculation on the $Ni(H_2Dtc)_2$ and Ni–$(PhDta)_2$ complexes is available (123). In this calculation the participation of the $4s$ and $4p_x$ orbitals in the ground state was found to be small. Furthermore, there was no participation of the $4p_y$ and $4p_z$ orbitals.

This finding is significant in that the participation of the $4p_z$ orbital of the nickel in the in-plane π system was considered as a reason for the unavailability of this orbital for axial σ bonds in base-adduct formation. It is proposed that the higher the metal character and the lower the energy of empty molecular orbitals of b_{1u} symmetry, the greater the chances for the formation of axial bonds.

F. Voltametric Studies on the Dithiocarbamate Complexes

As indicated in a previous review (627) comparison of voltametric data on the $M(R_2Dtc)_n$ complexes is difficult for many cases because different solvents and reference electrodes have been used. Major, extensive studies have been reported by two groups. The Nijmegen group have reported studies in CH_2Cl_2 versus a saturated calomel electrode (480, 602, 604, 609). The Australian group, on the other hand, reported electrochemical studies in acetone versus a Ag|AgCl, $0.1M$ LiCl electrode (115, 116, 309, 311, 312, 313).

The results of these two major studies are in agreement on a relative basis. Thus the dependence of the oxidation or reduction potentials on the R_2 groups was found roughly the same for all dithiocarbamate complexes.

A tabulation of some of the results obtained by the Australian group, Hendrickson, Martin, and co-workers, is presented in Table XXV. In this table are shown the potentials for the Bz_2Dtc and Cx_2Dtc complexes. These complexes show the greatest difference in electrochemical behavior and envelope the other R_2Dtc complexes.

Several major trends are apparent in Table XXV. (1) The redox potentials for any given ligand vary greatly with the nature of the metal. This correlation argues strongly that the electrochemical processes are primarily metal centered. An outstanding result is the exceptional stability toward oxidation of the Cr(III) d^3, and Co(III) d^6 complexes. (2) The $E_{1/2}$ values depend on the substituent R and show consistently that the order of stability of the oxidized or reduced species varies from the dibenzyl complexes to the dicyclohexyl complexes. The former are easier to reduce and more difficult to oxidize, while the latter are the easier to oxidize and more difficult to reduce.

TABLE XXV
Selected IR Vibrational Frequencies and Redox Potentials of the R_2Dtc Complexes.

Complex	ν(M–S), cm^{-1}	ν(C=N), cm^{-1}	$E_{1/2}$(oxid), V^a	$E_{1/2}$(red), V^a	Ref.
Cr(Bz$_2$Dtc)$_3$			$+1.22^b$		116
Cr(Cx$_2$Dtc)$_3$			$+0.97^b$		116
Mn(Bz$_2$Dtc)$_3$			$+0.53^c$	$+0.07^c$	311
Mn(Cx$_2$Dtc)$_3$			$+0.25^c$	-0.22^c	311
Mn(Et$_2$Dtc)$_3$			$+0.41^c$	-0.07^c	311
Fe(Bz$_2$Dtc)$_3$			$+0.67$	-0.24	113
Fe(Cx$_2$Dtc)$_3$			$+0.41$	-0.55	115
Fe(Et$_2$Dtc)$_3$			$+0.54^c$	-0.37^c	115
Co(Bz$_2$Dtc)$_3$			$+1.24^b$		112
Co(Cx$_2$Dtc)$_3$			$+0.97$		116
Co(Et$_2$Dtc)$_3$	355 (s), 390 (m)	1485			270
Ni(Bz$_2$Dtc)$_2$			$+0.97^d$	-1.24^e	312
Ni(Cx$_2$Dtc)$_2$			$+0.84^d$	-1.49^e	312
Ni(Et$_2$Dtc)$_2$	368 (s), 410 (s)	1520 (s)	$+0.89^d$	-1.34^e	270, 312
Cu(Bz$_2$Dtc)$_2$			$+0.77^e$	-0.24^c	313
Cu(Cx$_2$Dtc)$_2$			$+0.572^c$	-0.48^c	313
Cu(Et$_2$Dtc)$_2$	345 (s), 370 (m)	1490 (s)	$+0.67^c$	-0.37^c	270, 313
Ru(Bz$_2$Dtc)$_3$			$+0.67^d$	-0.48^c	309
Ru(Cx$_2$Dtc)$_3$			$+0.48^d$	-0.72^c	309
Ru(Et$_2$Dtc)$_3$		1490 (s)	$+0.56^d$	-0.66^c	309
Cu(Et$_2$Dtc)$_2$ClO$_4$	385 (m), 410 (s)	1570 (s)			270
Ni(Et$_2$Dtc)$_3$ClO$_4$	380 (s), 390 (m)	1515 (s)			270
Fe(PyrrolDtc)$_3$	320 (s)	1485 (s)			270
Fe(PyrrolDtc)$_3$ClO$_4$	327 (s)	1530 (s)			

^{a} Cyclic voltametry in acetone 0.1 M, in Et$_4$NClO$_4$, potentials versus Ag | AgCl | 0.1 M LiCl acetone (313).
^{b} Values estimated from a graph (116).
^{c} Reversible process.
^{d} Irreversible process.
^{e} Quasi-reversible process.

For the M(R$_2$Dtc)$_n$ complexes where R = alkyl group, the Taft relation $E_{1/2} = p\Sigma\sigma + x$ holds (480). The low values of p are thought to indicate that the MO's involved in the electron transfer process have little ligand contribution.

III. 1,1-DITHIOLATES

A. 1,1-Dithiolate Ligands

Bifunctional CH acids of the type H_2CXY react with carbon disulfide in the presence of a base to give either dithioacid or, with strongly electron-with-

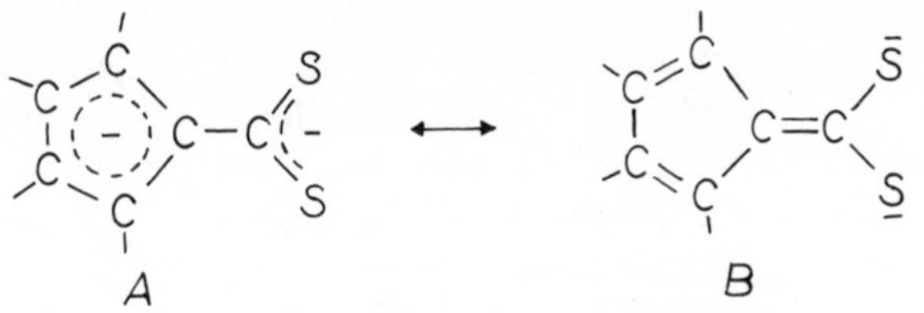

Fig. 52.   Resonance forms in the cyclopentadiene dithiolate dianion.

drawing X and Y groups, 1,1-ethylene dithiolate ligands. An account of the synthesis and characterization of 1,1-ethylene dithiolates has been presented for the work in this area prior to 1969 (153).

Since then relatively few new 1,1-ethylene dithiolate ligands have been reported. Outstanding among these new entries to the 1,1-dithiolate ligand group are the cyclopentadiene dithiolate (Fig. 52) (563), the 1,1-diperfluoro-methyl-2,2-ethylene dithiolate, $(CF_3)_2C=CS_2^{2-}$ (283), a series of dithiocarbimate ligands (64), the dithiocarbonate, CS_2O^{2-}, ligand (144, 239), and the alkyl-sul-fonyl-1,1-dithiolates $(R)(SO_2R)C:CS_2^{2-}$ (399).

B. 1,1-Dithiolate Complexes

Following the successful synthesis of the disodium salt of cyclopentadiene-dithiocarboxylic acid, Na_2CpD, Bereman and coworkers proceeded to explore the chemistry of this new ligand. The ligand is obtained by the reaction of NaCp with CS_2 in THF in a 2:1 molar ratio. The product can be obtained as either the THF or the CH_3CN solvate (563) and shows the $\nu(C=C)$ vibration at 1622 cm^{-1}. The $M(CpD)_2^{2-}$ complexes have been isolated as Et_4N^+ salts with M = Zn(II), Cu(II); (563); M = Co(II), Cd(II) (367), Ni(II), Pd(II) (25); and M = Fe(II). The $M(CpD)_3^{n-}$ complexes also have been reported for M = Mo(IV), $n = 2$ (368) and M = Fe(III), $n = 3$ (24).

The Cu(II) complex in the Zn(II) lattice and in frozen CH_3CN shows virtually identical EPR spectra ($g_{\parallel} = 2.094$, $g_{\perp} = 2.022$, $A_{\parallel} = 177.1 \times 10^{-4}$ cm^{-1}, $A_{\perp} = 47.6 \times 10^{-4}$ cm^{-1}). The EPR spectra in CH_3CN solu-tion show $\langle g \rangle = 2.046$ and $\langle a \rangle = 85.2 \times 10^{-4}$ cm^{-1}. It appears somewhat surprising that the $Cu(CpD)_2^{2-}$, a planar molecule, is not affected by the lattice forces of the Zn(II) complex lattice, the latter undoubtedly being a tetrahedral molecule. An x-ray powder diffraction study of the two crystalline complexes is necessary.

Bonding parameters derived from EPR data led the authors to conclude that the out-of-plane π bonding is "more covalent" than in the $Cu(R_2Dtc)_2$ complexes (563).

The $Co(CpD)_2^{2-}$ complex is a low-spin (2.55 BM, room temp. ?) unstable complex and is presumed to be planar. From an EPR analysis of this molecule, it was concluded that the unpaired electron resides in the d_{yz} orbital (367).

The magnetic moment (μ_{eff} = 4.83 BM) and Mössbauer spectra of the $Fe(CpD)_2^{2-}$ complex anion are typical of high-spin Fe(II). (IS = 0.69 mm sec^{-1}; metallic Fe reference); QS = 4.52 mm sec^{-1} 78°K). The large quadrupole splitting was rationalized in terms of a five-coordinate Fe(II) in the solid state. The $Fe(CpD)_3^{3-}$ is high spin (μ_{eff} = 5.93 BM). The Mössbauer spectrum of this complex is characterized by IS = 0.41 mm sec^{-1} (broad singlet with Γ of 1.71 mm/sec) at 78°K. The EPR spectrum shows g = 9.30, 4.31, and 0.771, which is considered to be consistent with a large zero-field splitting of the ^6S ground state (24).

The $(Et_4N)_2Mo(CpD)_3$ and $(Et_4N)MoO(CpD)_2$ complexes are both diamagnetic in the solid state. The interaction that accounts for the diamagnetism of the $MoO(CpD)_2^-$ complex is not maintained in solution where the EPR spectrum of this complex is typical for a Mo(V) species ($g_{\parallel}$ = 1.980, $g_\perp$ = 1.981, $A_{\parallel}$ = 52.61 x 10^{-4} cm^{-1}, $A_\perp$ + 23.35 x 10^{-4} cm^{-1}, in CH_3CN at 100°K; $\langle g \rangle$ = 1.980, $\langle a \rangle$ = 32.40 x 10^{-4} cm^{-1}, in CH_3CH solution, room temp.). The diamagnetic nature of the Mo(IV) complex with a d^2 configuration has been rationalized in terms of an orbitally "nondegenerate level lowest in energy" (368). The crystal structure of the monomeric planar $[Ni(CpD)_2^{2-}](Et_4N)_2^+$ complex has been determined (25). The mean Ni—S bond length 2.200(4) Å is similar to that found in other 1,1-dithiolate complexes. The rather large standard deviations in the C—C bonds in the ligand make the discussion of the structure of the ligand difficult, and the importance of resonance form a in Fig. 52 cannot be ascertained. The importance of this resonance form has been associated with the π-bonding ability of the ligand and is believed to be considerable on the basis of ESR studies (367, 563), coupling constants of the ring protons (24, 368, 369), and Mössbauer data (24, 369).

The 1,1-bis(trifluoromethyl)ethylene 2,2-dithiolate ligand is obtained (283) in oxidative addition reactions between $(PPh_3)_4Pt$ or $(PPh_3)_2IrCOCl$ and a mixture of 3,6-bis(2,2,2-trifluoro-1-trifluoromethyl ethylidene)s-tetrathian (a) and 3,5-bis(2,2,2 trifluoro-1-trifluoromethylidene)-1,2,4-trithiolan (b) (Fig. 53). The last two compounds were obtained by the reaction of bis(trifluoromethyl)-cyclodiazomethane and CS_2 (457). The products $(L)_2Pt[S_2C:C(CF_3)_2]$ and $(L)_2IrCOCl[S_2C:C(CF_3)_2]$ (L = Ph_3P or Ph_2MeP) showed ^{19}F NMR consistent with the above formulations, and the C=C stretching vibration in the new ligand at 1522 and 1519 cm^{-1}, respectively, for the two complexes. The use of

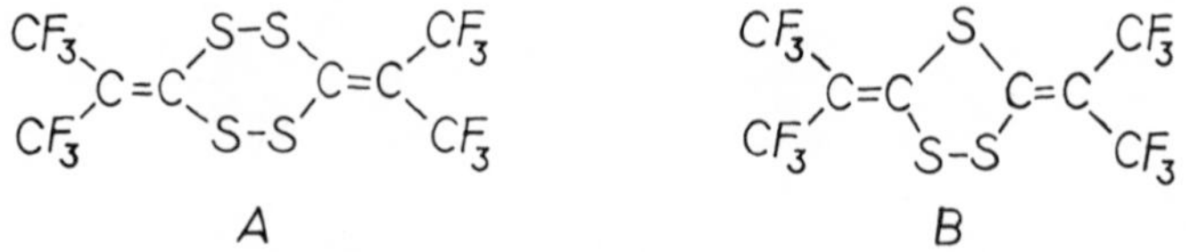

Fig. 53.   Oxidized precursors of the $(CF_3)_2C : CS_2^{2-}$ ligand.

this oxidative addition reaction for the synthesis of other complexes of the $S_2C:C(CF_3)_2^{2-}$ ligand opens an avenue for the isolation of potentially volatile sulfur chelates.

The synthesis of a large number of alkyl-sulfonyl 1,1-dithiolate ligands has been reported by the reaction between RSO_2CH_2R' and CS_2 in DMSO in the presence of NaH as a base (399). The coordination chemistry of these new ligands, $RSO_2R'C:CS_2^{2-}$, to the knowledge of the present author, has not been explored.

In a very interesting synthetic study McCleverty et al. obtained and characterized a wide series of mixed 1,1/1,2-dithio complexes of cobalt and iron (443). The complexes $[M(S_2C:X)(MNT)_2]^{3-}$ [M = Co or Fe; X = C(CN)$_2$, C(CN)(COOEt), C(CN)(CONH$_2$), CHNO$_2$, NCN] were isolated at the Ph_4P^+ salts. The exclusion of air and the presence of a mild reducing agent (sulfite ion) were necessary for the synthesis of these complexes from $[M(MNT)_2]_2^{2-}$ and the appropriate 1,1-dithiolate ligands in DMF solution (443). Mostly quasi-reversible one-electron oxidations were observed for the mixed-ligand complexes of this type:

$$[M(MNT)_2(S_2C:X)]^{3-} \rightleftharpoons [M(MNT)_2(S_2C:X)]^{2-} + e^- \qquad (49)$$

The spectral and magnetic properties of the complexes were reported.

C. 1,1-Dicarboalkoxy-2,2-Ethylene Dithiolate, R_2DED, Complexes

Compared to other 1,1-dithiolates, the R_2DED ligands are quite unique in their properties. These ligands appear to be more effective than the dithiocarbamate ligands in stabilizing metal ions in high formal oxidation states and are characterized by a conjugated π system (Fig. 54).

A number of complexes have been obtained with the Et_2DED ligand (henceforth DED) (Table XXVI) (503). Outstanding among these complexes are the $[Cu(III)(DED)_2]^-$, $[Fe(IV)(DED)_3]^{2-}$, and $[Mn(IV)(DED)_3]^{2-}$ anions. These complexes are remarkable in that they form very rapidly by air oxidation of the corresponding Cu(II), Fe(III), and Mn(III) complexes (99, 503). The

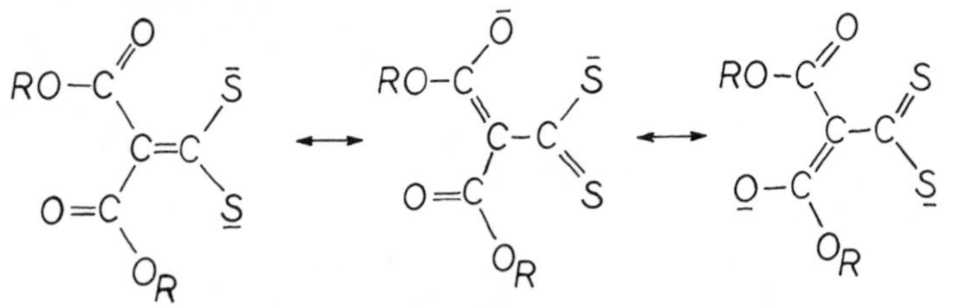

Fig. 54.   Resonance forms in the R_2DED ligands.

TABLE XXVI
$M(DED)_n^m$ -Complexes.

Complex[a]	$\nu(C{=}O)$, cm^{-1} [b]	$\nu(C{=}CS_2)$, cm^{-1} [b]	μ_{eff}^{corr}, BM (room temp.)	Ref.
$(BzPh_3P)_2Ni(DED)_2$	1720 (m), 1680 (s)	1445 (vs)	Diamagnetic	158
$(Ph_4P)_2Pd(DED)_2$	1711 (sh), 1674 (s), 1650 (sh)	1457 (vs)	Diamagnetic	158
$(Me_3PhN)_2Zn(DED)_2$	1708 (s), 1677 (s)	1465 (vs)	Diamagnetic	158
$(BzPh_3P)_2Cu(DED)_2$	1712 (s), 1683 (s)	1449 (vs)	1.66 (2)	157, 158, 334
$KCu(DED)_2$	1685 (m), 1670 (s), 1655 (s)	1495 (s), 1460 (s)	Diamagnetic	157, 158, 334
$(BzPh_3P)Cu(DED)_2$	1715 (s), 1680 (s)	1495 (s), 1460 (s)	Diamagnetic	157, 158, 334
$KAu(DED)_2$	1705 (s), 1670 (s)	1490 (s), 1450 (s)	Diamagnetic	157, 158, 334
$(BzPh_3P)Au(DED)_2$			Diamagnetic	157, 158, 334

$K_4Cu_8(DED)_6 \cdot 6H_2O$	1680 (s) 1610 (sh)	1480 (s) 1450 (s)	Diamagnetic	157, 158, 334
$(Ph_4P)_4Cu_8(DED)_6$	1695 (s)	1450 (s)	Diamagnetic	157, 158, 334
$(Ph_4P)_4Ag_8(DED)_6$	1708 (sh) 1691 (s)	1428 (s)	Diamagnetic	157, 158, 334
$(Me_3PhN)_3Co(DED)_3 \cdot 2H_2O$	1688 (s) 1630 (s) 1584 (sh)	1490 (m) 1457 (m)	Diamagnetic	157, 158, 334
$(Me_3PhN)_3Cr(DED)_3 \cdot 2H_2O$	1694 (m) 1652 (s)	1488 (w) 1408 (vs)	3.85	157, 158, 334
$(Me_3PhN)_3Rh(DED)_3 \cdot 2H_2O$			Diamagnetic	157, 158, 334
$(BzPh_3P)_2Fe(DED)_3$	1703 (s) 1646 (m)	1456 (s) 1438 (s)	2.92	158, 335, 336
$(BzPh_3P)_2Mn(DED)_3$	1708 (s) 1652 (m)	1455 (s) 1441 (s)	3.79	158
$(BzPh_3P)_2Sn(DED)_3$	1704 (s) 1673 (s) 1648 (s)	1458 (vs)	Diamagnetic	158

[a] DED = 1,1-dicarboethoxyethylene 2,2-dithiolate.
[b] Mineral oil mull.

$[Cu(II)(DED)_2]^{2-}$ complex can be isolated as a red-brown crystalline $BzPh_3P^+$ salt. However, solutions of this compound on standing in air rapidly undergo a disproportionation reaction to the Cu(III) complex and the $[Cu_8(DED)_6]^{4-}$ cluster (see below). All attempts to isolate the $[Fe(III)(DED)_3]^{3-}$ and the Mn(III) analogue have failed thus far. The exceptional stability of the oxidized complexes is reflected in the highly negative reduction potentials. Thus the Cu(III) and Fe(IV) complexes show reversible one-electron reduction at -0.78 and -1.20 V, respectively (cyclic voltametry on a Pt electrode in $CH_2Cl_2/$ Bu_4NClO_4 versus Ag|AgI). The Mn(IV) complex undergoes *irreversible* reduction at -1.4 V (503). The crystal structures of the $[Ni(II)(DED)_2]^{2-}$, $[Cu(III)(DED)_2]^-$ (157, 334), and $[Fe(IV)(DED)_3]^{2-}$ (335, 336) anions have been determined. The first two diamagnetic molecules are monomeric and centrosymmetric and contain four-coordinate planar MS_4 cores. A normal probability plot comparing intraligand distances in the Ni(II) and Cu(III) isoelectronic complexes shows that, collectively, the ligand parameters in the copper complex are systematically different than those in the nickel complex. The observed differences are consistent with a greater localization of charge on the sulfur atoms of the DED ligand in the copper complex (157).

Identical Ni–S and Cu–S bond distances are observed in these two structures, and the expected reduction of the ionic radius of copper is not reflected in a shorter Cu–S bond. Two possible reasons for this result are: (1) a concomitant increase of the ionic radii of the sulfur donors offsets the effect of a shorter Cu(III) radius, and (2) there exists significant covalency in the Cu–S bonds.

The magnetic properties ($\mu_{eff}^{corr} = 2.92$ BM, room temp.) (335, 336) and Mössbauer spectra (506) of the $Fe(DED)_3^{2-}$ complex anion were similar to those reported for the Fe(IV) dithiocarbamate complexes and indicative of an octahedrally coordinated iron in a d^4 configuration and a ^3T ground state. The crystal structure of this complex has been determined (Fig. 55). The geometry of the MS_6 moiety in the structure can be described as originating from a trigonal prism that suffers individual rotations of the chelating ligands around the C_2 axis. One of these rotations is more severe than the other two, causing a lowering of the overall symmetry to C_2 (158). The successful interpretation of the Mössbauer data (506) by a method based on ligand field theory is an additional indication that in the FeS_6 core of the $Fe(DED)_3^{2-}$ anion the iron exists in the 4+ oxidation state. A decrease in the spin–orbit coupling parameter by about 30% from the free ion value, (~ 500 cm^{-1}) as derived from the Mössbauer data analysis, is attributed to either a radial expansion or to overlap with sulfur orbitals. The similarity of the ligand structural parameters in the Ni(II) and Fe(IV) complexes (Table XXVII) and the virtually identical IR spectra of the Fe(IV) complexes to those of the x-ray isomorphous Sn(IV) analogues (503) suggest strongly that inter- or intraligand oxidation cannot

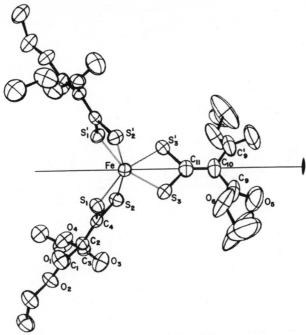

Fig. 55.   The molecular structure of the $[Fe^{IV}(DED)_3]^{2-}$ complex anion (158).

explain adequately the remarkable stability of the $Fe(DED)_3^{2-}$ complex. The structural and spectral data at this time suggest considerable covalency in the Fe(IV)–S bonds.

The $Mn(IV)(DED)_3(BzPh_3P)_2$ complex is x-ray isomorphous and, presumably, isostructural with the Sn(IV) and Fe(IV) complexes. It is obtained by the reaction of Mn(III) acetate with K_2DED in ethanol and has a magnetic moment of 3.79 ± 0.02 BM (room temp.), slightly lower than the spin-only value for a d^3 ion. The EPR spectrum of this complex diluted into the isomorphous diamagnetic Sn(IV) complex shows $g_\perp$ and $g_\parallel$ at 3.14 and 4.73, respectively and $A_\parallel = 80.5 \times 10^{-4}$ cm^{-1} (503).

The $M(DED)_3^{3-}$ complexes (M = Co, Rh, Cr) can be isolated as crystalline, x-ray isomorphous salts of the Me_3PhN^+ cation with two waters of hydration. The diamagnetism and electronic spectra of the Co(III) complex are consistent with a d^6 octahedrally coordinated ion. Cyclic voltametry shows quasi-reversible oxidation at +0.14 V (in CH_2Cl_2 versus Ag|AgI). A separation between the anodic and cathodic waves of 0.60 V suggests that a reversible rearrangement occurs upon oxidation of the $Co(DED)_3^{3-}$ anion. An investigation of the oxidation product is currently underway.

The $Cr(DED)_3^{3-}$ complex ($\mu_{eff}^{corr} = 3.85$ BM) shows the $^4A_2 \rightarrow {}^4T_2$ and

TABLE XXVII

Average Distances (Å) in the 1,1-Dicarboethoxyethylene 2,2-Dithiolate Complexes.
$[M_m(DED)_n]^p$

Distance	Fe(DED)$_3^{2-}$ [a,b]	Cu(DED)$_2^-$ [c]	Ni(DED)$_2^{2-}$ [c]	$[Cu_8(DED)_6]^{4-}$ [c]
$\overline{M-S}$	2.298 (8)	2.195 (5)	2.195 (2)	2.258 (10)
$\overline{S_1-S_2}$	2.786 (2)	2.766 (6)	2.789 (2)	3.043 (12)
	2.737 (3)[d]			
$\overline{C_1-O_1}$	1.215 (6)	1.18 (2)	1.208 (6)	1.20 (4)
$\overline{C_1-C_2}$	1.450 (8)	1.51 (2)	1.450 (7)	1.44 (4)
$\overline{C_2-C_3}$	1.487 (8)	1.50 (3)	1.486 (7)	1.56 (5)
$\overline{C_2-C_4}$	1.369 (7)	1.32 (2)	1.367 (6)	1.37 (4)
$\overline{C_3-O_3}$	1.190 (7)	1.15 (2)	1.190 (6)	1.17 (4)
$\overline{C_4-S_1}$	1.735 (5)	1.77 (1)	1.738 (5)	1.75 (3)
$\overline{C_4-S_2}$	1.735 (6)	1.76 (2)	1.749 (4)	1.75 (3)

[a] Intraligand distances reported for the symmetrical unconstrained ligands.
[b] Ref. 158.
[c] Ref. 335.
[d] Bite of the ligand bisected by the crystallographic C_2 axis.

$^4T_2 \rightarrow {}^4T_2(P)$ transitions at 15, 300, and 22,000 cm^{-1}. The energy of the $^4A_2 \rightarrow {}^4T_2$ transition, which is a direct measure of 10 Dq, when compared to that of other Cr(III) 1,1-dithio complexes, places the R_2DED ligand between i-MNT and R_2Dtc in the spectrochemical series.

For the last two ligands the $^4T_2 \rightarrow {}^4T_2$ transition in the Cr(III) complexes occurs at 15,100 and 15,500 cm^{-1}, respectively. A further indication of the remarkable ability of the R_2DED ligands to stabilize high oxidation states is the reversible oxidation of the Cr(III)(DED)$_3^{3-}$ complex at 0.22 V (503). However, at present it is still not possible to isolate a crystalline Cr(IV) complex by chemical methods.

1. Cu(I) and Ag(I) Cluster Compounds

The $Cu_8(DED)_6^{4-}$ cluster is obtained by the reaction between Cu(CH$_3$CN)$_4$·ClO$_4$·2H$_2$O and K$_2$DED in aqueous solution containing 5% CH$_3$CN. This cluster represents the second entry to the class of the copper(I) "cubanes" that

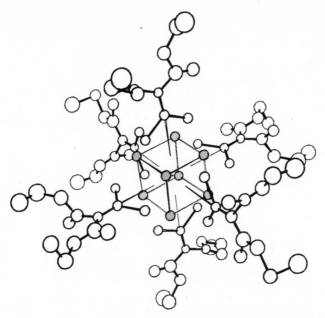

Fig. 56.   Molecular structure of the $Cu_8(DED)_6^{4-}$ "cubane" (336).

contain the Cu_8S_{12} core. The crystal structure of the Bu_4N^+ salt of the $[Cu_8(DED)_6]^{4-}$ cluster (335, 336) verified the suggested stoichiometry and revealed (Fig. 56) the basic Cu_8 cube inscribed in a distorted icosahedron of the 12 S atoms (T_h symmetry). The molecular parameters in the Cu_8S_{12} core are very similar to those reported for the same core in the $[Cu_8(i\text{-MNT})_6]^{4-}$ cluster (438). Structural comparisons among the cores of the i-MNT, Et_2DED, and dithiosquarate (DTS) clusters have led to the conclusion that in the Cu_8 cubes the $\overline{Cu-Cu}$ distances, which are found in the range between 2.790(11) and 2.844(20) Å, reflect weak attractive (bonding) Cu–Cu interactions (335).

The $Cu_8(R_2DED)_6^{4-}$ clusters also can be obtained with various other R groups (Me, Bz, t-Bu, etc.). With $R = t$-Bu the resulting cluster is very soluble in organic solvents, particularly as the Bu_4N^+ salt. Improved solubility properties have allowed for a study of the reactions of the $Cu_8(t\text{-}Bu_2DED)_6^{4-}$ cluster.

The studies of the author and coworkers have shown that this molecule undergoes two major reactions. The first of these reactions is proton addition addition (99, 503):

$$[Cu_8L_6]^{4-} \underset{OH^-}{\overset{H^+}{\rightleftharpoons}} [Cu_8L_6H]^{3-} \underset{OH^-}{\overset{H^+}{\rightleftharpoons}} [Cu_8L_6H_2]^{2-} \xrightarrow{2H^+} Cu_{10}L_8H_6 \qquad (50)$$
$$\qquad\quad\;\; A \qquad\qquad\qquad\qquad B \qquad\qquad\qquad C$$

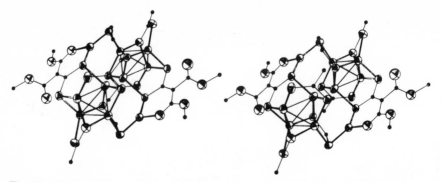

Fig. 57. Stereographic pair of the molecular structure of the $Cu_{10}(t\text{-}Bu_2\,DED)_2$-$(t\text{-}Bu_2\,DEDH)_6$ aggregate (165).

The crystal structure of the final product (**C**) has been determined (165) and reveals a multicopper aggregate that forms following the collapse of the Cu_8 core. Two individual tetrahedral Cu_4 units are connected by two unique Cu(I) ions (Fig. 57). The structural parameters of this interesting cluster are presented in Table XXVIII.

The site of protonation is unequivocally determined (165) as the "methine" carbon of the ligand (Fig. 58), which is transformed to a dithioacid ligand of the form $HCR_2\text{-}CS_2^-$. It is interesting to note that the mode of bridging co-ordination of the protonated $t\text{-}Bu_2\,DED$ ligands in the Cu_4 tetrahedral units is similar to that found with the $Et_2\,Dtc$ ligands in the structure of the $[Cu(Et_2\text{-}Dtc)]_4$ tetramer (322).

Following the addition of one or two protons, the Cu_8 cube does not collapse. Preliminary results on the crystal structure of the $[Cu_8(t\text{-}Bu_2\text{-}DED)_6H]^{3-}$ anion (Bu_4N^+ salt) show that Cu–Cu distances in the Cu_8 core are shorter than those in the Cu_8 core of the unprotonated cluster. If this difference is maintained in the final stages of refinement, it probably can be attributed to a decrease in the S–S repulsions in the S_{12} periphery expected to occur on protonation of two of the ligands. It appears that the Cu–Cu distances in the Cu_8 cubes reflect a delicate balance of (1) the coordination geometry require-ments of the copper atoms, (2) the Cu–Cu attractive interactions, and (3) the S–S repulsions in the S_{12} periphery.

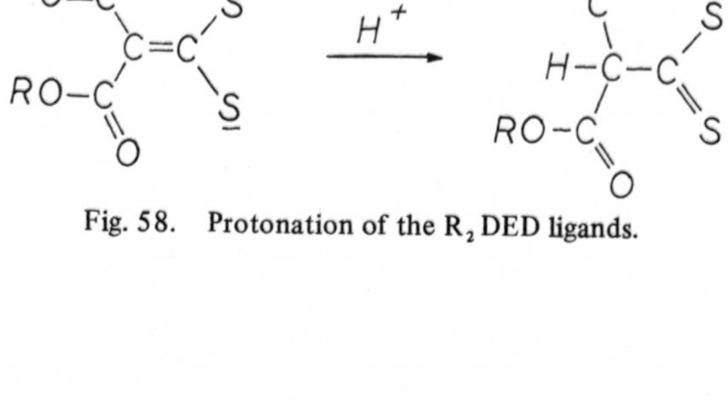

Fig. 58. Protonation of the $R_2\,DED$ ligands.

TABLE XXVIII
Selected Distances and Angles in the Cu_{10} (t-Bu_2 DED)$_2$ (t-Bu_2 –DEDH)$_6$ Aggregate.

Distances, Å		
Mean	Cu–Cu	2.819 (17)
Range	Cu–Cu	2.664 (2)–3.056 (2)
Inter Cu_4 unit[a]	Cu–Cu	2.872 (3)
Cu_4 unit-bridging Cu[b]	Cu–Cu	3.438 (2)
Mean	S–S (bite)	3.037 (4)
Bridging ligand	S–S (bite)	2.958 (4)
Mean	Cu–S	2.269 (32)
Range	Cu–S	2.230 (3)–2.322 (3)
Bridging Cu	Cu–S	2.187 (3)
		2.209 (3)
Bridging Cu	Cu–O	1.898 (8)

Angles, °		
Mean[c]	Cu–Cu–Cu	59.4 (58)
Range[c]	Cu–Cu–Cu	54.11 (5)–70.12 (6)
Mean	S–Cu–S	119.4 (72)
Range	S–Cu–S	106.0 (15)–131.8 (16)
Bridging Cu	S–Cu–S	148.1 (16)
Bridging Cu	S–Cu–O	99.6 (3)
		110.8 (3)

Σ S–Cu–S; Cu (1), 358.2; Cu (2), 358.9; Cu (3), 359.9; Cu (4), 356.3

[a] Closest approach across the center of symmetry.
[b] Closest approach.
[c] In the Cu_4 tetrahedra.

The other reaction that has been explored with the Cu_8(t-Bu_2 DED)$_6^{4-}$ cluster is the sulfur addition to the coordinated 1,1-dithio ligands, which proceeds according to Eq. 51:

$$[Cu_8 L_6]^{4-} \xrightarrow{\text{S}} [Cu_8 L_6 S_6]^{4-} \xrightarrow{\text{Cu(II) or Fe(III)}} [Cu_5 L_4 S_4]^- + [Cu_4(LS)_3]^{2-} \quad (51)$$

The properties of the $[Cu_5 L_4 S_4]^-$ cluster are quite remarkable. This intense blue diamagnetic cluster (λ_{max} 654 nm, $\epsilon \sim 46,100$) can be described formally as a mixed-valence Cu(I), Cu(III) cluster (164). The crystal structure of this molecule (Fig. 59) supports this *formalism* and reveals a Cu_5 rectangular pyramidal core (166). The basal copper atoms are three-coordinate and nearly planar and can be considered as Cu(I) units. The axial copper is four-coordinate and strictly planar, a coordination geometry appropriate for a d^8 Cu(III) center.

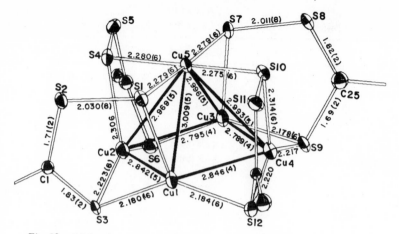

Fig. 59.   Molecular structure of the $[Cu_5(t\text{-Bu}_2 DEDS)_4]^-$ cluster (164).

The corresponding $Au(III)Cu_4$ cluster also has been synthesized and the crystal and molecular structure has been determined. The core is similar to the one shown in Fig. 59 with Au(III) at the apex of the pyramid. (164). The sulfur addition to the 1,1-dithiolate ligands has resulted in perthioethylene dithiolate ligands. Although sulfur addition to the 1,1-dithiolate ligands is well-documented (156), the formation of a perthiolate, $[S_3C=CXY]^{2-}$ ligand has not been demonstrated previously by an x-ray crystallographic study.

As might be expected this cluster shows interesting photochemical properties that seem to arise from partial photodissociation of the unique Cu(III) ion followed by a two-electron transfer from reducing substrates. Work in progress is aimed at understanding the interesting photochemistry of this cluster.

D. Dithiocarbimate Complexes

The chemistry of the dithiocarbimate ligands, $[RN:CS_2]^{2-}$, has not been studied to a great extent. Until recently only complexes of the $[(CN)N:CS_2]^{2-}$ ligand (CDC) and the bis-nickel complex of the $[PhN:CS_2]^{2-}$ ligand had been reported (153).

Recently the interesting reaction of coordinated organic isothiocyanates with excess isothiocyanates was reported to give dithiocarbimate ligands (64). The formation of these ligands is thought to occur by a "collapse" of $L_2M(RNCS)_2$ complexes followed by the extrusion of RNC and generation of the $L_2M(S_2C:NR)$ complexes [$L = Ph_3P$, $R = Ph$ or Me, $M = Pt(II)$]. Protonation of these complexes with trifluoroacetic acid in benzene affords the crystalline dithiocarbamates $[L_2M(S_2CNHR)]^+$.

Two studies of the pyrolysis of the $[Ph_2I^+]$ salts of $Co(CDC)_3^{3-}$ (533) and

$Ni(CDC)_2^{2-}$ (532) have been reported. Pyrolysis of $Co(CDC)_3^{3-}$ results in partial reduction of the metal ion to Co(II) and liberation of iodobenzene and phenyl disulfide. In pyridine solution phenylation of the ligand was observed and N-cyanodiphenyldithiocarbimate was isolated, $(PhS)_2 C:NCN$.

In a similar fashion pyrolytic autoarylation of the ligand was found in the pyrolysis of $(Ph_2 I)_2 Ni(CDC)_2$. The paramagnetic product ($\mu_{eff} = 2.38$ BM), $Ni(PhS_2 C:NCN)_2$, appears to contain a diphenylated ligand, and its low magnetic moment possibly suggests octahedral coordination of the Ni(II) ion in a polymeric structure. The description of this molecule as a polymeric $Ni(Ph_2 S_2 - C:NCN)(S_2 C:NCN)$ complex is favored by the authors on the basis of its reaction with $Ph_3 P$ to give $(Ph_3 P)_2 Ni(CDC)$ (532).

The organic chemistry of the dithiocarbimate dianions is rich and has been explored (178, 591). The synthesis and characterization of organotin CDC complexes has been reported in a patent (580) and a paper (569). In this later work the reaction of $K_2 CDC$ with $R_2 SnCl_2$ is reported to give the neutral $R_2 Sn(CDC)$ complexes (R = n-Bu, n-octyl) in high yields. The reaction of the same ligand with $Ph_2 SnCl_2$ resulted in ligand decomposition and formation of $[Ph_3 Sn]_2 S$ and $Ph_3 SnSCN$.

Casey and Thackeray reported on the synthesis of the $(\eta^5 -Cp)_2 V(CDC)$ complex by the reaction between $Na_2 CDC$ and $(\eta^5 -Cp)_2 VCl_2$ in an acetone–water 1:1 mixture (108). The magnetic moment of this complex (1.68 BM, 295°K) is similar to values obtained for other V(IV) complexes of the same type with the RXant and $R_2 Dtc$ ligands. The EPR and IR spectra of this compound also were reported.

E. Trithio- and Dithiocarbonates

Few synthetic studies have appeared in the literature that deal with CS_3^{2-} or $CS_2 O^{2-}$ complexes.

Dithiocarbonate complexes can be obtained by the reaction of xanthate complexes with K[RXant] ligands. Thus reaction between $Pt(RXant)_2$ with K(RXant) (R = Et or i-Pr) followed by addition of $Ph_4 AsCl$ generates $[Ph_4 As] - [Pt(S_2 COR)_3]$, which, following attempts at recrystallization, is converted to the $[Ph_4 A] [Pt(S_2 CO)(S_2 COR)]$ complex. The same main product is obtained in the reaction of the corresponding Pd(II) xanthate complex. It appears that the R group on the RXant ligand has an effect on the formation of the dithiocarbonate ligands. Thus with R = Me or Bz the $Pt(S_2 CO)_2^{2-}$ complex anion is obtained, which upon addition of phosphines gives the neutral $(L)_2 Pt(S_2 CO)$ complexes (L = PPh_3, $PMe_2 Ph$ or "$\frac{1}{2}Ph_2 PC_2 H_4 PPh_2$") (141, 144).

The paramagnetic "$Ni(NH_3)_3 CS_3$" complexes originally reported by Hofmann (328, 329) and reinvestigated by Seidel and Petzoldt (568) are now known to be the "double" salts $[Ni(NH_3)_6] [Ni(CS_3)_2]$. This conclusion was reached on the basis of vibrational analyses (91, 470), electrophoresis, and ultraviolet spectroscopy (523) of "$Ni(NH_3)_3CS_3$."

The infrared and Raman spectra of the $[M(CS_3)_2]^{2-}$ complexes have been measured (90) and assignments have been made. These assignments were extended to the $L_2M(CS_3)$ and $L_2M(CS_2O)$ complexes (M = Pd or Pt and L = Ph_2-MeP). The $\nu(C=S)$ vibration ($\sim$1030 cm^{-1}) and $\nu_{as}(C-S)$ at 850 cm^{-1} remained constant in these complexes. However, the $\nu_s(C-S)$ and (M–S) varied in the series $M(CS_3)_2^{2-}$, $L_2M(CS_3)$, and $L_2M(CS_2O)$.

More recently the infrared spectra of the $A_2[M(CS_3)_2]$ complexes were reported (470) (M = ^{58}Ni, ^{62}Ni, ^{64}Zn, and ^{68}Zn and A$^+$ = Ph_4P, Ph_4As, and Me_4N). A normal coordinate analysis of the $[^{58}Ni(CS_3)_2]^{2-}$ and its ^{62}Ni analogue was carried out. The bands at 1010 and 857 cm^{-1} were assigned to the C=S and C–S stretching modes, respectively (470), in agreement with the previous work (90). Another C–S stretching mode was assigned to a band at 507 cm^{-1}. The ^{58}Ni–S stretching vibrations were calculated at 366 and 384 cm^{-1}, respectively. Two bands near 260 and 205 cm^{-1} in the spectra of the $Zn(CS_3)_2^{2-}$ complex were affected by the ^{64}Zn–^{68}Zn substitution (3 $\sim$ 4 cm^{-1} shifts) and were tentatively assigned to Zn–S vibrations.

IV. REACTIONS OF DITHIOACID AND DITHIOLATE COMPLEXES

A. Substitution Reactions

The kinetics of the substitution reactions of the planar $Ni(L)_2^{2-}$ complexes with 1,1-dithiolate nucleophiles have been studied in aqueous solution at 25°C by a stopped-flow technique (502) (L = CPD^{2-}, NED^{2-}, CDC^{2-}, and i-MNT).

The formation constants of the $Ni(L)_2^{2-}$ complexes were obtained using a competition reaction between ethylenediamine and the dithiolate ligand. Values for log β_2 for the CPD^{2-}, NED^{2-}, CDC^{2-}, and i-MNT complexes of 20.3, 14.4, 13.7, and 12.2 were reported (502). The rates of formation of Ni(i-MNT)$_2^{2-}$ and Ni(CDC)$_2^{2-}$ were reported to be 4.5 x 10^4 and 9.1 x 10^4 sec^{-1}, respectively. The following mechanism was postulated for the formation of these complexes:

$$Ni(OH_2)_6^{2+} + L^{2-} \;\underset{}{\overset{k_0}{\rightleftharpoons}}\; Ni(OH_2)_6^{2+}L^{2-} \tag{52}$$

$$Ni(OH_2)_6^{2+}L^{2-} \;\overset{k_0}{\longrightarrow}\; Ni(OH_2)_4L + 2H_2O \tag{52b}$$

$$Ni(OH_2)_4L + L^{2-} \;\overset{fast}{\longrightarrow}\; NiL_2^{2-} + 4H_2O \tag{52c}$$

Reaction 52a is the diffusion-controlled formation of an ion pair and 52b represents the rate-determining loss of solvent water.

On the basis of a number of substitution reaction studies, the ease of displacement was reported to decrease in the order $i\text{-MNT}^{2-} > \text{NED}^{2-} > \text{CDC}^{2-} > \text{CPD}^{2-} > \text{MNT}^{2-}$. With the exception of NED^{2-} this order corresponds to the order in the formation constants of the Ni(L)_2^{2-} complexes (see above). The general ordering for the nucleophiles used was reported to be:

$$\text{CN}^- > \text{MNT}^{2-} > \text{NED}^{2-} > \text{CPD}^{2-} > \text{CDC}^{2-} \tag{53}$$

The mechanism proposed to account for the substitution reactions of the Ni(L)_2^{2-} complexes is shown in Fig. 60. A similar mechanism is proposed for substitution reactions with unidentate nucleophiles. Several of the five-coordinate intermediates shown in Fig. 60 were detected and their stabilities were estimated. It was suggested that the trans effect operates through the stability of the five-coordinate intermediate which in turn is correlated to the extent of involvement of the nickel $4p_z$ orbital in π bonding. The implication of the $4p_z$ orbital in π bonding should be considered *cum grano salis*, however, in view of recent theoretical calculations indicating that the involvement of the $4p_z$ orbital in the π system is very small even in the Ni(MNT)_2^{2-} complex (123). (See also Sect. 2E.)

A similar study of the rates and mechanisms of substitution reactions of the Pt(II) 1,1-dithiolate complexes, Pt(L)_2^{2-} ($\text{L} = i\text{-MNT}^{2-}$, NED^{2-}, CDC^{2-} and CPD^{2-}), also has been reported (352). The mechanism proposed for the substitution reactions of the Pt(L)_2^{2-} complexes with bidentate 1,1-dithio chelates is the same as that shown in Fig. 60.

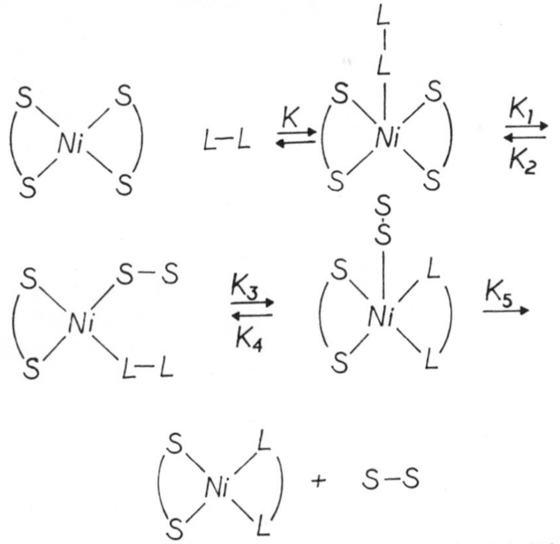

Fig. 60.   Proposed mechanism for substitution reactions in the 1,1-dithiolate complexes (502).

The main differences between the substitution reactions of the $Ni(L)_2^{2-}$ and $Pt(L)_2^{2-}$ complexes are interpreted in terms of the stabilities of the five-coordinate intermediates, which appear to be considerably greater with the Ni(II) complexes.

B. Sulfur-Addition Reaction

Since the first report of the sulfur-addition reaction to 1,1-dithio complexes (156), a number of structural studies have verified the originally proposed ring expansion of the MS_2C ring following this reaction. In addition, mechanistic studies on the sulfur insertion reaction and electronic spectral studies of the perthiocarboxylate Ni(II) complexes have been reported.

Structural Studies. Following the report of the first structure determination of a sulfur-rich 1,1-dithio complex, Fe(p-MePhDta)$_2$(p-MePhDtaS), by Coucouvanis and Lippard (160), (161) the structures of Ni(PhDtaS)$_2$ (48), Zn(PhDtaS)$_2$ (48), Ni(p-i-PrPhDta)(p-i-PrPhDtaS) (233), and Zn(p-i-PrPhDtaS)$_2$ (233) have been determined (Table XXIX). With no exception the "extra" sulfur atoms in all complexes are found inserted into the MSCS ring of the "parent", 1,1-dithio complexes, with the formation of MS_2CS five-membered chelate rings. The structure of the "mixed" ligand Ni(p-i-PrPhDta)(p-i-PrPhDtaS) is shown in Fig. 48. A structural feature common to all these complexes (48, 160, 233) is the rotation of the aromatic ring out of the plane of the five-membered chelate ring. It has been suggested that this rotation is caused by S⋯H contacts that arise when the S—C—S angle opens from $109°$ in the four-membered ring to $124°$ in the five-membered ring (233). The MS_4 centers in the Ni(II) complexes are planar and those in the Zn(II) complexes are distorted tetrahedral. Selected structural parameters are shown in Tables XXII and XXIX. An examination of the Ni—S bond lengths in the NiS_2CS five-membered rings reveals these bonds to be unusually short when compared to the Ni—S bonds in the NiSCS four-membered rings.

The question of whether sulfur insertion occurs at the C—S or the M—S bond was raised following the first isolation of the perthio-1,1-dithiolate complexes (156). Early studies of the sulfur-addition and sulfur-abstraction reactions using ^{33}S were reported for $Ni(CS_3)_2^{2-}$ (Eq. 54). It was concluded that the added sulfur was inserted into the C—S bond (156).

$$Ni(CS_3)_2^{2-} + 2*S \longrightarrow Ni(CS_4^*)_2^{2-}$$

$$Ni(CS_4^*)_2^{2-} + 2Ph_3P \longrightarrow Ni(CS_3)_2 + Ph_3PS* \tag{54}$$

TABLE XXIX
Selected Structural Parameters of the Sulfur-Rich 1,1-Dithio Complexes.

Complex	M–S, Å	S–S, Å	C–S, Å	MS$_2$CS–Ar, °[a]	S–C–S, °	Ref.
Ni(PhDtaS)$_2$	2.161 (5)	2.022 (3)	1.67 (1)	33	122.4 (5)	48
Ni(p-i-PrPhDta)(p-i-PrPhDtaS)	2.134 (4)[b]	2.016 (4)	1.688 (15)[b]		124[b]	233
	2.212 (6)		1.688 (10)		109	
Zn(PhDtaS)$_2$	2.331 (5)	2.007 (3)	1.69 (1)	40	126.2 (5)	48
Zn(p-i-PrPhDtaS)$_2$	2.321 (8)	2.008 (4)	1.687 (33)	26.9	126.4 (4)	233
Fe(p-MeDta)$_2$(p-MeDtaS)	2.211 (38)[b]	2.086 (8)	1.70 (3)[b]	30.2	121 (1)[b]	160, 161
	2.318 (25)		1.68 (3)		112 (1)	

[a] Twist of the aryl group relative to the MS$_2$CS unit.
[b] 'Sulfur-rich' ligand.

At a later date elegant studies by Fackler and coworkers on ^{34}S-enriched Ni(PhDta)$_2$S$_2$ reaffirmed the initial conclusions. Mass spectral intensities of the C$_6$H$_5$CS$^+$ fragment demonstrated that the photochemically initiated sulfur-addition reaction was highly specific (20, 232). (235).

Some scrambling was detected when the sulfur-addition reaction was carried out thermally. With zinc, total sulfur atom scrambling was observed. The rate of sulfur exchange in the Zn complexes was studied by PMR in solutions containing equal amounts of perthiocumate Zn(II) and dithiocumate Zn(II) complexes (233). Extrapolated to room temperature, the data suggested sulfur atom exchange lifetimes of $\sim 10^{-5}$ sec! The activation energy for this exchange process (~ 5 kcal mole^{-1}) is not very different from the activation energy of the sulfur abstraction from NiL$_2$S with Ph$_3$P (~ 8 kcal mole^{-1}).

Kinetic studies on the reaction of NiL$_2$S with Ph$_3$P established that the rate of sulfur abstraction depends on the concentration of both the complex and the phosphine (233).

It has been suggested that the use of zinc dithiolate complexes as vulcanization catalysts may be closely related to the ability of these complexes to form sulfur-rich complexes that have very labile disulfide sulfur atoms (230).

The synthesis of the perthiopivalate ligand, t-BuDtaS, has been reported and the Ni(t-BuDtaS)$_2$ and Zn(II) analogues have been obtained in crystalline form. A discussion of the infrared and electronic spectra of these complexes has been presented (246).

Attempts have been made to understand the electronic spectra and, particularly, the high-intensity bands in the visible region and the very low ligand field band ($\bar{\nu}_1 \sim 12$ kK) in the perthiocarboxylate complexes of Ni(II) (246). It is conceded that the use of ligand field treatment in this system, where mixing of electronic states occurs in the higher excited states, is only of limited value (246).

Similar conclusions were reached in another study, where an extensive tabulation of the electronic spectra of the dithio- and perthiocarboxylate Ni(II) complexes was presented (258). Recently it was reported by McCleverty and Morrison (440) that the reaction between Ni(H-p-MePhDtc)$_2$ and n-Bu$_4$N(Me$_2$-Dtc) affords Ni(Me$_2$Dtc)$_2$ and the yellow crystalline Ni(H-p-MePhDtcS)$_2$ complex, which contains the "sulfur-rich" trithiocarbamate ligand.

A subsequent examination of this material (633) did not show the expected N–H stretching vibrations in the IR spectrum. Furthermore, thin-layer chromatography of the yellow crystals revealed the presence of two components. Further studies on this compound are needed before an unambiguous characterization can be claimed.

The reaction of Na(R$_2$Dtc) (R = Me, Et) with K$_2$Cr$_2$O$_7$ in a 6:1 molar ratio in aqueous solution gives Cr(R$_2$Dtc)$_3$ and the very interesting blue Cr(R$_2$-Dtc)$_2$(R$_2$DtcO) complexes (339). The crystal structure of the ethyl homologue shows a six-coordinate distorted octahedral Cr(III) ion (Fig. 17) coordinated by two "normal" bidentate Et_2Dtc ligands. An oxodithiocarbamate ligand completes the coordination sphere and upon chelation forms a CrOSCS five-membered ring. The ring expansion has been considered a result of oxygen insertion (339) into the M–S bond. At this stage the method of preparation of this compound does not warrant such an interpretation.

C. Carbon Disulfide Elimination

Carbon disulfide elimination (Eq. 55) appears to be a general reaction of the thioxanthate complexes (153).

$$2M(S_2CSR)_n \longrightarrow 2CS_2 + [M(SR)(S_2CSR)_{n-1}]_2 \tag{55}$$

Lippard et al. have reported on the synthesis and structural characterization of the $Fe_2(SR)_2(S_2CSR)_4$ (162) and $Co_2(SR)_2(S_2CSR)_4$ (405, 409) complexes (see also Sections II.C.8a and II.C.8b). In these complexes the Fe_2S_2 and Co_2S_2 planar units contain mercaptide bridges.

The structure of the $[Ni(EtS)(DtSXant)]_2$ dimer has been reported and shows a Ni–Ni distance of 2.76 Å in a Ni_2S_2 unit that contains the bridging mercaptide ligands (614). This distance was proclaimed as "the first example of a binuclear planar Ni(II) complex with Ni–Ni bonds." At a later date Fackler and Zegarski reported on the structures of the $[Ni(BzS)(BzSXant)]_2$ (Fig. 45) and $[Pd(t\text{-}BuS)(t\text{-}BuSXant)]_2$ dimers (241). Each of these compounds contains distorted planar MS_4 coordination geometries with bridging mercaptide groups. The nickel complex displays a nonplanar $syn–endo$-Ni_2S_2 unit and the palladium complex displays a nonplanar $anti$-Pd_2S_2 unit. The Ni–Ni and Pd–Pd distances of 2.79(3) and 3.162(1) Å, respectively, are not considered to be the only attractive interactions that control the gross stereochemistry of the complexes. It has been suggested (241) that S⋯S attractive forces in the M_2S_2 units may be as important as M⋯M interactions in describing the fold of these units.

A similarly folded Pd_3S_3 unit is observed in the structure of the $[Pd(SEt)(EtSXant)]_3$ trimer. In this structure (241) (Fig. 49) a puckered six-membered ring of alternating Pd and S atoms displays the Pd atoms in an isosceles triangle with Pd–Pd distances of 3.655(2), 3.307(2), and 3.303(2) Å.

The kinetics of the carbon disulfide elimination reaction were studied using PMR and visible spectroscopy (16). This spontaneous reaction was found to be first order in the $M(RSXant)_2$ complexes (M = Ni, R = Et, t-Bu, Bz; M = Pd, R = t-Bu), both in the disappearance of the starting material and in the formation of the mercaptide-bridged species in $CHCl_3$ and THF. The pseudo-first-order kinetics observed in CS_2 are attributed to an equilibrium between the $M(RSXant)_2$ complexes and this solvent. Rate constants for the reaction are of the order of 10^{-3} to 10^{-1} min^{-1} depending on solvent, temperature, alkyl group, and metal [with R = Bz the rate for the Ni(II) complex is faster than that for the Pd(II) complex]. Activation parameters are in the ranges $\Delta H^{\ddagger} = 22.0–23.8$ kcal mole^{-1} and $\Delta S^{\ddagger} = +3$ to -7 eu. The possible pathways for the unimolecular decomposition of the M(RSXant)$_2$ complexes are shown in Fig. 61. The absence of EPR signal and of disulfide (RSSR) product rules out the homolytic scission of the C–S bond (K$_1'$, Fig. 61). Of the remaining a concerted pathway K$_1'$, is feasible.

The synthesis of the thioxanthate bridged-mercaptide complexes of Pt(II), Fe(III), and Co(III) have been reported (16, 405, 409).

A kinetic study of the reaction

$$2\text{Co(EtSXant)}_3 \longrightarrow [\text{Co(EtS)(EtSXant)}_2]_2 + 2\text{CS}_2 \qquad (56)$$

in CHCl$_3$ shows first-order kinetics. The rate-determining step in the CS$_2$ elimination process is thought to be cleavage of the MS$_2$C–SR bond (see Fig. 61) (405, 409).

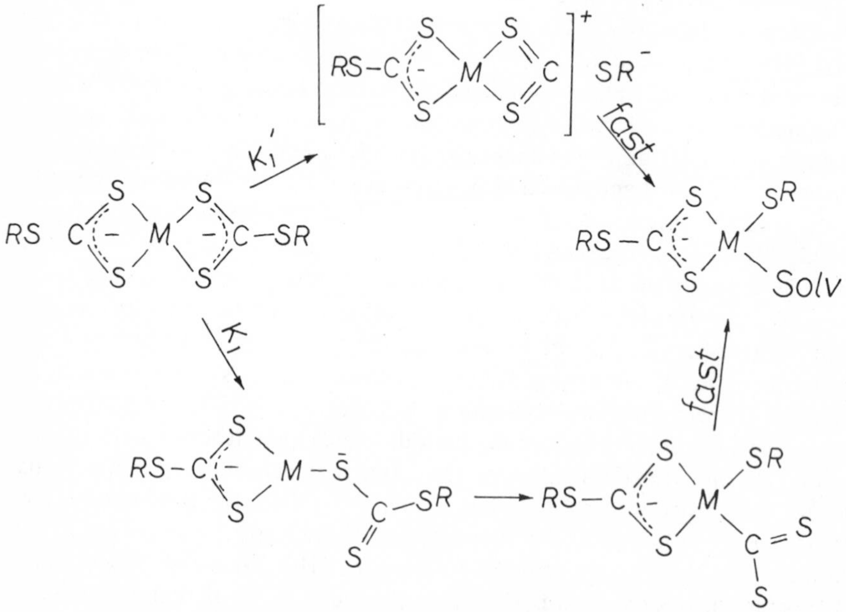

Fig. 61. Possible mechanisms for the CS$_2$ elimination reaction (16).

D. Reactions of 1,1-Dithiolates with Lewis Acids

The inert cations that accompany the anionic 1,1-dithiolate complexes $Ni(L)_2^{2-}$ (L = i-MNT, CDC, R_2DED, and S_2C:NPh) are readily replaced by the coordinately unsaturated $M'(Ph_3P)_2^+$ [M = Cu(I), Ag(I), Ag(I)] complex cations (100). The IR spectra of the crystalline polynuclear complexes $[Ni(L)_2]$ $[PPh_3)_nM']_2$ (n = 2 or 3) indicate that the anion–cation interactions occur primarily through the coordinated 1,1-dithiolate sulfer donors. In the $Ni(S_2C:NPh)_2[(PPh_3)_3Cu]_2$ complex the $(PPh_3)_3Cu^+$ cation is coordinated to the carbimate nitrogen of the ligand. Interactions of $(PPh_3)_2Cu^+$ with the $Ni(i$-$MNT)_2^{2-}$ complex occur with the π system of the nitrile groups of the i-MNT ligand. With $(PPh_3)_2Ag^+$ the anion–cation interaction is through the sulfur atoms of the i-MNT ligand.

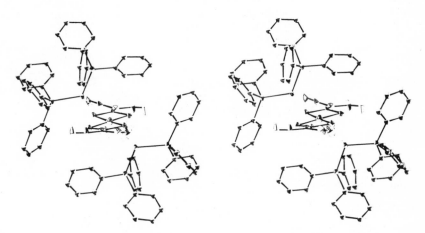

Fig. 62.   Stereoscopic view of $[Ag(PPh_3)_2]_2[Ni(i$-$MNT)_2]$ (154).

The structure of the $Ni(i$-$MNT)_2[Ag(PPh_3)_2]_2$ complex has been determined (154). The geometry of the NiS_4 group is square with a $\overline{Ni-S}$ bond length of 2.217(6) Å. The interaction of the $Ag(PPh_3)_2^+$ cations occurs at the NiS_4 moiety and the silver atoms are located above and below the NiS_4 plane in a chair configuration (Fig. 62). The Ag–Ni distances are 2.907(3) and 2.966(3) Å. The $\overline{Ag-S}$ bond length, 2.789(93) Å (range 2.689(5)–2.916(5) Å], is longer than single covalent bond lengths but appreciably shorter than the van der Waals contacts of 3.52 Å expected for a Ag–S interaction.

Abbreviations

Abbreviation	Compound	Structural formula
RDta⁻	Dithioacid anion	$(RCSS)^-$
R_2Dtc^-	Dithiocarbamate anion	$(R_2NCSS)^-$
R_4Tds	Tetraalkyl thiuram disulfide	$[R_2NC(S)S]_2$
R_4Tms	Tetraalkyl thiuram monosulfide	$[R_2NC(S)]_2S$
RXant⁻	O-alkyl or O-aryl dithiocarbonate anion (xanthate)	$(RO-CSS)^-$
RSXant⁻	Alkyl or aryl trithiocarbonate anion (thioxanthate)	$(RS-CSS)^-$
CpD	Cyclopentadiene dithiolate	$[C_5H_5=CSS]^{2-}$
NED	1-Nitro-2,2-ethylene dithiolate	$\{[(H)(NO_2)]C=CSS\}^{2-}$
CDC	Cyanodithiocarbimate	$\{NCN=CSS\}^{2-}$
CPD	1,1-Cyanophenyl-2,2-ethylene dithiolate	$\{[(CN)(C_6H_5)]C=CSS\}^{2-}$
RDtc	Dithiocarbimate anion	$[RN=CSS]^{2-}$
R_2DED	1,1-Dicarboalkoxy-2,2-ethylene dithiolate	$[(ROCO)_2C=CSS]^{2-}$
i-MNT	1,1-Dicyano-2,2-ethylene dithiolate	$[(NC)_2C=CSS]^{2-}$
MNT	1,2-Dicyano-1,2-ethylene dithiolate	$[S_2C_2(CN)_2]^{2-}$
TFD	Bis(perfluoromethyl)-1,2-dithietene	$[S_2C_2(CF_3)_2]^{2-}$
Sacsac	Dithioacetyl acetonate anion	$[(CH_3C(S))_2CH]^-$
Me	Methyl	$-CH_3$
Et	Ethyl	$-C_2H_5$
Pr	Propyl	$-C_3H_7$
i-Pr	iso-propyl	$-C_3H_7$
Bu	Butyl	$-C_4H_9$
Ph	Phenyl	$-C_6H_5$
Cx	Cyclohexyl	$-C_6H_{11}$
Bz	Benzyl	$-C_7H_7$
Morpho	Morpholine	OC_4H_4NH
Pyrrol	Pyrrolidine	C_4H_4NH
Py	Pyridine	C_5H_5N
Bipy	Bipyridine	$C_8H_6N_2$
Pip	Piperidine	$C_5H_{12}N$
PCy_3	Tricyclohexylphosphine	$P(C_6H_{11})_3$
η^5-Cp	Pentahaptocyclopentadienide	$C_5H_5^-$
Dithio-	Dithioacid anion (general)	$XCSS^-$

Acknowledgements

The author is grateful to numerous investigators for supplying reprints and preprints of their work. Thanks are due to Prof. M. Bonamico, Prof. J. Enemark, Prof. J. P. Fackler, Prof. R. Feltham, Prof. D. G. Holah, Dr. S. Husebye, Prof. S. J. Lippard, Prof. R. L. Martin, Prof. G. E. Manoussakis, Dr. W. J. Mitchell, Prof. L. Pignolet, Prof. T. A. Stephenson, Prof. A. H. White, Dr. J. Willemse, Dr. N. D. Yordanov, Dr. L. Zambonelli, and Prof. J. Zubieta. The author is particularly grateful to Mrs. Vida Brenner for typing the manuscript.

References

1. U. Aava and R. Hesse, *Ark. Kem., 30*, 149 (1968).
2. E. W. Abel and M. O. Dunster, *J. Chem. Soc., Dalton Trans., 1973*, 98.
3. H. Abrahamson, J. R. Heiman, and L. H. Pignolet, *Inorg. Chem., 14*, 2070 (1975).
4. V. Agarwala, Lakshmi, and P. B. Rao, *Inorg. Chim. Acta, 2*, 337 (1968).
5. R. L. Ake and G. M. Harris-Loew, *J. Chem. Phys., 52*, 1098 (1970).
6. S. Akerström, *Ark. Kemi, 14*, 387 (1959).
7. S. Akerström, *Ark. Kemi, 14*, 413 (1959).
8. V. G. Albano, P. L. Bellon, and G. Ciani, *J. Organomet. Chem., 31*, 75 (1971).
9. A. Albinati, A. Musco, G. Carturan, and G. Strukul, *Inorg. Chim. Acta, 18*, 219 (1976).
10. J. M. C. Alison and T. A. Stephenson, *J. Chem. Soc. Dalton Trans., 1973*, 254.
11. (a) E. C. Alyea, B. S. Ramaswamy, A. N. Bhat, and R. C. Fay, *Inorg. Nucl. Chem. Lett., 9*, 399 (1973). (b) E. C. Alyea and B. C. Ramaswamy, *Inorg. Nucl. Chem. Lett., 6*, 441 (1970).
12. A. Z. Amanov, G. A. Kukina, and M. A. Porai-Koshits, *Zh. Struct. Khim., 8*, 174 (1967).
13. H. Anacker-Eickhoff, P. Jennische, and R. Hesse, *Acta Chem. Scand. A, 29*, 51 (1975).
14. O. P. Anderson and S. Husebye, *Acta Chem. Scand. A, 24*, 3141 (1970).
15. J. A. M. Andrews, Ph.D. Thesis, Case Western Reserve University, 1971.
16. J. M. Andrews, D. Coucouvanis, and J. P. Fackler, Jr., *Inorg. Chem., 11*, 493 (1972).
17. C. C. Ashworth, N. A. Bailey, M. Johnson, J. A. McCleverty, N. Morrison, and B. Tabbiner, *Chem. Commun., 1976*, 743.
18. A. Avdeef and J. P. Fackler, Jr., *J. Coord. Chem., 4*, 211 (1975).
19. A. Avdeef, J. P. Fackler, Jr., and R. G. Fischer, Jr., *J. Am. Chem. Soc., 92*, 6927 (1970).
20. A. Avdeef, J. P. Fackler, Jr., and R. G. Fischer, Jr., *J. Am. Chem. Soc., 95*, 774 (1973).
21. K. W. Bagnall, D. Brown, and D. G. Holah, *J. Chem. Soc. Dalton Trans., 1968*, 1149.
22. R. Bally, *C. R. Ser. C, 271*, 1436 (1970).
23. R. Barral, D. Bocard, I. Serreede Roch, and L. Sajus, *Tetrahedron Lett., 17*, 1693 (1972).
24. R. D. Bereman, M. L. Good, B. J. Kalbacher, and J. Buttone, *Inorg. Chem., 15*, 618 (1976).
25. R. D. Bereman and D. Nalewajek, *Inorg. Chem., 15*, 2981 (1976).
26. R. D. Bereman and D. Nalewajek, *Inorg. Chem., 16*, 2687 (1977).
27. J. J. Bergendahl and E. M. Bergendahl, *Inorg. Chem., 11*, 638 (1972).
28. P. T. Beurskens, H. J. A. Blaauw, J. A. Cras, and J. J. Steggerda, *Inorg. Chem., 7*, 805 (1968).
29. P. T. Beurskens, W. P. J. H. Bosman, and J. A. Cras, *J. Cryst. Mol. Struct., 2*, 183 (1972).
30. P. T. Beurskens, J. A. Cras, T. W. Hummelink, and J. H. Noordik, *J. Cryst. Mol. Struct., 1*, 253 (1971).
31. P. T. Beurskens, J. A. Cras, T. W. Hummelink, and J. G. M. Van der Linden, *Rec. Trav. Chim., 89*, 984 (1970).
32. P. T. Beurskens, J. A. Cras, J. H. Noordik, and A. M. Sprujt, *J. Cryst. Mol. Struct., 1*, 93 (1971).
33. P. T. Beurskens, J. A. Crass, and J. J. Steggerda, *Inorg. Chem., 7*, 810 (1968).

34. P. T. Beurskens, J. A. Cras, and J. G. M. Van der Linden, *Inorg. Chem., 9*, 475 (1970).
35. A. N. Bhat, R. C. Fay, D. F. Lewis, A. F. Lindmark, and S. H. Strauss, *Inorg. Chem., 13*, 886 (1974).
36. (a) M. W. Bishop, J. Chatt, and J. R. Dilworth, *J. Organomet. Chem., 73*, C59 (1974).
   (b) J. Chatt and J. R. Dilworth, *J. Less-Common Met., 36*, 513 (1974).
37. M. W. Bishop, J. Chatt, J. R. Dilworth, G. Kaufman, S. Kim, and J. Zubieta, *Chem. Commun., 1977*, 70.
38. R. S. Bitcher, J. R. Ferraro, and E. Sinn, *Inorg. Chem., 15*, 2077 (1976).
39. H. J. Blaauw, Ph.D. Thesis, Nijmegen, 1965.
40. P. Bladon, R. Bruce, and G. R. Knox, *Chem. Commun., 1965*, 557.
41a. A. B. Blake, F. A. Cotton, and J. S. Wood, *J. Am. Chem. Soc., 86*, 3024 (1964).
41b. G. Bombieri, U. Croatto E. Forsellini, B. Zarli and R. Graziani, *J. Chem. Soc. Dalton Trans.*, 1972, 560.
42. M. Bonamico and G. Dessy, *Ric. Sci., 38*, 1106 (1968).
43. M. Bonamico, G. Dessy, and V. Fares, *Chem. Commun., 1969*, 1106.
44. M. Bonamico, G. Dessy, V. Fares, A. Flamini, and L. Scaramuzza, *J. Chem. Soc. Dalton Trans., 1976*, 1743.
45. M. Bonamico, G. Dessy, V. Fares, P. Porta, and L. Scaramuzza, *Chem. Commun., 1971*, 365.
46. M. Bonamico, G. Dessy, V. Fares, and L. Scaramuzza, *Ann. Chim., 60*, 644 (1970).
47. M. Bonamico, G. Dessy, V. Fares, and L. Scaramuzza, *Cryst. Struct. Commun., 2*, 201 (1973).
48. M. Bonamico, G. Dessy, V. Fares, and L. Scaramuzza, *J. Chem. Soc. A, 1971*, 3191.
49. M. Bonamico, G. Dessy, V. Fares, and L. Scaramuzza, *J. Chem. Soc., Dalton Trans., 1972*, 2515.
50. M. Bonamico, G. Dessy, V. Fares, and L. Scaramuzza, *J. Chem. Soc., Dalton Trans., 1974*, 1258.
51. M. Bonamico, G. Dessy, V. Fares, and L. Scaramuzza, *J. Chem. Soc., Dalton Trans., 1975*, 2250.
52. M. Bonamico, G. Dessy, V. Fares, and L. Scaramuzza, *J. Chem. Soc., Dalton Trans., 1975*, 2079.
53. M. Bonamico, G. Dessy, A. Mugnoli, A. Vaciago, and L. Zambonelli, *Acta Crystallogr., 19*, 886 (1965).
54. M. Bonamico, G. Dessy, A. Mugnoli, A. Vaciago, and L. Zambonelli, *Acta Crystallogr., 19*, 886 (1965).
55. (a) F. Bonati, G. Minghetti, and S. Cenini, *Inorg. Chim. Acta, 2*, 375 (1968). (b) F. Bonati and R. Ugo, *J. Organomet. Chem., 10*, 257 (1967).
56. A. M. Bond, A. T. Casey, and J. R. Thakeray, *Electronanal. Chem. Interf. Electrochem., 48*, 71 (1973).
57. A. M. Bond, A. T. Casey, and J. R. Thackeray, *Inorg. Chem, 12*, 887 (1973).
58. A. M. Bond, A. T. Casey, and J. R. Thackeray, *Inorg. Chem., 13*, 84 (1974).
59. A. M. Bond, A. T. Casey, and J. R. Thackeray, *J. Chem. Soc., Dalton. Trans., 1974*, 773.
60. A. M. Bond, A. T. Casey, and J. R. Thackeray, *J. Electrochem. Soc., 120*, 1502 (1973).
61. M. M. Borel and M. Ledesert, *Z. Anorg. Allg. Chem., 415*, 285 (1975).
62. W. P. Bosman and A. W. Gal, *Cryst. Struct. Commun., 4*, 465 (1975).
63. W. P. Bosman and A. Nieuwpoort, *Inorg. Chem., 15*, 775 (1976).
64. F. L. Bowden, R. Giles, and R. N. Haszeldine, *J. Chem Soc., Dalton Trans., 1974*, 578.

65. P. D. W. Boyd and R. L. Martin, *J. Chem. Soc., Dalton Trans., 1977*, 105.
66. R. A. Bozis, Univ. Microfilms, 72-14,064 [*Chem. Abstr., 77*, 108887y (1972)].
67. (a) G. C. Brackett, P. L. Richards, and W. S. Caughy, *J. Chem Phys., 54*, 4383 (1971).
    (b) G. C. Brackett, P. L. Richards, and H. H. Wickman, *Chem. Phys. Lett., 6*, 75 (1970).
68. D. C. Bradley and M. H. Chisholm, *J. Chem. Soc. A, 1971*, 2741.
69. D. C. Bradley and M. H. Gitlitz, *J. Chem. Soc. A, 1969*, 1152.
70. D. C. Bradley, I. F. Rendall, and K. D. Sales, *J. Chem. Soc., Dalton Trans., 1973*, 2228.
71. T. B. Brennan and I. Bernal, *Chem. Commun., 1970*, 138; *Inorg. Chim. Acta, 7*, 283 (1973).
72. T. Brennan and I. Bernal, *J. Phys. Chem., 73*, 443 (1969).
73. H. C. Brinkhoff, *Inorg. Nucl. Chem. Lett., 7*, 413 (1971).
74. H. C. Brinkhoff, Ph.D. Thesis, Nijmegen, 1970.
75. H. C. Brinkhoff, *Recl. Trav. Chim. Pays-Bas, 90*, 377 (1971).
76. H. Brinkhoff and A. M. Grotens, *Recl. Trav. Chim., 111*, 252 (1971).
77. H. C. Brinkhoff, J. A. Cras, J. J. Steggerda, and J. Willemse, *Recl. Trav. Chim., 88*, 633 (1969).
78. H. C. Brinkhoff, A. M. Grotens, and J. J. Steggerda, *Recl. Trav. Chim., 89*, 11 (1970).
79. N. J. Brøndmo, S. Esperas, H. Graver, and S. Husebye, *Acta Chem. Scand., 27*, 713 (1973).
80. N. J. Brøndmo, S. Esperas, and S. Husebye, *Acta Chem Scand., A29*, 93 (1975).
81. D. A. Brown, W. K. Glass, and M. A. Burke, *Spectrochim. Acta, 32A*, 137 (1976).
82. K. L. Brown, R. M. Golding, P. C. Healey, K. J. Jessop, and W. C. Tennant, *Aust. J. Chem., 27*, 2075 (1974).
83. D. A. Brown, B. J. Gordon, W. K. Glass, and C. J. O'Daly, *Proc. Int. Conf. Coord. Chem., 14th Toronto, 1972*, p. 646.
84. D. Brown and D. G. Holah, *Chem. Commun., 1968*, 1545.
85. D. Brown, D. G. Holah, and C. E. F. Rickard, *Chem. Commun., 1969*, 280.
86. D. Brown, D. G. Holah, and C. E. F. Rickard, *J. Chem Soc. A, 1970*, 423.
87. D. Brown, D. G. Holah, and C. E. F. Rickard, *J. Chem. Soc., Dalton Trans., 1970*, 786.
88. T. M. Brown and J. N. Smith, *J. Chem. Soc., Dalton Trans., 1972*, 1614.
89. A. H. Bruder, R. C. Fay, D. F. Lewis, and A. Sayler, *J. Am. Chem. Soc., 98*, 6932 (1976).
90. J. M. Burke and J. P. Fackler, Jr., *Inorg. Chem., 11*, 2744 (1972).
91. J. M. Burke and J. P. Fackler, Jr., *Inorg. Chem., 11*, 2744 (1972).
92. J. M. Burke and J. P. Fackler, Jr., *Inorg. Chem., 11*, 3000 (1972).
93. (a) R. J. Butcher, J. R. Ferraro, and E. Sinn, *J. Chem. Soc., Chem. Commun., 1976*, 910. (b) R. J. Butcher, J. R. Ferraro, and E. Sinn, *Inorg. Chem., 15*, 2077 (1976).
94. R. J. Butcher and E. Sinn, *J. Chem. Soc., Dalton Trans., 1975*, 2517.
95. R. J. Butcher and E. Sinn, *J. Chem. Soc., Dalton Trans., 1975*, 2518.
96. (a) R. J. Butcher and E. Sinn, *J. Am. Chem. Soc., 98*, 5159 (1976). (b) R. J. Butcher and E. Sinn, *J. Am. Chem. Soc., 98*, 2440 (1976).
97. H. Büttner and R. D. Feltham, *Inorg. Chem., 11*, 971 (1972).
98. V. M. Byr'ko, M. B. Polinskaya, A. I. Buser, and T. A. Kuz'mina, *Russ. J. Inorg. Chem. Int. Ed., 18*(6), 820 (1973).
99. M. L. Caffery, Ph.D. Thesis, University of Iowa, 1973.
100. M. L. Caffery and D. Coucouvanis, *J. Inorg. Nucl. Chem, 37*, 7081 (1975).
101. L. Cambi and A. Cagnasso, *Atti Accad. Nat. Lincei, 13*, 809 (1931); L. Cambi and L. Szëgo, *Ber. Dtsch. Chem. Ges., 64*, 2591 (1931).

102. L. Capacci, A. C. Villa, M. Ferrari, and M. Nardeli, *Ric. Sci., 37*, 993 (1967).

103. R. L. Carlin and A. E. Siegel, *Inorg. Chem., 9*, 1587 (1970).

104. A. Caron and J. Donohue, *Acta Crystallogr., 14*, 548 (1961).

105. A. T. Casey, D. J. Mackey, R. L. Martin, and A. H. White, *Aust. J. Chem., 25*, 477 (1972).

106. A. T. Casey and J. R. Thackeray, *Aust. J. Chem., 25*, 2085 (1972).

107. A. T. Casey and J. R. Thackeray, *Aust. J. Chem., 27*, 757 (1974).

108. A. T. Casey and J. R. Thackeray, *Aust. J. Chem., 28*, 471 (1975).

109. G. Cauquis and D. Lachenal, *Electroanal. Chem., 43*, 205 (1973).

110. G. Cauquis and D. Lachenal, *Inorg. Nucl. Chem. Lett., 9*, 1095 (1973).

111. E. Cervone, F. Diomedi-Camassei, M. L. Luciani, and C. Furlani, *J. Inorg. Nucl. Chem., 31*, 1101 (1969).

112. R. Chant, A. R. Hendrickson, R. L. Martin, and N. M. Rohde, *Aust. J. Chem., 26*, 2533 (1973).

113. R. Chant, A. R. Hendrickson, R. L. Martin, and N. M. Rohde, *Inorg. Chem., 14*, 1894 (1975).

114. G. E. Chapps, S. W. McCann, H. H. Wickman, and R. C. Sherwood, *J. Chem. Phys., 60*, 990 (1974).

115. R. Chant, A. R. Hendrickson, R. L. Martin, and N. M. Rohde, *Inorg. Chem., 14*, 1894 (1975).

116. R. Chant, A. R. Hendrickson, R. L. Martin, and N. M. Rohde, *Aust. J. Chem., 26*, 2533 (1973).

117. H. W. Chen, and J. P. Fackler, Jr., *Inorg. Chem. 17*, 22 (1978).

118. G. J. J. Chen, J. W. McDonald, and W. E. Newton, *Inorg. Chem., 15*, 2612 (1976).

119. G. J. J. Chen, J. W. McDonald, and W. E. Newton, *Inorg. Chim. Acta, 19*, L67 (1976).

120. J. H. B. Chenier, J. A. Howard, and J. C. Tait, *Can. J. Chem., 55*, 1644 (1977).

121. C. Chieh, *Can. J. Chem., 55*, 65 (1977).

122. C. Chieh and L. P. C. Leung, *Can. J. Chem., 54*, 3077 (1976).

123. G. Ciullo and A. Sgamellotti, *Z. Phys. Chem., 100*, 67 (1976).

124. M. Colapietro, A. Domenicano, L. Scaramuzza, and A. Vaciago, *Chem. Commun., 1968*, 302.

125. M. Colapietro, A. Domenicano, L. Scaramuzza, A. Vaciago, and L. Zambonelli, *Chem. Commun., 1967*, 583.

126. M. Colapietro, A. Domenicano, and A. Vaciago, *Chem. Commun., 1968*, 572.

127. M. Colapietro, A. Vaciago, D. C. Bradley, M. B. Hursthouse, and I. F. Rendall, *Chem. Commun., 1970*, 743.

128. M. Colapietro, A. Vaciago, D. C. Bradley, M. B. Hursthouse, and I. F. Rendall, *Chem. Commun., 1970*, 743; *J. Chem. Soc., Dalton Trans., 1972*, 1052.

129. D. J. Cole-Hamilton and T. A. Stephenson, *J. Chem. Soc., Dalton Trans., 1974*, 739.

130. D. J. Cole-Hamilton and T. A. Stephenson, *J. Chem. Soc., Dalton Trans., 1974*, 754.

131. D. J. Cole-Hamilton and T. Stephenson, *J. Chem. Soc., Dalton Trans., 1974*, 1818.

132. D. J. Cole-Hamilton and T. A. Stephenson, *J. Chem. Soc., Dalton Trans., 1976*, 2396.

133. H. M. Colquhoun, T. J. Greenhough, and M. G. H. Wallbridge, *Chem Commun., 1976*, 1019.

134. R. Colton, R. Levitus, and G. Wilkinson, *J. Chem. Soc., Dalton Trans., 1960*, 5275.

135. R. Colton and G. G. Rose, *Aust. J. Chem., 23*, 1111 (1970).

136. R. Colton and G. R. Scollary, *Aust. J. Chem., 21*, 1427 (1968).

137. N. G. Connely, M. Green, and T. A. Kuc, *Chem. Commun., 1974*, 542.

138. G. Contreras and H. Cortes, *J. Inorg. Nucl. Chem., 33*, 1337 (1971).

139. B. J. Corden and P. H. Rieger, *Inorg. Chem., 10*, 263 (1971).

140. M. C. Cornock, R. C. Davis, D. Leaver, and T. A. Stephenson, *J. Organomet. Chem.*, *107*, C43 (1976).

141. M. C. Cornock, R. O. Gould, C. L. Jones, J. D. Owen, D. F. Steele, and T. A. Stephenson, *J. Chem. Soc., Dalton Trans., 1977*, 496.

142. M. C. Cornock, R. O. Gould, C. L. Jones, and T. A. Stephenson, *J. Chem. Soc., Dalton Trans., 1977*, 496.

143. M. C. Cornock, D. R. Robertson, T. A. Stephenson, C. L. Jones, G. H. W. Milburn and L. Sawyer, *J. Organometal Chem.*, submitted for publication.

144. M. C. Cornock, D. F. Steele, and T. A. Stephenson, *Inorg. Nucl. Chem. Lett.*, *10*, 785 (1974).

145. M. C. Cornock and T. A. Stephenson, *Inorg. Nucl. Chem. Lett.*, *12*, 381 (1976).

146. M. C. Cornock and T. A. Stephenson, *J. Chem. Soc., Dalton Trans., 1977*, 501.

147. M. C. Cornock and T. A. Stephenson, *J. Chem. Soc., Dalton Trans., 1977*, 501.

148. F. A. Cotton, D. L. Hunter, R. Ricard, and R. Weiss, *J. Coord. Chem.*, *3*, 259 (1973).

149. F. A. Cotton and J. A. McCleverty, *Inorg. Chem.*, *3*, 1398 (1964).

150. F. A. Cotton and J. G. Norman, Jr., *J. Am. Chem. Soc.*, *94*, 5697 (1972).

151. F. A. Cotton and G. Wilkinson, *Adv. Inorg. Chemistry*, 2nd ed., Interscience, New York.

152. D. Coucouvanis, Ph.D. Thesis, Case Institute of Technology, 1967.

153. D. Coucouvanis, *Prog. Inorg. Chem.*, *11*, 233 (1970).

154. D. Coucouvanis, N. C. Baenziger, and S. M. Johnson, *Inorg. Chem.*, *13*, 1191 (1974).

155. D. Coucouvanis and J. P. Fackler, Jr., *Inorg. Chem.*, *6*, 2047 (1967).

156. D. Coucouvanis and J. P. Fackler, Jr., *J. Am. Chem. Soc.*, *89*, 1346 (1967).

157. D. Coucouvanis, F. J. Hollander, and M. L. Caffery, *Inorg. Chem.*, *15*, 1853 (1976).

158. D. Coucouvanis, F. J. Hollander, and R. Pedelty, *Inorg. Chem.*, *16*, 2691 (1977).

159. D. Coucouvanis, F. J. Hollander, D. Petridis, A. Simopoulos, and A. Kostikas, to be published.

160. D. Coucouvanis and S. J. Lippard, *J. Am. Chem. Soc.*, *90*, 3281 (1968).

161. D. Coucouvanis and S. J. Lippard, *J. Am. Chem. Soc.*, *91*, 307 (1969).

162. D. Coucouvanis, S. J. Lippard, and J. A. Zubieta, *Inorg. Chem.*, *9*, 2775 (1970).

163. D. Coucouvanis, S. J. Lippard, and J. A. Zubieta, *J. Am. Chem. Soc.*, *92*, 3342 (1970).

164. D. Coucouvanis, R. Pedelty, S. Kanodia, and M. Chu, manuscript in preparation.

165. D. Coucouvanis, D. Swenson, N. C. Baenziger, R. Pedelty, and M. L. Caffery, *J. Am. Chem. Soc.*, *99*, 8097 (1977).

166. D. N. Coucouvanis, D. Swenson, R. Pedelty, S. Kanodia, W. Richardson, and N. C. Baenziger, paper presented at the 174th Meeting of the American Chemical Society, Chicago, Illinois, Sept. 1977.

167. R. K. Cowsik, G. Rangarajan, and R. Srinivasan, *Chem. Phys. Lett.*, *8*, 136 (1970).

168. J. A. Cras, *Proc. Int. Conf. Coord. Chem., 16th, Dublin, 1974.*

169. J. A. Cras, J. H. Noordik, P. T. Beurskens, and A. M. Verhoeven, *J. Cryst. Mol. Structure*, *1*, 155 (1971).

170. J. A. Cras, P. J. H. A. M. van de Leemput, J. Willemse, and W. P. Bosman, *Recl. Trav. Chim. Pays-Bas, 96*, 78 (1977).

171. J. A. Cras, and J. Willemse, *J. Inorg. and Nucl. Chem.*, to be published.

172. J. A. Cras, J. Willemse, A. W. Gal, and B. G. M. C. Hummelink-Peters, *Recl. Trav. Chim., 92*, 641 (1973).

173. P. B. Critchlow and S. D. Robinson, *J. Chem. Soc., Dalton Trans., 1975*, 1367.

174. E. J. Cukauskas, B. S. Deaver, and E. Sinn, *Inorg. Nucl. Chem. Lett.*

175. E. J. Cukauskas, B. S. Deaver, and E. Sinn, *J. Chem. Phys., 67*, 1257 (1977).

176. E. J. Cukauskas, B. S. Deaver, Jr., and E. Sinn, *Chem Commun., 1974*, 698.

177. A. D'Addario D. G. Holah, and K. Knox, unpublished work. A. D'Addario, Ph.D. Thesis, 1970, University Microfilms, Ann Arbor, Michigan.
178. J. J. D'Amice and R. H. Campbell, *J. Org. Chem., 32*, 2567 (1967).
179. G. D'Ascenzo and W. W. Wendlandt, *J. Thermal Anal., 1*, 423 (1969).
180. G. R. Davies, J. A. J. Jarvis, B. T. Kilbourn, R. H. B. Mais, and P. G. Owston, *J. Chem. Soc. A, 1970*, 1275.
181. G. R. Davies, R. H. B. Mais, and P. G. Owston, *Chem. Commun., 1968*, 81.
182. R. Davis, M. N. S. Hill, C. E. Holloway, B. F. G. Johnson, and K. H. Al-Obaidi, *J. Chem. Soc. A, 1971*, 994.
183. W. K. Dean and J. W. Moncrief, *J. Coord. Chem , 6*, 107 (1976).
184. W. K. Dean and P. M. Treichel, *J. Organomet. Chem., 66*, 87 (1974).
185. M. K. DeArmond and W. J. Mitchell, *Inorg. Chem., 11*, 181 (1972).
186. M. H. M. De Croon, H. L. M. van Gaal, and A. van der Ent, *Inorg. Nucl. Chem. Lett., 10*, 1081 (1974).
187. L. J. DeHayes, H. C. Faulkner and W. H. Doub, Jr., and D. T. Sawyer, *Inorg. Chem., 14*, 2110 (1975).
188. J. M. De Lisle and R. M. Golding, *Proc. R. Soc. (Lond.), A296*, 457 (1967).
189. J. L. F. K. DeVries and R. H. Herber, *Inorg. Chem., 11*, 2458 (1972).
190. J. L. F. K. DeVries, C. P. Keijzers, and E. deBoer, *Inorg. Chem., 11*, 1343 (1972).
191. J. L. F. K. DeVries, J. M. Trooster, and E. deBoer, *Inorg. Chem., 10*, 81 (1971).
192. J. L. F. K. DeVries, J. M. Trooster, and E. deBoer, *Inorg. Chem., 12*, 2730 (1973).
193. J. C. Dewan, D. L. Kepert C. L. Raston, D. Taylor, and A. H. White, *J. Chem. Soc., Dalton Trans, 1973*, 2082.
194. M. M. Dhingra, P. Ganguli, V. R. Marathe, S. Mitra, and R. L. Martin, *J. Magn. Resonance, 20*, 133 (1975).
195. R. Dingle, *Inorg. Chem., 10*, 1141 (1971).
196. J. Dirand, L. Ricard, and R. Weiss, *Inorg. Chem., Lett., 11*, 661 (1975).
197. J. Dirand-Colin, L. Ricard, and R. Weiss, *Inorg. Chim. Acta., 18*, L21 (1976).
198. J. Dirand, L. Ricard, and R. Weiss, *J. Chem. Soc., Dalton Trans., 1976*, 278.
199. J. Dirand, L. Ricard, and R. Weiss, *Trans. Met Chem., 1*, 2 (1975).
200. J. Dirand-Colin, M. Schappacher, L. Ricard, and R. Weiss, *J. Less-Common Met., 54*, 91 (1976).
201. J. Dobrowolski, Z. Badkowska, and I. Kwiatkowska, *Rocz. Chem., 50*, 1043 (1976).
202. D. M. Doddrell, M. R. Bendall, and A. K. Gregson, *Aust. J. Chem., 29*, 55 (1976).
203. A. Domenicano, L. Torelli, A. Vaciago, and L. Zambonelli, *J. Chem. Soc., Dalton Trans (1968)*, 1351.
204. A. Domenicano, A. Vaciago, L. Zambonelli, P. L. Loader, and L. M. Venanzi, *Chem. Commun., 1966*, 476.
205. D. J. Duffy and L. H. Pignolet, *Inorg. Chem., 11*, 2843 (1972).
206. (a) D. J. Duffy, and L. H. Pignolet, *Inorg. Chem., 13*, 2045 (1974). (b) G. Durgaprasad, D. N. Sathyanarayana, and C. C. Patel, *Can. J. Chem., 47*, 631 (1969).
207. K. Dymock, G. J. Palenik, J. Slezak, C. L. Raston, and A. H. White, *J. Chem. Soc., Dalton Trans., 1976*, 28.
208. E. A. V. Ebsworth, G. Rocktäschel, and J. C. Thompson, *J. Chem. Soc. A, 1967*, 362.
209. B. L. Edgar, D. J. Duffy, M. C. Palazzotto, and L. H. Pignolet, *J. Am. Chem. Soc., 95*, 1125 (1973).
210. F. W. Einstein, E. Enwall, N. Flitcroft, and J. M. Leach, *J. Inorg. Nucl. Chem., 34*, 885 (1972).
211. A. G. El A'mma and R. S. Drago, *Inorg. Chem., 16*, 2975 (1977).
212. R. Eley, N. V. Duffy, and D. L. Uhrich, *J. Inorg. Nucl. Chem., 34*, 3681 (1972).

213. R. R. Eley, R. R. Myers, and N. V. Duffy, *Inorg. Chem.*, *11*, 1128 (1972).

214. E. Elfwing, H. Anacker-Eickhoff, P. Jennische, and R. Hesse, *Acta Chem. Scand., A*, *30*, 335 (1976).

215. J. H. Enemark and R. D. Feltham, *Coord. Chem. Rev.*, *13*, 339 (1974).

216. J. H. Enemark and R. D. Feltham, *J. Chem. Soc., Dalton Trans., 1972*, 718.

217. R. Engler, M. Dräger, and G. Gattow, *Z. Anorg. Allg. Chem., 403*, 81 (1974).

218. R. Engler, G. Kiel, and G. Gattow, *Z. Anorg. Allg. Chem., 404*, 71 (1974).

219. L. M. Epstein and D. K. Straub, *Inorg. Chem.*, *8*, 560 (1969).

220. S. Esperas and S. Husebye, *Acta Chem. Scand., 26*, 3293 (1972).

221. S. Esperas and S. Husebye, *Acta Chem. Scand., 27*, 706 (1973).

222. S. Esperas and S. Husebye, *Acta Chem. Scand. A, 29*, 185 (1975).

223. S. Esperas, S. Husebye, and S. E. Svaeren, *Acta Chem. Scand., 25*, 3539 (1971).

224. I. P. Evans, A. Spencer, and G. Wilkinson, *Chem. Soc., Dalton Trans., 1973*, 204.

225. J. A. Evans, M. J. Hacker, R. D. W. Kemmit, D. R. Russell, and J. Stocks, *Chem. Commun., 1972*, 72.

226. A. H. Ewald, R. L. Martin, E. Sinn, and A. H. White, *Inorg. Chem., 8*, 1837 (1969).

227. A. E. Ewald and E. Sinn, *Aust. J. Chem., 21*, 927 (1968).

228. J. Fabian, *Theor. Chim. Acta, 12*, 200 (1968).

229. C. Fabiani, R. Spagna, A. Vaciago, and L. Zambonelli, *Acta Crystallogr., B27*, 1499 (1971).

230. J. P. Fackler, Jr., *Adv. Chem. Ser., 150, Inorg. Compd. Unusual Prop. Symp.*, 394 (1976).

231. J. P. Fackler, Jr., A. Avdeef, and R. G. Fischer, Jr., *J. Am. Chem. Soc., 95*, 774 (1973).

232. J. P. Fackler, Jr., and J. A. Fetchin, *J. Am. Chem. Soc., 92*, 2912 (1970).

233. J. P. Fackler, Jr., J. A. Fetchin, and D. C. Fries, *J. Am. Chem. Soc., 94*, 7323 (1972).

234. J. P. Fackler, Jr., J. A. Fetchin, and W. C. Seidel, *J. Am. Chem. Soc., 91*, 1217 (1969).

235. J. P. Fackler, Jr., J. A. Fetchin, and J. A. Smith, *J. Am. Chem. Soc., 92*, 2910 (1970).

236. J. P. Fackler, Jr. and D. G. Holah, *Inorg. Nucl. Chem. Lett., 2*, 251 (1966).

237. J. P. Fackler, Jr., I. J. B. Lin, and J. Andrews, *Inorg. Chem., 16*, 450 (1977).

238. J. P. Fackler, Jr., D. P. Schussler, and H. W. Chen, *Synth. React. Inorg. Met-Org. Chem., 8*(1), 27 (1978).

239. J. P. Fackler, Jr. and W. C. Seidel, *Inorg. Chem., 8*, 1631 (1969).

240. J. P. Fackler, Jr., W. C. Seidel, and J. A. Fetchin, *J. Am. Chem. Soc., 90*, 2707 (1968).

241. J. P. Fackler, Jr. and W. J. Zegarski, *J. Am. Chem. Soc., 95*, 8566 (1973).

242. L. Fanfani, A. Nunzi, P. F. Zanazzi, and A. R. Zanzari, *Acta Crystallogr. B28*, 1298 (1972).

243. F. J. Farrell and T. G. Spiro, *Inorg. Chem., 10*, 1606 (1971).

244. B. N. Figgis and G. E. Toogood, *J. Chem. Soc., Dalton Trans., 1972*, 2177.

245. B. W. Fitzsimmons and A. C. Sawbridge, *J. Chem. Soc., Dalton Trans., 1972*, 1678.

246. A. Flamini, C. Furlani, and O. Piovesana, *J. Inorg. Nucl. Chem., 33*, 1841 (1971).

247. S. R. Fletcher, J. F. Rowbottom, A. C. Skapski, and G. Wilkinson, *Chem. Commun., 1970*, 1572.

248. S. R. Fletcher and A. C. Skapski, *J. Chem. Soc., Dalton Trans., 1972*, 1073.

249. S. R. Fletcher and A. C. Skapski, *J. Chem. Soc., Dalton Trans., 1972*, 1079.

250. C. Flick and E. Gelerinter, *Chem. Phys. Lett., 23*, 422 (1973).

251. O. Foss, *Acta Chem. Scand., 7*, 227 (1953).

252. W. O. Foye, "Report of New England Association of Chemistry Teachers", *J. Chem. Educ., 46*, 841 (1969).

253. W. O. Foye, J. R. Marshall, and J. Mickles, *J. Pharm. Sci., 52*, 406 (1963).

254. W. O. Foye and J. Mickles, *J. Med. Pharm. Chem.*, *5*, 846 (1962).
255. K. A. Fraser and M. M. Harding, *Acta Crystallogr.*, *22*, 75 (1967).
256. M. Freni, D. Giusto, and P. Romiti, *J. Inorg. Nucl. Chem.*, *33*, 4093 (1971).
257. D. C. Fries and J. P. Fackler, Jr., *Chem. Commun.*, *1971*, 276.
258. C. Furlani, A. Flamini, A. Sgamelloti, and C. Bellito, *J. Chem. Soc., Dalton Trans.*, *1973*, 2404.
259. R. H. Furneaux, *Inorg. Nucl. Chem. Lett.*, *12*, 501 (1976).
260. L. R. Gahan and M. J. O'Connor, *Chem. Commun.*, *1974*, 67.
261. L. R. Gahan and M. J. O'Connor, *Chem. Commun.*, *1974*, 67.
262. I. F. Gainulin, N. S. Garifyanov, and V. V. Trachevskii, *Jzv. Akad. Nauk. S.S.S.R. Ser. Khim.*, *1969*, 2176; L. N. Duglav and Z. I. Usmanov, *Zh. Strukt. Khim.*, *16*, 312 (1975).
263. A. W. Gal, Ph.D. Thesis, Nijmegen, Netherland, 1975.
264. A. W. Gal, G. Beurskens, J. A. Cras, P. T. Beurskens, and J. Willemse, *Recl. Trav. Chim.*, *95*, 157 (1976).
265. A. W. Gal, A. F. J. M. Van der Ploeg, F. A. Vollenbroek, and W. Bosman, *J. Organomet. Chem.*, *96*, 123 (1975).
266. P. Ganguli, V. R. Marathe, S. Mitra, and R. L. Martin, *Chem. Phys. Lett.*, *26*, 529 (1974).
267. K. W. Given, B. M. Mattson, M. F. McGuiggan, G. L. Miessler, and L. H. Pignolet, *J. Am. Chem. Soc.*, *99*, 4855 (1977).
268. K. W. Given, B. M. Mattson, and L. H. Pignolet, *Inorg. Chem.*, *15*, 3152 (1976).
269. K. W. Given and L. H. Pignolet, *Inorg. Chem.*, *16*, 2982 (1977).
270. R. M. Golding, C. M. Harris, K. J. Jessop, and W. C. Tennant, *Aust. J. Chem.*, *25*, 2567 (1972).
271. R. M. Golding, P. C. Healy, P. Colombera, and A. H. White, *Aust. J. Chem.*, *27*, 2089 (1974).
272. R. M. Golding, P. Healy, P. Newman, E. Sinn, W. C. Tennant, and A. H. White, *J. Chem. Phys.*, *52*, 3105 (1970).
273. R. M. Golding, P. C. Healy, P. W. G. Newman, E. Sinn, and A. H. White *Inorg. Chem.*, *11*, 2435 (1972).
274. R. M. Golding, L. L. Kok, K. Lehtonen, and R. K. Nigam, *Aust. J. Chem.*, *28*, 1915 (1975).
275. R. M. Golding and K. Lehtonen, *Aust. J. Chem.*, *27*, 2083 (1974).
276. R. M. Golding, K. Lehtonen, and B. J. Ralph, *Aust. J. Chem.*, *28*, 2393 (1975).
277. R. M. Golding, B. D. Lukeman, and E. Sinn, *J. Chem. Phys.*, *56*, 4147 (1972).
278. R. M. Golding, A. D. Rae, B. J. Ralph, and L. Sulligoi, *Inorg. Chem.*, *13*, 2499 (1974).
279. R. M. Golding, E. Sinn, and W. C. Tennant, *J. Chem. Phys.*, *56*, 5296 (1972).
280. R. M. Golding, A. H. White, and P. Healy, *Trans. Faraday Soc.*, *67*, 1672 (1971).
281. H. Graver and S. Husebye, *Acta Chem. Scand.*, *A29*, 14 (1975).
282. R. Graziani, B. Zarli, A. Cassol, G. Bombieri, E. Forsellini, and E. Todello, *Inorg. Chem.*, *9*, 2116 (1970).
283. M. Green, R. B. L. Osborn, and F. G. A, Stone, *J. Chem. Soc. A, 1970*, 944.
284. A. K. Gregson and D. M. Doddrell, *Chem. Phys. Lett.*, *31*, 125 (1975).
285. C. M. Guzy, J. B. Raynor, and M. C. R. Symons, *J. Chem. Soc. A, 1969*, 2987.
286. H. Hagihara, Y. Watanabe, and S. Yamashita, *Acta Crystallogr.*, *B24*, 960 (1968).
287. H. Hagihara and S. Yamashita, *Acta Crystallogr.*, *21*, 350 (1966).
288. H. Hagihara, N. Yoshida, and Y. Watanabe, *Acta Crystallogr.*, *B25*, 1775 (1969).
289. H. Hagihara, N. Yoshida, and Y. Watanabe, *Acta Crystallogr.*, *B25*, 1775 (1969).
290. V. Hahnkam, G. Kiel, and G. Gattow, *Z. Anorg. Allg. Chem.*, *368*, 127 (1966).

291. G. R. Hall and D. N. Hendrickson, *Inorg. Chem., 15*, 607 (1976).

292. K. Hara, W. Mori, M. Inoue, M. Kishita, and M. Kubo, *Bull. Chem. Soc. Jap., 42*, 576 (1969).

293. R. D. Harcourt and G. Winter, *J. Inorg. Nucl. Chem., 37*, 1039 (1975).

294. R. D. Harcourt and G. Winter, *J. Inorg. Nucl. Chem., 39*, 360 (1977).

295. C. S. Harreld and E. O. Schlemper, *Acta Crystallogr., B27*, 1964 (1971).

296. G. Harris, *Theor. Chim. Acta, 5*, 379 (1966).

297. G. Harris, *Theor. Chim. Acta, 10*, 119 (1968).

298. G. Harris-Loew, *J. Chem. Phys., 48*, 2191 (1968); G. Harris-Loew and R. L. Ake, *J. Chem. Phys., 51*, 3143 (1969).

299. P. J. Hauser, J. Bordner, and A. F. Schreiner, *Inorg. Chem., 12*, 1347 (1973).

300. P. J. Hauser, A. F. Schreiner, and R. S. Evans, *Inorg. Chem., 13*, 1925 (1974).

301. P. J. Hauser, A. F. Schreiner, J. D. Gunter, W. J. Mitchell, and M. K. DeArmond, *Theor. Chim. Acta, 24*, 78 (1972).

302. P. C. Healy and E. Sinn, *Inorg. Chem., 14*, 109 (1975).

303. P. C. Healy and A. H. White, *J. Chem. Soc., Dalton Trans., 1972*, 1163.

304. P. C. Healy and A. H. White, *J. Chem. Soc., Dalton Trans., 1972*, 1369.

305. P. C. Healy and A. H. White, *J. Chem. Soc., Dalton Trans., 1972*, 1883.

306. P. C. Healy and A. H. White, *J. Chem. Soc., Dalton Trans., 1973*, 284.

307. P. R. Heckley and D. G. Holah, *Inorg. Nucl. Chem. Lett., 6*, 865 (1970).

308. P. R. Heckley, D. G. Holah, and D. Brown, *Can. J. Chem., 49*, 1151 (1971).

309. A. R. Hendrickson, J. M. Hope, and R. L. Martin, *J. Chem. Soc., Dalton Trans., 1976*, 2032.

310. A. R. Hendrickson and R. L. Martin, *Chem. Commun., 1974*, 873.

311. A. R. Hendrickson, R. L. Martin, and N. M. Rohde, *Inorg. Chem., 13*, 1933 (1974).

312. A. R. Hendrickson, R. L. Martin, and N. M. Rohde, *Inorg. Chem., 14*, 2980 (1975).

313. A. R. Hendrickson, R. L. Martin, and N. M. Rohde, *Inorg. Chem, 15*, 2115 (1976).

314. A. R. Hendrickson, R. L. Martin, and D. Taylor, *Aust. J. Chem., 29*, 269 (1976).

315. A. Hendrickson, R. L. Martin, and D. Taylor, *Chem. Commun., 1975*, 843.

316. A. R. Hendrickson, R. L. Martin, and D. Taylor *J. Chem. Soc. Dalton Trans., 1975*, 2182.

317. D. N. Hendrickson, private communication.

318. H. M. Hendriks, W. P. Bosman, and P. T. Beurskens, *Cryst. Struct. Commun., 3*, 447 (1974).

319. K. Henrick, C. L. Raston, and A. H.,White, *J. Chem. Soc. Dalton Trans., 1976*, 26.

320. A. Hermann and R. M. Wing, *Inorg. Chem., 11*, 1415 (1972).

321. R. Hesse, *Advances in the Chemistry of the Coordination Compounds*, Macmillan, New York, 1961, p. 314.

322. R. Hesse, *Ark. Kemi, 20*, 481 (1963).

323. R. Hesse and P. Jennische, *Acta Chem. Scand., 26*, 3855 (1972).

324. R. Hesse and L. Nilson, *Acta Chem. Scand., 23*, 825 (1969).

325. D. M. Hill, L. F. Larkworthy, and M. W. O'Donoghue, *J. Chem. Soc. Dalton Trans., 1975*, 1726.

326. H. A. O. Hill, D. Williams, and N. Zarb-Adami, *J. Chem. Soc. Faraday Trans., 2*, 72(9), 1494 (1976).

327. J. L. Hoard and J. V. Silverton, *Inorg. Chem., 2*, 235 (1963).

328. K. A. Hofmann, *Z. Anorg. Chem., 11*, 379 (1896).

329. K. A. Hofmann, *Z. Anorg. Chem., 14*, 263 (1897).

330. D. G. Holah and C. N. Murphy, *Can. J. Chem., 49*, 2726 (1971).

331. D. G. Holah and C. N. Murphy, *Can. J. Chem., 49*, 2726 (1971).

332. D. G. Holah and C. N. Murphy, *Inorg. Nucl. Chem. Lett.*, *8*, 1069 (1972).
333. D. G. Holah and C. N. Murphy, *J. Thermal. Anal.*, *3*, 311 (1971).
334. F. J. Hollander, M. L. Caffery, and D. Coucouvanis, *J. Am. Chem. Soc.*, *95*, 1125 (1973).
335. F. J. Hollander and D. Coucouvanis, *J. Am. Chem. Soc.*, *96*, 5646 (1974).
336. F. J. Hollander and D. Coucouvanis, *J. Am. Chem. Soc.*, *99*, 6268 (1977).
337. F. J. Hollander, R. Pedelty, and D. Coucouvanis, *J. Am. Chem. Soc.*, *96*, 4032 (1974).
338. M. Honda, M. Komura, Y. Kawasaki, T. Tanaka, and R. Okawara, *J. Inorg. Nucl. Chem.*, *30*, 3231 (1968).
339. J. M. Hope, R. L. Martin, D. Taylor, and A. H. White, *Chem. Commun.*, *1977*, 99.
340. B. F. Hoskins and B. P. Kelly, *Chem. Commun.*, *1970*, 45.
341. B. F. Hoskins and B. P. Kelly, *Inorg. Nucl. Chem. Lett.*, *8*, 875 (1972).
342. B. F. Hoskins, R. L. Martin, and N. M. Rohde, *Aust. J. Chem.*, *29*, 213 (1976).
343. B. F. Hoskins and C. D. Pannan, *Aust. J. Chem.*, *29*, 2337 (1976).
344. B. F. Hoskins and C. D. Pannan, *Chem. Commun.*, *1975*, 408.
345. B. F. Hoskins and A. H. White, *J. Chem. Soc. A*, *1970*, 1668.
346. M. R. Hunt, A. G. Kruger, L. Smith, and G. Winter, *Aust. J. Chem.*, *24*, 53 (1971).
347. S. Husebye, *Acta Chem. Scand.*, *24*, 2198 (1970).
348. S. Husebye, *Acta Chem. Scand.*, *27*, 756 (1973).
349. S. Husebye and G. Helland-Madsen, *Acta Chem. Scand.*, *24*, 2273 (1970).
350. S. Husebye and S. E. Svaeren, *Acta Chem. Scand.*, *27*, 763 (1973).
351. J. Hyde and J. Zubieta, *J. Inorg. Nucl. Chem.*, *39*, 289 (1977).
352. M. J. Hynes and A. J. Moran, *J. Chem. Soc., Dalton Trans.*, *1973*, 2280.
353. Y. Iimura, T. Ito, and H. Hagihara, *Acta Crystallogr.*, *B28*, 2271 (1972).
354. O. A. Ileperuma and R. D. Feltham, *Inorg. Chem.*, *14*, 3042 (1975).
355. O. A. Ileperuma and R. D. Feltham, *Inorg. Chem.*, *16*, 1876 (1977).
356. O. A. Ileperuma and R. D. Feltham, *J. Am. Chem. Soc.*, *98*, 6039 (1976).
357. H. Iwasaki, *Chem. Lett.*, *1972* 1105.
358. H. Iwasaki, *Acta Crystallogr.*, *B29*, 2115 (1973).
359. H. Iwasaki and H. Hagihara, *Acta Crystallogr.*, *B28*, 507 (1972).
360. M. J. Janssen, *Rec. Trav. Chim.*, *79*, 454 (1960).
361. P. Jennische and R. Hesse, *Acta Chem. Scand. A*, *27*, 3531 (1973).
362. P. Jennische, A. Olin, and R. Hesse, *Acta Chem. Scand.*, *26*, 2799 (1972).
363. K. A. Jensen, B. M. Dahl, and P. H. Nielsen, *Acta Chem. Scand.*, *26*, 2241 (1972).
364. B. F. G. Johnson, K. H. Al-Obaidi, and J. A. McCleverty, *J. Chem. Soc. A*, *1969*, 1668.
365. D. L. Johnston, W. L. Rohrbaugh, and W. D. Horrocks, Jr., *Inorg. Chem.*, *10*, 1474 (1971).
366. R. N. Jowitt and P. C. H. Mitchell, *J. Chem. Soc. A*, *1969*, 2632.
367. B. J. Kalbacher and R. Bereman, *Inorg. Chem.*, *12*, 2997 (1973).
368. B. J. Kalbacher and R. D. Bereman, *Inorg. Chem.*, *14*, 1417 (1975).
369. B. J. Kalbacher and R. D. Bereman, *J. Inorg. Nucl. Chem.*, *38*, 471 (1976).
370. A. E. Kalinin, A. I. Gusev, and Y. T. Struchkov, *J. Struct. Chem. (Engl. Transl.)*, *14*, 804 (1973).
371. A. E. Kalinin, A. I. Gusev, and Y. T. Struchkov, *Zh. Strukt. Khim.*, *14*, 859 (1973).
372. P. Karagiannidis, D. Kessisoglou, and G. Manoussakis, *Chim. Chron., New Ser.*, *6*, 487 (1977).
373. P. Karagiannidis, S. T. Papastefanou, and G. Manoussakis, *Chimika Chron. New Ser.*, *6*, 385 (1977).
374. T. Katada, S. Kato, and M. Mizuta, *Chem. Lett.*, *1975*, 1037.

375. S. Kato, A. Hori, H. Shiotani, M. Mizuta, N. Hayashi, and T. Takakuwa, *J. Organomet. Chem.*, *82*, 223 (1974).

376. S. Kato, T. Kato, T. Yamauchi, Y. Shibahasi, E. Kakuda, M. Mizuta, and Y. Ishii, *J. Organomet. Chem.*, *76*, 215 (1974).

377. S. Kato and M. Mizuta, *Bull. Chem. Soc. Jap*, *45*, 3492 (1972).

378. S. Kato, M. Mizuta, and Y. Ishii, *J. Organomet. Chem.*, *55*, 121 (1973).

379. S. Kato, M. Wakamatsu, and M. Mizuta, *J. Organomet. Chem.*, *78*, 405 (1974).

380. E. O. Kazimir, Ph.D. Thesis, Fordham University, Bronx, N.Y.

381. C. P. Keijzers and E. deBoer, *J. Chem. Phys.*, *57*, 1277 (1972).

382. C. P. Keijzers and E. deBoer, *Mol. Phys.*, *6*, 1743 (1975).

383. C. P. Keijzers and E. deBoer, *Mol. Phys.*, *29*, 1007 (1975).

384. C. P. Keijzers, H. J. M. DeVries, and A. Van der Avoird, *Inorg. Chem.*, *11*, 1338 (1972).

385. C. P. Keijzers, P. L. A. C. M. Van der Meer, and E. deBoer, *Mol. Phys.*, *6*, 1733 (1975).

386. R. Kellner, P. Prokopowski, and H. Malissa, *Anal. Chim. Acta*, *68*, 401 (1974).

387. W. Kemula, A. Hulanicki, and W. Nawrot, *Rocz. Chem.*, *36*, 1717 (1962); *38*, 1065 (1964).

388. D. L. Kepert, *Inorg. Chem.*, *11*, 1561 (1972).

389. J. V. Kingston and G. Wilkinson, *J. Inorg. Nucl. Chem.*, *28*, 2709 (1966).

390. W. L. Klotz and M. K. DeArmond, *Inorg. Chem.*, *14*, 3125 (1975).

391. N. Kobayashi and T. Fujisawa, *Bull. Chem. Soc. Jap.*, *49*, 2780 (1976).

392. N. Kobayashi, A. Osawa, and T. Fujisawa, *Bull. Chem. Soc. Jap.*, *47*, 2287 (1974).

393. E. König and S. Kremer, *Theor. Chim. Acta*, *20*, 143 (1971).

394. A. Kopwillem, *Acta Chem. Scand.*, *26*, 2941 (1972).

395. A. Kostikas, D. Petridis, A. Simopoulos, and M. Pasternak, *Solid State Commun.*, *13*, 1661 (1973).

396. R. Kramolowsky, *Angew. Chem.*, *8*, 202 (1969).

397. S. F. Krzeminski and D. K. Straub, *J. Chem. Phys.*, *58*, 1086 (1973).

398. E. J. Kupchik and C. T. Theisen, *J. Organomet. Chem.*, *11*, 627 (1968).

399. D. Laduree, P. Rioult, and J. Vialle, *Bull. Soc. Chim. Fr., 1973* 637.

400. S. Lahiry and V. K. Anand, *Chem. Commun., 1971*, 1111.

401. G. N. La Mar and F. A. Walker, *J. Am. Chem. Soc.*, *95*, 6950 (1973).

402. G. M. Larin, V. V. Zelentsov, Y. V. Rakitin, and M. E. Dyatkina, *Russ. J. Inorg. Chem.*, *17*, 1110 (1972).

403. J. G. Leipoldt and P. Coppens, *Inorg. Chem.*, *12*, 2269 (1973).

404a. D. F. Lewis and R. C. Fay, *J. Am. Chem. Soc.*, *96*, 3843 (1974).

404b. D. F. Lewis and R. C. Fay, *Inorg. Chem.*, *15*, 2219 (1976).

405. D. F. Lewis, S. J. Lippard, and J. A. Zubieta, *J. Am. Chem. Soc.*, *94*, 1563 (1972).

406. D. F. Lewis, S. J. Lippard, and J. A. Zubieta, *Inorg. Chem.*, *11*, 823 (1972).

407. T. Li and S. J. Lippard, *Inorg. Chem.*, *13*, 1791 (1974).

408. E. Lindner and R. Grimmer, *J. Organomet. Chem.*, *25*, 493 (1970).

409. S. J. Lippard, *Acc. Chem. Res.*, *6*, 282 (1973).

410. G. Losse and E. Wottgen, *J. Prakt. Chem.*, *13*, 260 (1961).

411. R. D. MacDonald and G. Winter, *Inorg. Nucl. Chem. Lett.*, *10*, 305 (1974).

412. D. J. Machin and J. F. Sullivan, *J. Less-Common Met.*, *19*, 413 (1969).

413. L. Malatesta, *Gazz. Ital.*, *68*, 195 (1938), and references therein.

414. L. Malatesta, *Gazz. Chim. Ital.*, *69*, 752 (1939).

415. A. Malliaris and D. Niarchos, *Inorg. Chem.*, *6*, 1340 (1976).

416. M. Maltese, *J. Chem. Soc., Dalton Trans., 1972*, 2664.

417. W. Manchot and S. Davidson, *Chem. Ber.*, *62*, 681 (1929).

418. G. E. Manoussakis and P. Karayannidis, *Inorg. Nucl. Chem. Lett., 6*, 71 (1970).
419. G. Manoussakis and P. Karayannidis, *J. Inorg. Nucl. Chem., 31*, 2978 (1969).
420. G. E. Manoussakis, M. Lalia-Kantouri, and R. B. Huff, *J. Inorg. Nucl. Chem., 37*, 2330 (1975).
421. G. E. Manoussakis and C. A. Tsipis, *J. Inorg. Nucl. Chem., 35*, 743 (1973).
422. G. E. Manoussakis and C. A. Tsipis, *Z. Anorg. Allg. Chem., 398*, 88 (1973).
423. G. E. Manoussakis, C. Tsipis, and C. Hadjikostas, *Can. J. Chem., 53*, 1530 (1975).
424. V. R. Marathe and S. Mitra, *Chem. Phys. Lett., 21*, 62 (1973).
425. A. L. Marchese, M. Scudder, A. M. Van den Bergen, and B. O. West, *J. Organ. Chem., 121*, 63 (1976).
426. G. Marcotrigiano, G. C. Pellacani, and C. Preti, *J. Inorg. Nucl. Chem., 36*, 3709 (1974).
427. J. M. Martin, P. W. G. Newman, B. W. Robinson, and A. H. White, *J. Chem. Soc., Dalton Trans., 1972*, 2233.
428. R. L. Martin, N. M. Rohde, G. B. Robertson, and D. Taylor, *J. Am. Chem. Soc., 96*, 3647 (1974).
429. R. L. Martin and A. H. White, *Inorg. Chem., 6*, 712 (1967).
430. R. L. Martin and A. H. White, *Nature, 223*, 394 (1969).
431. R. L. Martin and A. H. White, *Trans. Met. Chem., 4*, 113 (1968).
432. R. Mattes and G. Pauleickhoff, *Spectrochim. Acta, 30A*, 379 (1974).
433. R. Mattes and W. Stork, *Spectrochim. Acta Part A, 30A*, 1385 (1974).
434. B. M. Mattson, J. R. Heiman, and L. H. Pignolet, *Inorg. Chem., 15*, 564 (1976).
435. B. M. Mattson and L. H. Pignolet, *Inorg. Chem., 16*, 488 (1977).
436. P. L. Maxfield, *Inorg. Nucl. Chem. Lett., 6*, 693 (1970).
437. F. Mazzi and C. Tadini, *Z. Kristallogr. Kristallgeomer., Kristallphys. Kristallchem., 118*, 378 (1963).
438. L. E. McCandlish, E. C. Bissell, D. Coucouvanis, J. P. Fackler, Jr., and K. Knox, *J. Am. Chem. Soc., 90*, 7357 (1968).
439. J. A. McCleverty, S. McLuckie, N. J. Morrison, N. A. Bailey, and N. W. Walker, *J. Chem. Soc., Dalton Trans., 1977*, 359.
440. J. A. McCleverty and N. J. Morrison, *Chem. Commun., 1974*, 1048.
441. J. A. McCleverty and N. J. Morrison, *J. Chem. Soc., Dalton Trans., 1976*, 541.
442. J. A. McCleverty and N. J. Morrison, *J. Chem. Soc., Dalton Trans., 1976*, 2169.
443. J. A. McCleverty, D. G. Orchard, and K. Smith, *J. Chem. Soc. A, 1971*, 707.
444. B. J. McCormick, *Can. J. Chem., 47*, 4283 (1969).
445. B. J. McCormick, *Inorg. Chem., 7*, 1965 (1968).
446. B. J. McCormick and E. M. Bellott, *Inorg. Chem., 9*, 1779 (1970).
447. J. W. McDonald, J. L. Corbin, and W. E. Newton, *Inorg. Chem., 15*, 2056 (1976).
448. J. W. McDonald, J. L. Corbin, and W. E. Newton, *J. Chem. Soc., Dalton Trans., 1974*, 1044.
449. K. T. McGregor, D. J. Hodgson, and W. E. Hatfield, *Inorg. Chem., 12*, 731 (1973).
450. (a) H. A. Meinema and J. G. Noltes, *J. Organomet. Chem., 25*, 139 (1970). (b) E. R. Menzel and J. R. Wasson, *J. Magn. Resonance, 23*, 285 (1976).
451. S. Merlino, *Acta Cyrstallogr., Sect. B, 24*, 1441 (1968).
452. S. Merlino, *Acta Crystallogr., Sect. B, 25*, 2270 (1969).
453. S. Merlino and F. Sartori, *Acta Crystallogr., Sect. B, 28*, 972 (1972).
454. P. B. Merrithew and P. G. Rasmussen, *Inorg. Chem., 11*, 325 (1972).
455. E. R. Meuzek and J. R. Wasson, *J. Magn. Resonance, 23*, 285 (1976).
456. G. L. Miessler, G. Stuk, T. P. Smith, K. W. Given, M. C. Pallazzotto, and L. H. Pignolet, *Inorg. Chem., 15*, 1982 (1976).

457. R. B. Minasyan, E. M. Rokhlin, N. P. Gambaryan, Y. V. Zeifman, and I. L. Knunyants, *Bull. Acad. Sci., USSR, 1965*, 746.
458. P. C. H. Mitchell and R. D. Scarle, *J. Chem. Soc., Dalton Trans., 1975*, 111.
459. P. C. H. Mitchell and R. D. Scarle, *J. Chem. Soc., Dalton Trans., 1975*, 2552.
460. R. W. Mitchell, J. D. Ruddick, and G. Wilkinson, *J. Chem. Soc. A, 1971*, 3224.
461. W. J. Mitchell and M. K. DeArmond, *J. Lumin., 4*, 137 (1971).
462. W. J. Mitchell and M. K. DeArmond, *J. Mol. Spectrosc., 41*, 33 (1972).
463. S. Mitra, A. H. White, and C. L. Raston, *Aust. J. Chem., 29*, 1899 (1976).
464. F. G. Moers, R. W. M. ten Hoedt, and J. P. Langhout, *Inorg. Chem., 12*, 2196 (1977).
465. F. G. Moers, R. W. M. ten Hoedt, and J. P. Langhout, *J. Organomet. Chem., 65*, 93 (1974).
466. F. W. Moore and M. L. Larson, *Inorg. Chem., 6*, 998 (1967).
467. J. S. Morris and E. O. Schlemper, cited in private communication to I. Bernal [*Inorg. Chim. Acta, 7*, 283 (1973)].
468. E. L. Muetterties, *Inorg. Chem., 12*, 1963 (1973).
469. E. L. Muetterties, *Inorg. Chem., 13*, 1011 (1974).
470. A. Müller, P. Christophiemk, I. Tossidis, and C. K. Jorgensen, *Z. Anorg. Allg. Chem., 401*, 274 (1973).
471. W. G. Mumme and G. Winter, *Inorg. Nucl. Chem. Lett., 7*, 505 (1971).
472. C. G. R. Nair and K. K. M. Yusuff, *J. Inorg. Nucl. Chem., 39*, 281 (1977).
473. P. W. G. Newman, C. L. Raston, and A. H. White, *J. Chem. Soc., Dalton Trans., 1973*, 1332.
474. P. W. G. Newman and A. H. White, *J. Chem. Soc., Dalton Trans., 1972*, 1460.
475. P. W. G. Newman and A. H. White, *J. Chem. Soc., Dalton Trans., 1972*, 2239.
476. W. E. Newton, G. J. J. Chen, and J. W. McDonald, *J. Am. Chem. Soc., 98*, 5387 (1976).
477. W. E. Newton, J. L. Corbin, D. C. Bravard, J. E. Searles, and J. W. McDonald, *Inorg. Chem., 13*, 1100 (1974).
478. W. E. Newton, J. L. Corbin, and J. W. McDonald, *J. Chem. Soc., Dalton Trans., 1974*, 1044.
479. Y. Nigo, I. Masuda and K. Shinra, *Chem. Commun., 1970*, 476.
480. A. Nieuwpoort, Ph.D. Thesis, Nijmegen, Netherlands, 1975.
481. A. Nieuwpoort, H. M. Claessen, and J. G. M. van der Linden, *Inorg. Nucl. Chem. Lett., 11*, 869 (1975).
482. A. Nieuwpoort and J. J. Steggerda, *Recl. Trav. Chim. Pays-Bas, 95*, 250 (1976).
483. A. Nieuwpoort and J. J. Steggerda, *Recl., J. R. Neth. Chem. Soc., 95*, 289 (1976).
484. A. Nieuwpoort and J. J. Steggerda, *Recl., J. R. Neth. Chem. Soc., 95*, 290 (1976).
485. A. Nieuwpoort and J. J. Steggerda, *Recl., J. R. Neth. Chem. Soc., 95*, 294 (1976).
486. L. Nilson and R. Hesse, *Acta Chem. Scand., 23*, 1951 (1969).
487. J. H. Noordik, *Cryst. Struct. Commun., 2*, 81 (1973).
488. J. H. Noordik and P. T. Beurskens, *J. Crystal. Mol. Struct., 1*, 339 (1971).
489. J. H. Noordik, Th. W. Hummelink, and J. G. M. Van der Linden, *J. Coord. Chem., 2*, 185 (1973).
490. C. O'Connor J. D. Gilbert, and G. Wilkinson, *J. Chem. Soc. A, 1969*, 84.
491. B. H. O'Connor and E. N. Masten, *Acta Crystallogr., 21*, 828 (1966).
492. I. Ojima, T. Onishi, T. Iwamoto, N. Inamoto, and K. Tamaru, *Inorg. Nucl. Chem. Lett., 6*, 65 (1970).
493. M. C. Palazzotto, D. J. Duffy, B. L. Edgar, L. Que, Jr., and L. H. Pignolet, *J. Am. Chem. Soc., 95*, 4537 (1973).
494. M. C. Palazzotto and L. H. Pignolet, *Inorg. Chem., 13*, 1781 (1974).

495. M. C. Pallazzotto and L. H. Pignolet, *Chem. Commun., 1972*, 6.
496. D. C. Pantaleo and R. C. Johnson, *Inorg. Chem., 9*, 1248 (1970).
497. N. Pariyadath, W. E. Newton, and E. I. Stiefel, *J. Am. Chem. Soc., 98*, 5388 (1976).
498. E. A. Pasek and D. K. Straub, *Inorg. Chim. Acta, 21*, 23 (1977).
499. E. A. Pasek and D. K. Straub, *Inorg. Chim. Acta, 21*, 29 (1977).
500. E. A. Pasek and D. K, Straub, *Inorg. Chem., 11*, 259 (1972).
501. G. S. Patterson and R. H. Holm, *Inorg. Chem., 11*, 2285 (1972).
502. R. G. Pearson and D. A. Sweigart, *Inorg. Chem., 9*, 1167 (1970).
503. R. Pedelty-Beetner, Ph.D. Thesis, University of Iowa, 1977.
504. G. C. Pelizzi and C. Pelizzi, *Inorg. Chim. Acta, 4*, 618 (1970).
505. D. Petridis, A. Simopoulos, and A. Kostikas, *Phys. Rev. Lett., 27*, 1171 (1971).
506. V. Petrouleas, A. Kostikas, A. Simopoulos, and D. Coucouvanis, *J. Phys., Conf. Int. Appl. Eff. Mössbauer, Corfu* (1976).
507. A. Pignedoli and G. Peyronel, *Gazzetta72*, 745 (1962).
508. L. H. Pignolet, *Inorg. Chem., 13*, 2051 (1974).
509. L. H. Pignolet, *Top. Curr. Chem., 56*, 91 (1975).
510. L. H. Pignolet, R. A. Lewis, and R. H. Holm, *Inorg. Chem., 11*, 99 (1972).
511. L. H. Pignolet, D. J. Duffy, and L. Que, Jr., *J. Am. Chem. Soc., 95*, 295 (1973).
512. L. H. Pignolet, R. A. Lewis, and R. H. Holm, *J. Am. Chem. Soc., 93*, 360 (1971).
513. L. H. Pignolet and B. M. Mattson, *Chem. Commun., 1975*, 49.
514. L. H. Pignolet, G. S. Patterson, J. F. Weiher, and R. H. Holm, *Inorg. Chem., 13*, 1263 (1974).
515. J. R. Pilbrow, A. D. Toy, and T. D. Smith, *J. Chem. Soc. A, 1969*, 1029.
516. A. T. Pilipenko and G. I. Gridchina, *Tr. Kom. Anal. Khim. Akad. Nauk. SSSR, 3*, 178 (1951).
517. O. Piovesana and G. Cappuccilli, *Inorg. Chem., 11*, 1543 (1972).
518. O. Piovesana and C. Furlani, *Chem. Commun., 1971*, 256.
519. O. Piovesana, C. Furlani, A. Flamini, A. Sgamellotti, and C. Bellitto, *Atti Accad. Naz. Lincei. Cl. Sci. Fis., Mat. Nat. Rend., 54*, 763 (1973).
520. O. Piovesana and L. Sestili, *Inorg. Chem., 13*, 2745 (1974).
521. O. Piovesana, L. Sestili, C. Bellitto, A. Flamini, M. Tomassini, P. F. Zanazzi, and A. R. Zanzari, *J. Am. Chem. Soc., 99*, 5190 (1977).
522. O. Piovesana, L. Sestili, C. Bellitto, A. Flamini, M. Tomassini, P. F. Zanazzi, and A. R. Zanzari, *J. Am. Chem. Soc.*, submitted for Publication.
523. J. N. Pons, J. Roger, and M. Stern, *Compt. Rend. 276*, 855 (1973).
524. J. Potenza and D. Mastropaolo, *Acta Crystallogr., 29B*, 1830 (1973).
525. J. Powell and A. W. L. Chan, *J. Organomet. Chem., 35*, 203 (1972).
526. C. Preti and G. Tosi, *J. Inorg. Nucl. Chem., 38*, 1746 (1976).
527. C. Preti and G. Tosi, *Z. Anorg. Allg. Chem., 418*, 188 (1975).
528. H. Pritzkow and P. Jennische, *Acta Chem. Scand. A, 29*, 60 (1975).
529. L. Que, Jr. and L. H. Pignolet, *Inorg. Chem., 13*, 351 (1974).
530. E. A. Pasek and D. K. Straub, *Inorg. Chem., 11*, 259 (1972).
531. L. Que, Jr. and L. H. Pignolet, *Inorg. Chem., 113*, 351 (1974).
532. K. K. Ramaswamy and R. A. Krause, *Inorg. Chem., 9*, 1136 (1970).
533. K. K. Ramaswamy and R. A. Krause, *Inorg. Chem., 9*, 2649 (1970).
534. C. L. Raston and A. H. White, *Aust. J. Chem., 29*, 523 (1976).
535. C. L. Raston and A. H. White, *Aust. J. Chem., 29*, 739 (1976).
536. C. L. Raston and A. H. White, *J. Chem. Soc., Dalton Trans., 1974*, 1790.
537. C. L. Raston and A. H. White, *J. Chem. Soc., Dalton Trans., 1975*, 2405.
537a. C. L. Raston and A. H. White, *J. Chem. Soc., Dalton Trans.*, 1975 (2425).

538. C. L. Raston and A. H. White, *J. Chem. Soc., Dalton Trans., 1975*, 2410.
539. C. L. Raston and A. H. White, *J. Chem. Soc., Dalton Trans., 1975*, 2418.
540. C. L. Raston and A. H. White, *J. Chem. Soc., Dalton Trans., 1975*, 2422.
541. C. L. Raston and A. H. White, *J. Chem. Soc., Dalton Trans., 1976*, 32.
542. C. L. Raston and A. H. White, *J. Chem. Soc., Dalton Trans., 1976*, 791.
542a C. L. Raston and A. H. White, *Aust. J. Chem. 30*, 2091 (1977).
543. C. L. Raston, A. H. White, and A. C. Willis, *J. Chem. Soc., Dalton Trans., 1975*, 2429.
544. (a) C. L. Raston, A. H. White, and G. Winter, *Aust. J. Chem., 29*, 731 (1976). (b) A. Ray, D. N. Sathyanarayana, G. D. Prasad, and C. C. Patel, *Spectrochim. Acta, 29A*, 1579 (1973).
545. W. Regenass, S. Fallab, and S. Erlenmeyer, *Helv. Chim. Acta, 38*, 1448 (1955).
546. L. Ricard, private communication.
547. L. Ricard, J. Estienne, and R. Weiss, *Inorg. Chem., 12*, 2182 (1973).
548. L. Ricard, J. Estienne, and R. Weiss, *Chem. Commun., 1972*, 906.
549. L. Ricard, J. Estienne, P. Karagiannidis, P. Toledano, J. Fischer, A. Mitschier, and R. Weiss, *J. Coord. Chem., 3*, 277 (1974).
550. L. Ricard, P. Karagiannidis, and R. Weiss, *Inorg. Chem., 12*, 2179 (1973).
551. L. Ricard, C. Martin, R. Wiest, and R. Weiss, *Inorg. Chem., 14*, 2300 (1975).
552. L. Ricard and R. Weiss, *Inorg. Nucl. Chem. Lett., 10*, 217 (1974).
553. R. Richards, C. E. Johnson, and H. A. O. Hill, *J. Chem. Phys., 53*, 3118 (1970).
554. R. Richards, C. E. Johnson, and H. A. O. Hill, *Trans. Faraday Soc., 65*, 2847 (1969).
555. T. L. Riechel, L. J. DeHayes, and D. T. Sawyer, *Inorg. Chem., 15*, 1900 (1976).
556. S. D. Robinson and A. Sahajpal, *Inorg. Chem., 16*, 2718 (1977).
557. S. D. Robinson and A. Sahajpal, *J. Organomet. Chem., 99*, C-65 (1975).
558. J. F. Rowbottom and G. Wilkinson, *Inorg. Nucl. Chem. Lett., 9*, 675 (1973).
559. J. F. Rowbottom and G. Wilkinson, *J. Chem. Soc., Dalton Trans., 1972*, 826.
560. T. Sakurai, H. Okabe, and H. Isoyama, *Bull. Jap. Pet. Inst., 13*, 243 (1971).
561. R. Y. Saleh and D. K. Straub, *Inorg. Chem., 13*, 1559 (1974).
562. R. Y. Saleh and D. K. Straub, *Inorg. Chem., 13*, 3017 (1974).
563. P. C. Savino and R. D. Bereman, *Inorg, Chem., 12*, 173 (1973).
564. R. L. Schlupp and A. H. Maki, *Inorg. Chem., 13*, 44 (1974).
565. M. Schmidt, H. Schumann, F. Gliniechi, and J. F. Jaggard, *J. Organomet. Chem., 17*, 277 (1969).
566. P. W. Schneider, D. C. Bravard, J. W. McDonald, and W. E. Newton, *J. Am. Chem. Soc., 94*, 8640 (1972).
567. A. F. Schreiner and P. J. Hauser, *Inorg. Chem., 11*, 2706 (1972).
568. H. Seidel and D. Petzoldt, *Naturwissenschaften, 56*, 283 (1969).
569. R. Seltzer, *J. Org. Chem., 33*, 3896 (1968).
570. G. M. Sheldrick and W. S. Sheldrick, *J. Chem. Soc. A. 1970*, 490.
571. D. Shopov and N. D. Yordanov, *Proc. XIV Int. Conf. Coord. Chem., Toronto, 1972*.
572. L. F. Shvydka, Yu. I. Usatenko, and F. M. Tulyupa, *Zh. Neorg. Khim., 18*, 756 (1973).
573. T. H. Siddall, *Inorg. Nucl. Chem. Lett., 7*, 545 (1971).
574. E. Sinn, *Coord. Chem. Rev., 12*, 185 (1974).
575. E. Sinn, *Inorg. Chem., 15*, 369 (1976).
576. A. Simopoulos, D. Petridis, A. Kostikas, and H. H. Wickman, *Chem. Phys., 2*, 452 (1973).
577. J. N. Smith and T. M. Brown, *Inorg. Nucl. Chem. Lett., 6*, 441 (1970).
578. N. Sonoda and T. Tanaka, *Inorg. Chim. Acta, 12*, 261 (1975).
579. B. Spivack, Z. Dori, and E. I. Stiefel, *Inorg. Nucl. Chem. Lett., 11*, 501 (1975).

580. W. A. Stamm and C. E. Greco, U.S. Patent 3,316, 284 (April 25, 1967).

581. (a) D. F. Steele and T. A. Stephenson, *Inorg. Nucl. Chem. Lett.*, *9*, 777 (1973). (b) H. J. Stoklosa and J. R. Wasson, *Inorg. Nucl. Chem. Lett.*, *10*, 377 (1974). (c) W. L. Steffen and R. C. Fay, *Inorg. Chem.*, *17*, 779 (1978).

582. G. St. Nikolov, *Inorg. Nucl. Chem. Lett.*, *7*, 1213 (1971).

583. G. St. Nikolov, N. Jordanov, and I. Havezov, *J. Inorg. Nucl. Chem.*, *33*, 1055 (1971).

584. G. St. Nikolov, N. Jordanov, and I. Havezov, *J. Inorg. Nucl. Chem.*, *33*, 1059 (1971).

585. G. St. Nikolov and N. Tyutyulkov, *Inorg. Nucl. Chem. Lett.*, *7*, 1209 (1971).

586. H. J. Stoklosa and J. R. Wasson, *Inorg. Nucl. Chem. Lett.*, *10*, 377 (1974).

587. D. R. Swift, Ph.D. Thesis, Case Western Reserve University, 1970.

588. Y. Takeda, N. Watanabe, and T. Tanaka, *Spectrochim. Acta*, *32A*, 1553 (1976).

589. T. Tanaka and N. Watanabe, *Org. Magn. Resonance*, *6*, 165 (1974).

590. K. I. The, L. Vande-Griend, W. A. Whitla, and R. G. Cavell, *J. Am. Chem. Soc.*, *99*, 7379 (1977).

591. R. J. Timmons and L. S. Wittenbrook, *J. Org. Chem.*, *32*, 1566 (1967).

592. V. F. Toropova, G. K. Budnikov, E. P. Medyantseva, and N. A. Ulakhovich, *Zh. Obsch, Khim.*, *46*, 638 (1976).

593. I. Tossidis, A. Singollitou, G. Manoussakis, *Inorg. Nucl. Chem. Lett.*, *11*, 283 (1975).

594. I. Tossidis, A. Singollitou-Kourakou, and G. Manoussakis, *Inorg. Nucl. Chem. Lett.*, *12*, 357 (1976).

595. A. D. Toy, S. H. H. Chaston, J. R. Pilbrow, and T. D. Smith, *Inorg. Chem.*, *10*, 2219 (1971).

596. C. A. Tsipis, C. C. Hadjikostas, and G. E. Manoussakis, *Inorg. Chim. Acta*, *23*, 163 (1977).

597. P. J. H. A. M. Van de Lemput, J. A. Cras, and J. Willemse, private communication.

598. P. J. H. A. M. Van de Leemput, J. A. Cras, J. Willemse, P. T. Beurskens, and E. Menger, *Recl. Trav. Chim. Pays-Bas*, *95*, 191 (1976).

599. P. J. H. A. M. Van de Leemput, J. Willemse, and J. A. Cras, *Recl. Trav. Chim. Pays-Bas*, *95*, 53 (1976).

600. J. G. M. Van der Aalsvoort and P. T. Beurskens, *Cryst. Struct. Commun.*, *3*, 653 (1974).

601. J. G. M. Van der Linden, *J. Inorg. Nucl. Chem.*, *34*, 1645 (1972).

602. J. G. M. Van der Linden, Ph.D. Thesis, Nijmegen (1972).

603. J. G. M. Van der Linden, *Recl., Trav. Chim.*, *90*, 1027 (1971).

604. J. G. M. Van der Linden and H. G. J. Van de Roer, *Inorg. Chim. Acta*, *5*, 254 (1971).

605. J. G. M. Van Rens and E. de Boer, *Chem. Phys. Lett.*, *31*, 377 (1975).

606. J. G. M. Van Rens, E. Van der Drift, and E. de Boer, *Chem. Phys. Lett.*, *14*, 113 (1972).

607. J. G. M. Van Rens, M. P. A. Viegers, and E. de Boer, *Chem. Phys. Lett.*, *28*, 104 (1974).

608. H. Van Willigen and J. G. M. Van Rens, *Chem. Phys. Lett.*, *2*, 283 (1968).

609. Z. B. Varadi and A. Nieuwpoort, *Inorg. Nucl. Chem. Lett.*, *10*, 801 (1974).

610. G. S. Vigee and J. J. Selbin, *Inorg. Nucl. Chem.*, *31*, 3187 (1969).

611. G. S. Vigee and C. L. Watkins, *J. Inorg. Nucl. Chem.*, *34*, 3936 (1972).

612. J. F. Villa, D. A. Chatfield, M. M. Bursey, and W. E. Hatfield, *Inorg. Chim. Acta*, *6*, 332 (1972).

613. A. C. Villa, A. G. Manfredotti, C. Guastini, and M. Nardelli, *Acta Crystallogr., Sect. B*, *28*, 2231 (1972).

614. A. C. Villa, A. G. Manfredotti, M. Nardelli, and C. Pelizzi, *Chem. Commun., 1970*, 1322.

615. J. F. Villa and W. E. Hatfield, *Inorg. Chem., 10*, 2038 (1971).
616. K. Von Jürgen, R. Lenck, S. N. Olafsson, and R. Kramolowsky, *Angew. Chem., 23*, 811 (1976).
617. A. Wahlberg, *Acta Chem. Scand., A30*, 433 (1976).
618. (a) A. Wahlberg, *Acta Chem. Scand., A30*, 614 (1976). (b) Y. Watanabe and K. Yamahata, *Sci. Pap. Inst. Phys. Chem. Res. (Tokyo), 64*, 71 (1975).
619. L. K. White and R. Belford, *J. Am. Chem. Soc., 98*, 4428 (1976).
620. H. H. Wickman, *J. Chem. Phys., 56*, 976 (1972).
621. H. H. Wickman and F. R. Merritt, *Chem. Phys. Lett., 1*, 117 (1967).
622. H. H. Wickman, A. M. Trozzolo, H. J. Williams, G. W. Hulls, and F. R. Merritt, *Phys. Rev., 155*, 563 (1967).
623. J. G. Wijnhoven, *Cryst. Struct. Commun., 2*, 637 (1973).
624. J. G. Wijnhoven, T. E. M. Van den Hark, and P. T. Beurskens, *J. Cryst. Mol. Struct., 2*, 189 (1972).
625. R. L. Wilfong and E. E. Maust, *Bur. Mines Rep. Invest., 1*, 7963 (1974).
626. J. Willemse and J. A. Cras, *Rec. Trav. Chim. Pays-Bas, 91*, 1309 (1972).
627. J. Willemse, J. A. Cras, J. J. Steggerda, and C. P. Keijzers, *Struct. Bonding, 28*, 83 (1976), Ref. 136.
628. J. Willemse, J. A. Cras, and P. J. H. A. M. Van de Leemput, *Inorg. Nucl. Chem. Lett., 12*, 255 (1976).
629. J. Willemse, J. A. Cras, J. G. Wijnhoven, and P. T. Beurskens, *Recl. Trav. Chim. Pays-Bas, 92*, 1199 (1973).
630. J. Willemse, F. W. Pijpers, and J. J. M. Backus, *Recl. Trav. Chim. Pays-Bas, 94*, 185 (1975).
631. J. Willemse, P. H. F. M. Rouwette, and J. A. Cras, *Inorg. Nucl. Chem. Lett., 8*, 389 (1972).
632. J. Willemse and J. J. Steggerda, *Chem. Commun., 1969*, 1123.
633. J. Winniczek and J. P. Fackler, Jr., personal communication.
634. R. A. Winograd, D. L. Lewis, and S. J. Lippard, *Inorg. Chem., 14*, 2601 (1975).
635. G. Winter, *Aust. J. Chem., 29*, 559 (1976).
636. K. Yamanouchi, J. H. Enemark, J. W. McDonald, and W. E. Newton, *J. Am. Chem. Soc., 99*, 3529 (1977).
637. N. D. Yordanov, V. I. Iliev, and D. M. Shopov, *Proc. Bulgarian Acad. Sci., 29*, 1653 (1976).
638. N. D. Yordanov and D. Shopov, *Inorg. Chim. Acta, 5*, 679 (1971).
639. N. D. Yordanov and D. Shopov, *J. Chem. Soc., Dalton Trans., 1976*, 883.
640. N. D. Yordanov and D. Shopov, *J. Inorg. Nucl. Chem., 38*, 137 (1976).
641. N. D. Yordanov and D. Shopov, *Proc. Int. Conf. Coord. Chem., 16th, Ireland, 1974*.
642. N. D. Yordanov, V. Terziev, and D. Shopov, *Proc. Bulg. Acad. Sci., 27*, 1529 (1974).
643. J. Zubieta, personal communication.
644. J. A. Zubieta and J. Hyde, unpublished results; Ref. 37 in Ref. 10.
645. J. A. Zubieta and G. B. Maniloff, *Inorg. Nucl. Chem. Lett., 12*, 121 (1976).

Subject Index

Cumulative Index, Volumes 1-26